Concrete Composite Columns

Woodhead Publishing Series in Civil and Structural Engineering

Concrete Composite Columns

Behavior and Design

Chong Rong

Department of Civil Engineering, Xi'an University of Architecture and Technology, Xi'an, Shaanxi, P.R. China

ELSEVIER

WP WOODHEAD PUBLISHING

An imprint of Elsevier

Woodhead Publishing is an imprint of Elsevier
50 Hampshire Street, 5th Floor, Cambridge, MA 02139, United States
The Boulevard, Langford Lane, Kidlington, OX5 1GB, United Kingdom

ISBN: 978-0-323-85171-8 (print)

ISBN: 978-0-323-85335-4 (online)

For information on all Woodhead Publishing publications
visit our website at https://www.elsevier.com/books-and-journals

Publisher: Matthew Deans
Acquisitions Editor: Glyn Jones
Editorial Project Manager: John Leonard
Production Project Manager: Surya Narayanan Jayachandran
Cover Designer: Matthew Limbert

Typeset by MPS Limited, Chennai, India

Contents

Preface

Concrete composite structures have high mechanical behaviors and have been widely used in practical engineering, including steel reinforced concrete composite structures, concrete filled steel tubular structures, FRP confined concrete composite structures, FRP–steel reinforced concrete composite structures, and so on. In the concrete composite structure, especially for the concrete composite column, different building materials can produce complementary effects, which can improve mechanical properties of each material, so as to further improve mechanical behaviors of the concrete composite components. Therefore, the concrete composite columns have high bearing capacity, good ductility, and good seismic performance. The existing studies showed core concrete is the most complex material in composite columns, and the section type is an important factor influencing overall mechanical behaviors. Under different combination types, different lateral confining stress will confine the core concrete, and its mechanical properties are also obviously different. If there is no strength degradation, mechanical properties of the concrete show stress path independence. But once the concrete enters the strength degradation stage, it shows stress path dependence. Meanwhile, the characteristics of each confining material are different. For example, FRP can provide continuous confining force but shows brittleness. And steel can provide vertical force and constant confining force, but the local buckling problem is more complex. It is necessary to analyze the interaction characteristics of different materials, based on a combination of theoretical analysis and experimental study. And it has high significance to use a reasonable design to limit inferior properties of building materials. Based on the above background, the author makes a detailed analysis and review of confined concrete in the concrete composite columns and discusses the design method and application of concrete composite columns, so as to give new research ideas to concrete composite columns, which is the original intention of the author to write this book.

This book is mainly based on the research contents of the author during his doctoral period at Xi'an University of Architecture and Technology and joint training period at the University of Wollongong, and combined with some research contents of the author during his postdoctoral period. This book analyzes concrete composite columns from the perspective of confining concrete, which comprehensively introduces the theoretical model of concrete under multiaxial stress conditions, axial mechanical behaviors of various concrete composite columns, seismic behaviors of the steel frame confined concrete composite columns, and its application in practical engineering.

The book covers a wide range of contents and is divided into four parts. The first part focuses on the existing concrete composite columns, including Chapters 1 and 2. Chapter 1 introduces a variety of concrete composite columns, including steel–concrete composite columns, FRP confined concrete columns (FCC), and FRP–steel–concrete composite columns. Afterward, it makes supplemental experiments to analyze two types of concrete composite columns. In Chapter 2, the lateral confining force in concrete composite columns is analyzed, and a simplified calculation model of stress and strain is proposed, then the design-oriented constitutive model is established. The second part focuses on theoretical analysis of the confined concrete, including Chapters 3 and 4. In Chapter 3, the Twin Shear Strength Theory is analyzed and improved to propose the failure criterion of the concrete under multiple stress. Chapter 4 establishes analysis-oriented constitutive models of actively and passively confined concrete, including the stress calculation model based on the improved failure criterion, and strain calculation model based on the improved energy balance method. The third part focuses on compressive behaviors of novel confined concrete columns, including Chapters 5, 6, and 7. Chapter 5 analyzes axial compressive behaviors of steel frame confined concrete column (SFC), including axial compression test and analysis, finite element numerical analysis, and design-oriented constitutive model. Chapter 6 analyzes axial compressive behaviors of confined recycled aggregate concrete (RAC) column, including RAC filled in steel tube column, FRP confined RAC column, and double confined RAC column. Chapter 7 analyzes the compressive behaviors of the hybrid double-skin tubular rectangular column, including the axial compression test and eccentrical compression test. The fourth part focuses on the behavior design method and application of the confined concrete column, which sets SFC as an example, including Chapters 8, 9, and 10. Chapter 8 presents an experimental study on the seismic behavior of SFC. Chapter 9 discusses the section design method of SFC, including the section analysis method, analysis of influence factors, calculation method, and damage index model. Chapter 10 discussed the engineering application and structural analysis of SFC, including the static elastic–plastic analysis, pushover analysis, seismic design method, and structural performance design.

Associate Professor Chong Rong edited this book. Professor Shi Qingxuan gave the main guidance. Two postgraduate students (Tian Wenkai and Qv Yunsong) drew all figures in this book. At the same time, most contents of this book are supported by the National Natural Science Foundation of China (52108171). I express my sincere thanks.

It should be pointed out that the content of concrete composite columns is very rich. However, this book is only analyzed from the perspective of confined concrete. It is inevitable to miss something in the book. In this regard, I will gradually enrich and improve in future research. Because of the limited level of the author, there are inevitable omissions and mistakes in the book. I sincerely hope that readers will criticize and correct those issues.

Chong Rong

Department of Civil Engineering, Xi'an University of Architecture and Technology, Xi'an, Shaanxi, P.R. China

Review and further analysis of concrete composite columns

1

1.1 Introduction

In recent decades, high property materials are widely applied in civil engineering, including steel tube, section steel, and fiber reinforced plastic (FRP) composite. In addition, the existing studies have rapidly analyzed the mechanical behaviors of the concrete composite columns composed of concrete and high property materials. Based on many existing experimental and theoretical studies about the concrete composite columns, it is clearly recognized that the steel profile materials and FRP profile materials can be the confining materials and filling materials in concrete columns to improve the mechanical behavior, seismic behavior, and durability of these columns. In the existing researches on the application of FRP and steel in concrete composite columns, the constitutive models of various confined concrete have been proposed to identify and simulate the mechanical characteristics of concrete under complex stress states (e.g. steel tube confined concrete, FRP-confined concrete, and stirrup confined concrete). What is more important is that the rational utilization of mechanical properties of each material in the combination of materials (i.e. concrete, steel, and FRP) have increasingly become the focus of research. In other words, three key questions are important for the selection of the reasonable combination mode and the optimum section form of concrete composite columns: (1) What are materials in composite structure? (2) What are the mechanical properties of each material? (3) How to combine these materials?

In these chapters, we comprehensively reviewed the existing experimental and theoretical researches about various concrete composite columns, including concrete-filled steel tube columns, steel-reinforced concrete-filled steel tube columns, FRP-confined concrete composite columns, and FRP-confined steel-reinforced concrete columns. Based on the further analysis of mechanical behaviors of these concrete composite columns with various section forms, we identify the mechanical properties of each material under the combining effect at first. Then we point out the insufficient matter of the existing section form and demonstrate the conclusions from two experimental studies. Finally, we focus on how to combine the characteristics of the three materials to make their advantages complementary, so as to determine a section form of a composite column with excellent behavior. These design and study ideas show the purpose and significance of this book: the mechanical properties of various confined concrete, the section design method of concrete composite columns, the behavior design method, and the application of concrete composite columns.

Concrete Composite Columns. DOI: https://doi.org/10.1016/B978-0-323-85171-8.00001-X

1.2 Steel-concrete composite columns

Compared with the traditional reinforced concrete structures, the steel-concrete composite columns have well mechanical behaviors, plasticity deformation, and toughness. Therefore, the application of the steel-concrete composite structures can reduce the cross-sectional area of the contribution members, which leads to large usage space in a building and reduce the structure's self-weight. Meanwhile, the steel tube can be the outer formwork in the concrete pouring process and reinforcement cage may not be set in some concrete composite structures. It is easy to understand that these engineering structures are efficient and environmentally protected due to the short construction period and the easy building process. Compared with the traditional steel structures, the steel-concrete composite structures have well durability, fire resistance, and stability. As mentioned above, the concrete composite structure is outstanding in safety, economy, environmental protection, and durability. Moreover, the high-behavior structures have been widely applied in super high-rise building projects, long-span bridge projects, large-scale water conservancy projects, ports and coasts, and deep-sea projects. We have widely known that the column is the main bearing member of the concrete composite structure. The steel-concrete composite column also plays an important role in the composite structure system, and the quality of the composite column directly determines the overall mechanical behavior and economic index of the composite structure. In general, the existing steel-concrete composite column mainly includes steel-reinforced concrete column, concrete-filled steel tube column, and steel-reinforced concrete-filled steel tube column. For these concrete composite columns, the steel tube can provide confining effect on the core concrete, which leads to more complex stress mechanism in the core concrete. Meanwhile, the mechanical properties and stress mechanism of the confined concrete are keys to the behavioral improvement of the composite. Therefore, this section focuses on the research status of concrete-filled steel tube column and steel reinforced concrete-filled steel tube column. Further, we put forward some development methods and study ideas for steel-concrete composite columns, based on the review and the analysis.

1.2.1 Concrete-filled steel tube columns

The existing studies propose the concrete-filled steel tube columns based on the steel column and reinforced concrete column. Based on the loading characteristics, we can divide the concrete-filled steel tube columns into three types: the concrete-filled outer steel tube column, the concrete-filled inner steel tube column, and the steel tube confined concrete column. The section forms in Fig. 1.1 show main structural characteristics. We can divide the differences of the three concrete composite columns into two following contents: the position of the steel tube in section design and the load bearing situation of the steel tube. The two differences cause the different stress states of the steel tube and the core concrete, which will lead to significant differences in the overall stress state, the durability, and the fire resistance.

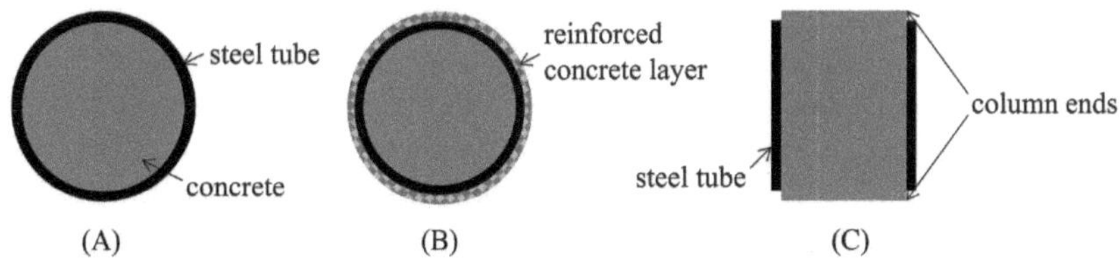

Figure 1.1 Three kinds of concrete-filled steel tube columns (A) concrete-filled outer steel tube column; (B) concrete-filled inner steel tube column; (C) steel tube confined column.

Therefore, the contents, the methods, and the focuses are different based on the existing studies about the three types of concrete-filled steel tube columns.

From the 1960s, the concrete-filled outer steel tube was valued and studied in developed countries, such as the Soviet Union, Western Europe, North America, and Japan. In the 1980s, the development of pumping concrete technology promoted the application and research of the concrete-filled outer steel tube in the world. At this time, many researchers developed a series of studies about the concrete-filled outer steel tubular concrete composite columns in China (Cai, 1999). As shown in Fig. 1.1, it is easy to understand that the outer steel tube bears the expansion pressure of concrete and the axial pressure of the external load. Therefore, the steel tube is in the two-direction stress state, including the longitudinal compression and the circumferential tension. Under the stress state, the steel tube can confine the core concrete, and the core concrete can improve the stability of the outer steel tube. A lot of existing researches focused on experimental studies and the calculation methods of bearing capacity for short columns, long columns, and eccentrically loaded columns. Zhong and Wang (1980) used strength theory to derive the calculation formulas of ultimate bearing capacity and stable bearing capacity of the composite column, and further analyzed the influence factors of bearing capacity, including the concrete creep, the shrinkage, and the temperature. The experimental study proposed by Tang et al. (1982) analyzed over 100 concrete-filled outer steel tube cylinders under axial or eccentric compression. The study parameters included section steel ratio, concrete strength, and column slenderness ratio. The experimental results show that the confinement coefficient of steel tubes is the important parameter affecting the bearing capacity and deformation capacity of the short concrete composite columns. Meanwhile, the study by Tang et al. (1982) derived the calculation formula for ultimate strength of concrete-filled steel tube short columns based on three proposed assumptions. In the experimental research on axial compression of long columns and eccentric compression of short columns (Cai & Jiao, 1984), the stability of concrete-filled steel outer tube columns was analyzed, and the formulas for corresponding bearing capacity calculation were proposed. Then, the experimental study by Gu et al. (1991) presented the mechanical behaviors of high-strength concrete-filled steel tubular columns, including axial and eccentric compression behaviors of long or short columns. The results show that the mechanical behaviors of high-strength concrete composite columns are similar to that of the ordinary strength concrete composite columns. The experimental study presented by Yu et al. (2007) on 17 self-compacting concrete-filled steel tube short columns under axial compression mainly studied the influence factors of the bearing capacity, including the opening and slotting in the middle of the steel tube. The test results show that the opening of small holes and slots has little effect on the bearing capacity and the deformation but may reduce the combined elastic modulus in the elastic stage.

Fig. 1.1 shows the section form of concrete-filled inner steel tube column which is composed of the inner steel tube, the core concrete, and the outer reinforced concrete layer. In this type of concrete-filled steel tube column, the stability of steel tube is improved by two-side concrete layers. The experimental study presented by Zhao et al. (1996) analyzed the seismic behavior of small-scale high-strength concrete-filled inner steel tube columns, including a total of 38 specimens under the low cyclic loading tests. The test results show that the composite column has high shear capacity and ductility. Further, the study by Zhao et al. (1996) suggested the design limit values of axial compression ratio, volume stirrup ratio, section ratio of steel tube, and confinement coefficient. Li et al. (1999) presented experimental studies on the concrete-filled inner steel tube columns. The study results can draw the following conclusions: the outer concrete lay can bear 90% of the lateral force; the axial force and the steel ratio have little effect on the lateral force of the inner concrete-filled steel tube column; the differences in strength grade between inner concrete column and outer concrete layer can improve the ductility of the concrete composite column. Later, the study by Li et al. (1999) discussed the influence factors of the axial force distribution of the composite column, including the axial compression ratio, the section ratio of outer concrete layer, and the confinement coefficient of steel tube. Lin et al. (2001) presented experimental studies on seven concrete-filled inner steel tube concrete composite columns, including the axial compressive test and the eccentric compressive tests, and proposed the calculation formula of the axial compression ratio limit and the normal section bearing capacity of concrete composite columns by theoretical analysis. The study of Cai et al. (2002) presented the axial compression tests of 10 high-strength concrete-filled inner steel tube columns and discussed the effects of steel ratio, longitudinal reinforcement ratio, stirrup ratio, and section form on the bearing capacity and ductility of the concrete composite columns. Based on the theoretical analysis presented by Nie et al. (2005), the dividing volume stirrup ratio was derived by the foundation that concrete inside and outside the steel tube reaches the limit state at the same time. Based on the study results by Nie et al. (2005), we can know that if the stirrup ratio of the outer concrete layer is less than the dividing volume stirrup ratio, the outer concrete layer will damage quickly because of insufficient confining effect, which leads to the lower ultimate bearing capacity of the composite column than the calculated value. Therefore, the relationship between the section ratio of the inner concrete and the stirrup ratio of the outer concrete layer needs to be considered in the design, to ensure the safety of the members and the most rational utilization of materials.

For the steel tube confined concrete column, the steel tube does not penetrate up and down in joints, only bears the expansion pressure of inner concrete, and is in the unidirectional stress state of circumferential tension, as shown in Fig. 1.1. Therefore, the steel tube confined concrete column is less affected by the local buckling and the stability of steel tube. In 1985, Tomii, Sakino of Japan, and Xiao of China put forward the concept of steel tube confined concrete, which was called "tubed column" (Tomii et al., 1985). In the following 10 years, Martin et al. (2000), Mei et al. (2001), and Peter et al. (2002), respectively proposed the axial compression bearing capacity model of steel tube confined concrete short cylinders. In 2004, Xiao et al. (2004) expounded on the advantages and development prospects of the steel tube confined concrete columns in China, and they carried out the pseudo-static tests on 4 circular section columns and two square section columns with steel casing locally confined at two column ends. The test results show that steel tube confined concrete columns have great seismic ductility. The experimental studies about steel tube confined concrete short columns by Han et al. (2005) showed that the concrete composite columns have great energy consumption capacity even under high axial compression ratio. The experimental study presented by Zhou and Liu (2010) tested a series of steel tube confined concrete specimens to discuss the effect of diameter–thickness ratio and shear span ratio on the axial behavior and seismic behavior of circle and square section columns. Zhou and Liu (2010) further established the design theories and methods of steel tube confined reinforced concrete and steel tube confined concrete, and proposed the reasonable structural measured, which can be the foundation for the popularization and application of steel tube confined concrete columns.

As shown above, the three types of concrete-filled steel tube concrete composite columns have great axial compressive behavior and great seismic behavior, while their different structural forms lead to different stress characteristics and respective disadvantages. For concrete-filled outer steel tube columns, the outer steel tube has some disadvantages, including the local buckling, less fire resistance, and the less corrosion resistance. Besides, the concrete-filled inner steel tube columns have the complex construction process, because of many formworks needed in construction. For steel tube confined concrete columns, the design methods of beam-column joints and the construction process are inconvenient due to the no penetrating steel tube. Therefore, the three types of concrete composite columns have their own characteristics. We should reasonably combine or further improve these characteristics for application in practical engineering.

1.2.2 Concrete-filled special-shaped steel tube column

Concrete-filled special-shaped steel tube column is a novel column that combines the advantages of reinforced concrete special-shaped column structure and concrete-filled steel tubular structure. The new structure form can meet the needs of humanized high-rise building design in the future. Based on the major difference in structural measures to enhance the confining effect on core concrete, the main section forms of concrete-filled special-shaped steel tube column include:

(1) Ordinary T-shaped, L-shaped, and cross-shaped concrete-filled steel tube columns, as shown in Fig. 1.2. Compared with reinforced concrete special-shaped

columns, this kind of special-shaped concrete-filled steel tubular columns can improve the ductility and seismic behavior of members. However, the disadvantage is that the steel tube wall has a little confining effect on core concrete, and the confining effect on the improvement of bearing capacity is much less than that of circular section.

(2) T-shaped, L-shaped, and cross-shaped concrete-filled steel tubular columns with ribs, as shown in Fig. 1.3. Similar to the square steel tube, the longitudinal ribbed stiffeners are welded on the middle or weak area of the column section to enhance the lateral stiffness of the steel tube wall, delay the local buckling of the steel tube wall under vertical load, and improve the confining effect on core concrete, so as to improve the bearing capacity and ductility of the members.

(3) Concrete-filled special-shaped steel tube column with binding bars, as shown in Fig. 1.4. In this type of the concrete composite column, some studies set the binding bars at certain spacing on each side to delay the local buckling of the steel tube wall and enhance the confining effect of the steel tube on core concrete, so as to improve the bearing capacity and ductility of the members. However, the special-shaped section form may not satisfy with the esthetic requirements of the building. The plate surface of the column is no longer neat and flat, due to the

(A) (B) (C)

Figure 1.2 The concrete-filled special-shaped steel tube columns. (A) T-shaped column; (B) L-shaped column; (C) cross-shaped column.

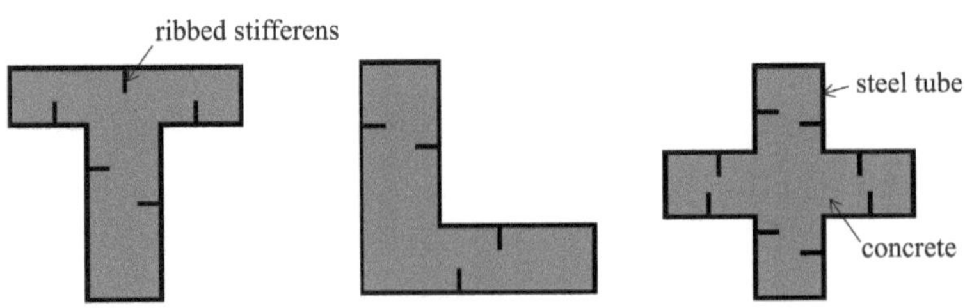

(A) (B) (C)

Figure 1.3 The concrete-filled special-shaped steel tube columns with ribbed stiffeners. (A) T-shaped column; (B) L-shaped column; (C) cross-shaped column.

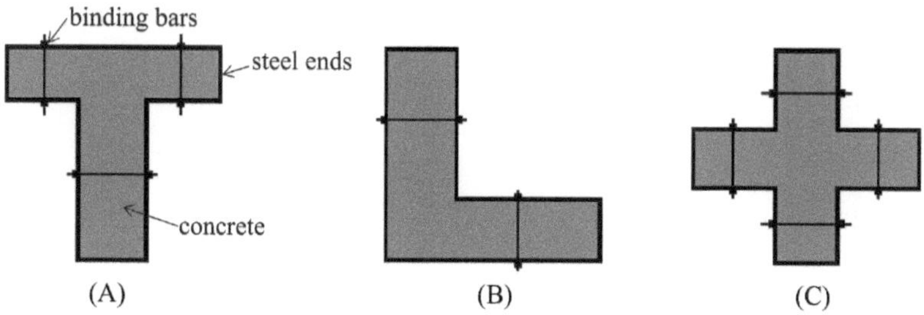

Figure 1.4 The concrete-filled special-shaped steel tube columns with binding bars. (A) T-shaped column; (B) L-shaped column; (C) cross-shaped column.

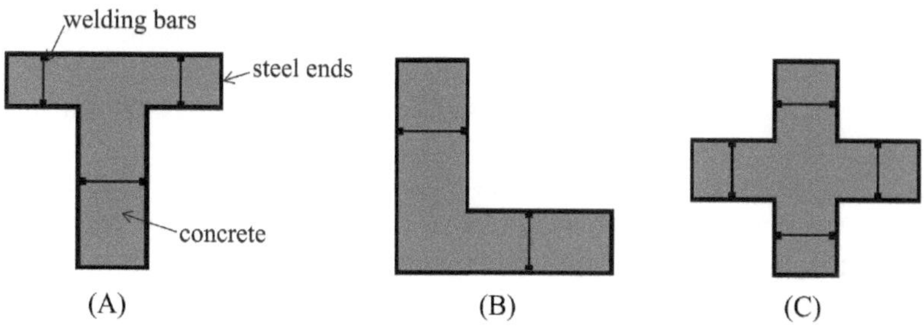

Figure 1.5 The concrete-filled special-shaped steel tube columns with welding bars. (A) T-shaped column; (B) L-shaped column; (C) cross-shaped column.

existence of tie rods and bolts. Therefore, it is necessary to deal with the relationship between convex bolts and building decoration. Meanwhile, there are many holes in the steel surface, which cause significant damage to the steel tube wall. Under the axial loading, due to the stress concentration, the influence of the hole edge on the mechanical properties of the members should be further studied.

(4) Concrete-filled special-shaped steel tube column with welding bars, as shown in Fig. 1.5. Some studies proposed this type of concrete composite column to solve the problem that the protruding binding bars on the slab surface may affect the aesthetics of the building. The mechanical behaviors of the concrete composite columns are similar to that of concrete-filled special-shaped steel tube column with binding bars. The welding bars delay the local buckling of steel tubes and provide a strong confining effect on core concrete. However, the quality of welds during construction is hard to meet the design and application requirements, due to the large number of plate welding joints and complex welding residual stress. Therefore, the behaviors of the concrete composite columns need to be further studied.

(5) The spliced composite column by welding two rectangle or square columns, as shown in Fig. 1.6. There is no concave corner weak area in this kind of concrete composite column. In addition, the width−thickness ratio of the member section is small. Therefore, the steel tube can provide an constructive confining effect on core

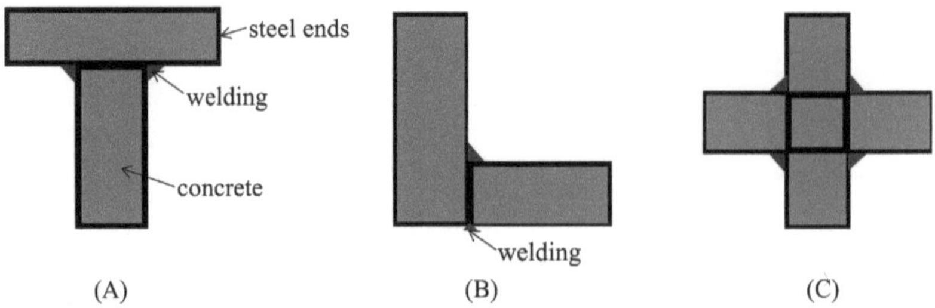

Figure 1.6 The spliced composite column by welding two rectangle or square columns. (A) T-shaped column; (B) L-shaped column; (C) cross-shaped column.

concrete to improve the bearing capacity. However, the two steel tubes are too difficult to weld to ensure the weld quality in this section form. Meanwhile, the two parallel close steel plates between the two longitudinal welds show small deformation under the bending load. Therefore, the section form cannot play the full material properties of the steel parts.

The existing researches on concrete-filled special-shaped steel tube columns focused on the strength bearing capacity and the stable bearing capacity under axial loading, the strength bearing capacity under eccentric loading (including unidirectional and bidirectional eccentric loading), and the stable bearing capacity under unidirectional eccentric loading.

For the axial compression mechanical behavior, Du et al. (2008a) presented compression experimental studies on 6 T-shaped concrete-filled steel tubular columns, and variation parameters included the limb thickness, the limb width, the web width, and the thickness of steel tube. The experimental results show that the failure modes of all specimens show the waist drum form, and there are no local compression failure and instability failure. In the study by Du et al. (2008a), they proposed the calculation method of axial compression bearing capacity by parameter analysis, yet the method does not reasonably consider the confining effect provided by the steel tube. Subsequently, Du et al. (2008b) carried out an axial compression test on 20 T-shaped concrete-filled steel tube short columns welded by square steel tube and rectangular steel tube, and the variation parameters included confining effect parameter, web width, and slenderness ratio. The failure modes of specimens can be divided into the shear failure and the local convex failure. The test results show that the deformation of the rectangular steel tube and the square steel tube are concordant in the entire loading process, and the two independent composite components of the T-shaped composite column can superimpose to calculate the bearing capacity. The study presented by Long and Cai (2006) analyzed the effect of binding bars on axial mechanical behavior of L-shaped concrete-filled steel tube short columns by axial compression tests. The test results show that the smaller width–thickness ratio of steel tube leads to better ductility of the specimen; the large spacing between two binding bars only can increase the ductility of the specimen; the small spacing between two binding bars can improve both the bearing capacity and ductility of the specimen; the diameter of binding bar has little effect on the bearing capacity and ductility. In addition, the study by Long and Cai (2006) showed that most of the existing standards are not suitable for the calculation of bearing capacity of L-shaped concrete-filled steel tubular columns with binding bars. The experimental studies presented by Zuo et al. (2011) also showed that the setting of binding bars can change the yield mode of steel tube and delay the local buckling of steel tube, so as to improve the bearing capacity and ductility of specimens. Based on the computational comparisons in the study by Zuo et al. (2011), the existing standards do not provide accurate calculation formula of bearing capacity, due to the negligence of confining effect provided by binding bars.

For the eccentric compression mechanical behavior, Shen et al. (2009) presented an experimental study on the behavior of the L-shaped concrete-filled steel tubular columns, the test results show all specimens occurred the overall instability failure because of the increase of lateral deflection, and the lateral deformation curve is close to the sinusoidal half-wave curve. In addition, the longitudinal strain results of the mid-span section of steel tube show that the section deformation belongs to the plane section deformation. Therefore, the fiber model can be used in section analysis of the L-shaped composite column. Zuo et al. (2010) carried out the eccentric compressive tests on 8 L-shaped concrete-filled steel tubular columns with blinding bars, the variation parameters included eccentricity, load angle, horizontal spacing, and diameter of blinding bar. The test results show that the deformation of the specimen includes bending deformation and local buckling of the steel tube; the blinding bars delay the local buckling of steel tube, improve the bearing capacity and the ductility of the eccentric member; the section deformation of the composite column under eccentric compression conforms to the plane section assumption. Both of the above two studies proposed the calculation method by the fiber model and verified the proposed method by applying the experimental results. Based on the theoretical calculation model, a simplified calculation method of biaxial compression-bending capacity is obtained by regression analysis.

Based on the above reviews, the keys to theoretical analysis of the mechanical behaviors of concrete-filled special-shaped steel tube columns are the equivalent constitutive models of concrete and steel under the corresponding section structure. In addition, we should study the following contents in detail: the definition of the wall width–thickness ratio between the ordinary thick special-shaped concrete-filled steel tube column and the thin-walled composite column; the effective section of the thin-walled special-shaped concrete-filled steel tube column; and the calculation method of bearing capacity of the concrete composite columns after local buckling occurred on the steel tube. However, few existing studies focus on the experimental and theoretical analysis of the two-direction compression-bending stability bearing capacity of the special-shaped concrete-filled steel tubular columns.

For the seismic behavior, Lin et al. (2009) presented an experimental study for 7 L-shaped concrete-filled steel tubular columns, and the variation parameters included loading method, axial compression ratio, and effect of the welding ribs. The test results show that the L-shaped concrete-filled steel tubular column has good energy dissipation capacity and ductility. In addition, they examined the applicability of the constitutive models of the steel and the concrete by the numerical analysis based on the fiber model. Wang and Lu (2005) carried out the low cyclic loading test on 12 ordinary concrete-filled special-shaped steel tube columns (i.e. 6 T-shaped specimens and 6 L-shaped specimens), the variation parameters included axial compression ratio, core concrete strength, and thickness of steel tube. The test results showed that with the increasing axial compression ratio, the ultimate bearing capacity increases slightly or even decreases, and the ductility decreases; thicker steel tube can improve the ultimate load and ductility; the core concrete strength has an obvious effect on the ultimate load, yet it does not influence the ductility. We can clearly find that the confining effect of steel tube in ordinary special-shaped concrete composite columns is too weak to improve the mechanical properties of core concrete. The strengthening measures in the concrete composite columns are necessary.

1.2.3 Steel reinforced concrete-filled steel tube column

Steel reinforced concrete-filled steel tube column is a new type of composite column which composes of the outer steel tube, inner section steel, and concrete-filled between the steel parts. Fig. 1.7 shows the existing section forms. Based on the existing studies, we can summarize the main characteristics of these concrete composite columns as follows: (1) combining the advantages of solid-web steel reinforced and concrete-filled steel tube columns, its great mechanical behavior can lead to the section size of columns, so as to increase the service space in buildings; (2) the confining effect provided by the outer steel tube can improve the mechanical properties and deformation capacity of core concrete, which leads to great seismic behavior of the concrete composite columns; (3) since the longitudinal and stirrup are not necessary, the construction behavior of the steel reinforced concrete-filled steel tube column is more convenient than steel reinforced concrete column; (4) the composite column has the fire resistance; (5) the composite column is suitable for high axial pressure.

Since 2002, Wang et al. (2003, 2004) carried out a series of experimental studies on the mechanical behaviors of over 33 steel reinforced concrete-filled steel tube concrete composite columns. The analysis results show the following conclusions:

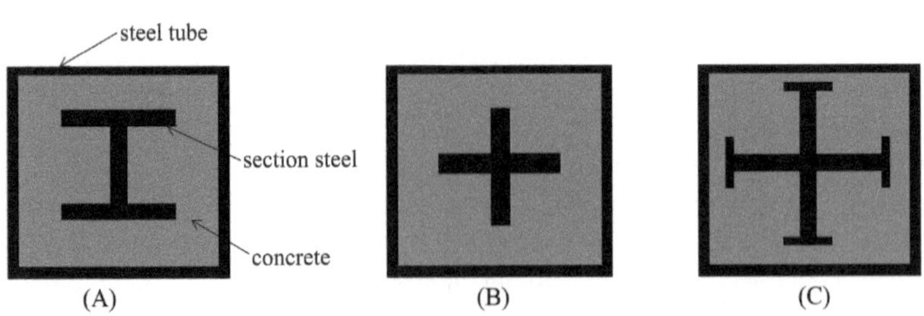

Figure 1.7 The section forms of steel reinforced concrete-filled steel tube column. (A) I-steel; (B) cross-shaped steel; (C) special cross-shape steel.

the higher concrete strength, the higher confinement index, or the higher steel embedded index can increase the axial compressive bearing capacity and ductility of the concrete composite columns; the eccentric bearing capacity and ductility increase with the increase of confinement index or steel embedded index, confinement index or steel embedded index, it decreases with the increase of slenderness ratio and eccentricity; the concrete composite columns are suitable for the structural design in high seismic intensity region, and the axial compression ratio should not be limited, due to the great ductility; the major factors of the mechanical properties of compression-bending concrete composite columns are confinement index, steel embedded index, axial compression ratio and slenderness ratio, where the last factor is the adverse one; with the constant shear span ratio, the major factors of the ductility of the composite column are concrete strength, steel embedded index and axial compression ratio, where the increase of axial compression ratio can decrease energy dissipation capacity and ductility of the composite column. Yang (2006) also presented axial compressive experiments on the concrete composite columns with I-section steel. The experimental results show that the outer steel tube and inner I-section steel improve the bearing capacity and ductility of the concrete composite columns; the elastic limit of the composite column under axial load increases with the increase in concrete strength.

The experimental studies presented by Zhu et al. (2011) analyzed seismic behaviors of the steel reinforced concrete-filled square steel tube columns, based on 39 test specimens. The study results show that qualified self-compacting concrete can be poured without vibration; the bearing capacity and ductility of the concrete composite columns decrease with the increase in concrete strength, or increase with the increase of width–thickness ratio and steel embedded index; the axial compression ratio is the major factor affecting the ductility of the composite column, which should be limited. The above conclusions show that square steel tube may not provide adequate confining effect on core concrete, which leads to a low contribution of concrete to the seismic behavior of the composite column. Based on the experimental study proposed by Yang (2014), the concrete composite columns with cross-section steel have good energy consumption and seismic behavior. The increase in steel embedded index can improve the bearing capacity and ductility; the increase in axial compression leads to a quick strength degradation rate and bad ductility.

Wen (2010) presented seismic experiments on over 32 joint specimens, including steel beam-composite column joints and reinforced concrete beam-composite column joints. The experimental results showed that all joint specimens have plump hysteretic curves, good bearing capacity, and good seismic behavior. Although the axial compression ratio has little effect on these joint specimens, the study by Wen (2010) suggested that the design avoids the buckling failure of the concrete composite column under high axial compression ratio. It also proposed the calculation method of shear bearing capacity in the core area of heavy-duty column joints.

As mentioned above, the steel reinforced concrete-filled steel tube columns have great mechanical behavior and seismic behavior. The steel tube can provide confining effect on core concrete, which leads to no reinforcement required and behavior improvement of the concrete composite columns. Under the confining effect provided by the steel tube, the mechanical properties of core concrete are complex, and the constitution model is the key to the theoretical analysis of the composite column.

1.2.4 Theoretical analysis of core concrete in steel tube

We have found that the steel tube confined concrete is the most complex and important part of the mechanical analysis of steel-concrete composite columns. At present, the mechanical models mainly belong to the separation mechanical model, which can clearly analyze the interaction relationship between steel and concrete. Besides, the separation mechanical model can identify the contribution and detailed mechanical properties of each material, to have clear physical and mechanical significance. A series of existing researches presented theoretical studies on the mechanical model of the core concrete in concrete-filled steel tube columns, such as the empirical relationship model, non-linear elastic model, plastic model (e.g., plastic fracture model and plastic damage model), endochronic theoretical model and so on.

In several studies (Choi & Xiao, 2009; Han et al., 2005; Liang & Fragomeni, 2009), the empirical relationship models were established by the fitting analysis method. The explicit function expressions expressed these models by fitting the stress–strain curves of steel tube confined concrete obtained by test results or database in existing references. In establishment process of the empirical relationship model, we can easily obtain the mechanical response parameters of concrete-filled steel tube column. However, the fitting process of test data has obvious disadvantages: the parameters in the empirical relationship model do not have strict physical and mechanical meaning; the parameter selection may be influenced by researcher's cognition, and the accuracy of the selection is not determined; the accuracy of the model depends on the sample quantity and the sample quality of the database. Therefore, the empirical relationship model is just suitable for specific objects, which affects its popularization and application.

The non-linear elastic model shows the following characteristics: the changing relationship between the stress and the deformation of core concrete shows a non-linear trend; the strain returns along the curve without residual strain in the unloading process. The non-linear elastic model can reflect the main deformation characteristics of concrete under loading. The simple and intuitive expression equation of the model can provide accurate prediction results for the core concrete under monotonic proportional loading. However, we can clearly find that the model does not reflect the differences in deformation characteristics under unloading and loading. Therefore, the non-linear elastic model is not suitable for core concrete under cyclic loading and non-proportional loading.

The plastic model of core concrete in the steel tube is too complex to apply in manual computation. In general, the plastic mechanical model is realized by computational programming, which has been developed swiftly and violently with the development of computer technology. In particular, the large-scale general finite element software in the finite element analysis study can calculate the mechanical properties of any type of confined concrete. ANSYS and ABAQUS are two types of general finite element software used in existing researches.

For the application of plastic theory models, the plastic fracture model is gradually developed in the finite element analysis of the confined concrete (Rajeshkumar et al., 2013; Soundararajan & Shanmugasundaram, 2010). The plastic fracture model can effectively describe the physical phenomena (e.g., plastic flow, strain hardening, and yield stress changing of concrete) and has a precise mathematical construction form. In addition, the basic assumption is that the concrete will destroy once the stress state reaches the failure surface. However, the basic assumption makes the plastic fracture model unable to analyze effectively the mechanical behavior of the concrete after strain softening.

Against the background of the plastic fracture model, some studies (Han et al., 2008; Wang et al., 2013) proposed the plastic damage model by the combination of plastic mechanics and damage mechanics, and the model includes the plasticity part and the damaged part. In some existing researches (Ellobody et al., 2006; Lee & Fenves, 1998), the selection of the plastic damage model of core concrete in steel tube can be guided by the following ideas: one is software application method, the constitutive model is directly defined by the parameter change in the existing finite element software (e.g., ABAQUS, ANSYS, etc.); another is theoretical derivation method, the constitutive model is established by the combination between the related mechanical theory and the stress characteristics of core concrete in a given member. The study presented by Zhong (1999) proposed the plastic damage model of core concrete in concrete-filled steel tube column, and the stress increment included three parts in the model: the elastic part, the plastic part, and the fracture damage part. In the calculation process of the proposed model, the plastic stress is solved by the loading surface in the principal stress space based on the classical plastic theory, and the fracture damage stress is solved by the potential function in the strain space. Meanwhile, the strain increment composes of elastic strain, plastic strain, and micro-crack strain. The plastic damage model provided by Zhong (1999) can describe the characteristics of concrete in the hardening stage and softening stage. However, two kinds of loading surfaces must be defined, which leads to a large amount of calculation and complex application conditions.

The study presented by Han and Zhong (1993) provided the endochronic theoretical model, which introduced an internal variable, namely endochronic, to reflect the irreversible damage accumulation. In the proposed model, it divided the deformation of concrete under loading into two parts: the volume change part and the shape change part, then each part includes the elastic deformation and the plastic deformation. Although the endochronic theoretical model differs from the classical plasticity theory, it can simulate the complex properties of concrete. However, the parameters in the proposed model are too many and not independent of each other. Moreover, some parameters lack a clear physical meaning.

Based on the above reviews, the existing theoretical analysis model of the core concrete in concrete-filled steel tube column mainly has the following disadvantages: the expression equation forms vary greatly; the reflection of plastic flow and damaged deterioration of concrete in these models differs from the actual stress situation;

interaction mechanism between concrete and steel tube on their interface is not studied deeply in existing theoretical analysis; the calculation accuracy and efficiency of the models need to be further improved. Therefore, it is necessary to have further study on the theoretical constitutive model for the core concrete in steel tube, especially for the confined concrete in other kinds of concrete composite columns.

1.3 FRP-confined concrete column

FRP is a high property and linear elastic material, which has been deeply studied and widely applied in civil engineering. The FRP-confined concrete column, which comprises an FRP outer tube and concrete-filled in, has gained much attention in existing studies, due to the complex properties of concrete. Similar to the concrete-filled steel tube column, the FRP tube can provide lateral confining pressure on core concrete and improve the mechanical properties and the ductility of core concrete.

However, there are many differences between FRP-confined concrete and steel tube confined concrete: (1) the FRP-confined concrete belongs to passively confined type, the lateral pressure increases with the increase of the lateral expansion of concrete under axial compression, due to the linear elastic property of FRP, whereas the steel tube confined concrete likely belongs to actively confined, the lateral pressure provided by steel tube tends to be constant after entering the yield state; (2) the ultimate failure of FRP-confined concrete happens suddenly with a stress reductions, due to the rupture of FRP tube, whereas failure of steel tube confined concrete does not correspond to a significant change in the stress–strain curve, and the failure form is the local buckling of the steel tube; (3) FRP tube only provides the lateral confining pressure on core concrete, whereas the steel tube provides both the confining pressure on core concrete and the vertical bearing capacity for the composite column.

As shown above, the FRP-confined concrete has great property improvement and complex confining mechanism. Therefore, the experimental and theoretical studies of FRP-confined concrete columns are reviewed and analyzed in the following sections.

1.3.1 The mechanical properties

It is well known that due to the confining effect, concrete is the most complex material in composite structures. The FRP-confined concrete column has a simple interaction mechanism between the FRP tube and core concrete. Therefore, the FRP-confined concrete column should be the basic composite column to comprehensively study the mechanical property of confined concrete. A series of existing researches (Jian & Ozbakkaloglu, 2015; Piekarczyk et al., 2011; Realfonzo & Napoli, 2011; Xie & Ozbakkaloglu, 2015) have shown that the FRP tube can only provide efficient confining pressure distributed around the concrete section to improve the mechanical properties and the durability of the composite column.

Besides, the major factors of the mechanical properties of FRP-confined concrete are the basic properties of unconfined concrete and FRP, the section shape, and the section size. For the basic material properties, high elasticity modulus and large thickness of the FRP tube can provide confining pressure, and the lower-strength concrete can reflect better confining effect under the same confining pressure. Moreover, the section shape can influence the distribution and the stress value of the lateral pressure, including circle shape, square shape, rectangle shape, and special shape, as shown in Fig. 1.8.

As we knew, FRP-confined concrete belongs to passively confined type. Although the load action forms are similar between the passive type and the active type, the former makes the core concrete have more complex axial mechanical properties. In order to compare mechanical properties of the two types of confined concrete, the normalization curves of axial stress versus axial and lateral strains are shown in Fig. 1.9, and we summarize the data from the experimental studies

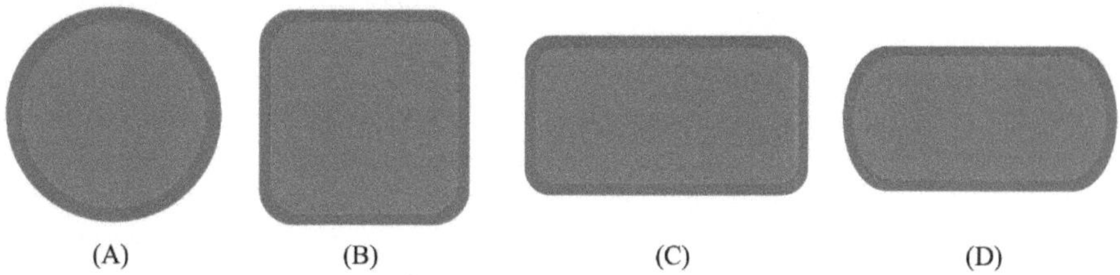

Figure 1.8 The section shapes of FRP-confined concrete columns. (A) circle section; (B) square section; (C) rectangle section; (D) special-shape section.

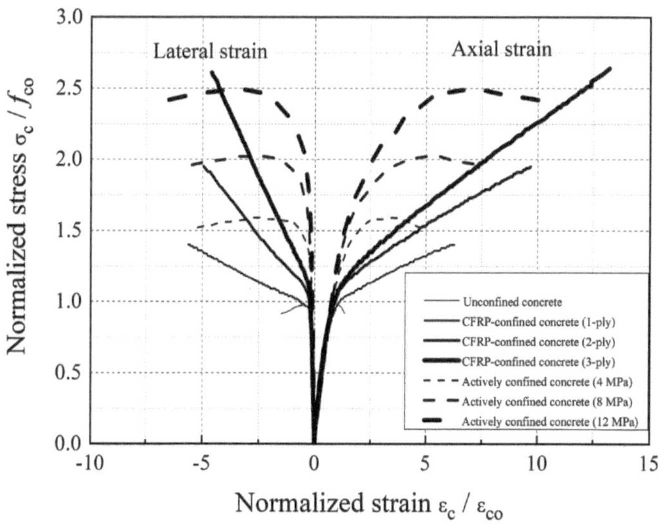

Figure 1.9 The normalization curves of axial stress versus axial and lateral strains of FRP-confined concrete and actively confined concrete. The stress—strain relationship.

(Candappa et al., 2001; Lam & Teng, 2003, 2004; Xie & Ozbakkaloglu, 2015). The stress–strain curve trend of actively confined concrete is similar to that of unconfined concrete, including the increasing stage and decreasing stage. The constant confining pressure can improve the peak stress, peak strain, and the changing trend of curves. In general, the stress–strain curve of FRP-confined concrete shows a bilinear increasing trend, where the first stage is influenced by the properties of filled concrete, and the second stage is influenced by the combination of the stiffness of the FRP tube, the strength grade, and the lateral expansion rate of filled concrete. In addition, the stress and the strain of a peak point between the two stages are slightly larger than the peak point of unconfined concrete, and the peak point of unconfined concrete simply to be the peak point in most existing studies.

For the circle section of FRP-confined concrete, the lateral confining pressure (p) provided by the FRP tube is uniformly distributed around the circumference, as shown in Fig. 1.10. The lateral confining pressure can be computed by $p = f_{frp} / D$, where f_{frp} is the pulling force provided by the FRP tube. The study by Fahmy and Wu (2010) presented an experimental study on various types of FRP-confined concrete cylinders with the same stiffness as FRP tube. The study found that the mechanical properties of confined concrete are irrelevant to the FRP type. And they proposed the axial strength computational model based on lateral stiffness. Furthermore, Ozbakkaloglu and Jian (2013) reviewed and analyzed the experimental studies of over 250 researches on FRP-confined concrete cylinders, and they identified the accuracy of the presented test data by variation trends of parameters in an extensive database. The authors suggested that the actual rupture strains of FRP are smaller than the FRP coupon tensile test results and proposed the ultimate stress model of FRP-confined concrete based on the proposed strain reduction coefficient of FRP tube and the strength of unconfined concrete. Furthermore, Vincent and Ozbakkaloglu (2013) followed the previous study with a study on the effect of the strain reduction coefficient and the strength of unconfined concrete. Then, they summarized the following conclusions based on the analysis results: (1) the strain reduction coefficient has a little relationship with the thickness of the FRP tube; (2) the reduction coefficient decreases with the increase of unconfined concrete strength; (3) the existing stress model cannot provide accurate prediction results for the high-strength concrete. In addition, the study of Dong et al. (2015) presented an

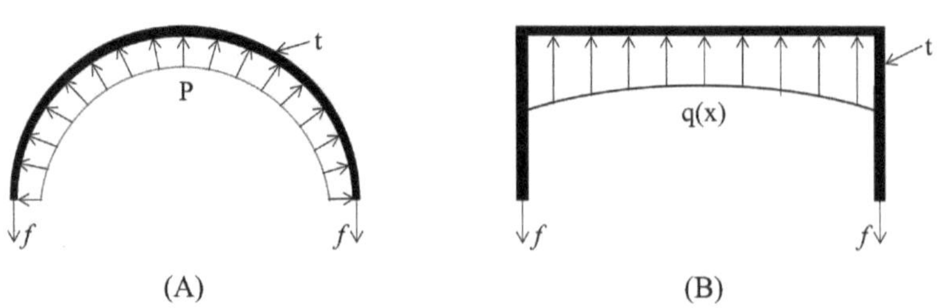

Figure 1.10 The distribution of confining pressure on core concrete. (A) circle section; (B) square or rectangle section.

experimental study to conclude that the peak point between the two curve stages is related to the stiffness of the FRP tube, and the ultimate stress is related to the actual rupture strain, which means that the reduction coefficient should be introduced into the stress–strain models of FRP-confined concrete. The study presented by Bouchelaghem et al. (2011) analyzed the effect of the FRP types, and the experiment results indicate that carbon FRP (CFRP) has high stiffness, a small rupture strain and a large strain reduction coefficient, and a high material utilization rate. In summary, the uniform lateral pressure provided by the FRP tube confines the core concrete in circle section. The confining effect depends on the lateral confining stress, the unconfined concrete strength, and the section size. All study results of FRP-confined concrete cylinders can be the research foundation for all types of section forms.

Except for the circle section column, the square section or rectangular section columns also have been widely applied in practical engineering. However, the non-uniformly pressure confines the core concrete in these two sections, as shown in Fig. 1.11. Therefore, the confining mechanism and mechanical properties of the FRP-confined concrete columns are relatively complex. A series of experimental and theoretical studies (Lo et al., 2015; Saleem et al., 2017; Zhang et al., 2010) have been presented over the past two decades on mechanical behaviors of square and rectangular FRP-confined concrete columns. The characteristics of mechanical properties can be summarized as follows: (1) the ultimate axial strain of the two sectional columns is larger than that of circle section columns; (2) the core concrete in the two sectional columns is confined by non-uniform confining pressure, and the section can be divided into two parts, that is, the effective confined part which is the shaded area in Fig. 1.11 and the invalid confined area which is the blank area in Fig. 1.11; (3) the lateral strain of the FRP tube on section corner is the region leads to the ultimate failure of specimens; (4) the sectional corner has given chamfer, and the effective confinement area increases with the increase of chamfer radius, which leads to larger confining effect; (5) the small length-to-width ratio of FRP tube leads to more effective confining effect. Furthermore, a study Wu and Wei (2010) presented an experimental study on the effect of the length-to-width ratio. This study's results show that the axial ultimate strain of the rectangle column is not influenced by the length-to-width ratio, and the peak stress enhancement of

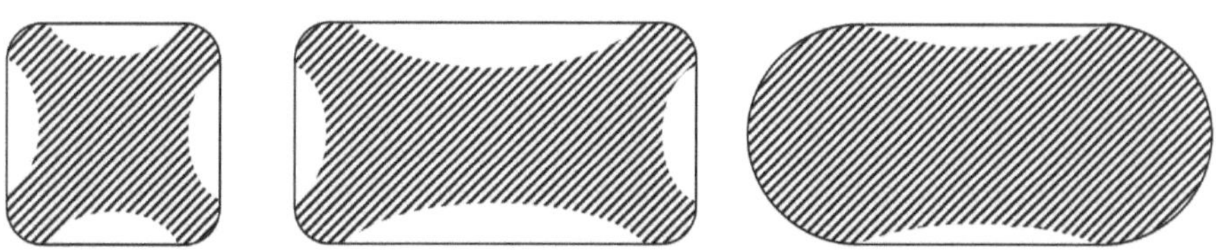

(A) (B) (C)

Figure 1.11 The distribution of confining pressure on non-circle sections. (A) square section; (B) rectangle section; (C) special-shaped section.

core concrete is negligible for a length-to-width ratio over 2, so there is no confining effect in the concrete composite columns with a large length-to-width ratio.

As mentioned above, the mechanical properties of FRP-confined square and rectangular concrete are complex due to the non-uniform confining effect. And the axial stress–strain curves of rectangular concrete under various FRP-confinement effects provided by Ozbakkaloglu (2013) are shown in Fig. 1.12. The characteristics of the stress–strain curves of FRP-confined rectangular concrete are summarized as follows: (1) under the large confining effect, the curves show a bilinear trend, which is similar to the curve of circle-shaped concrete composite columns; (2) under the low confining pressure, the curve shows a trilinear trend, an addition horizontal stage between the two increasing stages; (3) under the insufficient confining effect, an addition increasing stage between the two increasing stages. In general, under the confining pressure provided by non-circular sectional FRP tube, most stress–strain curves of core concrete show the third type of curve trend.

It is easy to find that the sectional corner is the most important part of the rectangle section of the concrete composite columns. Ozbakkaloglu (2013) presented a further study on the section corners, and found that the radius of the section chamfer plays an important role in the mechanical properties of the confined concrete: with the increase in the chamfer radius, the effective confining area increases, whereas the axial ultimate strain decreases. Moreover, the study found that the reinforcement at the section corner can lead to a more uniform confining pressure and larger confining effect provided by the FRP tube, which can improve the ultimate

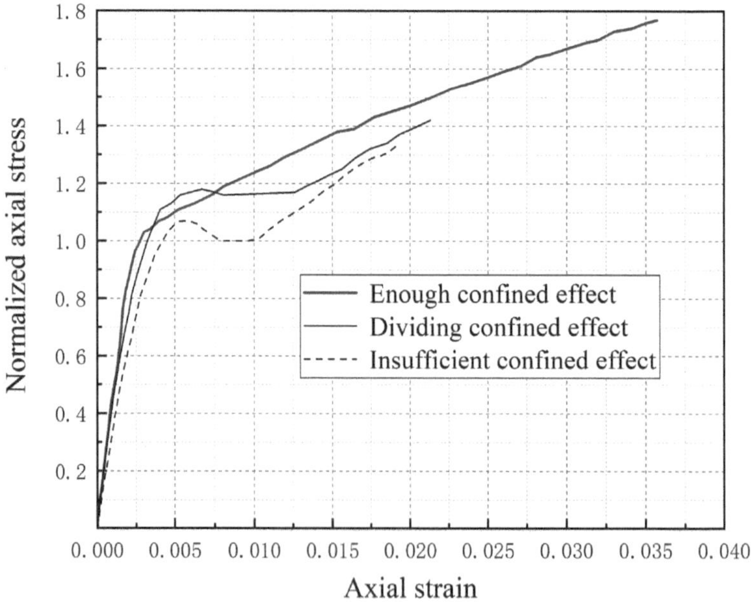

Figure 1.12 Axial stress-strains of square or rectangular FRP-confined concrete. The stress–strain relationship.

stress without the decrease in the ultimate strain. The most important finding is that the FRP rupture area changes from the corner to the central area. Corresponding to the above analysis conclusions, the study by Tan et al. (2013) proposed a new section form which expands the corner area into a whole short side, and the section composed of two semicircle sides and two long sides (namely capsule shape), as shown in Fig. 1.8D. With confining effect of the semicircle side, the stress–strain curve shows a bilinear trend, which means the FRP tube can effectively confine the core concrete. However, the engineering application value is low.

1.3.2 The stress–strain model for FRP-confined concrete

Based on existing studies, we clearly know that lateral confining pressure can improve the strength and ductility of concrete. And a series of experimental and theoretical studies have been presented on the influencing factors of the stress–strain curve of FRP-confined concrete. Based on a variety of parameters, several stress–strain models were established to reflect the axial compression behavior of FRP-confined concrete. Initially, most of the existing studies (Khaloot et al., 1991; Saadatmanesh, 1994) were adjusted directly from the constitutive model of active confined concrete, but these models are inaccurate. Later, some studies (Mirmiran et al., 1999; Saafi et al., 1999) proposed the stress–strain models for FRP-confined concrete. However, these models were established by fitting analysis of limited experimental data. Based on the study proposed by Lam and Teng (2004), most of the existing models belong to the database of empirical models, so they cannot be suitable for an extensive database. Therefore, it is necessary to summarize the existing researches, find the research direction, and establish a unified model which is suitable for most kinds of FRP-confined concrete.

Based on the study proposed by Ozbakkaloglu et al. (2013), the existing proposed models from 1982 to 2011 are reviewed and assessed, and these constitutive models of FRP-confined concrete can be divided into: (1) the design-oriented model, the model comprises a series of equations to reflect the stress–strain curve, which is mainly used in engineering design; (2) the analysis-oriented model, the model is established by numerical iteration procedure, based on the lateral strain model and the stress station of concrete, and the model is mainly used to establish the non-linear finite element model for complex structure analysis. For the design-oriented model, the study (Khalili & Fardis, 1982) first proposed the parabola models referenced from the actively confined concrete. And then some study (Karbhari & Gao, 1997; Nanni & Bradford, 1995; Xiao & Wu, 2000) developed these models into simple bilinear models with two linear rising stages. Finally, some studies (Lillistone & Jolly, 2000; Moran & Pantelides, 2002; Toutanji, 1999) proposed several accurate models in recent years, which are the bilinear model comprising parabolic first stage and linear second stage. However, most of the design-oriented models focused on the stress–strain curves with the rising second stage. Only a few models (Binici, 2008; Lam & Teng, 2003; Wu et al., 2006) proposed the design for the softening stage of the stress–strain curves under the insufficient confining effect. For the analysis-oriented model, Teng et al. (2007) proposed a general design idea and assumed that the stress path is independent, and the axial strain of

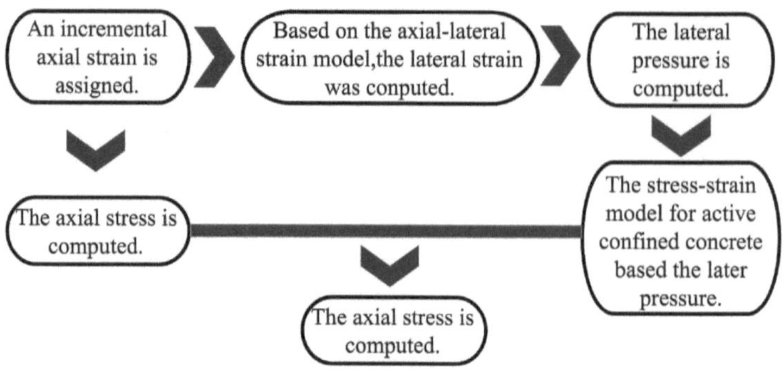

Figure 1.13 The computational process of analysis-oriented model. A flowchart.

FRP-confined concrete is the same as that of the active confined concrete under the same lateral confining pressure. The computational process proposed by Teng et al. (2007) is shown in Fig. 1.13. Following the studies (Hany et al., 2016; Lo et al., 2015; Xiao et al., 2010) about the finite element analysis model, the accuracy mainly depends on the axial-lateral strain model, axial stress model, failure criterion, and stress–strain model of active confined concrete. However, the study proposed by Xiao et al. (2010) suggested that the high-strength concrete shows a path-dependent effect. Then, the study proposed by Ozbakkaloglu and Akin (2012) proved that the path-dependent effect is mainly shown in concrete with insufficient confined effect, and the characteristics of the high-strength concrete have no particularity.

As aforementioned, the axial stress model and the strain model are the key factors in both the design-oriented model and the analysis-oriented model. Shi et al. (2017) deduced the axial stress model based on the twin shear strength theory, and found that the coefficient of confining effect is related to the internal friction angle and the cohesive force of core concrete, but the two parameters are difficult to measure. However, the internal friction angle and the cohesive force can be reflected by the lateral confining pressure and the unconfined concrete strength. In general, the existing studies got the confining coefficient by fitting experimental results in most researches. And it is a convenient method to establish the constitutive model by fitting an extensive database. However, the fitting constitutive model has neither clear physical meaning nor complete mathematic theory as its basis. Against the background, some studies proposed the models based on the innovation in the basic theory, which have definitude of physical significance. Wu and Zhou (2010) established the stress model based on the Hoek-Brown failure criterion. In the study presented by Rong et al. (2018), the failure criterion of concrete under multiaxial stress conditions was improved from the double shear strength theory, which was used to establish the unified stress model (Rong & Shi, 2018). Furthermore, Lim et al. (2016) established the evaluation method for the stress–-strain model. They proposed the predictive model for the ultimate state of FRP-confined concrete by the genetic programming algorithm. In summary, the existing studies are developed from the database fitting to the theoretical deduction. The theoretical development means the research in civil engineering gradually developed into a scientific problem.

1.3.3 The further study of FRP-confined concrete under insufficient confining effect

Based on the above studies, it is clearly known that the relationship between confining pressure and concrete can be used to reflect the major factors for the mechanical properties of FRP-confined concrete (e.g., section ship, stiffness of FRP tube, concrete strength grade, and column size). At present, the strong or weak confining effect on FRP-confined concrete with square and rectangle section have been studied in detail. However, most existing studies ignore the mechanical behavior of the FRP-confined concrete cylinder under weak confining effect. It is necessary to study the properties of FRP-confined concrete cylinders under various types of confining pressure. Therefore, in this section, the experiment (Rong et al., 2020) is set to study various confining effects provided by the FRP tubes.

In an experiment (Rong et al., 2020), one pair of unconfined concrete cylinders and three pairs of glass FRP (GFRP) confined concrete cylinders were tested by axial compression. The cylinder had a diameter of 150 mm and a height of 300 mm. The peak stress of the unconfined concrete was 44.6 MPa. The GFRP tubes were prefabricated via a wet-layup process by wrapping resin-impregnated glass fiber sheets around a foam core, with an overlapping length of 150 mm. The GFRP tubes were composed of one, three, or five plies of fibers; each layer had a nominal thickness of 0.165 mm based on the weight of fibers. The GFRP tubes had an average elastic modulus of 82.11 Gpa and a tensile strength of 1223.4 MPa. For each GFRP tube, four lateral strain gauges with a gauge length of 20 mm were attached at the mid-height of the tube, all four strain gauges were evenly located outside the overlapping zone. Each specimen had two linear variable displacement transducers (LVDTs) placed diagonally to measure the overall shortening. For some specimens, two additional LVDTs diagonal placed were used to measure the axial deformation of the 100 mm mid-height region. We installed two LVDTs on two rings which were fixed to the outer GFRP tube of the specimens using screws. Finally, the loading rate was 0.2 mm/min for all the GFRP confined specimens. In addition, Fig. 1.14 shows the loading process.

Before compression tests stop, all specimens showed suddenly circumferential rupture of FRP tube with an enormous sound. Fig. 1.15 shows the axial stress versus axial and lateral strains curves of GFRP confined concrete. Based on the experimental results and analysis, we can draw the following conclusions: (1) the GFRP tube has less effect on the first stage of the stress–strain curves, due to the small lateral strain and the linear increase of lateral confining pressure; (2) the stress–-strain curve shows the bilinear trend under large confining pressure, as shown in the specimen with 5-ply GFRP; (3) the stress–strain curve also shows the bilinear trend under lower confining pressure, as shown in the specimen with 3-ply GFRP;

Figure 1.14 The loading process of GFRP confined concrete cylinders. A picture.

however, there is a short horizontal transition between two stages, which is similar to the stress–strain curves of GFRP confined high-strength concrete (Ozbakkaloglu & Jian, 2013); (4) the stress–strain curve shows the trilinear trend under insufficient confining pressure, as shown in the specimen with 1-ply GFRP, which is similar to the curves of the rectangular section; (5) the third stage of the stress–strain curve of the specimen with 1-ply GFRP is a horizontal line, mainly because the circle GFRP tube with low lateral stiffness leads to non-obvious increasing trend; however, the continuous increase of lateral stiffness of rectangle FRP tube leads to obvious re-increasing stage. As aforementioned, the major factors of FRP-confined concrete are the lateral stiffness, section shape, and the properties of unconfined concrete.

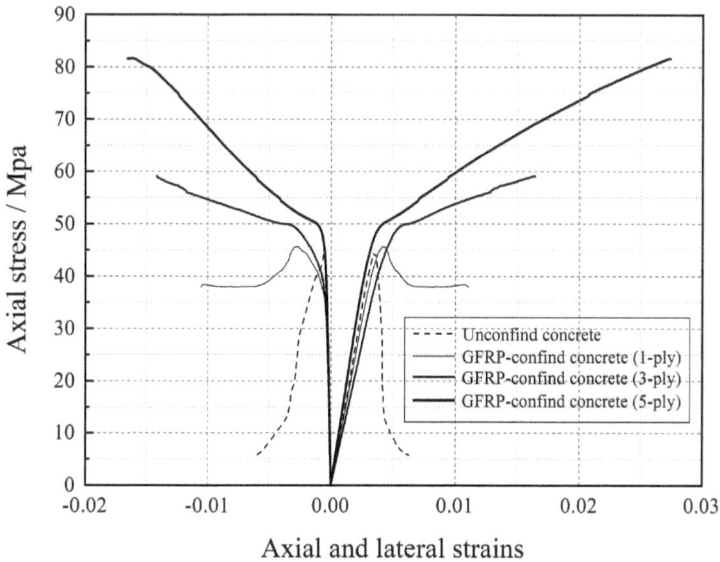

Figure 1.15 The axial stress versus axial and lateral strains curves of GFRP confined concrete. A stress−strain relationship figure.

1.4 FRP-steel-concrete composite column

In recent years, it developed many new types of concrete composite columns from the FRP-confined concrete columns, which comprise FRP section, steel section, and concrete section. Although each new composite column has its own design idea, all of them have great mechanical behavior and seismic behavior, and their application can reduce the structure weight. We can divide the section form into two types: the hybrid double-skin tubular columns (DSTCs) proposed by Teng et al. (2004), which comprises a steel tube inside, an FRP tube outside, and concrete in between; the FRP-confined steel-concrete composite column (i.e., the FRP-confined concrete-filled steel tubular columns and FRP-confined concrete-encased steel columns (CCFTs)) steel section can provide both the axial bearing capacity and lateral confining pressure. In this section, we first review and analyze the existing researches on FRP-steel-confined concrete composite columns. Then, we set an additional experiment to study the interaction relationship between FRP tube and steel tube.

1.4.1 Hybrid double-skin tubular column

The DSTC is a new type of concrete composite column that comprises the FRP outer tube, the steel empty inner tube, and the confined concrete inside, as proposed by Teng et al. (2004) at Hong Kong Polytechnic University. Several experimental and theoretical studies on DSTCs (Abdekarim & Elgawady, 2016; Yu, Hu, et al., 2016) showed that the cross-sectional area can be greatly increased with the constant bearing capacity; therefore, the increase in slenderness ratio can lead to high seismic behavior of the concrete composite columns. The section shapes of DSTCs may be combined with circular or square-shaped FRP tubes and steel tubes, and the circle columns are shown in Fig. 1.16A and B, and the square columns are shown in Fig. 1.16C and D and the rectangular DSTC is shown in Fig. 1.16E. The steel tube is circular in most researches, therefore, the steel tubes in DSTCs were the empty circular tube in this paper, unless otherwise specified.

Yu, Zhang, et al. (2017) presented an experimental study on the circular DSTCs, which comprise circular large-strain FRP tubes and circular steel tubes, the following conclusions can be drawn: (1) there is a decreasing stage after the second increasing stage in the load-displacement curve, due to the local buckling of steel tube; (2) the thickness–diameter ratio of steel tube is the important influence factor, the smaller thickness–diameter ratio leads to the short descending stage; (3) if thickness–diameter ratio of steel tube is large, the FRP will not rupture at the failure of columns; (4) when the steel tube of the DSTC is full of concrete, the local buckling of steel tube could be restricted by the two-side concrete, meanwhile, the great confining effect make the load-displacement curve show the bilinear trend, but the axial ultimate strain is small. The study presented by Yu and Teng (2013) analyzed the comparisons between the circular DSTCs and square DSTCs, and they drew the following conclusions: (1) the mechanical behavior of the square DSTCs is similar to that of the square FRP-confined columns; (2) the circle DSTCs have well mechanical property, because of the great confining effect provided by circle FRP tube; (3) the hollow ratio has an important effect on circular DSTCs, but the non-obvious effect on square DSTCs. In addition, the study presented by Fanggi and Ozbakkaloglu (2015) analyzed the square DSTCs with different shaped steel

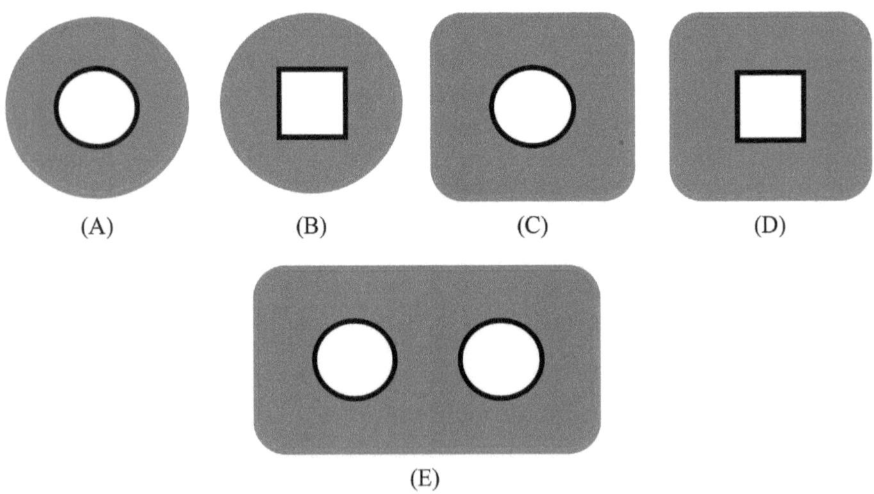

Figure 1.16 The hybrid double-skin tubular columns. (A) circle section with circle steel tube; (B) circle section with square steel tube; (C) square section with circle steel tube; (D) square section with square steel tube; (E) rectangle section.

tubes, and the following conclusions could be drawn: (1) hollow ratio has a significant effect on the mechanical properties of all specimens; (2) the ultimate strain increases with the increasing diameter–thickness ratio of steel tube; (3) the elastic modulus of concrete in DSTCs is larger than that of the unconfined concrete or the concrete in FRP-confined concrete columns; (4) the ultimate stress and strain of DSTCs with circle steel tube are larger than that of DSTCs with square steel tube. Then, Fanggi and Ozbakkaloglu (2013) studied the circular DSTCs with different shaped steel tubes, the study results suggested that the square shape and low strength of the inner steel tube are disadvantaged to mechanical behaviors of DSTCs.

A series of experimental studies have been presented on the influence factors of DSTCs. In the study presented by Ozbakkaloglu and Fanggi (2014), the influence factors included the thickness of FRP tube, unconfined concrete strength, sectional hollow ratio, yield strength, and diameter–thickness ratio of steel tube, and the following conclusions could be drawn: (1) the thicker FRP tube and the lower unconfined concrete strength could show the better confining effect; (2) the lower yield strength of steel tube may reduce the mechanical property of the confined concrete in DSTCs, but the effect is less significant; (3) the increasing thickness of steel tube could improve the mechanical behavior of the columns; (4) the mechanical mechanism of the solid DSTCs differs from that of the empty DSTCs; (5) the filled concrete in DSTCs can limit the local buckling of steel tube, improve the ultima stress of confined concrete with the little decrease of the ultima strain. The study presented by Abdekarim and Elgawady (2016) shows that the mechanical behavior of the DSTCs is related to the lateral stiffness of FRP tube, and lower lateral stiffness leads to the lower strength and the larger ultima strain of DSTCs. Based on the study presented by Zhou et al. (2017), the confining effect is significant for the DSTCs with lightweight aggregate concrete, and the hollow ratio has little effect on the confining effect. Based on the above analysis, it is well known that the local buckling is an important factor. Therefore, Peng (2017) introduced ribbed steel tubes to improve the mechanical behaviors of the DSTCs. The section form of the special composite column is shown in Fig. 1.17, and the following conclusion can be drawn: (1) the steel ribs delays the local buckling of the steel tube, so as to improve the mechanical behaviors; (2) under the same steel ratio, the increasing

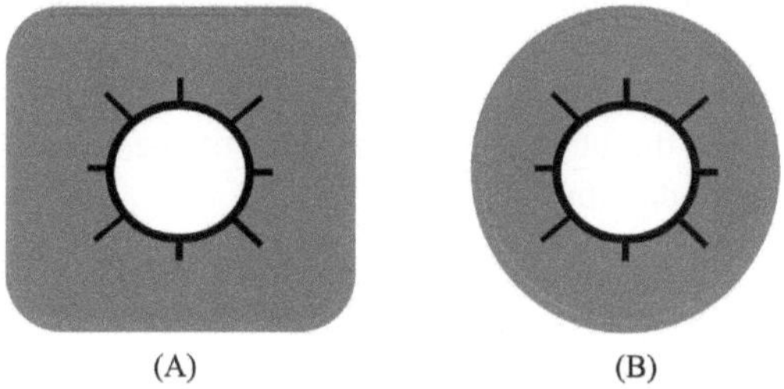

Figure 1.17 The hybrid double-skin tubular columns with ribbed slabs. (A) square section; (B) circle section.

number of steel ribs can improve the mechanical behaviors; (3) the corner distribution of the steel ribs has a better effect in square DSTCs, because of the non-uniform confining effect provided by square FRP tube.

Furthermore, the study presented by Wong et al. (2008) discussed the internal mechanical mechanisms between concrete and FRP or steel tube in experimental studies that are based on the mechanical comparisons of the FRP-confined concrete columns, the FRP-confined hollow concrete columns, and the DSTCs, and following conclusions can be drawn: (1) the stiffness and the peak stress in first stage of concrete in the FRP-confined hollow concrete column are larger than that of concrete in the DSTCs, because the stress concentration around the hollow lead to the larger confining pressure provided by FRP tube; (2) the second stage of the stress−strain curve shows descending trend, the failure of the FRP-confined hollow concrete column occurred on the core concrete, and there is no hoop rupture of FRP throughout the loading process; (3) the stress−strain curves of DSTCs are similar to that of FRP-confined concrete column, because the inner steel tube can lead to internal force redistribution, and the redistributed force makes the lateral confining pressure uniform; (4) the latter stage of the stress−strain curves of DSTCs show descending trend, due to the local buckling of steel tube; (5) both the hoop rupture of FRP and the local buckling of steel tube can be the failure mode of DSTCs. Yu et al. (2010) presented the finite element analysis on the mechanical properties of the confined concrete in DSTCs, the study showed that the stiffness ratio of FRP, lateral strain, and hollow ratio have an important effect, and the constitutive models of core concrete were proposed by parameter fitting.

Except for the axial monotonic compression studies, there are existing experimental studies on cyclic loading compression for the DSTCs. The experimental study presented by Yu et al. (2012) shows that the stress−strain curves of the DSTCs under cyclic loading are the same as that under axial monotonic loading, however, the loading loss in some DSTC specimens is because of the bonding slip between concrete and steel tube. The experimental study presented by Abdekarim and Elgawady (2016) analyzed the effect of diameter−thickness ratio of steel tube and the wrapping method of FRP, and the following conclusions can be drawn: (1) the large diameter−thickness ratio of steel tube easily leads to the bonding slip, so as to affect the mechanical behavior; (2) the FRP wrapping with 45 oblique angles can increase the ultima strain and decrease the ultima stress of core concrete; (3) the FRP tubes, which are wrapped in both the hoop direction and the oblique direction, could increase the ultima strain and the ultima stress, and the failure modes can be divided into two stages: the first stage is the rapture of the hoop FRP and second is the rapture of the oblique FRP, because of the axial strain redistribution of the FRP in the height direction.

1.4.2 FRP-confined concrete-filled steel tube column

In recent years, concrete-filled steel tube columns have been widely used in practical engineering structures. As we knew for the composite column, the core concrete can only limit the inward buckling of the steel tube. However, the strength and ductility degradation will be mainly controlled by the outward buckling of the steel tube at both ends of the column. Against these disadvantages, existing researches proposed the FRP-confined concrete-filled steel tube column, where the outer local buckling of the steel tube can be restricted by the FRP tube. The study presented by Xiao (2009) used steel or FRP to confine both ends of a steel tube, the reinforced method can significantly increase the bearing capacity and ductility of the concrete-filled columns. The more important thing is that the confining effect of FRP is better than that of steel tube. Furthermore, FRP-confined concrete-filled steel tube columns were studied by several scholars (Choi et al., 2005; Mao & Xiao, 2006), and they suggested the design sections of the new concrete composite columns, as shown in Fig. 1.18.

Hu et al. (2011) presented an experimental study on CFRP confined concrete-filled thin-walled steel tube column, and the following conclusions can be drawn: (1) the wrapping CFRP could limit or even remove the local buckling of steel tube; (2) the wrapping FRP can increase the ultimate stress and ultimate strain of core concrete; (3) the failure mode is the hoop rupture of FRP tube; (4) the stress–strain curves of core concrete can be divided into three stages: the first stage is controlled by concrete, the second stage is controlled by steel tube and FRP, and the third stage is controlled by FRP. In addition, Hu (2011) had a further study on FRP-confined thin-walled steel tube in detail, and the study showed that FRP has well-confining effect on both hollow steel tubes and concrete-filled steel tube columns. In the study by Tao et al. (2007), both circular CCFT and square CCFT were studied and the following conclusions can be drawn: (1) the sectional shape has a significant effect on the mechanical behavior; (2) the wrapping FRP has well-confining effect on the circular section; (3) with the increasing thickness of FRP, the ductility of circular CCFT is decreasing, while that of square CCFT is increasing. The study of Park et al. (2010) showed that the increasing thickness of FRP can increase the strength and ductility of the concrete composite columns. However, the wrapping FRP has little effect on the bearing capacity of the rectangular-shaped columns. Tao et al. (2007) presented an experimental study on the FRP-confined concrete-filled steel tube columns under axial cyclic loading, the study results can be concluded as follows: (1) the failure mode of the composite column under axial cyclic loading is the same as that under axial monotonous

(A) (B)

Figure 1.18 The FRP-confined concrete-filled steel tube columns. (A) square section; (B) circle section.

loading; (2) the enveloping curve under cyclic loading is the same as the stress--strain curve under monotonic loading; (3) the cyclic loading process leads to the damage accumulation of core concrete, therefore, the stress response and the strain response under cyclic loading is depended to the loading method and it can be predicted by the stress–strain curve under monotonic loading.

The confined concrete is the key and complex part of FRP-confined concrete-filled steel tube columns. Therefore, existing researches focused on the mechanical property of the confined concrete and the interaction mechanism between FRP and steel tube. Teng et al. (2013) proposed the stress–strain models of the concrete in the composed columns, based on the stress–strain models of actively confined concrete, the axial strain-lateral strain models, and the lateral pressure computational method of steel tube. However, the models did not consider the unconfined concrete strength, the local hoop strain of FRP, and the stiffness ratio between steel tube and FRP tube. Based on that, Teng et al. (2015) proposed two kinds of analysis-oriented models: (1) one model directly incorporates the confining pressure of steel tube into the total lateral confining pressure; (2) another model introduces the lateral stiffness ratio between FRP and steel tube, which considers the interaction relationship between FRP tube and steel tube. Furthermore, the study of Dong et al. (2016) showed that the local buckling of steel tube leads to the interface separation between steel tube and core concrete, which ultimately leads to the failure of the concrete composite columns, whereas the interface separation could be restricted by wrapping FRP tube. Therefore, Dong et al. (2016) proposed the constitutive models of the core concrete in the two tubes, based on the axial-lateral strain model, the constitutive model of confined concrete, the biaxial constitutive model of steel tube, and the stiffness of FRP tube. And each stress state or strain state can be computed by the proposed model. Based on the above studies, it is found that enough wrapping FRP could both provide an effective confining effect on core concrete and prevent interface separation between steel tube and core concrete, and the steel tube can prevent the sudden failure after the brittle rupture of FRP tube.

Based on existing experiments, the failure model of FRP-confined concrete-filled steel tube columns shows the hoop rupture of FRP tube, and the local buckling of steel tube is also an important influence factor. In order to further restrict the local buckling of the steel tube, foam plastic materials were used to set a layer between the FRP tube and steel tube in the study by Xiao et al. (2005), as shown in Fig. 1.19A. The study by Xiao et al. (2005) shows that the plastic layer could delay the rupture of FRP and the ductility of the composite column would be greatly improved.

Furthermore, the comparisons of the load-strain curves of concrete-filled steel tube (CFT), the FRP-confined composite column (CCFT), and the confined composite column with plastic layer (CCFT-G) are shown in Fig. 1.20. The following conclusions can be drawn from the curve comparisons: (1) the first stage in the curves of the three columns is similar; (2) the plastic layer has a buffer effect, therefore, the lateral strain of the FRP tube is smaller than that of steel tube, and rupture of the FRP tube is delayed, which leads to the increasing ultimate strain; (3) the ultimate stresses of CCFT and CCFT-G are similar. In addition, the thickness of the

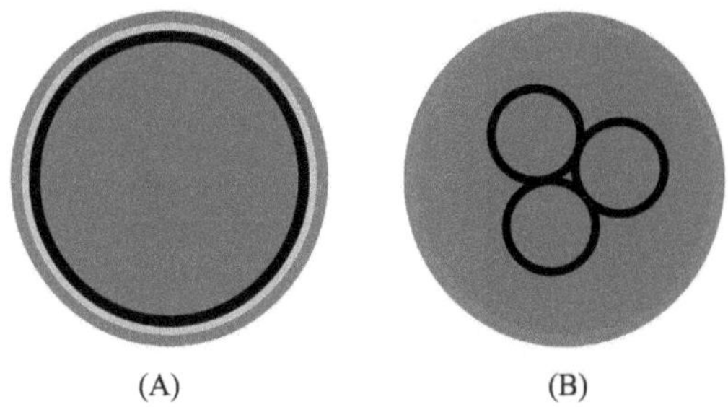

Figure 1.19 The section of new-type FRP-confined concrete-filled steel tubes. (A) FRP-confined composite column; (B) FRP-confined concrete-filled multitube column.

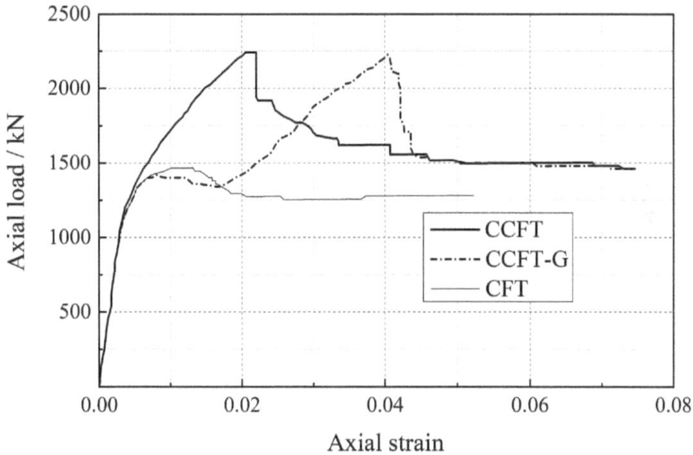

Figure 1.20 The load-strain curves of CFT, CCFT, and CCFT-G. The load-strain relationship.

plastic layer is just 1 mm and provides ignorable stress, so the lateral confining pressure of FRP in CCFT-G is similar to that in CCFT. Then it can be concluded that stress is loading-path independence and strain is loading-path dependence.

In general, it is difficult to produce oversize steel tubes for mega-columns in practical engineering. Therefore, Yu, Chan, et al. (2017) proposed a new type of concrete composite columns with a number of small-size steel tubes, namely the FRP-confined concrete-filled multitube column (MTCC), as shown in Fig. 1.19B. The mechanical behavior characteristics of the concrete composite columns could be summarized as follows: (1) the load-strain curves show the bilinear trend, which are similar to FRP-confined concrete columns; (2) there is no local buckling of the inner steel tube, the combination between the concrete and FRP could completely limit the steel tubes; (3) with the same steel ratio, the ductility of FRP-confined concrete-filled multitube column is better than that of FRP-confined concrete-filled steel tube column.

1.4.3 FRP-confined steel reinforced concrete column

The steel-reinforced concrete columns also have been widely used in practical engineering. Therefore, the concrete composite columns confined by FRP tube have been widely studied in many existing researches (Karimi et al., 2012; Zakaib & Fam, 2012). The common section shapes are shown in Fig. 1.21.

The study presented by Yu, Lin, et al. (2016) shows that I-shaped steel reinforced concrete column can be confined by FRP, and the section deformation agrees with the plain section assumption under eccentric loading. Karimi et al. (2013) presented an experiment on the slenderness ratios of FRP-confined I-shaped steel reinforced concrete columns. It suggested that the slenderness ratio has an obvious effect when the value is over 19.8, and the limit value increases with the increasing confining effect provided by FRP tube. In the study on FRP-confined I-shaped steel reinforced concrete columns (Huang et al., 2012), Huang et al. drew the following conclusions: the lateral deformation of the composite column is related to the lateral stiffness and axial stiffness of the I-steel columns; the circle columns have well mechanical properties than the square concrete composite columns; the I-shaped steel section could not provide obvious confining effect in the concrete composite columns. Based on the above conclusions, Huang et al. (2021) studied the concrete composite columns with cross-shaped steel, and the following conclusions can be drawn: (1) the cross-section steel can provide confining effect on the weak area of core concrete in the square section; (2) FRP can effectively limit the local buckling of cross-section steel; (3) the mechanical behaviors of the square concrete composite columns can be improved by confining effect of FRP and cross-section steel; (4) the failure mode of the composite column is the hoop rupture of FRP at the midpoint of both the boundary and the height direction. In addition, the above researches showed that the analysis-oriented model of FRP-confined concrete can be used to establish the finite element model of FRP-confined steel reinforced concrete column.

1.4.4 The further study of FRP-confined concrete-filled steel tube

As aforementioned, it is clearly known that the key parameters of the failure of FRP-confined concrete-filled steel tube columns are the local buckling of steel tube

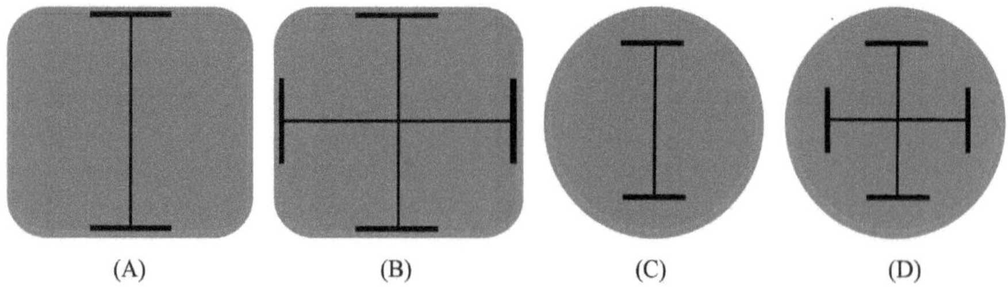

Figure 1.21 The FRP-confined steel reinforced concrete columns. (A) I-steel in square section; (B) cross-shaped steel in square section; (C) I-steel in circle section; (D) cross-shaped steel in circle section.

and the interaction relationship between FRP tube and steel tube. However, most of the existing studies focused on concrete composite columns with thin-wall steel tube. Therefore, this section presents an experimental study to analyze the interaction relationships between FRP tube and steel tube, and the studied influence factor is the FRP wrapping method (Rong et al., 2020).

In the experiment design, the cylinder had a bottom diameter of 180 mm and a height of 500 mm. The peak stress of unconfined concrete was 53.6 MPa and the peak strain was 0.0045. The steel tube had an outer diameter of 180 mm and a thickness of 4.5 mm. The FRP tubes were prefabricated via a wet-layup process by wrapping resin-impregnated aramid fiber, with an overlapping length of 180 mm. The FRP wrapping methods are shown in Table 1.1, each specimen was given a name, and the naming rules are as follows: CFT means the concrete-filled steel tube column; CCFT means the FRP-confined concrete-filled steel tube column; the number after letter "T" means the number of FRP plies; and the letter after the number means the wrapping method of FRP tube. In addition, each layer had a nominal thickness of 0.18 mm based on the weight of fibers, and the sketch is shown in Fig. 1.22. The FRP used in the study had an average elastic modulus of 184 GPa and a tensile strength of 2741.6 MPa. For each FRP tube, four lateral strain gauges with a gauge length of 20 mm were attached at the mid-height of the tube, all four strain gauges were evenly located outside the overlapping zone. For each specimen, two LVDTs diagonal placed were used to measure the overall axial shortening. Finally, the loading rate was 0.5 mm per minute for all the FRP-confined specimens.

After the compression test, the failure modes of the specimens are shown in Fig. 1.23. It is found that the failure modes of all specimens were similar, due to the local buckling of steel tube and rapture of FRP tube. And for the CCFTs and CCFT-Bs, the failure process was that the local buckling first occurred on the ends of steel tube, followed by the hoop wrapping on the ends of FRP tube, and finally the whole FRP wrapped. For the CCFT-Ts, the failure process was that the local buckling first occurred on the void area of steel tube, followed by the hoop wrapping on the central FRP tube, and finally the two end FRP tubes wrapped.

Table 1.1 The wrapping method of AFRP around the steel tube.

Item	Plies of FRP	Wrapping method of FRP
CFT	0	—
CCFT2	2	Overall wrapping
CCFT3	3	Overall wrapping
CCFT2B	2	Two-sectional wrapping
CCFT3B	3	Two-sectional wrapping
CCFT2T	2	Three-sectional wrapping
CCFT3T	3	Three-sectional wrapping

Source: From Rong, C., Shi, Q. X., & Zhao, H. C. (2020). Behavior of FRP composite columns: review and analysis of the section forms. *Advances in Concrete Construction*, *9*(2), 125–137.

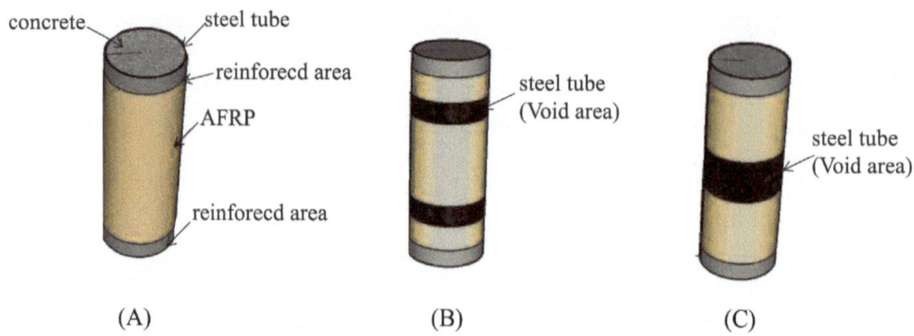

Figure 1.22 The sketch of new CCFTs section. (A) CCFT; (B) CCFT2T or CCFT3T; (C) CCFT2B or CCFT3B.

Figure 1.23 The failure modes of the specimens. A picture.

The load-strain curves of all specimens are shown in Fig. 1.24, where the curve of "Steel + FRP-confined concrete" is obtained by the direct addition of confined concrete and steel tube, and the load of the confined concrete is computed by the stress–strain model proposed by Teng et al. (2007). Based on the curve comparison, the following conclusions can be drawn: (1) the confining effect of FRP is non-obvious because the FRP is directly wrapped in the steel tube, and the local buckling of steel tube is against the transmission of the lateral pressure; (2) the local buckling of steel tube is the key to the failure of columns, and if the confining effect of FRP is not enough, the thickness changing can be neglected; (3) the load-strain curves of CCFT-B are similar to that of CCFT, and the void area in the middle has no effect on the failure modes because the local buckling is at the ends; (4) the ductility of the CCFT-T is well, due to the void area at the ends can be the reserve area for local buckling of the steel tube, and the expansion effect provided by the steel tube on the FRP is slowed down; (5) For the CCFT-T, the thicker FRP tube leads to better ultimate strength, the thinner FRP tube leads to better ductility.

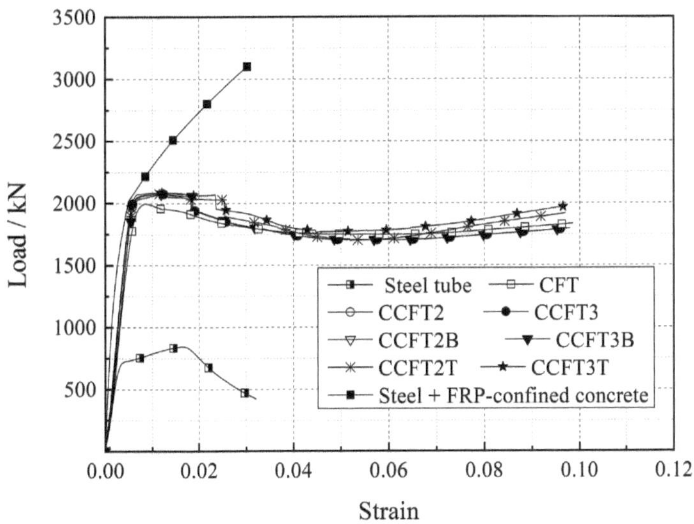

Figure 1.24 The load-strain curves of all test specimens. The load-strain relationship.

Table 1.2 The characteristics of each material.

Item	Characteristic	Disadvantage
FRP	Linear elasticity	Brittle failure
Steel	Linear elasticity before yielding, constant stress after yielding	Decreased mechanical behavior because of the local buckling
Concrete	The property depends on confining stress	The mechanical behavior is complex, the influence factors are confining types and confining effect

1.5 Discussion for the composite column

Based on the above reviews and additional experimental studies, properties of the compositive materials and section form are two main influence factors for the mechanical behavior of concrete composite columns. The former factor is the foundation part and the latter factor is a design method to make the material properties further improved. Thus, the section form should be reasonably designed by following the complementary effect of each material to play maximize utilization and optimize the structure behavior.

There were many existing researches on the mechanical behavior of the concrete composite columns with different compositive materials and various sectional forms. We deeply analyzed the material properties, as shown in Table 1.2. However, the following contents are ignored in most existing researches: (1) the local buckling of section steel and steel tube; (2) the stress variation of concrete under multiaxial stress condition; (3) the stress mechanism of FRP-confined concrete under insufficient confining effect; (4) the preventive measures and monitoring methods for suddenly rupture of FRP tube; (5) the reasonable compound mode between different materials. It is necessary to study the complementary effect of materials and find a reasonable design method for the practical engineering structure.

Based on existing studies, some measures can be concluded as follows: (1) the concrete should be confined to prevent brittle failure; (2) the circular section can provide greatly confining effect, and it can use in the core area of the non-circular section; (3) in non-circular section, FRP should be whole wrapped, and the corner should be reinforced; (4) the steel could put in inefficient confining region to add the confining effect; (5) the local buckling of steel part is a key defect in the composite column, and it can be restricted by the combination of concrete and FRP. Therefore, this paper proposes a reasonable design of section form, and it can be shown in the flowchart in Fig. 1.25.

Based on the reasonable design flowchart of section form and the above review in this paper, the following research points should be studied in future studies: (1) the detailed properties of concrete under complex stress condition; (2) the detailed mechanical behavior of confined concrete, including various types and developing methods of confining compression; (3) the reasonable section form to make the materials play the advantages and improve the disadvantage; (4) the seismic behavior of new composite structures; (5) the application of new composite structures in

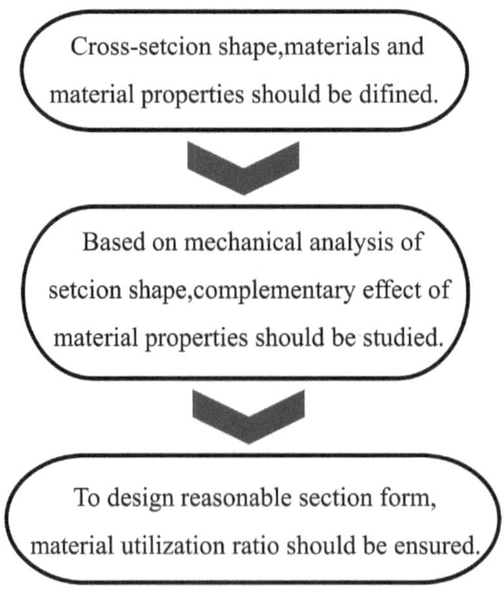

Fig. 1.25 The reasonable section design method. A flowchart.

practice engineering. All the mentioned study contents have been presented in the following chapters.

1.6 Conclusions

To study the relationships and the interaction mechanism between FRP, steel, and concrete, this chapter presented the review and analysis of the existing experimental and theoretical researches of the concrete composite columns, including the concrete-filled steel tube columns, the steel reinforced concrete-filled steel tube columns, the FRP-confined concrete columns, and the FRP-confined steel-concrete columns. Meanwhile, two additional experiments were set to definitely study the mechanical properties of FRP-confined concrete and the interaction relationships between the steel tube and FRP tube. This paper can be the theoretical foundation to study and design the concrete composite columns with excellent behavior in future researches. Based on this, the following conclusions can be drawn:

(1) The mechanical properties of confined concrete are only related to the confining types, relative confining stiffness, and the unconfined concrete properties have little effect on them. Therefore, the unified stress–strain model of confined concrete can be established based on the confining characteristics.

(2) The stress–strain curves of steel tube confined concrete show two stages, and the curve trend is similar to unconfined concrete. The curves of FRP-confined concrete under enough confining effect show bilinear trend. Otherwise, the stress–-strain curves show a trilinear trend.

(3) The non-uniform confining effect of the square and the rectangular section can lead to increasing stiffness in the latter loading stage.

(4) If there is no damage in concrete, the stress path is loading path independent. The stress path independence and the strain path dependence are shown in Fig. 1.20.

(5) It is necessary to establish the uniform stress–strain model of confined concrete. We suggest the model should be established as a science project, based on the combination of theoretical analysis and experimental data. The stress model can be derived based on the failure criterion and the strain model can be derived by energy method.

(6) The FRP-confined steel-concrete composite columns have excellent mechanical properties, seismic behavior, and durability. The enveloping curves of the concrete composite columns under axial cyclic loading are similar to the stress–strain curves under axial monotonic loading.

(7) The failure of most concrete composite columns is mainly controlled by the local buckling of the steel. And the local buckling of steel tube cannot be restricted by the materials filled by single-side. For example, the failure of DSTCs is because of the interior local buckling of the steel tube, and the failure of concrete-filled steel tube is because of the outer local buckling of the steel tube.

(8) The local buckling of the steel tube can be limited by the outer FRP tube and the inner confined concrete. However, the FRP tube has an obvious effect on the thin-walled steel tube, but little effect on the thick-wall steel tube.

(9) Based on some studies (Huang et al., 2012, 2021; Karimi et al., 2013), there are two methods to restrict the local buckling. For the circle section, the concrete should be filled between steel tube and FRP tube. For the other shaped section, the steel should be placed in the weak confining area of the concrete section to add the confining effect.

Acknowledgments

This work was supported by the National Natural Science Foundation of China [grant numbers 52108171; 51878540]. The authors gratefully acknowledge two Master's students (Qv Yunsong and Tian Wenkai) of Xi′an University of Architecture and Technology for the drawing works.

References

Abdekarim, OI, & Elgawady, MA (2016). Behavior of hollow FRP-concrete-steel columns under static cyclic axial compressive loading. *Engineering Structures*, *123*, 77−88.

Binici, B (2008). Design of FRPs in circular bridge column retrofits for ductility enhancement. *Engineering Structures*, *30*(3), 766−776.

Bouchelaghem, H, Bezazi, A, & Scarpa, F (2011). Compressive behaviour of concrete cylindrical FRP-confined columns subjected to a new sequential loading technique. *Composites Part B: Engineering*, *42*(7), 1987−1993.

Cai, J, Xie, X, Yang, C, He, J, & Chen, H (2002). An experimental research on the composite column with core of high-strength concrete filled steel tube under axial compression loading. *Journal of South China University of Technology (Natural Science Edition)*, *30* (6), 81−85.

Cai, S (1999). Recent development of steel tube-confined concrete structures in China. *China Civil Engineering Journal*, *32*(4), 16−25.

Cai, S, & Jiao, Z (1984). Behavior and ultimate strength of short concrete-filled steel tubular columns. *Journal of Building Structure*, *6*(5), 13−29.

Candappa, DC, Sanjayan, JG, & Setunge, S (2001). Complete triaxial stress-strain curves of high-strength concrete. *Journal of Materials in Civil Engineering*, *3*(3), 209−215.

Choi, KK, & Xiao, Y (2009). Analytical studies of concrete-filled circular steel tubes under axial compression. *Journal of Structural Engineering*, *136*(5), 565−573.

Choi, K, Xiao, Y, & He, W (2005). Confined concrete-filled tubular columns. *Journal of Structural Engineering*, *131*(3), 488−497.

Dong, CX, Kwan, AKH, & Ho, JCM (2015). Effects of confining stiffness and rupture strain on performance of FRP confined concrete. *Engineering Structures*, *97*, 1−14.

Dong, CX, Kwan, AKH, & Ho, JCM (2016). Axial and lateral stress-strain model for concrete-filled steel tubes with FRP jackets. *Engineering Structures*, *126*, 365−378.

Du, G, Xu, L, Xu, H, & Wen, F (2008a). Experiment research on behaviour of T-shaped concrete filled steel tube stubs. *Journal of Huazhong University of Science and Technology (Urban Science Edition)*, *25*(3), 188–190.

Du, G, Xu, L, Xu, H, & Wen, F (2008b). Study on behavior of composite T-shaped concrete filled steel tubular short columns under axial compression. *Journal of Xi'an University of Architecture and Technology (Natural Science Edition)*, *40*(4), 549–554.

Ellobody, E, Young, B, & Lain, D (2006). Behaviour of normal and high strength concrete-filled compact steel tube circular stub columns. *Journal of Construction Steel Research*, *62*, 706–715.

Fahmy, MFM, & Wu, Z (2010). Evaluating and proposing models of circular concrete columns confined with different FRP composites. *Composites Part B: Engineering*, *41*(3), 199–213.

Fanggi, BAL, & Ozbakkaloglu, T (2013). Compressive behavior of aramid FRP-HSC-steel double-skin tubular columns. *Construction and Building Materials*, *48*(19), 554–565.

Fanggi, BAL, & Ozbakkaloglu, T (2015). Square FRP-HSC-steel composite columns: Behavior under axial compression. *Engineering Structures*, *92*, 156–171.

Gu, W, Cai, S, & Feng, W (1991). Research on the behavior and loading capacity of slender steel tubular columns filled with high-strength concrete. *Building Science*, *3*, 3–8.

Han, I, Yao, G, & Zhao, X (2005). Tests and calculations for hollow structural steel (HSS) stub columns filled with self-consolidating concrete (SCC). *Journal of Constructional Steel Research*, *61*(9), 1241–1269.

Han, LH, Liu, W, & Yang, YF (2008). Behaviour of concrete-filled steel tubular stub columns subjected to axially local compression. *Journal of Constructional Steel Research*, *64*, 377–387.

Han, LH, & Zhong, ST (1993). The use of endochronic theory in describing the constitutive relations of the core concrete of concrete filled steel tubular (CFST) under mulitiple-loaded conditions. *Journal of Harbin Architecture and Civil Engineering Institute*, *26* (2), 48–55.

Han, L, Yao, G, & Chen, Z (2005). Experimental behavior of steel tube confined concrete (STCC) column. *Steel and Composite Structures*, *5*(6), 459–484.

Hany, NF, Hantouche, EG, & Harajli, MH (2016). Finite element modeling of FRP-confined concrete using modified concrete damaged plasticity. *Engineering Structures*, *125*, 1–14.

Hu, YM (2011). *Behavior and modeling of FRP-confined hollow and concrete-filled steel tubular columns*. Hong Kong Polytechnic University.

Hu, YM, Yu, T, & Teng, JG (2011). FRP-confined circular concrete-filled thin steel tubes under axial compression. *Journal of Composites for Construction*, *15*(5), 850–860.

Huang, L, Yu, T, & Zhang, SS (2021). FRP-confined concrete-encased cross-shaped steel columns: Effects of key parameters. *Composite Structures*, *272*(4), 114252.

Huang, L, Zhang, SS, Yu, T, & Wang, ZY (2012). *Concrete-encased steel columns confined with large rupture strain FRP composites: Axial compression tests*. In: Proceedings, the 24th Australian Conference on the Mechanics of Structures and Materials (ACMSM24), Perth.

Jian, CL, & Ozbakkaloglu, T (2015). Influence of concrete age on stress-strain behavior of FRP-confined normal- and high-strength concrete. *Construction and Building Materials*, *82*(4), 61–70.

Karbhari, V M, & Gao, Y (1997). Composite jacketed concrete under uniaxial compression-verification of simple design rquations. *Journal of Materials in Civil Engineering*, *9*(4), 185–193.

Karimi, K, Tait, MJ, & El-Dakhakhni, WW (2012). Influence of slenderness on the behavior of a FRP-encased steel-concrete composite column. *Journal of Composites for Construction*, *16*(1), 100–109.

Karimi, K, Tait, MJ, & El-Dakhakhni, WW (2013). Analytical modeling and axial load design of a novel FRP-encased steel-concrete composite column for various slenderness ratios. *Engineering Structures*, *46*, 526–534.

Khalili, HH, & Fardis, MN (1982). FRP-encased concrete as a structural material. *Magazine of Concrete Research*, *121*(34), 191–202.

Khaloot, AR, Irshaid, A, & Ahmad, SH (1991). Behavior of concrete spirally confined by fiberglass filaments. *Magazine of Concrete Research*, *156*(43), 143–148.

Lam, L, & Teng, JG (2003). Design-oriented stress-strain model for FRP-confined concrete. *Construction and Building Materials*, *17*(6), 471–489.

Lam, L, & Teng, JG (2004). Ultimate condition of fiber reinforced polymer-confined concrete. *Journal of Composites for Construction*, *8*(6), 540–548.

Lee, J, & Fenves, GL (1998). Plastic-damage model for cyclic loading of concrete structures. *Journal of Engineering Mechanics*, *124*(8), 892–900.

Li, H, Wang, Z, & Wu, B (1999). Analysis on axial loading distribution and limitation of nominal axial compression ratio of laminated column with high strength concrete filled steel tube. *World Information on Earthquake Engineering*, *15*(2), 1–8.

Liang, Q, & Fragomeni, S (2009). Nonlinear analysis of circular concrete-filled steel tubular short columns under axial loading. *Journal of Constructional Steel Research*, *65*, 2186–2196.

Lillistone, D, & Jolly, CK (2000). An innovative form of reinforcement for concrete columns using advanced composites. *Structural Engineer*, *78*(23), 20–28.

Lim, JC, Karakus, M, & Ozbakkaloglu, T (2016). Evaluation of ultimate conditions of FRP-confined concrete columns using genetic programming. *Computers and Structures*, *162*, 28–37.

Lin, Y, Cheng, W, Xu, M, & Zuo, J (2001). Experimental study on limit values of axial compression ratio of SRC columns with circular steel tube. *China Civil Engineering Journal*, *34*(6), 23–28.

Lin, Z, Shen, Z, & Luo, J (2009). Hysteretic behavior of concrete-filled L-section steel tubular columns under cyclic loading. *Progress in Steel Building Structures*, *11*(2), 12–17.

Lo, SH, Kwan, AKH, & Ouyang, Y (2015). Finite element analysis of axially loaded FRP-confined rectangular concrete columns. *Engineering Structures*, *100*, 253–263.

Long, Y, & Cai, J (2006). Experimental investigation into axial compressive behavior of L-shaped concrete filled steel tubular stub columns with binding bars. *Journal of South China University of Technology (Natural Science Edition)*, *34*(11), 87–92.

Mao, XY, & Xiao, Y (2006). Seismic behavior of confined square CFT columns. *Engineering Structures*, *28*(10), 1378–1386.

Martin, D, Shea, O, & Russsell, Q (2000). Design of circular thin-walled concrete filled steel tubes: Discussion and closure. *Journal of Structural Engineering*, *126*(11), 1295–1303.

Mei, H, Kiiousis, P, & Ehsani, M (2001). Confinement effects on high-strength concrete. *ACI Structural Journal*, *98*(4), 548–553.

Mirmiran, A, Singhvi, A, & Monti, G (1999). FRP-confined concrete model. *Journal of Composites for Construction*, *3*(3), 62–65.

Moran, DA, & Pantelides, CP (2002). Stress-strain model for fiber reinforced polymer confined concrete. *Journal of Composites for Construction*, *6*(4), 233–240.

Nanni, A, & Bradford, N M (1995). FRP jacketed concrete under uniaxial compression. *Construction & Building Materials*, *9*(2), 115–124.

Nie, J, Bai, Y, Li, S, Zhao, J, & Yan, X (2005). Analyses on composite column with inside filled steel tube under axial compression. *China Civil Engineering Journal*, *38*(9), 9–13.

Ozbakkaloglu, T (2013). Compressive behavior of concrete-filled FRP tube columns: Assessment of critical column parameters. *Engineering Structures*, *51*(51), 188–199.

Ozbakkaloglu, T, & Akin, E (2012). Behavior of FRP-confined normal- and high-strength concrete under cyclic axial compression. *Journal of Composites for Construction*, *16*(4), 451–463.

Ozbakkaloglu, T, & Fanggi, BAL (2014). Axial compressive behavior of FRP-concrete-steel double-skin tubular columns made of normal-and high-strength concrete. *Journal of Composites for Construction*, *18*(1), 46–47.

Ozbakkaloglu, T, & Jian, CL (2013). Axial compressive behavior of FRP-confined concrete: Experimental test database and a new design-oriented model. *Composites Part B: Engineering*, *55*(12), 607–634.

Ozbakkaloglu, T, Jian, CL, & Vincent, T (2013). FRP-confined concrete in circular sections: Review and assessment of stress-strain models. *Engineering Structures*, *49*(2), 1068–1088.

Park, JW, Hong, YK, & Choi, SM (2010). Behaviors of concrete filled square steel tubes confined by carbon fiber sheets (CFS) under compression and cyclic loads. *Steel and Composite Structures*, *10*(2), 187–205.

Peng, KD (2017). *Compression tests on square hybrid FRP-concrete-steel tubular columns with a rib-stiffened steel inner tube*. University of Wollongong.

Peter, M, John, F, & Bonacci, M (2002). Composite response of high-strength concrete confined by circular steel tube. *ACI Structural Journal*, *128*(8), 1073–1081.

Piekarczyk, J, Piekarczyk, W, & Blazewicz, S (2011). Compression strength of concrete cylinders reinforced with carbon fiber laminate. *Construction and Building Materials*, *25*(5), 2365–2369.

Rajeshkumar, B, Anil, KP, & Helen, S (2013). FE analysis of concrete filled cold formed steel sections using Ansys. *International Journal of Advanced Computer Engineering*, *3* (1), 11–18.

Realfonzo, R, & Napoli, A (2011). Concrete confined by FRP systems: Confinement efficiency and design strength models. *Composites Part B: Engineering*, *42*(4), 736–755.

Rong, C, & Shi, QX (2018). Axial-strength model for FRP-confined concrete based on the improved twin shear strength theory. *Composite Structures*, *202*, 102–110.

Rong, C, Shi, QX, Zhang, T, & Zhao, HC (2018). New failure criterion models for concrete under multiaxial stress in compression. *Construction and Building Materials*, *161*, 432–441.

Rong, C, Shi, QX, & Zhao, HC (2020). Behaviour of FRP composite columns: Review and analysis of the section forms. *Adavances in Concrete Construction*, *9*(2), 125–137.

Saadatmanesh, H (1994). Strength and ductility of concrete columns externally reinforced with fiber composite straps. *Aci Structural Journal Technical Paper*, *91*(4), 91–S43.

Saafi, M, Toutanji, HA, & Li, Z (1999). Behavior of concrete columns confined with fiber reinforced polymer tubes. *Aci Materials Journal*, *96*(4), 500–509.

Saleem, S, Hussain, Q, & Pimanmas, A (2017). Compressive behavior of PET, FRP confined circular, square, and rectangular concrete columns. *Journal of Composites for Construction*, *21*(3), 04016097.

Shen, Z, Lin, Z, & Luo, J (2009). *Hysteretic behaviors of steel frames with concrete-filled Lshaped steel tube columns* (pp. 65–99).

Shi, QX, Rong, C, & Zhang, T (2017). A constitutive model for confined concrete in composite structures. *Steel and Composite Structures*, *24*(6), 689–695.

Soundararajan, A, & Shanmugasundaram, K (2010). Finite element analysis on the flexural behaviour of concrete filled steel tube beams. *Journal of Theoretical and Applied Mechanics*, *48*(2), 505–516.

Tan, KH, Bhowmik, T, & Balendra, T (2013). Confinement model for FRP-bonded capsule-shaped concrete columns. *Engineering Structures*, *51*(2), 51–59.

Tang, G, Zhao, B, & Zhu, H (1982). Study on the fundamental structural behavior of concrete filled steel tubular columns. *Journal of Building Structure*, *3*(1), 13–31.

Tao, Z, Han, LH, & Zhuang, JP (2007). Axial loading behavior of CFRP strengthened concrete-filled steel tubular stub columns. *Advances in Structural Engineering*, *10*(1), 37–46.

Teng, JG, Hu, YM, & Yu, T (2013). Stress-strain model for concrete in FRP-confined steel tubular columns. *Engineering Structures*, *49*(2), 156–167.

Teng, JG, Huang, YL, & Lam, L (2007). Theoretical model for fiber-reinforced polymer-confined concrete. *Journal of Composites for Construction*, *11*(2), 201–210.

Teng, JG, Lin, G, & Yu, T (2015). Design-oriented stress-strain model for concrete under combined FRP-steel confinement. *Journal of Composites for Construction*, *20*(4), 04015084.

Teng, JG, Yu, T, & Wong, YL (2004). *Hybrid FRP-concrete-steel double-skin tubular columns: Stub column tests* (pp. 1390–1400).

Tomii, M, Sakino, K, & Xiao, Y (1985). *Earthquake resisting hysteretic behavior of reinforced concrete short columns confined by steel tube* (pp. 119–125).

Toutanji, HA (1999). Stress-Strain characteristics of concrete columns externally confined with advanced composite sheets. *Aci Materials Journal*, *96*, 397–404.

Vincent, T, & Ozbakkaloglu, T (2013). Influence of concrete strength and confinement method on axial compressive behavior of FRP confined high- and ultra high-strength concrete. *Composites Part B: Engineering*, *50*(7), 413–428.

Wang, D, & Lu, X (2005). Experimental study on seismic behavior of concrete-filled steel T-section, L-section columns. *Journal of Building Structures*, *26*(4), 39–44.

Wang, Q, Zhao, D, & Guan, P (2003). Study on the mechanical properties of axially loaded steel tubular columns filled with steel-reinforced high-strength concrete. *Journal of Building Structures*, *24*(6), 45–49.

Wang, Q, Zhao, D, & Guan, P (2004). Study of the mechanical properties of axially loaded section steel and steel tube short columns filled with steel-reinforced concrete. *China Journal of Highway and Transport*, *17*(3), 36–40.

Wang, R, Han, LH, & Hou, CC (2013). Behavior of concrete filled steel tubular(CFST)members under lateral impact: Experiment and FEA model. *Journal of Constructional Steel Research*, *80*, 188–201.

Wen, Y (2010). *Research on seismic behavior of joints of heavy-load steel tubular columns filled with steel-reinforced concrete*. Dalian University of Technology.

Wong, YL, Yu, T, & Teng, JG (2008). Behavior of FRP-confined concrete in annular section columns. *Composites Part B: Engineering*, *39*, 451–466.

Wu, G, Lu, ZT, & Wu, ZS (2006). Strength and ductility of concrete cylinders confined with FRP composites. *Construction and Building Materials*, *20*(3), 134–148.

Wu, YF, & Wei, YY (2010). Effect of cross-sectional aspect ratio on the strength of CFRP-confined rectangular concrete columns. *Engineering Structures*, *32*(1), 32–45.

Wu, YF, & Zhou, YW (2010). Unified strength model based on hoek-brown failure criterion for circular and square concrete columns confined by FRP. *Journal of Composites for Construction, 14*(2), 175–184.

Xiao, QG, Teng, JG, & Yu, T (2010). Behavior and modeling of confined high-strength concrete. *Journal of Composites for Construction, 14*(3), 249–259.

Xiao, Y (2009). Applications of FRP composites in concrete columns. *Advances in Structural Engineering, 7*(4), 335–343.

Xiao, Y, He, W, & Choi, K (2005). Confined concrete-filled tubular columns. *Journal of Structural Engineering, 131*(3), 488–497.

Xiao, Y, He, W, & Mao, X (2004). Development of confined concrete filled tubular (CCFT) columns. *Journal of Building Structures, 25*(6), 59–66.

Xiao, Y, & Wu, H (2000). Compressive behavior of concrete confined by carbon fiber composite jackets. *Journal of Materials in Civil Engineering, 12*(2), 139–146.

Xie, T, & Ozbakkaloglu, T (2015). Behavior of steel fiber-reinforced high-strength concrete-filled FRP tube columns under axial compression. *Engineering Structures, 90*, 158–171.

Yang, B (2006). *Research on the behavior of steel tube column filled with steel-reinforced concrete*. Inner Mongolia University of Science and Technology.

Yang, D (2014). *Research on the seismic behavior of steel tube column filled with steel-reinforced concrete*. Inner Mongolia University of Science and Technology.

Yu, T, Chan, CW, Teh, LH, & Teng, JG (2017). Hybrid FRP-concrete-steel multitube concrete columns: Concept and behavior. *Journal of Composites for Construction, 21*(6), 04017044.

Yu, T, Hu, YM, & Teng, JG (2016). Cyclic lateral response of FRP-confined circular concrete-filled steel tubular columns. *Journal of Constructional Steel Research, 124*, 12–22.

Yu, T, Lin, G, & Zhang, SS (2016). Compressive behavior of FRP-confined concrete-encased steel columns. *Composite Structures, 154*, 493–506.

Yu, T, & Teng, JG (2013). Behavior of hybrid FRP-concrete-steel double-skin tubular columns with a square outer tube and a circular inner tube subjected to axial compression. *Journal of Composites for Construction, 17*(2), 271–279.

Yu, T, Teng, JG, & Wong, YL (2010). Stress-strain behavior of concrete in hybrid FRP-concrete-steel double-skin tubular columns. *Journal of Structural Engineering, 136*(4), 379–389.

Yu, T, Zhang, B, Cao, YB, & Teng, JG (2012). Behavior of hybrid FRP-concrete-steel double-skin tubular columns subjected to cyclic axial compression. *Thin-Walled Structures, 61*(6), 196–203.

Yu, T, Zhang, SS, Huang, L, & Chan, C (2017). Compressive behavior of hybrid double-skin tubular columns with a large rupture strain FRP tube. *Composite Structures, 171*, 10–18.

Yu, Z, Ding, F, & Cai, C (2007). Experimental behavior of circular concrete-filled steel tube stub columns. *Journal of Constructional Steel Research, 63*(2), 165–174.

Zakaib, S, & Fam, A (2012). Flexural performance and moment connection of concrete-filled GFRP tube-encased steel I-sections. *Journal of Composites for Construction, 16*(5), 604–613.

Zhang, DJ, Wang, YF, & Ma, YS (2010). Compressive behaviour of FRP-confined square concrete columns after creep. *Engineering Structures, 32*(8), 1957–1963.

Zhao, G, Zhang, D, & Huang, C (1996). Study of earthquake resistant behaviour of high strength concrete column reinforced with concrete filled steel tube. *Journal of Dalian University of Technology*, *36*(6), 759–766.

Zhong, S (1999). *Concrete filled steel tubular high rise buildings*. Heilongjiang Science and Technology Press.

Zhong, S, & Wang, Y (1980). Research on analytical theory of axial compression member of concrete filled steel tubes. *Journal of Building Structure*, *1*, 61–71.

Zhou, X, & Liu, J (2010). *Behaviors and design of steel tube confined concrete columns*. Science Press.

Zhou, Y, Liu, X, Xing, F, Li, D, Wang, Y, & Sui, L (2017). Behavior and modeling of FRP-concrete-steel double-skin tubular columns made of full lightweight aggregate concrete. *Construction and Building Materials*, *139*, 52–63.

Zhu, M, Liu, J, & Wang, Q (2011). Experimental study of seismic behavior of square steel tubes filled with steel-reinforced high-strength concrete. *China Civil Engineering Journal*, *44*(7), 55–63.

Zuo, Z, Cai, J, Qian, Q, Zhang, J, & Chen, Q (2011). Experimental study on T-shaped CFT stub columns with binding bars subjected to axial compression. *China Civil Engineering Journal*, *44*(11), 43–51.

Zuo, Z, Cai, J, & Zhu, C (2010). Bearing capacity of L-shape CFT stub columns with binding bars subjected to eccentric compression load. *Engineering Mechanics*, *27*(7), 161–167.

Design-oriented constitutive model

2

2.1 Introduction

The lateral pressure provided by confining materials (e.g., steel tube, hoops, and FRP) can restrain the internal micro-cracks development of core concrete, so as to improve the mechanical properties of the core concrete and improve the overall performance of the composite column (Han & Yang, 2007; Soliman, 2011). Meanwhile, the inner concrete also has feedback effects on the confining materials, for example, restraining the local buckling of steel tube or section steel, improving the utilization of FRP, and improving the stability of the whole structure (Samani & Attard, 2012; Takahama & Oniki, 2014). Therefore, the composite method between different materials can improve the mechanical behavior, the ductility, and the durability of the composite column. At present, the concrete composite structure has been widely used in practice engineering, including the high-strength hoops confined concrete column, the steel reinforced concrete column, the concrete filled in steel tube, the steel reinforced concrete filled in steel tube, the FRP confined column, the FRP confined concrete filled in steel tube, the FRP confined reinforced concrete column, and so on. In most of the concrete composite columns, the core concrete is under the confining station, which is the most complex part. And the lateral confining effect can improve the strength and the toughness of the core concrete, which differs greatly from the properties of the initial concrete (Lee, 2007; Ren et al., 2014; Toutanji & Saafi, 2002; Yu et al., 2010). It is necessary to study the properties of concrete under various confining effects for the application of concrete composite columns.

The stress–strain curve can reflect the basic nonlinear property of material, including the elastic module, the strength, the ductility, and the stress condition in the total loading process. Therefore, the constitutive model for the concrete under various confining effects is important to study the nonlinear behavior of composite members or composite structures. As we knew in Chapter 1, the stress–strain curves of confined concrete show different trends in the different confining materials. The existing studies have proposed a series of constitutive models for the concrete confined by given materials or in a given section form (Han & Yang, 2007; Soliman, 2011). In these theoretical studies, the constitutive models have little difference in the first increasing stage, and the mathematical expressions of the relationship between stress and strain mainly focus on the unitary quadratic equation, the unitary cubic equation, and even the linear equation. In these proposed equations, the independent variable is strain ratio, the dependent variable is stress ratio, and the equation coefficients are simple. The studies proposed these equations by easily fitting the test results. However, there are many differences in the later stage

Concrete Composite Columns. DOI: https://doi.org/10.1016/B978-0-323-85171-8.00002-1

of the existing constitutive models, because of the complex properties of concrete in the plastic condition and the damage condition (Shi et al., 2011). The stress–strain curves also show various developing trends, including the declining trend, the first decreasing and then leveling trend, first declining and then rising trend, and the rising trend. Therefore, most of the research focus on mathematical expressions of the later stage of the constitutive models.

Scott (1982) presented a combination of experimental and theoretical studies on the concrete confined by the overlapping hoops. In addition, they studied the stress–strain behavior of the core concrete at low and high strain rates in detail. The experiment results showed that the stress and strain are both greatly improved and their increase ratio is similar. Meanwhile, the declining stage of the stress–strain curves shows a linear trend that is gentler than that of the unconfined concrete. Therefore, the study (Scott, 1982) assumed that the improvement of both the peak stress and the peak strain is the same, which is related to the lateral confining pressure. And the decreasing stage of the constitutive model for the confined concrete can be simply fitted in a linear form. However, the two proposed hypotheses are not accurate, due to the test errors or the low confining pressure. The peak stress and the peak strain cannot be calculated together. We should express the declining stage of the constitutive model as a nonlinear equation. Furthermore, we should focus on the following content in the model establishment: the reasonable calculation of the peak stress and the peak strain of the various confined concrete; and the development trends for the stress–strain curves of the various confined concrete, especially for the FRP confined concrete under weaker confining effect. Mander et al. (1988) assumed that the additional part of the compressive strength directly relates to the lateral confining pressure, and they introduce the effective confining coefficient to reflect the added value of the compressive strength. Then they established the unified constitutive model for the low-strength hoops confined concrete, including the increasing and decreasing stages. Meanwhile, the decreasing stage of the confined concrete shows a slow trend, which means that the lateral confining effect can improve the ductility of core concrete. However, the constitutive model is only suitable for the concrete confined by hoops, which can provide low confining pressure. We easily found that the different lateral confining pressure leads to different compression improvement effects. That means the effective confining coefficient is not a constant value, and the model proposed by Mander et al. (1988) will provide high prediction results for the concrete under the high confining effect. Afterward, Daniel and Patrick (1995) presented an experimental study on the high-strength hoops confined concrete and established the constitutive model for the confined concrete based on the analysis of the confining effect. In that study (Daniel & Patrick, 1995), they emphatically analyzed the declining stage of stress–strain curves and studied the influence factors of the curve trend. Then they determined the decreasing stage of the constitutive model as an exponential function form, based on the introduced parameters. In addition, they used the iterative calculation method in the proposed model to improve accuracy of the proposed model. However, the large number of fitting parameters and the complex calculation method make the proposed model not suitable for the application in practical engineering design. In recent decades, concrete-filled steel tube columns have been further studied and widely applied. Han presented a series of studies on the concrete-filled steel tube columns and established the constitutive model of the steel tube confined concrete (Han & Yang, 2007). The proposed constitutive model expresses the decreasing stage by modifying the key parameters of the decreasing stage in the constitutive model of unconfined concrete. In addition, the stress–strain behaviors of the steel confined concrete in the square and circle sections have many differences, due to the different distributions of the lateral confining pressure. The declining stage of the stress–strain curves of steel tube confined concrete in square section declines fast, whereas that of the confined concrete in circle section decreases slowly and even increases, due to the stiffening properties of steel tube. Therefore, we should choose the calculation process for the different section forms.

At present, many existing studies have proposed a series of constitutive models for the given confined concrete (Shi et al., 2014; Toutanji, 2001), including the hoops confined concrete, the steel tube confined concrete, the FRP confined concrete, and so on. Even for the same type of confined concrete, the equation forms and the coefficient selections differ from each other. Though these proposed models can provide accurate prediction results for their test results and have their own theoretical values, the varied constitutive models will be detrimental to the application of the confined concrete composite columns. In order to increase the practicability of the constitutive models and expand the application of confined concrete, it is necessary to establish unified simple models which can be suitable for most types of confined concrete. Based on the reviews and analysis of confined concrete in Chapter 1, the confined concrete can be divided into two types: the actively confined concrete and the passively confined concrete. Each type of confined concrete has its own curve trend, so as to have the particular equation form. Under the given section form, we should select and modify the coefficients, based on the influence factors of the composite columns, including the lateral stiffness of the confining materials, the compressive strength of the unconfined concrete, the section shape and the section size of the composite columns, and so on. All influence factors of the mechanical properties can boil down to several common parameters, which can be used to propose mathematical equations. Therefore, we will discuss the lateral confining effect and the simple confining mechanism at first. This chapter follows the mechanism analysis with the inner stress condition of the core concrete under the lateral confining pressure, based on the Twin Shear Unified Strength Theory. After that, we propose the unified calculation equations of the axial compressive properties. Finally, we establish the unique constitutive models for different confined concrete and verify the proposed models.

2.2 Lateral confining pressure around core concrete

The lateral confining pressure is the key to influencing the developing trend of stress–strain curves of confined concrete. Besides, the outer confining materials provide the lateral confining pressure by the following process: first, the core concrete occurs the lateral expansion under axial compression; second, the lateral expansion leads to the circumferential tension state of the outer confining materials; third, the tension force translates into the lateral confining pressure on core concrete, because of the balance of forces on the section. Therefore, the section shape directly influences the distribution of the confining pressure, the section including the circle section and the non-circle section (e.g., square, rectangle, and special shape). Except for the section shape, the stress state of the confining materials also has a significant effect on the distribution of the lateral confining pressure in the non-circle section. For example, the steel tube provides both axial bearing capacity and lateral confining effect in the concrete-filled steel tube column, and the steel tube shows obvious local buckling in the compression process. Therefore, the lateral pressure in concrete-filled steel tube is less than that in hoops confined concrete with the same steel ratio. Meanwhile, the square FRP confined concrete column has a great confining effect because of the four circle chamfers, and the radius of the chamfer has an obvious effect on the confining pressure. Therefore, we will discuss the lateral pressure in four types of confined concrete columns: the confined concrete cylinder, the concrete-filled square steel tube column, the square hoops confined concrete column, and the FRP confined concrete column.

2.2.1 Confined concrete cylinder

For the composite columns in the circle section, we show the international relationship between the outer confining materials and the core concrete in Fig. 2.1. The stress state of the core concrete includes the uniform circumferential pressure and the axial compression stress. However, the stress state in the lateral confining materials is various. In this section, we discuss three common types of confined concrete composite columns: the steel tube confined concrete; the hoops confined concrete; the FRP confined concrete. For the steel tube confined concrete, the stress

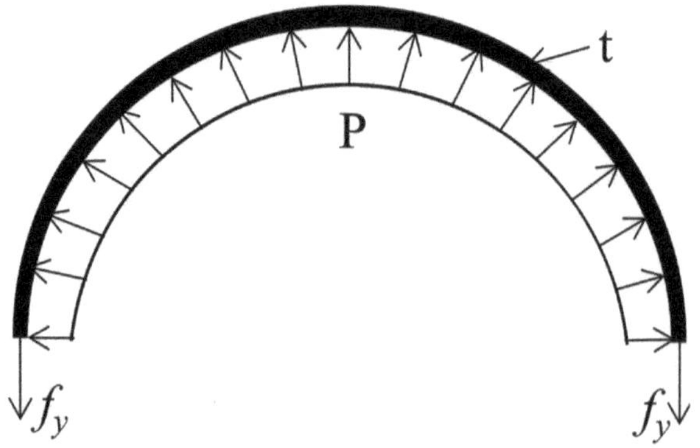

Figure 2.1 Diagrammatic sketches of confining pressure in circle section. One figure to describe the confining pressure in circle section.

state in the steel tube includes the axial compressive stress, the outer radial stress, and the circumferential tensile stress. The axial compressive stress can influence the bearing capacity of the steel tube and has little effect on core concrete, and the local buckling will affect the stress (Shi et al., 2017).

The outer radial stress and the circumferential tensile stress are the two major factors to influence the confining pressure around the core concrete, where the former is equal to the confining pressure and the latter provides the total lateral force, the two stresses keep a balanced relationship. Therefore, we can divide the stress state in the steel tube into the axial stress (σ_3), the redial stress (σ_2), and the circumferential tensile stress (σ_1). Compared with the other stresses, the redial stress is too small to ignore in stress analysis, $\sigma_2 \approx 0$. According to the Von Mises yield criterion, the mathematical expression can be simplified to that:

$$\sigma_1^2 + \sigma_3^2 - \sigma_1\sigma_3 = \sigma_s^2 \tag{2.1}$$

where σ_s is the steel yield stress. Then, the relationship between σ_1 and σ_3 can be expressed as:

$$\sigma_3 = \frac{\sigma_1 \pm \sqrt{\sigma_1^2 - 4(\sigma_1^2 - \sigma_s^2)}}{2} \tag{2.2}$$

Based on Eq. (2.2), we easily find that when the circumferential tensile stress reaches the yield stress ($\sigma_1 = \sigma_s$), the axial stress can show two states (i.e., $\sigma_3 = \sigma_s$ or $\sigma_3 = 0$), and one state ($\sigma_3 = 0$) will not appear in the practical engineering. Therefore, it can be got that when the circumferential tensile stress reached the yield stress, circumferential tensile stress is equal to the steel yield stress (f_y) directly.

According to the force balanced relationship, the lateral pressure on core concrete section can be computed as follows: $p = 2tf_y/D$, where D is the diameter of the column section, t is the thickness of the steel tube, and f_y is the yield stress of steel tube. For the hoops confined concrete, the stress state in the hoops includes the outer radial stress and the circumferential tensile stress, where the axial compressive stress is transmitted between concrete parts and can be ignored, because of the discontinuous distribution of the hoops along the column height. Just because of the discontinuous distribution, the confining effect provided by hoops shows the longitudinal uneven distribution. Therefore, the lateral pressure on core concrete section can be computed as follows: $p = k_s\rho_s f_y/2$, where k_s is the reduction factor of the lateral confining pressure in longitudinal uneven distribution, ρ_s is the steel ratio of the hoops, and f_y is the yield stress of the hoops. For the FRP confined concrete, the stress state in the FRP tube includes the axial compression stress, the outer radial stress, and the circumferential tensile stress. However, the axial compressive stress is too small to ignore in the stress analysis. Therefore, the core concrete is uniformly confined by the FRP tube, the lateral pressure can be computed as follows: $p = 2tf_y/D = 2tE_{frp}\varepsilon_l/D$, where t is the thickness of FRP tube, E_{frp} is the elastic modulus of FRP tube, ε_l is the lateral strain of FRP tube, and D is the section diameter.

2.2.2 Concrete-filled square steel tube column

Compared with the steel tube confined concrete in circle section, the internal confining mechanism of the confined concrete in square section is more complex, due to the lateral uneven confining pressure. It is easily found that the lateral concrete expansion is the largest in the corner, and the lateral concrete expansion is decreasing from the corner to the middle part. Therefore, the surface which is away from corners has a weak interaction relationship between the steel tube and the core concrete. And the failure mode of concrete-filled square steel tube column is caused by the yield of the steel tube at the corner and the local buckling of the steel plate in the middle part of each side. In addition, the uneven confining pressure can lead to the changeable confining effect on the square concrete section, and the mechanical properties of core concrete are not unified.

In order to analyze the stress condition in square confined concrete, Varma et al. (2005) studied the high-strength square concrete filled steel tube beam and columns by using the fiber model. For the square section under the lateral uneven confining pressure in their study, they divided the core concrete into the effective confining area and the noneffective confining area. However, the analysis method is only suitable for the fiber model or the numerical analysis in the finite element software. It is difficult to divide the confining areas in the mathematical expression of the constitutive model of square confined concrete. Based on the stress analysis of the confined concrete in the circle section, the lateral confining pressure directly relates to the confining materials and the lateral expansion behavior of concrete.

Since the uneven confining effect is caused by the uneven lateral expansion and the uneven lateral expansion can be seen as a special expansion behavior, we can transform the uneven lateral expansion into a nominal uniform lateral expansion by the stress transformation method. To reflect the lateral uneven lateral expansion, the distribution curve along the side length changed with the width-to-thickness ratio L/t of the tube wall, as shown in Fig. 2.2. We can find that the confining stress on the concrete part which is away from corners closes to zero, because of the local buckling of steel tube and the weak interaction effect on the two material surfaces in the section part.

The normal confining pressure is distributed along the side length following the law of quadratic curve, and then the equation is $q(x) = 4qx^2/L^2$, where q is the maximum lateral confining force at the corner, and $q = 2tf_y/L$; x is the distance from an assumed coordinate to the midpoint on one side. Based on the theoretical study on stress–strain model for confined concrete by Mander et al. (1988), the uneven distribution can be the homogeneous

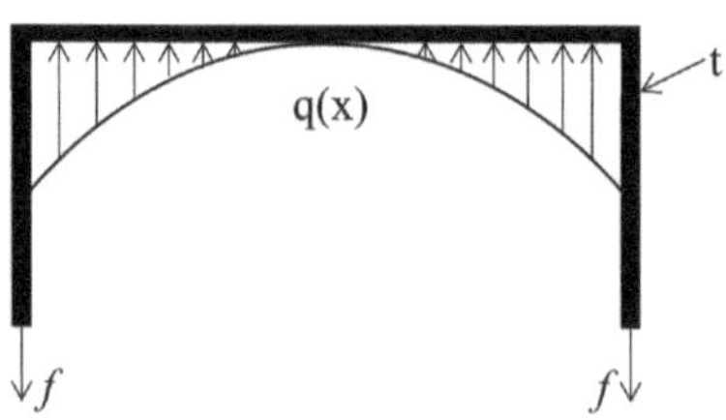

Figure 2.2 Diagrammatic sketches of confining pressure in concrete-filled square steel tube column. One figure to describe the confining pressure in the square section.

distribution of the lateral pressure along the surface of core concrete, and the nominal lateral uniform pressure is equivalent to the average value of the stress curve equation. Based on the integral process, the nominal lateral pressure can be computed as follows:

$$p = \frac{2}{L}\int_0^{L/2} q(x) = q/3 \tag{2.3}$$

2.2.3 Square hoops confined concrete column

For the square hoops confined concrete, the distribution of lateral pressure is similar to that of square steel tube confined concrete, the confining pressure decreases with increasing distance away from the corner at each side, and the confining pressure at the corner directly relates to the yield stress of the hoops. However, the hoops can provide obvious confining pressure on the midpoint of each side. It means that the square hoops can provide a large confining effect on the core concrete than the square steel tube with the same steel ratio, as shown in Fig. 2.3.

We can find the main reason for the stress state of hoops. As we discussed in above paragraph, the hoops are mainly extruded in the lateral and axial directions. And the vertical compressive stress is transmitted from the both-side concrete, due to the discontinuity distribution of the hoops. Compared with the steel yield stress, the vertical compressive stress is too small to ignore in the stress analysis. Therefore, the stress state in the hoops mainly includes the lateral pressure from the core concrete and the tensile stress from the hoops, which is at the same stress plane. And the failure model is the hoops yield damage at the corner and the crushing destruction of the corner concrete. Without the local buckling effect, the square hoops can provide a larger confining effect than square steel tube. In addition, the distribution of lateral pressure provide by the hoops differs from that provided by the steel tube. Except for the stress state in the column section, the lateral pressure shown the uneven distribution along the height direction. The concrete which is the same height as hoops is better confined than that in the section between two hoops, namely "arch effect", as shown in Fig. 2.3. The arch effect increases with the

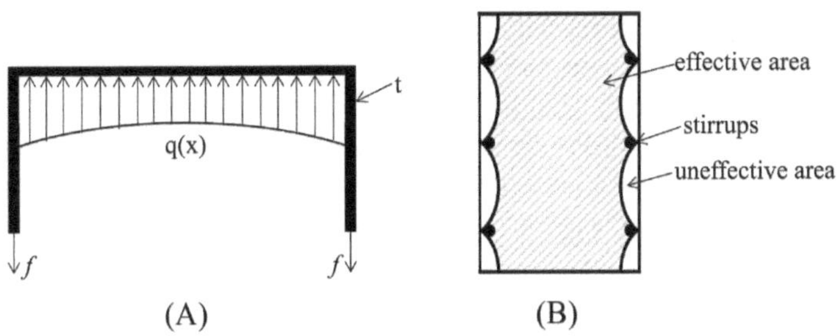

Figure 2.3 Diagrammatic sketches of confining pressure in hoops confined concrete column. (A) cross section; (B) longitudinal section.

increase in the spacing between two hoops. Based on the existing studies (Mander et al., 1988; Shi et al., 2011, 2013) on the theoretical constitutive models of the hoops confined concrete, the reduction of confining effect caused by the arch effect can be mainly reflected in the stress–strain curves.

In order to simplify the calculation, we use the hoops volume ratio and the corresponding parameter fitting in the constitutive model to reflect the influence effect, because of the little change in the diameter and the spacing between the hoops. For the similar distribution of the lateral pressure, we can consider the hoops as the square steel tube with the same steel volume ratio. Then, the steel ratio of hoops can be calculated as follows: $\rho_v = 4tL/L^2 = 4t/L$, where L is the side length, and t is the thickness of steel tube. Meanwhile, the stress reduction of the arch effect and the uneven lateral pressure in this section can be ignored by the following reasons: (1) the steel ratio of hoops is low; (2) these influence factors can be reflected in volume ratio indirectly; (3) the spacing between hoops is small in hoops confined concrete; (4) the hoops are embedded in concrete, so as to have great cooperative deformation ability; (5) the confining effect in hoops confined concrete is lower than that in steel tube confined concrete or FRP confined concrete. Therefore, the lateral confining pressure provided by hoops can be assumed as the homogeneous pressure, we can compute the analysis pressure as follows:

$$p = 2tf_y/L = \rho_y f_y/2 \tag{2.4}$$

where t is the nominal thickness of the hypothetical steel tube, f_y is the yield stress of hoops, L is section length, and ρ_v is the steel volume ratio.

2.2.4 FRP confined concrete column

For square FRP confined concrete, the distribution of lateral pressure provided by the FRP tube is similar to that of square hoops confined concrete, and the stress state of the FRP tube is also similar to that of the hoops. The stress state of the FRP tube mainly includes the lateral pressure provided by the core concrete and the tensile stress provided by the FRP tube, and the axial compressive stress is much smaller than other stresses. However, the confining effect provided by the FRP tube is large and the bonding relationship between the core concrete and the FRP tube is weak. The reduction effect of the square section is obvious. Meanwhile, the failure model of the square FRP confined concrete shows the crushing concrete and the FRP rupture at the corner. It means that the lateral stiffness of the square column is the largest on the diagonal, and there is a large stress concentration in the FRP tube at the corner. Therefore, we introduce the section shape coefficient and the stress concentration coefficient in the equation of lateral confining pressure to reflect the reduction of confining effect. The equation is as follows:

$$p = k_s k_f (2tf_f rp/L) \tag{2.5}$$

where t is the thickness of FRP tube, f_{frp} is the tensile stress of FRP tube, L is the section length, k_s is the section shape coefficient, and k_f is the stress concentration coefficient.

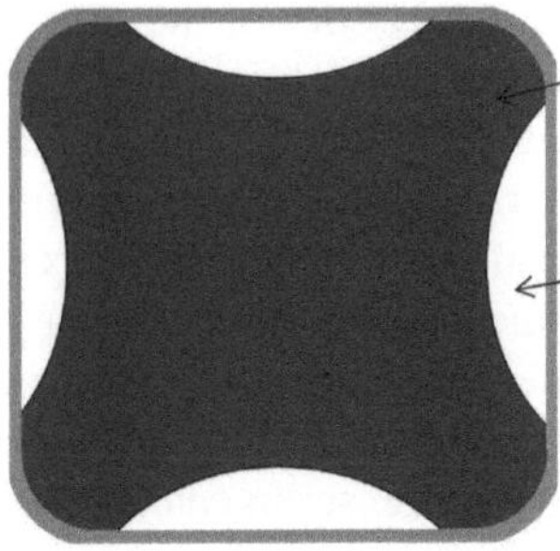

Figure 2.4 Dividing sections of the square section. The section includes the effective area and the ineffective area.

Based on the experimental and theoretical studies (Teng et al., 2002; Wang & Restrepo, 2001), the distribution of the lateral pressure is the arch curve, and the concrete in the circular chamfer part is under the maximal confining effect. Therefore, the side of the effective confining area is similar to a quadratic parabola, and the tangent of the endpoint is at an angle of 45° to the column surface. In the equation, k_s is the section shape coefficient that adopts the ratio of the effective confining area to the total section area, it showed the dividing sections in Fig. 2.4. Finally, the equation is as follows:

$$k_s = (b^2 - 4r^2 + \pi r^2 - 2a^2/3)/(b^2 - 4r^2 + \pi r^2) \tag{2.6}$$

where r is the radius of the circular chamfer, b is the side length of the section, and a is the level line length on one side. The detailed schematic diagram is shown in Fig. 2.4.

As discussed in the above section, the lateral stiffness of the square column is the largest on the diagonal, and there is a large stress concentration at the corner of the FRP tube. Therefore, in the experimental study of the FRP confined square column, the FRP at the corner usually ruptures first. Campione and Miraglia presented an experimental study on the mechanical behavior of the FRP confined concrete and focused on the strength and strain analysis. Based on the relationship analysis between the stress concentration coefficient and the circular chamfer size, the calculation equation is as follows:

$$k_f = [2r(1 - \sqrt{2}k_i/2)/b + \sqrt{2}k_i/2] \tag{2.7}$$

where $k_i = 0.2121$.

2.3 Stress calculation and strain calculation

The design-oriented constitutive models of confined concrete have some obvious characteristics, including simple expression, clear physical meaning, accurate prediction, and universal application. The simple solution method of model establishment is: (1) to analyze the characteristics of the stress–strain curve of the target confined concrete; (2) to determine the characteristic points of the stress–strain curve; (3) to fit the stress–strain equation. We can find that the characteristic stress calculation and characteristic strain calculation of confined concrete are the keys to proposed design-oriented constitutive models. Based on the reviews on the stress–strain curves of various confined concrete in Chapter 1, the characteristic points include: the peak points of steel tube confined concrete and hoops confined concrete, the peak points, and ultimate points of FRP confined concrete. Therefore, we should study the stress calculation and the strain calculation for these characteristic points.

This section mainly focuses on the stress calculation method, based on theoretical analysis and data fitting. However, the characteristic strain of concrete is difficult to determine in the theoretical method, due to the large discreteness of concrete. Most studies provided the characteristic strain prediction by the fitting method in an extensive database. Therefore, we propose the characteristic strain calculation by modifying the existing models in design-oriented constitutive model. We will present the further theoretical analysis and mathematical derivation for the stress calculation and the strain calculation in Chapters 3 and 4.

2.3.1 Characteristic stress calculation of confined concrete

In all kinds of confined concrete, the stress condition in core concrete is compression in the three axial directions, and the key influence factor is the lateral confining pressure. As mentioned above, we can calculate the lateral confining pressure of each kind of concrete composite column in Section 2.2. After the determination of lateral confining pressure, we can calculate the axial stress of confined concrete by using the same foundation. For the complex stress state in the confined concrete under multiaxial compressive pressure, we use the Twin Shear Unified Strength Theory to be the failure criterion, as proposed by Yu et al. (2007). The theory holds that the yield failure of each material is related not only to the maximum principal shear stress but also to the secondary principal shear stress. That is, whether the dangerous point in the member is in the state of unidirectional stress or complex stress, the yield failure occurs when the sum of the two larger principal shear stresses at the point reaches the sum of the ultimate double shear stress of the material. The sum of the ultimate double shear stress can be obtained from the unidirectional tensile yield failure experiment. This strength theory can not only explain the yield failure of plastic materials but also explain the tensile failure, shear failure, compression failure, and various biaxial and triaxial failures of materials. Therefore, the theory can apply to confined concrete under multiaxial compression conditions. The theory has great academic significance and strong application value, and the mathematical expressions are as follows:

$$F = \sigma_1 - \frac{\alpha}{1+b}(b\sigma_2 + \sigma_3) = \sigma_t, \sigma_2 \leq \frac{\sigma_1 + \alpha\sigma_3}{1+\alpha} \tag{2.8}$$

$$F' = \frac{1}{1+b}(\sigma_1 + b\sigma_2) - \alpha\sigma_3 = \sigma_t, \sigma_2 > \frac{\sigma_1 + \alpha\sigma_3}{1+\alpha} \tag{2.9}$$

where σ_1, σ_2, and σ_3 are three principal stresses, $\alpha = \sigma_t/\sigma_c$ is the ratio of tensile strength to the compressive strength of material, σ_t and σ_c are the uniaxial tensile strength limit and uniaxial compression strength limit of material, and b is a weighted parameter.

In order to consider the mechanical properties of concrete, we should introduce the cohesive force c of concrete and internal friction angle φ. Based on the Mohr theory, if the material is damaged along the plane under the action of two-dimensional stress, there must be a combination of the normal stress (σ) and the shear stress (τ) in this plane. The normal stress (σ) and the shear stress (τ) in the failure plane can be calculated by the force balance:

$$\sigma = \sigma_1 cos\varphi + \sigma_3 sin\varphi \tag{2.10}$$

$$\tau = (\sigma_1 - \sigma_3)cos\varphi \times sin\varphi \tag{2.11}$$

If Eqs. (2.10) and (2.11) are substituted into Eqs. (2.8) and (2.9), we can gain the following equations.

$$F = \tau_{13} + b\tau_{12} + sin\varphi(\sigma_{13} + b\sigma_{12}) = (1+b)c \times cos\varphi, F \geq F' \tag{2.12}$$

$$F' = \tau_{13} + b\tau_{23} + sin\varphi(\sigma_{13} + b\sigma_{23}) = (1+b)c \times cos\varphi, F < F' \tag{2.13}$$

where τ_{12}, τ_{23}, and τ_{13} are three principal shear stresses, $\tau_{12} = (\sigma_1 - \sigma_2)/2$, $\tau_{23} = (\sigma_2 - \sigma_3)/2$, $\tau_{13} = (\sigma_1 - \sigma_3)/2$, and σ_{12}, σ_{23} and σ_{13} are the principal stresses corresponded to three principal shear stresses, $\sigma_{12} = (\sigma_1 + \sigma_2)/2$, $\sigma_{23} = (\sigma_2 + \sigma_3)/2$, $\sigma_{13} = (\sigma_1 + \sigma_3)/2$.

Further, the subtraction equation is set between Eqs. (2.12) and (2.13), and the equation selection can be easily found in the equation: $F' - F = b(1 - \sin\varphi)(\sigma_1 - \sigma_2) \geq 0$. Based on the discriminant formula, the stress condition in confined concrete should be analyzed by using Eq. (2.13). It is easily known that the stress condition in confined concrete is that $0 > \sigma_1 = \sigma_2 > \sigma_3$. Bade on the special stress condition, the relationship between the first principal stress (σ_1) and the third principal stress (σ_3) is proposed by simplifying the Eq. (2.13), which can be as follows:

$$(1+b)/2(\sigma_1 - \sigma_3) + (1+b)/2(\sigma_1 + \sigma_3)sin\varphi = (1+b)c \times cos\varphi$$

It can be simplified as follows:

$$-\sigma_3 = \frac{2c \times \cos\varphi}{1 - sin\varphi} - \frac{1 + \sin\varphi}{1 - sin\varphi}\sigma_1 \tag{2.14}$$

However, the cohesive force (c) and the internal friction angle (φ) of concrete are not only hardly measured but also changed with the lateral confining effect. Therefore, the two parameters are not suitable for the analysis of stress condition in confined concrete. It is necessary to use other parameters to replace the cohesive force (c) and the internal friction angle (φ). For the unconfined concrete under uniaxial compression, the stress condition in concrete can be shown as $\sigma_3 = -f_{co}$, $\sigma_1 = \sigma_2 = 0$. It is easily found that $f_{co} = (2c\cos\varphi)/(1 - \sin\varphi)$, $k=(1 + \sin\varphi)/(1 - \sin\varphi)$. For the confined concrete under uniaxial compression, the stress condition in concrete is that $\sigma_3 = -f_{cc}$, $\sigma_1 = \sigma_2 = -f_l$. Therefore, the third principal stress in the failure plane can be expressed by using Eq. (2.14), as follows:

$$f_{cc} = f_{co} + kf_l \tag{2.15}$$

where f_{co} is the strength of the unconfined concrete, f_{cc} is the axial stress in the failure plane, f_l is the lateral confining pressure, k is defined as the confining influence coefficient which is related to the cohesive force (c), and the internal friction angle (φ).

It should be noted that the axial stress calculation equation is irrelevant to weighted parameter (b) of the intermediate principal stress. Based on the mathematical derivation, when the second principal stress is equal to the third principal stress, then $\tau_{12}+b\tau_{23}=(1 + b)\ (\sigma_1 - \sigma_3)$ and $\sigma_{12} + b\sigma_{23} = (1 + b)\ (\sigma_1 + \sigma_3)$. If the two conversion methods are applied in Eq. (2.13), the $(1 + b)$ will be removed in the equation simplification process. This probably means that the confining pressure is much smaller than the axial compression stress in confined concrete; therefore, the influence of the second principal stress can be ignored. The actual stress condition of confined concrete is mutually complementary with the derivation result based on the analysis of the Twin Shear Unified Strength Theory.

It should be noted that the expression form of Eq. (2.15) is the same as the strength calculation model of confined concrete in most existing studies. However, the lateral confining coefficients have different theoretical significance and different expression methods in the same expression equation. Most of the existing studies determined the lateral confining coefficient as a constant value (Han & Yang, 2007; Mander et al., 1988; Yu et al., 2010), and we can gain the constant value by fitting the test results of particular confined concrete. However, the study in this book considers that the lateral confining coefficient has a relationship with the cohesive force (c) and internal friction angle (φ), which are varied by original properties of the unconfined concrete and different values of the lateral confining pressure. The value of the lateral confining coefficient is not related to the type of confined concrete and will change the value under the different lateral confining pressure. Therefore, mathematical expressions of the lateral confining coefficient should be a nonlinear equation, which is composed of the unconfined concrete strength and the lateral confining pressure. Based on the stress condition of concrete under triaxial compression, the internal friction angle decreases with the increase in lateral compression stress, so the lateral confining effect is reduced under increasing confining pressure. The more intuitive expression is that the lateral confining coefficient (k) decreases with the increase in lateral confining pressure. In order to study the lateral confining coefficient, several axial compressive test results of various confined concrete are collected and analyzed in this section, including steel tube confined concrete, hoops confined concrete, and FRP confined concrete (Han & Yang, 2007; Ozbakkaloglu & Lim, 2013; Shi et al., 2011, 2013). Based on these existing experimental studies, we select almost 350 groups of representative compression test results. To balance the effect of the lateral confining effect and the concrete properties, we introduce the relative confining ratio (ω) in this section, which is the ratio of the lateral confining pressure to the compressive strength of unconfined concrete, $\omega = f_l / f_{co}$. In addition, we can gain the lateral confining coefficient (k) by reverse derivation of Eq. (2.15), based on the known value of the unconfined concrete strength, the axial compressive strength, and the lateral pressure. Based on the data analysis, the relationship between the lateral confining coefficient (k) and the relative confining ratio (ω) is shown in Fig. 2.5.

The Fig. 2.5 shows that the value of k decreases with the increase of ω, and the decreasing speed changes from fast to slow, then tends to zero at last. Meanwhile, the k is deducing from 5 and closing to 1. The characteristic of the relationship between k and ω agrees with the existing study results. When ω is less than 0.1, the confining pressure is much smaller than the concrete strength, and the strength improvement much larger than the lateral confining pressure, k is about 4 which is similar to the study results of hoops confined concrete by Mander et al. (1988). When ω is over 0.8, the confining pressure is approximately equal to the concrete strength, or even larger than that, the strength improvement is a little larger than the lateral confining pressure, which is close to the triaxial compressive state. Furthermore, the functional relationship between k and ω can be fitted by the summarized data. We obtain the equation for k as follows:

$$k = 1 + 1/(3.5\omega^{0.6973}) \tag{2.16}$$

where ω is the relative confining ratio, $\omega = f_l / f_{co}$.

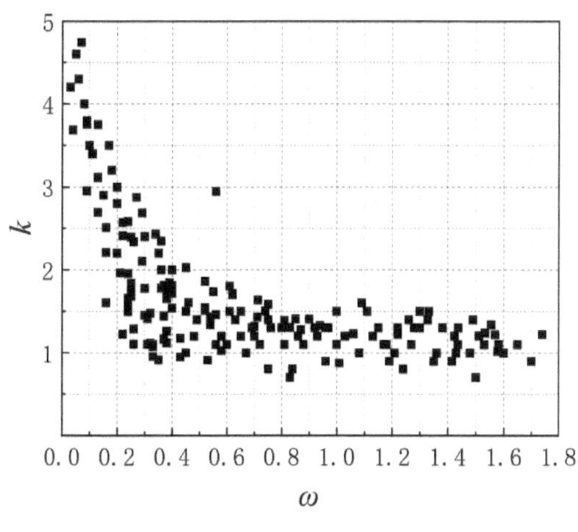

Figure 2.5 The relationship between $\boldsymbol{k}$ and $\boldsymbol{\omega}$. This is a scatter diagram.

Except for the above theoretical influence factors, the size effect is another influencing factor for the compression behavior of confined concrete. Based on the existing studies (Sakino et al., 2004; Shan et al., 2016), the comparisons between small size-specimen and big size-specimen show the following conclusions: the increase in specimen size reduces the compressive strength of unconfined concrete, due to the increasing concrete defects; the increase of the specimen size significantly reduces the confining effect of the outer confining materials; the lateral strain on the outer confining materials is small in the big size-specimen, which means the increase in the specimen size reduces the expansion of the concrete section. Therefore, the size coefficient should be introduced to reflect the strength reduction. Based on the experimental and theoretical studies by Sakino et al. (2004), the size coefficient (γ_u) can be computed as follows: $\gamma_u = 1.67D^{-0.112}$, where the parameter D is the outer diameter of the composite column in circle section.

As discussed in the above section, we can reflect the differences between the various confined concrete by the developing trend of the lateral confining pressure. And the strength improving mechanism is similar, due to the load path independence of the stress (Teng et al., 2002). Therefore, the stress equations for the different confined concrete can be expressed in the same equation form, based on the unconfined concrete strength and the lateral confining pressure. We can obtain the stress equation of confined concrete by the combination of Eqs. (2.15) and (2.16), as follows:

$$f_{cc} = \gamma_u f_{co} + kf_l \tag{2.17}$$

where γ_u is the size coefficient (Sakino et al., 2004), $\gamma_u = 1.67D^{-0.112}$, D is the diameter in the composite column with circle section, the range of D is from 50 to 800 mm. For the composite columns with the square section, the parameter D is the value of the side length (L) on the outer surface of the cross sections. Furthermore, f_{co} is the uniaxial compressive strength of unconfined concrete. For the composite columns with the square section, f_{co} can be taken as the standard compressive strength. But for the composite columns with the square section, the compressive strength is larger than that of the circle section, and f_{co} should be replaced by the uniaxial compressive strength of cylinder (f_{cy}). The conversion relationship between the two kinds of compression strengths is as follows: $f_{cy} = mf_c$, where m is the conversion coefficient. The conversion coefficient (m) has been analyzed by Guo et al. (1982), as shown in Table 2.1.

Table 2.1 The value of m.

Concrete strength	C20–C40	C50	C60	C70	C80
m	0.8	0.83	0.86	0.875	0.89

Source: From Guo, Z. H., Zhang, X. Q., Zhang, D. C., & Wang, R. Q. (1982). Experimental investigation of the complete stress–strain curve of concrete. *Journal of Building Structures*, *3*, 1–12.

Therefore, the characteristic stress calculation process is as follows: (1) to identify the basic properties of the concrete and the confining materials, including the section shape, the strength of unconfined concrete, the elastic modulus of FRP, the elastic modulus and the yield stress of steel; (2) to calculate the lateral confining pressure around the core concrete by the proposed method in Section 2.2; (3) to calculate the characteristic stresses of each kind of the confined concrete, based on Eqs. (2.16) and (2.17), and Table 2.1.

In order to verify the calculation equations for the characteristic stresses, we compare the calculation results with the experiment result, and the experiment database is summarized from the existing experimental studies (Han & Yang, 2007; Li et al., 2015; Ozbakkaloglu & Lim, 2013; Patel et al., 2012; Sakino et al., 2004; Zhang et al., 2007), including steel tube confined concrete, hoops confined concrete, and FRP confined concrete. As we knew, the axial stress of FRP confined concrete columns and hoops confined concrete columns are easily gained by the bearing capacity and the section area. But for the concrete-filled steel tube columns, the bearing capacity should be combined by the core concrete and the steel tube. In order to simplify the data collection, we adopted the bearing capacity (N_u) for concrete-filled steel tube columns in the calculation verification. It is known that the axial stress of steel tube is equal to yield stress at the peak load, based on Eq. (2.2). Therefore, the equation of bearing capacity of concrete-filled steel tube column is as follows:

$$N_u = \sigma_s A_s + f_{cc} A_c \tag{2.18}$$

where N_u is the bearing capacity, σ_s is the yield stress of the steel tube, A_s is the section area of the steel tube, f_{cc} is the compressive strength of the steel tube confined concrete, and A_c is the section area of the core concrete.

The comparisons between the calculation results and the experiment results are shown in Fig. 2.6. Since the peak stress of the core concrete in the steel tube is

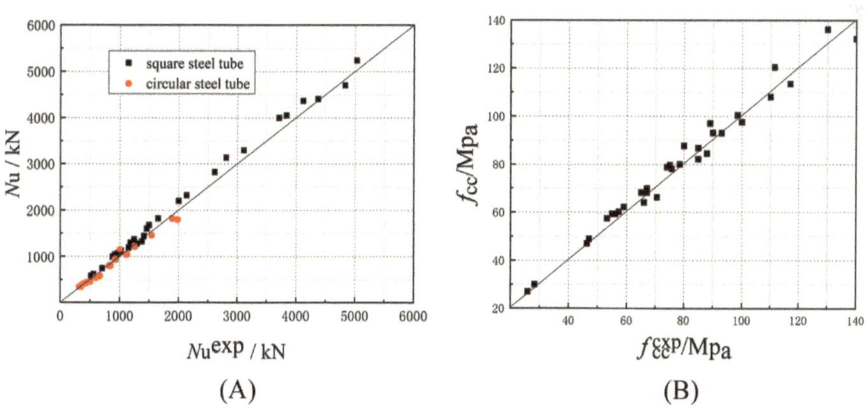

Figure 2.6 Comparison between calculation results and experiment results of composite columns. (A) concrete filled steel tube column; (B) FRP or hoops confined concrete.

hard to gain, we divide the verification into the comparisons of the concrete-filled steel tube column and the comparisons of the FRP confined concrete and the hoops confined concrete.

Based on the comparison in Fig. 2.6, the data points are evenly distributed on both sides of the angular bisector which is $x = y$. It means that the proposed stress calculation equation can provide accurate prediction results for the confined concrete. However, this is an intuitive judgment for the comparison in Fig. 2.6. We should verify the comparison results in the theoretical method. In order to reflect the coincidence between experimental results and computational results, the following evaluation parameters are computed: the correlation coefficient (R), which reflects the average ratio of calculation result to experiment result; the residual mean (Q), which reflects the average value of calculation result errors; the residual standard deviation (S), which reflects the dispersion degree of the calculation result and the experiment result. And three evaluation parameters can be computed by the following equations:

$$R = \sum\nolimits_1^n (com_i/exp_i)/nQ = \left[\sum\nolimits_1^n (exp_i - com_i\right]/nS = \sqrt{\left(\sum\nolimits_1^z (Q_i - Q)^2\right)/(z-1)} \tag{2.19}$$

where exp_i is the experimental data, com_i is the computation data, and n is the quantity of the data.

Based on above equations, the evaluation parameters of concrete filled steel tube columns are R=0.997, Q=52.41, and S=8.71, as well as the evaluation parameters of FRP confined concrete and hoops confined concrete are R=0.970, Q=0.22, and S=0.09. The values of Q and S of concrete-filled steel tube columns are larger. It is mainly because we adopt the bearing capacity of the composite column in the comparison and the large value leads to a large error computation value. To sum up, the calculation results by the proposed equation can agree better with the experiment results in this chapter.

2.3.2 Characteristic strain calculation of confined concrete

It has been known that steel tube confined concrete and hoops confined concrete belong to actively confined concrete and FRP confined concrete belongs to passively confined concrete. The two types of confined concrete show different curve trends. And the characteristic strain should also be studied in different methods, due to the different stress conditions.

For the steel tube confined concrete and the hoops confined concrete, we can divide the axial compressive pressure into two parts: one part is the self-compression strength and another part is the addition strength by the lateral confining pressure. Furthermore, the stress state in the confined concrete has a hydrostatic pressure part and the net axial pressure, which expresses that: $f_{cc} = f_c' + f_l$. And f_c' is the pressure without hydrostatic pressure, it can be seen as the improving properties of the core concrete, which expresses that $f_c' = f_{co} + (k-1)f_l$. The two pressures (i.e., f_l and f_c') can be used in the strain analysis and the peak strain calculation of confined concrete. Based on the above stress state analysis, we can divide the axial compressive strain of the confined concrete into two corresponding parts. One part of strain is directly influenced by the lateral confining pressure, which can be calculated by the hydrostatic pressure which is equal to the lateral confining pressure, called as hydrostatic pressure part. The other part of strain is influenced by the stress without the hydrostatic pressure (f_c'), which is higher than the strength of the unconfined concrete. This improving part is because the concrete properties can be improved under the lateral confining effect, called the well-performance concrete part. Therefore, there are two parts to the peak strain calculation, one part is the strain ε_c of the core concrete after property improvement under the confining effect, and the other part is the strain ε_p generated by the hydrostatic pressure. We can calculate the first strain, based on the defined stress (f_c') and the stress calculation of the unconfined concrete. And based on the hydrostatic pressure, we can calculate the second strain. Based on the experiment data analysis in existing studies (Shi et al., 2017), the calculation equation of peak strain (ε_{cc}) of confined concrete is as follows:

$$\varepsilon_{cc} = \varepsilon_c + \varepsilon_p \tag{2.20}$$

where $\varepsilon_c = 700 + 172 \times f_c'^{0.5} \times 10^{-6}$, $f_c' = f_c + (k-1)p$, $\varepsilon_p = 800 \times (2\omega)^{0.2} \times 10^{-6}$.

To verify the strain calculation equation, we set the comparisons between the experiment results and computation results of the peak strain of the steel confined

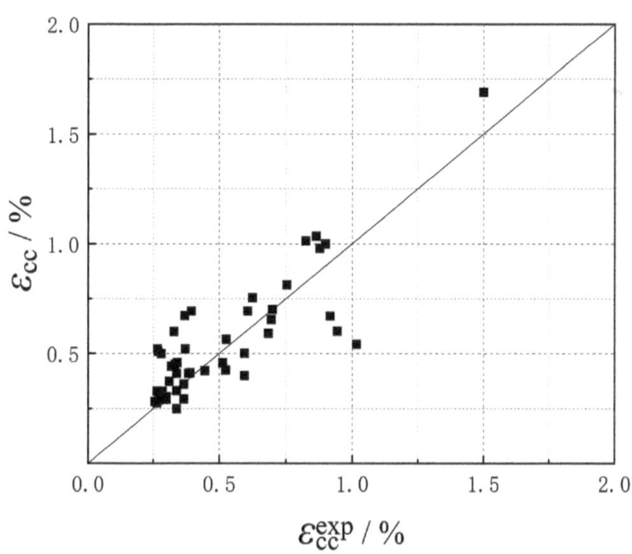

Figure 2.7 Comparison between experiment results and computation results of steel tube or hoops confined concrete. This is just one figure.

concrete and the hoops confined concrete in Fig. 2.7. And we summarize the experiment data from the existing studies (Guo et al., 1982; Han & Yang, 2007; Shi et al., 2017). Simultaneously, the evaluation parameters for goodness of the data fit are R=0.831, Q=0.041 and S=0.16, respectively. The curve comparisons in Fig. 2.7 and the evaluation parameters show that the prediction results by the proposed equation agree well with the experiment results.

It is well known that the stress–strain curves of FRP confined concrete show two kinds of curve trends. One kind of the stress–strain curve includes a softening stage, which is similar to the curve of unconfined concrete, and the peak point is a unique characteristic point. Another kind of the stress–strain curve shows a bilinear increasing trend. The first increasing stage is mainly influenced by the unconfined concrete properties, and the second increasing stage is mainly influenced by the combination of the FRP properties and the concrete properties. In addition, the characteristic point between the two increasing stages is named the peak point, which has higher stress and higher strain than the peak point of the unconfined concrete. Since the elastic modulus of the FRP tube is smaller than that of steel tube, the peak points of the FRP confined concrete and the steel tube confined concrete have significant differences. Meanwhile, the point closes to the failure of FRP confined concrete is named the ultimate point which has the maximal stress and the maximal strain. For the two characteristic points, the stress can be easily calculated by the method in the above section. However, the characteristic strain calculation equation should be studied in detail, because of the changing lateral pressure provided by the FRP tube.

At present, the peak strain is calculated by directly increasing the properties of unconfined concrete, which modify the peak strain of the unconfined concrete by an improving coefficient (a), the equation is $\varepsilon_{cp}=a\varepsilon_{co}$. However, if the lateral strain of the FRP tube is always overestimated at the peak point, then the existing stress equation will provide larger prediction results or even distortion prediction results for the peak point. In order to get the accurate prediction, we should gain the strain equation based on the strain addition, and the equation form is similar to the stress equation. And we can relate the strain addition to several influence factors, including the strength of unconfined concrete and the relative confining ratio (ω). The peak strain calculation equation is as follows:

$$\varepsilon_{ck} = \varepsilon_{co} + (1.02\omega/f_{co})\varepsilon_{co} \tag{2.21}$$

where ε_{ck} is the peak strain of FRP confined concrete, ε_{co} is the peak strain of unconfined concrete, f_{co} is the peak stress of unconfined concrete, and ω is the relative confining ratio, i.e., $\omega = f_l/f_{co}$.

For the ultimate strain, the accurate prediction results are gained with much difficulty, because of many influence factors, including the large confining effect, the imprecise fracture strain of the FRP tube, and the specimen size. In general, the average absolute error of the existing ultimate strain equations is over 30%. It is necessary to establish the ultimate strain equations to correctly reflect the variation of the ultimate strain with the lateral confinement characteristics of the FRP tube. It has been

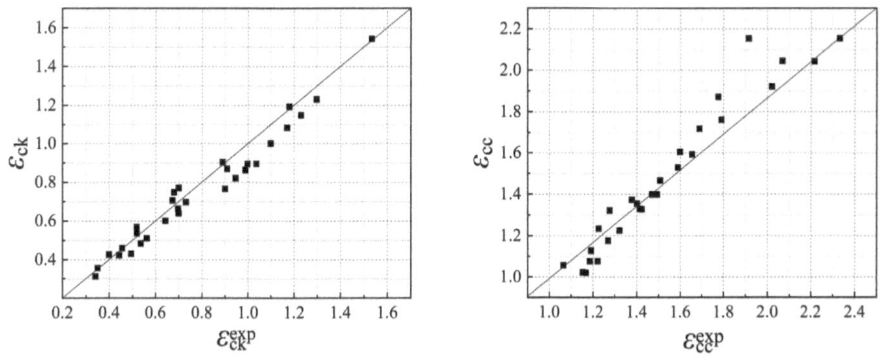

Figure 2.8 Comparison between experiment results and computation results of FRP confined concrete. (A) the peak strain; (B) the ultimate strain.

found that the influence factors of the ultimate strain are the same as that of the ultimate stress. Therefore, based on the data regression calculation with the influence factors (i.e., the compression strength of unconfined concrete, the ultimate lateral confining pressure, and the section size), the ultimate strain equation is as follows:

$$\varepsilon_{cc} = \varepsilon_{co} + 62.4\varepsilon_{co} f_{l,rup}/f_{co} \tag{2.22}$$

where, $f_{l,rup}$ is the maximal lateral confining pressure of the FRP tube before rupture.

Based on Eqs. (2.21) and (2.22), the calculation result verifications of the peak strain and the ultimate strain of FRP confined concrete are shown in two figures in Fig. 2.8. As shown in Fig. 2.8, the data points are unevenly distributed on both sides of the angular bisector which is $x=y$, and it means that the dispersibility of data is large. There are several following reasons: (1) the characteristic strain is hard to predict; (2) the experiment results have some test errors; (3) the rupture strain is inaccurate in some references. Therefore, the detailed evaluation parameters are not used for these data comparisons. In order to verify the proposed equations, we introduce the mean absolute error (*MR*) in this section, $MR = \sum(|com_i - exp_i|/exp_i)/n$. Based on the data computation, the mean absolute error (*MR*) of the peak strain is less than 15%, and the mean absolute error (*MR*) of the ultimate strain is less than 18%. Therefore, the two characteristic strain equations can provide good prediction results.

2.4 Design-oriented constitutive models of confined concrete

Based on the reviews and analysis of the composite columns presented in Chapter 1, we can divide the core concrete into the actively confined concrete (e.g., the steel tube confined concrete and the hoops confined concrete) and the passively confined concrete (e.g., the FRP confined concrete). Therefore, we should establish the constitutive models of the two types of confined concrete, respectively. Meanwhile, the design-oriented constitutive models should have simple expression and universal application characteristics. The proposed model will be modified by existing models. And the parameter selection and determination are proposed by the fitting methods.

2.4.1 Steel tube or hoops confined concrete

In this section, we set the steel tube or hoops confined concrete as the examples to study the design-oriented constitutive model of the actively confined concrete. For the stress–strain curves of the unconfined concrete, it can be found that the curve first shows a quickly increasing trend and then a slowly increasing trend before the peak point. The developing law reflects that there are few internal micro-cracks before the peak strain. Therefore, the confined pressure just has an obvious effect on the values of peak stress and peak strain and has little effect on the trend of the curves before the peak point. Then, the rising stage of the design-oriented constitutive model of the steel tube or hoops confined concrete can adopt the mathematical equation of the unconfined concrete. For the decreasing stage after the peak point, the inner micro-cracks of concrete appear and expand continuously, which leads to the stress loss. Besides, the lateral confining effect can restrict the inner developing micro-cracks, which causes the slowly decreasing trend of the confined concrete. Based on the above analysis, the decreasing stage of the design-oriented constitutive models of the steel tube or hoops confined concrete can be modified by that of the unconfined concrete.

At present, the classic constitutive model of unconfined concrete is the piecewise curve equation proposed by Guo et al. (1982) of Tsinghua University. The equation adopts a dimensionless coordinate ($x = \varepsilon/\varepsilon_{co}$, $y = \sigma/\sigma_{co}$) to establish the equation, as follows:

$$y = \left\{ \begin{array}{l} \alpha_a x + (3 - 2\alpha_a)x^2 + (\alpha_a - 2)x^3 x \leq 1 \\ x/\left[\alpha_d(x-1)^2 + x\right] x > 1 \end{array} \right\} \tag{2.23}$$

The increasing and the decreasing stages are continuous at the peak point of the stress–strain curve and conform to the geometric characteristics of the typical uniaxial compressive curve of the unconfined concrete and have a corresponding physical significance for each parameter in the increasing and the decreasing stages. We set the equations and draw the theoretical curves with different parameters in Fig. 2.9. The increasing parameter (α_a) of the curve equation is the ratio of the initial elastic modulus to the peak secant modulus, and the decreasing parameter (α_d) of the curve equation mainly influences the decline of the curve, and their equations are shown as follows:

$$\left\{ \begin{array}{l} \alpha_a = 2.4 - 0.0125 f_{co} \\ \alpha_d = 0.157 f_{co}^{0.785} - 0.905 \end{array} \right\} \tag{2.24}$$

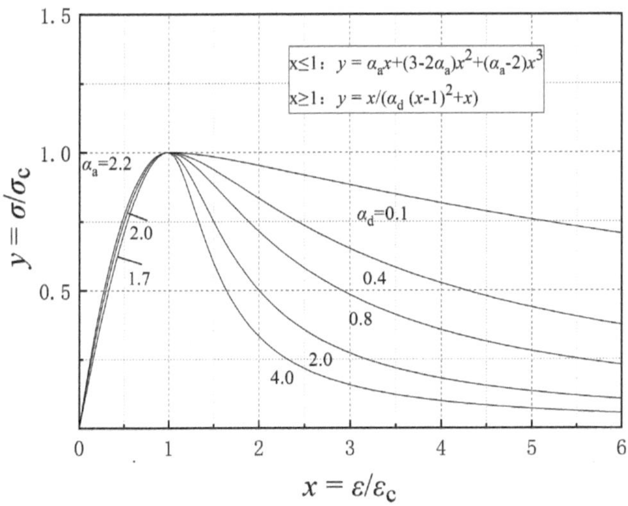

Figure 2.9 The theoretical stress–strain curves of concrete. This is just one figure.

As we knew, the different confined concretes have their own constitutive models in the existing studies, which makes it more difficult to use the mechanical properties of the confined concrete. Based on the lateral confining pressure analysis, we can establish a unified model for all types of confined concrete. This is because coefficients in the model can be modified by the properties of lateral confining materials. Based on the coefficient analysis of the constitutive model by Guo et al. (1982) and the analysis of lateral confining effect, the unified constitutive model of the confined concrete can be modified directly from the constitutive model of the unconfined concrete. Based on the lateral confining pressure calculation method and the constitutive model of unconfined concrete proposed by (Guo et al., 1982), we introduce some parameters and modify these parameters to establish the unified constitutive model.

Based on the above analysis, before the peak point in the stress–strain curves, the lateral confining pressure just has a greater effect on the peak stress and the peak strain, yet has little effect on the curve trend of the rising stage. Therefore, the curve equation of the rising stage is directly consistent with the constitutive model of unconfined concrete. In addition, we can gain the peak stress and the peak strain of the confined concrete in Section 2.3, then we substitute them into Eq. (2.23) to calculate the stress–strain curve of the increasing stage. After the peak point in the stress–strain curve, the internal micro-cracks of concrete increase which leads to the stress reduction, and the lateral confining pressure can restrict the internal increasing micro-cracks of concrete, so as to make the decreasing stage of the stress–strain curve tend to be slow. It can be found in Eq. (2.23) that the decreasing parameter (α_d) controls the curve trend of the decreasing stage. However, the declining speed shows first fast and then slow, which is not suitable for the actual curve trend of the stress–strain curves of confined concrete.

Then, we should introduce a new coefficient (namely the slope coefficient n) in the equation of the declining stage to reflect the slow declining trend of the stress–strain curves of confined concrete. For hoops confined concrete, it is well known that the confining effect along the length direction is not homogeneous, due to the "arch effect". Therefore, the confining effect on the concrete in the spacing area between hoops is less significant, and the overall confining effect should be reduced. In Section 2.2, we introduce the steel volume ratio ρ_v of the hoops confined concrete to calculate the lateral confining pressure. And the curve equation of the increasing stage of the hoops confined concrete is the same as that of the steel tube confined concrete. It is right to calculate the curve equation of the increasing stage, due to the less internal micro-cracks of concrete. After the peak point of the stress–strain curve, the internal increasing micro-cracks of concrete are not effectively restricted, because of the nonuniform confining effect along the length direction, especially for the concrete in the spacing area between hoops. Therefore, the decreasing stage of the stress–strain curve of the hoops confined concrete will show quicker declining speed than that of the steel tube confined concrete, which is similar to that of unconfined concrete. In order to reflect these characteristics of hoops confined concrete, we introduce the effective constraint coefficient (Ψ) to consider the nonuniform confining effect along the length direction. In order to meet the applicability of the proposed model, we add the effective constraint coefficient (Ψ) in the Mander model (Mander et al., 1988) and modify the declining parameter (α_d) by fitting the experiment database for the hoops confined concrete. Based on the above analysis, we can propose the mathematical expression of the constitutive model for the active confined concrete by modifying Eq. (2.23). The expression is shown as follows:

$$y = \left\{ \begin{array}{l} \alpha_a x + (3 - 2\alpha_a)x^2 + (\alpha_a - 2)x^3 x \leq 1 \\ x/[\psi\alpha_d(x-1)^n + x]x > 1 \end{array} \right\} \tag{2.25}$$

where x is the non-dimensional value of strain to the peak strain, $x = \varepsilon_c/\varepsilon_{cc}$; y is the non-dimensional value of stress to the peak stress, $y = \sigma/f_{cc}$; α_a is the rising parameter, $\alpha_a = 2.4 - 0.0124 f_{cc}$; α_d is the declining parameter, which is an important factor for the decreasing trend of the curve after peak point, and the equation will be studied and established in the following paragraph; Ψ is the effective confining coefficient, for the steel tube confined concrete $\Psi = 1$, for hoops confined concrete $\Psi = 0.021/\rho_v$, ρ_v is the steel volume ratio of hoops; and n is the declining slope coefficient (Shi et al., 2014), $n = 1 + \exp(-3\omega)$.

Based on the review of the confined concrete, the influence factors of the decreasing trend of the stress–strain curves after the peak point are the unconfined concrete properties and the lateral confining pressure. Meanwhile, based on the collection and the analysis of the experiment results of the steel tube confined concrete and the hoops confined concrete (Han & Yang, 2007; Mander et al., 1988), it can be found that the characteristics in the decreasing stage of the stress–strain curve of the confined concrete are mainly related to the relative confining ratio (ω). The lateral relative confining ratio can directly reflect the confining effect on core concrete. The larger lateral relative confining ratio can lead to well deformation capacity of confined concrete, then the decreasing stage of the constitutive model will appear later, and even the declining stage does not appear. Based on the above analysis, there are several influence factors that should be considered in detail, including the relative confining ratio, the ultimate strain of confined concrete, the compression strength of unconfined concrete, and the yield stress of confining materials. Based on the theoretical analysis and the data fitting of experiment results, we can propose the computation equation of the decreasing parameter (α_d), as follows:

$$\left\{ \begin{array}{l} \alpha_d = 3d/(17(d-1)^n) \\ d = \dfrac{0.0033}{\varepsilon_{cc}}\left[1 + 2.18 \times 10^5 (2\omega\sigma_c\varepsilon_{cc}/f_y)^{1.2}\right] \end{array} \right\} \tag{2.26}$$

The proposed model can predict the stress−strain curves of the actively confined concrete, including the steel tube confined concrete and the hoops confined concrete. And the two typical types of actively confined concrete have been analyzed. Therefore, we use two examples to verify the model by the comparisons between calculation curves and experiment curves (Han & Yang, 2007; Shi et al., 2014). Finally, the curve comparisons are shown in Fig. 2.10. As shown in Fig. 2.10, the fitting degrees of curves are great by comparison to the computation results. Therefore, the proposed constitutive model of actively confined concrete in this study can fit better with the stress−strain curves from experiment results.

2.4.2 FRP confined concrete

It is well known that the FRP confined concrete shows two kinds of curve trends under different confining effects. The stress−strain curve of FRP confined concrete under weak confining effect has three stages: first rising stage, second declining stage, and third little rising stage. The stress−strain curve of FRP confined concrete under strong confining effect has two stages: first rising stage and second rising stage. Because of the unconventional behavior under the two kinds of confining effect, two kinds of constitutive models should be established for FRP confined concrete. Furthermore, it is important to judge the limits value between the strong effect and the weak effect of the FRP tube for the core concrete. The study presented by (Wu et al., 2006) introduced the limit value of confining effect (λ) to judge whether there is a softening section in the stress−strain curve, and the influence factor is the ultimate confining ratio ($f_{l,a}/f_{co}$), where the $f_{l,a}$ is the ultimate confining stress of the FRP tube, f_{co} is the compression strength of the unconfined concrete. The judgment method is that when $f_{l,a}/f_{co} < \lambda$, there is a softening stage in the curve; otherwise, the curve shows a bilinear increasing trend. At the same time, through the analysis of an extensive database, it is suggested that the limit value

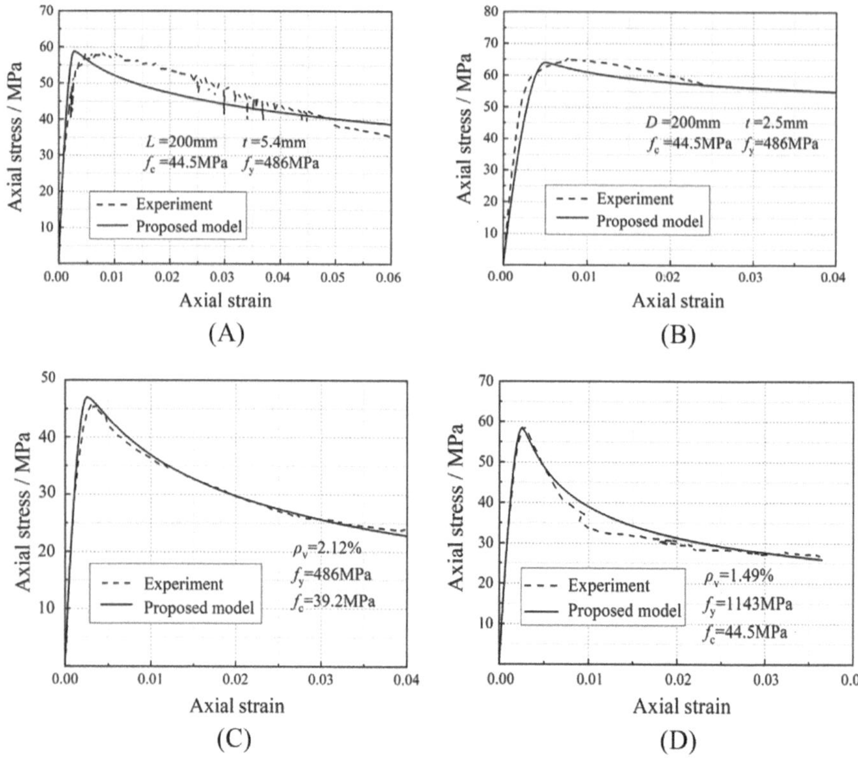

Figure 2.10 The verification of the proposed constitutive model of actively confined concrete. (A) square steel tube; (B) circle steel tube; (C) hoops; (D) hoops.

takes 0.13 for FRP confined concrete columns with ordinary elastic modulus (elastic modulus less than or equal to 250 GPa).

For the weak confining effect, we can divide the whole stress−strain curve of the FRP confined concrete in three stages. The first stage is the rising stage, and the stress−strain relationship is basically linear. FRP does not provide obvious lateral confining effect, because of the little expansion of core concrete. When the load gradually increases, the internal cracks of concrete continue to appear and expand, the slope of the stress−strain curve decreases, and the volume strain of concrete turns to expand. And the volume expansion can increase the confining effect provided by the FRP tube, so the peak stress of concrete is improved. Because of the weak confining of the FRP tube, the stress is not kept after crossing the peak point and the curve turns to the second stage, i.e., the softening stage. The stress−strain curve in the softening stage is approximately a straight line, and we can determine the slope and the limit point by the lateral restraint characteristics of the FRP tube. It means that even if the peak points of the curve are the same, the slope of the softening stage and the limit point will be different. After the limit point, the increasing confining effect of the FRP tube can restrict the internal cracks in the core concrete, the curve turns to the third stage, i.e., the little rising stage. In addition, we can gain the limit point by using the judgment methods, compute the remnant stress by the judgment equation ($f_{cr} = f_{l,a}/\lambda$), and compute the corresponding strain by the equation of the second curve stage. Meanwhile, the third stage is simplified as a leveling line. Based on the above analysis, the equations of the constitutive model of the FRP confined concrete with a softening stage as follows:

$$\sigma c = \begin{Bmatrix} f_{cc}\left[\alpha_a(\varepsilon_c/\varepsilon_{cc}) + (3-2\alpha_a)(\varepsilon_c/\varepsilon_{cc})2 + (\alpha_a-2)(\varepsilon_c/\varepsilon_{cc})3\right]0 \le \varepsilon_c < \varepsilon_{cc} f_{cc} + E_2(\varepsilon_c - \varepsilon_{cr})\varepsilon_{cc} \le \varepsilon_c < \varepsilon_{cr} \\ f_{cr}\varepsilon_{cr} \le \varepsilon_c < \varepsilon_{cu} \end{Bmatrix} \tag{2.27}$$

where the E_2 is the softening elastic modulus, which is influenced by the section size, the properties of the FRP tube, and unconfined concrete, then $E_2 = -11Rf_{co}{}^3/E_{frp}t_{frp}$.

For the strong confining effect, we can divide the whole stress–strain curve of the FRP confined concrete into two linear increasing stages. In fact, there is a nonlinear variation stage between the two linear increasing curves. However, the stress and the strain at the nominal peak point are difficultly determined discretely. And the stage has little effect on the whole curve, because of the short length. Therefore, the nonlinear variation can be ignored. The equations of the constitutive model of the FRP confined concrete under the strong confining effect are as follows:

$$\sigma_c = \begin{Bmatrix} f_{cc}\varepsilon_c/\varepsilon_{ce}0 \le \varepsilon_c < \varepsilon_{cc} \\ f_{cc} + \dfrac{f_{cu} - f_{cc}}{\varepsilon_{cu} - \varepsilon_{cc}}(\varepsilon_c\text{-}\varepsilon_{cc})\varepsilon_{cu} \le \varepsilon_c \le \varepsilon_{cc} \end{Bmatrix} \tag{2.28}$$

The two proposed models can predict the stress–strain curves of the passively confined concrete, including the weak confining effect and the strong confining

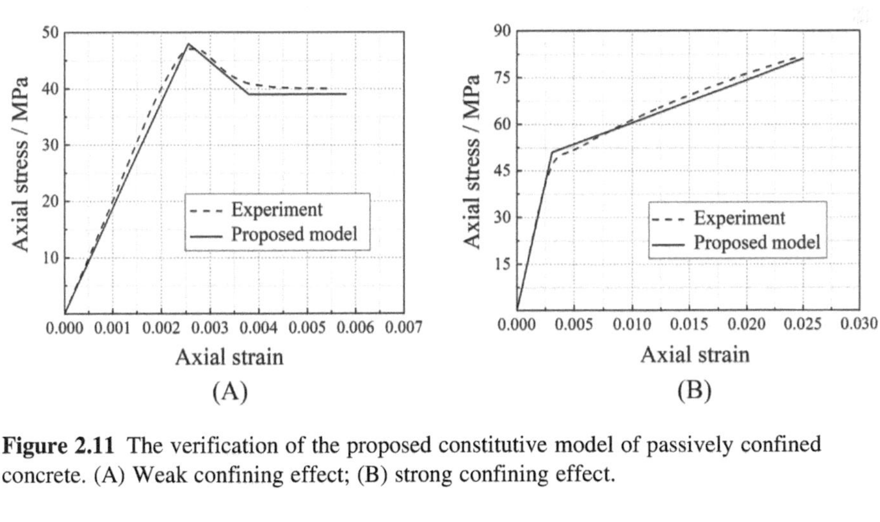

Figure 2.11 The verification of the proposed constitutive model of passively confined concrete. (A) Weak confining effect; (B) strong confining effect.

effect. And the two typical types of passively confined concrete have been analyzed. Therefore, we use two examples to verify the model by the comparisons between calculation curves and experiment curves (Wei et al., 2017; Wu et al., 2006). Finally, the curve comparisons are shown in Fig. 2.11. As shown in Fig. 2.11, the fitting degrees of curves are great by comparison to the computation results. Therefore, the two proposed constitutive models in this section can provide accurate prediction of stress−strain curves for passively confined concrete.

2.5 Conclusions

The design-oriented constitutive model is important for the application of composite columns in practical engineering. To establish the simple and convenient constitutive model, the actively confined concrete and the passively confined concrete should be studied, respectively. This section analyzes the stress state based on the Twin Shear Unified Strength Theory and proposes the characteristic stress equation for the confined concrete. Then we propose the characteristic strain equation of each type of confined concrete by the theoretical analysis and the fitting method for the experiment results. Finally, we establish the design-oriented constitutive models of the actively confined concrete and the passively confined concrete. In summary, the following conclusions can be drawn:

(1) The lateral confining pressure is the most important influence factor in the mechanical behavior of confined concrete. This section proposes the calculation equations of lateral confining pressure for three kinds of confined concretes, including steel tube confined concrete, hoops confined concrete and FRP confined concrete.

(2) Based on the lateral confining pressure and the unconfined concrete properties, we analyze the stress condition of the confined concrete, and introduce the relative confining ratio (ω) to reflect the confining effect. Then we propose an equation to compute the peak stress of confined concrete, based on the Twin Shear Unified Strength Theory. Based on the analysis and the verification, the equation is suitable for various confined concretes, since the equation is based on the lateral confining pressure.

(3) Based on the characteristic stress and the properties of confining materials, we propose the calculation equation of characteristic strain by the fitting method.

(4) Based on the constitutive model of unconfined concrete, we establish the constitutive model of actively confined concrete (i.e., steel tube confined concrete and hoops confined concrete) by modifying the main parameters and introducing some important parameters. The comparison between calculation curves and experiment curves shows a good fit for the proposed model.

(5) Based on the characteristics of passively confined concrete, the limit value of confining effect (λ) to judge whether there is a softening section in the stress−strain curve. Then we establish the two constitutive models for two types of passively confined concrete. The comparison between calculation curves and experiment curves shows a good fit for the proposed model.

(6) The foundations of the proposed models are the basic mechanical properties of unconfined concrete and lateral confining materials. The constitutive model proposed in this study is simple and convenient for application in practical engineering. Furthermore, this solution method is suitable for both actively and passively confined concrete, therefore, the proposed model belongs to a unified constitutive model of various confined concrete.

Acknowledgments

This work was supported by the National Natural Science Foundation of China [grant numbers 52108171; 51878540]. The authors gratefully acknowledge two Master's students (Mr. Qv Yunsong and Mr. Tian Wenkai) of Xi'an University of Architecture and Technology for the drawing works.

References

Daniel, C, & Patrick, P (1995). Stress-strain model for confined high strength concrete. *Journal of Structural Engineering*, *121*, 468–477.

Guo, ZH, Zhang, XQ, Zhang, DC, & Wang, RQ (1982). Experimental investigation of the complete stress-strain curve of concrete. *Journal of Building Structures*, *3*, 1–12.

Han, LH, & Yang, YF (2007). *The technology of modern steel tube confined concrete structures*. China Building Industry Press.

Lee, SJ (2007). Capacity and the moment-curvature relationship of high-strength concrete filled steel tube columns under eccentric load. *Steel and Composite Structures*, *7*(2), 135–160.

Li, N, Lu, YY, Li, S, & Liang, HJ (2015). Statistical-based evaluation of design codes for circular concrete-filled steel tube columns. *Steel & Composite Structures*, *18*, 519–546.

Mander, JB, Priestley, MJN, & Park, R (1988). Theoretical stress-strain model for confined concrete. *Journal of Structural Engineering*, *114*, 1804–1826.

Ozbakkaloglu, T, & Lim, JC (2013). Axial compressive behavior of FRP-confined concrete: experimental test database and a new design-oriented model. *Composite Part B: Engineering*, *55*(12), 607–634.

Patel, VI, Liang, QQ, & Hadi, MNS (2012). Nonlinear inelastic behavior of circular concrete-filled steel tubular slender beam-columns with preload effect. *Concrete Institute of Australia*, *13*, 395–402.

Ren, QX, Hou, C, Lam, D, & Han, LH (2014). Experiments on the bearing capacity of tapered concrete filled double skin steel tubular (CFDST) stub columns. *Steel and Composite Structures*, *17*, 667–686.

Sakino, K, Nakahara, H, Morino, S, & Nishiyama, I (2004). Behavior of centrally loaded concrete-filled steel-Tube short columns. *Journal of Structural Engineering*, *130*, 180–188.

Samani, AK, & Attard, MM (2012). A stress-strain model for uniaxial and confined concrete under compression. *Engineering Structures*, *41*, 335–349.

Scott, BD (1982). Stress-strain behavior of concrete confined by overlapping hoops at low and high strain rates. *ACI Structural Journal, 79*, 13–27.

Shan, B, Cai, J, Xiao, Y, & Giorgio, M (2016). Experimental research on large-scale square square columns confined with wrapped GFRP under axial compressive load. *Journal of Hunan University (Natural Sciences), 43*(3), 75–82.

Shi, QX, Rong, C, & Zhang, T (2017). A constitutive model for confined concrete in composite structures. *Steel and Composite Structures, 24*(6), 689–695.

Shi, QX, Tian, Y, Wang, N, & Hou, W (2011). Comparison study of axial behavior of high-strength concrete confined by normal- and high-strength lateralties. *Andvanced Science Letters, 4*, 2681–2685.

Shi, QX, Wang, N, Tian, JB, & Shi, JL (2014). A practical stress-strain model for high-strength stirrups confined concrete. *Journal of Building Materials, 17*, 216–222.

Shi, QX, Wang, N, Wang, QW, & Men, JJ (2013). Uniaxial compressive stress-strain model for high-strength concrete confined with high-strength lateral ties. *Engineering Mechanics, 30*, 131–137.

Soliman, EKS (2011). Behavior of long confined concrete column. *Ain Shams Engineering Journal, 23*, 141–148.

Takahama, U, & Oniki, T (2014). Lateral strain-to-axial strain relationship of confined concrete. *Journal of Structural Engineering, 13*, 130–138.

Teng, JG, Chen, JF, Smith, ST, & Lam, L (2002). *FRP strengthened RC structure*. John Wiley & Sons, Ltd..

Toutanji, H (2001). Design equations for concrete columns confined with hybrid composite materials. *Andvanced Composite Materials, 10*, 127–138.

Toutanji, H, & Saafi, M (2002). Stress-strain behavior of concrete columns confined with hybrid composite materials. *Material Structure, 35*, 338–347.

Varma, AH, Sause, R, & Ricles, JM (2005). Development and validation of fiber model for high strength square concrete filled steel tube beam-columns. *ACI Structural Journal, 102*, 73–84.

Wang, YC, & Restrepo, JI (2001). Investigation of concentrically loaded reinforced concrete columns confined with glass fiber reinforced polymer jackets. *ACI Structural Journal, 98*(3), 377–385.

Wei, Y, Wu, G, Wu, ZS, Lv, ZT, & Guo, ZX (2017). Study on stress-strain relationship for concrete rectangular columns sufficiently confined with FRP. *Building Structure, 37* (12), 75–78.

Wu, G, Wu, ZS, & Lv, ZT (2006). Study of the stress-strain relationship of FRP-confined circular concrete column with a strain-softening response. *China Civil Engineeing Journal, 39*(11), 7–14.

Yu, MH, Li, JC, & Ma, GW (2007). *Structural Plasticity*. Springer-Verlag.

Yu, T, Teng, JG, Wong, YL, & Dong, SL (2010). Finite element modeling of confined concrete-I Drucker-Prager type plasticity model. *Engineering Structures, 32*, 665–679.

Zhang, SM, Liu, JP, & Ma, L (2007). Experimental research and bearing capacity analysis of axially compressed stub columns of circular tube confined high-strength concrete. *China Civil Engineering Journal, 40*, 24–31.

Failure criterion of concrete under multiaxial compression

3

3.1 Introduction

Concrete composite structures have been wildly used in practical engineering and deeply studied in existing researches. The concrete composite members have excellent behavior, including high bearing capacity, high ductility, high fire resistance, and high durability. Therefore, the concrete composite structures are indispensable parts of high-rise buildings, arch dams, bridges, and other structures. For various building materials in composite structures, the mechanical properties of concrete are the most complex, especially for the concrete under the lateral confining effect. Under the lateral confining effect provided by different confining materials, the concrete reflects various mechanical properties which have been reviewed and analyzed in the last two chapters. There are many existing studies (Kaya et al., 2015; Øverli, 2016; Shkolnik, 2008; Yan & Lin, 2008) on the behavior of concrete under multiaxial stress. The experimental and theoretical study results show that the compressive strength of the concrete increases with the increase in lateral compressive pressure. Based on these studies, many stress models have been proposed for concrete under multiaxial stress condition, and these proposed models are applied in the constitutive model establishment for the actively or the passively confined concrete. It should be noted that most of the proposed stress calculation models are based on the compressive strength of unconfined concrete and lateral compressive pressure, and the establishment process is the parameters fitting in huge amounts of test data. However, it is just suitable for rough prediction in structure design and engineering analysis. It is difficult to deeply study the stress condition of confined concrete by the two ordinary influence factors (Ozbakkaloglu, 2013). It is necessary to study the strength model of confined concrete in detail.

There are many existing studies on the influence factors of the mechanical behavior of concrete. The study presented by Wang et al. (2016) analyzed the effect of concrete types, based on the experimental analysis of ultra-high-strength concrete under triaxial compression. The study presented by Zhou et al. (2011) analyzed the effects of free water content and strain rate on the mechanical behavior of concrete. The study by Rossi et al. (2013) analyzed the effect of compressive stress levels on the concrete strain. Moreover, the specimen size, the loading rate, and the loading types were deeply studied in the studies by Wang et al. (2017) and Zheng et al. (2007). However, most of the existing studies are still based on experimental analysis and fitting methods. It is necessary to study the stress condition of concrete by mechanics theoretical analysis and mathematical derivation, especially for confined concrete.

Concrete Composite Columns. DOI: https://doi.org/10.1016/B978-0-323-85171-8.00003-3

For the existing theoretical studies, the relationship between macroscopic and microscopic mechanical properties of concrete is analyzed, based on mathematical theory and numerical studies (Tang & Zhang, 2009; Xiao et al., 2011). These special macro-level and micro-level study methods show that the change in the behavior of the composite material is systematic, which means that the different micro-characteristics can be expressed by a similar macroscopic pattern. Therefore, the mechanical properties of confined concrete can be analyzed and calculated by macroscopic mechanics, and any type of concrete can be studied in the same mechanical method. Based on the macroscopic mechanical and microscopic mechanical studies of confined concrete presented by Zhang (1998) and Zhu et al. (2005), the basic mechanical influence factors of concrete under multiaxial stress can be summarized in the following six aspects: the tensile-compressive strength difference, the hydrostatic stress (σ_m), the shear stress (variation in the minimum stress), the principal stress, the middle principal stress (σ_2), and the variation in the middle principal stress. It can be found that the stress condition of core concrete in concrete composite construction is complex. However, the normal failure criterion model of concrete is focused on the simple stress condition. The normal mechanical foundation is not suitable for the confined concrete, and the designing method for complex structures is not yet economical and reasonable.

For all building materials, the failure criterion is the basic theoretical foundation to analyze the stress condition and propose the corresponding stress calculation methods. The failure criterion can be described by a mathematical function model to express the damage envelope surface. Therefore, the accurate failure criterion is very important to quantify and calculate the mechanical properties of confined concrete. Many early studies realized the importance of the failure criterion for concrete under multiaxial stress, and they have proposed a variety of failure criteria based on different stress characters (Asteris & Plevris, 2016; Menetrey & Willam, 1995; Piscesa et al., 2016; Plevris & Asteris, 2014; Shang & Song, 2013; Syrmakezis & Asteris, 2001; Yu, 2011), including the Mohr–Coulomb criterion, the Drucker–Prager criterion, the Willam–Warnker criterion, the Ottosen criterion, Asteris and Plevrisand criterion, and the Hsieh–Ting–Chen criterion. In these proposed failure criteria, the several stress parameters are combined together to reflect the original equations of tensile and compressive meridians, then used to establish the mathematical expression of the yield surface of the concrete in complex stress conditions. The stress parameters include the principal stress (σ_1, σ_2, σ_3), the stress invariant (I_1, J_2, J_3), the octahedral stress (σ_{oct}, τ_{oct}, θ) and the mean stress (σ_m, τ_m, θ). The mathematical derivation process of the proposed failure criterion is that:

(1) the basic stress parameters are defined first
(2) then the failure surface of the concrete in failure state is described
(3) finally, the calculation model of failure criterion is established.

Therefore, the accuracy of concrete failure criterion is influenced by the following two factors: the rationality of selected stress parameters, which should be the main influence factors of the concrete failure criterion and have definite physical significance; the reasonable selection of stress failure surface of concrete, which decides the stress boundary condition in the derivation process. However, the selected stress parameters in most existing models do not have obvious physical significance or selecting principles. Meanwhile, few failure criteria can precisely analyze the failure mechanism of confined concrete based on theoretical foundation. Therefore, it is necessary to analyze the mechanical properties of concrete under multiaxial stress, select the stress parameters which both have a clear physical significance, and reflect the failure mechanism of concrete under multiaxial stress. Further, the mathematical expression of failure criterion should be established by theoretical deduction, based on reasonable boundary conditions.

Against the background, the stress condition of concrete under multiaxial compression is deeply studied at first by the existing biaxial and triaxial concrete compressive experiments. Then, the development trend of maximum principal stress is analyzed, and the main influence factors are determined, based on the test results. After that, the Twin Shear Strength Theory is introduced, and the parameters in this theory are discussed. Based on deficiencies of the theory, the new stress parameters are introduced in the theoretical equation to reflect the development trend of maximum principal stress. In the new proposed equations, the unknown coefficients should be solved by the selected boundary condition in stress condition of concrete under multiaxial stress. Finally, the coordinate system transformations are used to gain the solving method of unknown coefficients, three kinds of failure criterion models for concrete under multiaxial compression are established, and the coefficients in each failure criterion are identified. Based on the above establishment methods, the proposed failure criterion models have the following characteristics:

(1) based on the analysis of stress condition of concrete under multiaxial stress, the selected stress parameters have clear physical meaning and reasonable theoretical value
(2) the particular mathematical derivation method makes the failure criterion models keep the reasonable theoretical value
(3) the establishment method do not rely on the data fitting, but rely on the exact solution for the unknown coefficients by selected boundary condition, so the failure criterion models have the theoretical solution which can be modified by selecting unique boundary condition to adapt in all types of concrete.

3.2 Stress condition of concrete under multiaxial compression

In order to select accurate stress parameters for the failure criterion models, the stress condition of concrete under multiaxial compressive pressure should be studied deeply. Therefore, the biaxial compressive test and triaxial compressive test of concrete are described, based on the study by Rong et al. (2018).

3.2.1 Describe of multiaxial compressive experiment

For convenient production and test, the concrete is mixed in the field and made into concrete cubes (100 × 100 × 100 mm) (Rong et al., 2018). And the concrete cube specimens are cured in standard conditions (i.e., the temperature of 20°C ± 2°C and humidity > 95 %) for 28 days and then cured in natural environment for three months, so as to eliminate the influence of different batching times and loading times on the concrete strength (Rong et al., 2018). After the curing process of concrete specimens, all specimens are tested in true triaxial compression machine. Before the multicompression tests, three layers of plastic membrane with glycerin should be set on the compression surface of concrete cube to eliminate the friction between the machine surface and specimen surface (Rong et al., 2018).

For the biaxial compression test and triaxial compression test (Rong et al., 2018), two kinds of proportional monotonic loading modes are used. For the biaxial compressive test, the stress increase of compression in two vertical directions. The stress ratio of two directions ranges from 0 to 1 (i.e., $\sigma_2{:}\sigma_3 = 0.00, 0.25, 0.50, 0.75$ and 1.00), and the first principal stress is equal to 0 in the biaxial compressive test. For the triaxial compressive test, three types of proportional stress ratio in loading processes should be considered, including two types: one type is the same stresses in two directions and another type is different stresses in three directions. Therefore, 14 triples of concrete cube specimens (i.e., $\sigma_1{:}\sigma_2{:}\sigma_3$ = 0:0:1, 0:1:1, 1:1:1, 0.1:0.1:1, 0.1:0.25:1, 0.1:0.3:1, 0.1:0.5:1, 0.1:0.75:1, 0.1:1:1, 0.25:0.25:1, 0.25:0.3:1, 0.25:0.75:1, 0.25:1:1 and 0.3:1:1) are tested in triaxial compression experiment. In this study presented by Rong et al. (2018), the three same specimens are tested in each kind of stress ratio, and the final data is averaged by the three test results. In addition, it should be noted that all three principal stresses are negative due to the compressive condition of concrete specimens. Therefore, the principal stresses could be expressed as $\sigma_1 \geq \sigma_2 \geq \sigma_3$. However, the absolute values of the three principle stresses are in reverse order. They can be divided into the minimum principle stress (σ_1), the middle principle stress (σ_2), and the maximum principle stress (σ_3).

3.2.2 The stress condition of concrete under multiaxial compression

The biaxial compression test results show that the failure form is closely relevant to the stress ratio (σ_2 / σ_3). The failure forms of specimens under low-stress ratio (i.e., $\sigma_2 / \sigma_3 = 0$ and 0.25) show several roughly parallel cracks, which is similar to the failure forms of concrete under uniaxial compression. The failure forms of specimens under higher stress ratio (i.e., $\sigma_2 / \sigma_3 = 0.5$) show several roughly parallel cracks on the concrete surface without loading and polygon cracks on the concrete surface with low loading. The failure forms of specimens under high-stress ratio (i.e., $\sigma_2 / \sigma_3 = 0.75$ and 1) show flake splitting cracks on the concrete surface without loading and polygon cracks on the other two surfaces. The biaxial compressive test results are shown in Table 3.1, where each specimen is given a name: BCS means biaxial compression strength and the following number is specimen number. The test results in Table 3.1 show that the maximum principal stress (σ_3) increases with the increase in middle principal stress (σ_2) at first and then decreases, and the

Table 3.1 Biaxial compressive strength of specimens under various stress ratios.

Concrete type	Stress ratio (σ_2 / σ_3)	$-\sigma_2$ (MPa)	$-\sigma_3$ (MPa)
BCS1	0.00	0.00	34.20
BCS2	0.25	10.72	42.91
BCS3	0.50	22.43	44.85
BCS4	0.75	31.57	42.09
BCS5	1.00	40.66	40.66

stress ratio (σ_2 / σ_3) at peak point is about 0.5. The test results will be analyzed in the following sections.

In order to describe the stress condition easily, the maximum stress (σ_3) is named axial principle stress and other lower stresses are named lateral principle stresses (σ_1 and σ_2) which can be divided into the large one and the small one. The triaxial compression test results show that the failure mode is closely relevant to the two lateral principle stresses. The failure modes of specimens under low lateral principle stress show several roughly parallel cracks, which is similar to the failure modes of concrete under uniaxial compression. The failure modes of specimens under one high lateral stress (i.e., σ_2 / σ_3 = 0.5, 0.75) show several roughly parallel cracks and polygon cracks on the two concrete surfaces. The failure modes of specimens under one high lateral stress (i.e., σ_2 / σ_3 = 0.5, 0.75 and σ_2 / σ_3 is less than 0.25) show several roughly parallel cracks and polygon cracks on the two concrete surface. The failure modes of specimens under two high lateral stresses (i.e., σ_2 / σ_3 is over 0.25) show polygon cracks on the two concrete surface with multiple broken concrete fragments. The failure modes of specimens under high triaxial compressive pressure (i.e., σ_1: σ_2: σ_3 = 1:1:1) show multiple concrete fragments accompanied by concrete powder on the failure surface. The triaxial compressive test results are summarized in Table 3.2, where each specimen is given a name: TCS means triaxial compressive strength, and the following number is the specimen number. Meanwhile, the uniaxial tensile strength (f_t) of the concrete is 2.07 MPa. Based on the test results in Table 3.2, the following conclusions can be drawn:

(1) the splitting failure is changing into flow failure with the increase of triaxial compressive strength
(2) with the similar minimum principle stress (σ_1, which is from TCS3 to TCS9), the maximum principle stress (σ_3) increases with the increase of middle principle stress (σ_2) and then decreases, where the stress ratio at peak point is about 0.3 which is smaller than that of biaxial compressvie test results
(3) based on comparisons of TCS11, TCS12 and TCS13, the middle principle stress has little effect on maximum principle stress (σ_3) under the high maximum principal shear stress ($\sigma_3 - \sigma_1$)
(4) based on the comparison between TCS13 and RCS 14, the high-stress ratio do not directly lead to high maximum principal stress (σ_3), but the value of minimum principal stress (σ_1) has obvious effect.

Table 3.2 Triaxial compressive strength of specimens under various stress ratios.

Concrete type	Stress ratio (σ_1:σ_2:σ_3)	$-\sigma_1$ (MPa)	$-\sigma_2$ (MPa)	$-\sigma_3$ (MPa)
TCS1	0.00:0.00:1.00	0.00	0.00	22.35
TCS2	0.00:1.00:1.00	0.00	27.41	27.41
TCS3	1.00:1.00:1.00	325.34	325.34	325.34
TCS4	0.10:0.10:1.00	5.45	5.57	52.83
TCS5	0.10:0.25:1.00	7.15	17.53	72.87
TCS6	0.10:0.30:1.00	7.82	21.38	71.77
TCS7	0.10:0.50:1.00	6.81	36.15	69.82
TCS8	0.10:0.75:1.00	6.08	45.77	61.65
TCS9	0.10:1.00:1.00	5.73	56.96	57.23
TCS10	0.25:0.25:1.00	37.73	38.29	150.36
TCS11	0.25:0.30:1.00	45.47	55.89	188.01
TCS12	0.25:0.75:1.00	46.57	135.71	187.28
TCS13	0.25:1.00:1.00	41.73	163.47	166.59
TCS14	0.30:1.00:1.00	25.05	83.35	83.96

Based on the analysis of biaxial compressive test results and triaxial compressive test results, it is found that the shear stress and hydrostatic stress have complex influencing mechanism and have important effects on failure mode. Both of them should be seriously considered in the establishment of the failure criterion. The stress condition of concrete under multiaxial pressure will be studied in the following section.

3.2.3 The stress condition of concrete under multiaxial compression pressure

As shown in Table 3.1, the maximum principal shear stress ($\sigma_3 - \sigma_1$) is not the unique influence factor on concrete failure. The middle principal stress also has obvious effect on the maximum principal stress of concrete under biaxial compression pressure. The relationship between middle principal stress and maximum principle stress is shown in Fig. 3.1. The relationship curve shows the trend that first increased and then decreased, and it shows univariate conic variation. At the σ_2 is about 23 MPa, the relationship curve reaches the peak point. These results mean that the middle principal stress is an important factor in the failure of concrete. Therefore, the middle principal stress should be considered in the failure criterion establishment.

Based on the analysis in Table 3.2, the effect of middle principal stress is ambiguous, thus the stress condition should be studied in detail. With the two same low principle stresses ($\sigma_1 = \sigma_2$), the relationship between minimum principle stress and maximum principle stress is shown in Fig. 3.2. The test results show that the rising speed of the curve has continuous reduction trend, which reflects that the effect of higher lateral pressure on the improvement of axial stress is lower. In addition, the hydrostatic stress has obvious effect on stress condition of the concrete failure

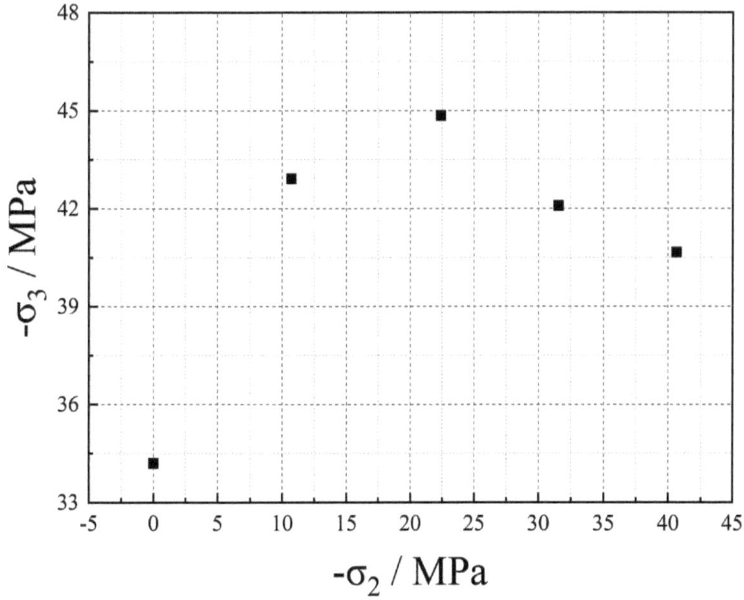

Figure 3.1 The relationship between $\sigma 3$ and $\sigma 2$ under the biaxial compression pressure.

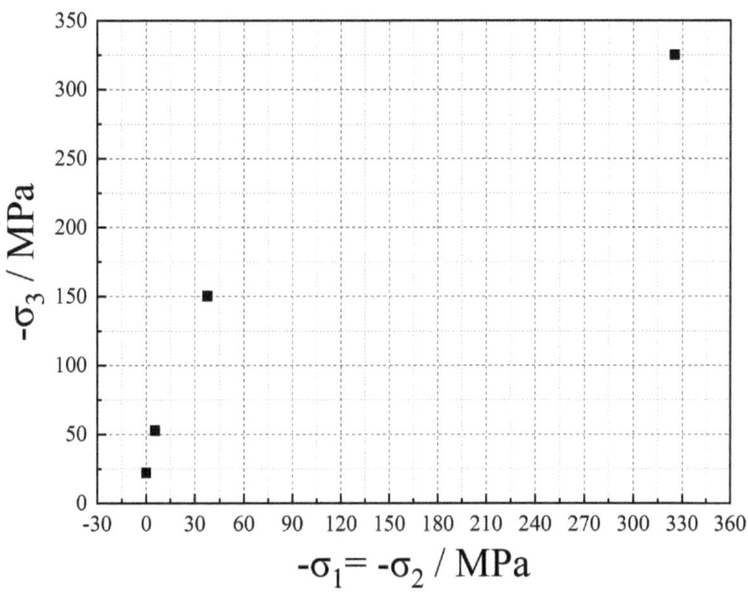

Figure 3.2 The relationships between σ_3 and σ_1 under triaxial compression ($\sigma_1 = \sigma_2$).

under lower multi-compressive pressure, and the principal shear stress has obvious effect on stress condition of the concrete failure under higher multi-compressive pressure. However, the effect provided by which kind of principal shear stress ($\sigma_2 - \sigma_1$, $\sigma_3 - \sigma_1$ or $\sigma_3 - \sigma_2$) is not yet determined. Therefore, the three kinds of principal shear stress are further studied. With the two same principal stresses ($\sigma_2 = \sigma_3$), the relationship between minimum principle stress (σ_1) and maximum principle stress (σ_3) is shown in Fig. 3.3. The curve trend in Fig. 3.3 is the same as that in Fig. 3.2. Based on the comparisons of Figs. 3.1–3.3, and Table 3.2, the following conclusions can be drawn:

(1) Figs. 3.2 and 3.3 show similar change trends which means that the maximum compressive strength cannot only be expressed by the middle principal stress
(2) the stress ratios (σ_2 / σ_3) at the peak point is changed with the minimum principle stress (σ_1), which means the effect of middle principal stress (σ_2) is also influenced by the minimum stress (σ_1)
(3) the middle principle shear stress (i.e., $\sigma_3 - \sigma_2$ or $\sigma_2 - \sigma_1$) is the main factor, yet the maximum principal shear stress ($\sigma_3 - \sigma_1$) is not the only crucial factor.

Based on the conclusions, the influence of principal shear stress (variation in the minimum stress), principal stress, middle principal stress (σ_2) and middle principal shear stress (i.e., $\sigma_3 - \sigma_2$ or $\sigma_2 - \sigma_1$) should be considered together. Therefore, it is necessary to introduce and modify a reasonable theoretical mechanical model to include all influences of these stress parameters.

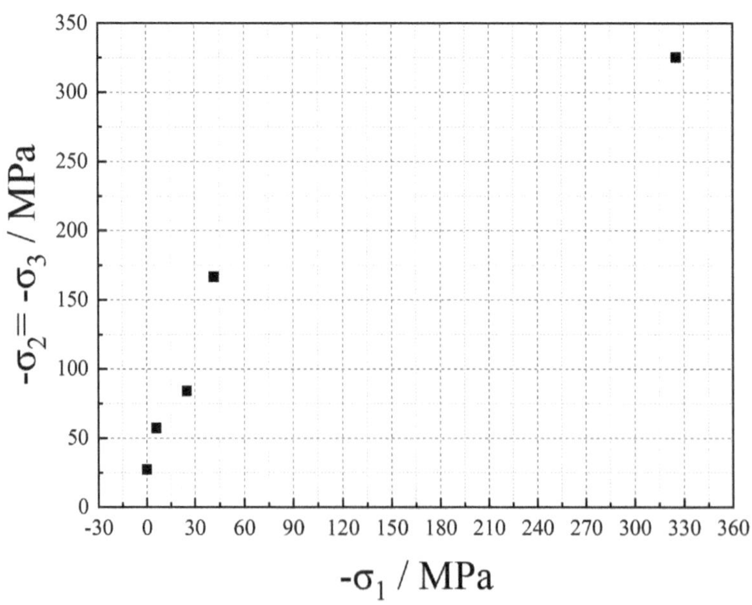

Figure 3.3 The relationships between σ_3 and σ_1 under triaxial compression ($\sigma_2 = \sigma_3$).

3.3 The theoretical foundation of failure criterion models

Based on the stress analysis in Section 3.2, the stress condition at the failure condition of concrete under multiaxial compression pressure is complex. There are many influence factors on the failure of concrete under multiaxial compressive pressure which can be concluded into the following four types: the principal stress, the principal shear stress, the middle principal stress (σ_2), and the hydrostatic stress (σ_m). However, the influence rules of the four influence factors are unobvious, due to the coupling effect between each other. Therefore, it is necessary to select the reasonable strength theoretical foundation and the corresponding calculation method to gain the failure criterion models of concrete under multiaxial compressive pressure.

3.3.1 Development history of strength theory

Strength theory is used to study the yield law or the failure law of materials (concrete, ceramics, rock, geotechnical, and metal) under complex stress conditions, and provide calculation failure criterion for strength design in the relevant engineering structure design. However, due to the complexity of strength theory, Leonardo da Vinci (1452 ~ 1519) and Galileo Galilei (1564 ~ 1642) began to study the tensile test of iron wire and stone in the early 15th and 16th centuries. Until the emergence of three strength theories (the first and second strength theories and Tresca yield criterion (1864)) in the 19th century, and marked by the M−C theory (1900) in the 20th century, researches have established a series of strength models from different ideas and for different materials. Especially, in more than 100 years since the 20th century, there are a vast number of research literature have established in this field. The study presented by Yu (2011) have systematically sorted and studied these theoretical references, then established the Twin Shear Strength Theory. Therefore, the existing studies on the strength theory field will not be described in this section, and the following contents are mainly discussed: the loopholes in the double shear strength theoretical system, the influence effect of the intermediate stress and the hydrostatic stress on multiaxial strength of concrete, the relationship between strength parameters that reflect materials, and the general adaptability of strength model.

In the development of strength theory, the studies on the effect of middle principal stress have been in the ascendant, but it has not been well solved for many years. In theory, if the material strength is independent of the middle principal stress, the strength theory will be greatly simplified and the test will be easy to carry out, then the strength theory with simple form can be easily used to solve practical problems in engineering application. At the end of the 19th century, Föppl presented the earliest experimental study on the effect of middle principal stress. In the 1919s, D. Sandel introduced the combination parameter of the shear stress and the hydrostatic stress to reflect the effect of middle principal stress. Until the 1960s, after the emergence of true triaxial testing machine, the middle principal stress effect is studied in experiments under true triaxial stress conditions. So far, serious strength models reflecting intermediate stress (and hydrostatic stress) are proposed. In these models, the famous Drucker–Prager criterion (1952) is proposed by modifying the Sandel model. Further, the study presented by Mogi (1967) confirmed the existence of the intermediate principal stress effect through experiments.

On the other hand, in order to facilitate study and application, some researches try to establish a unified strength model that can reflect various materials. For example, the combination strength theory (unified strength theory) is proposed in the 1940s. However, the Encyclopedia of China in 1985 pointed out that it is not a combination strength theory or a unified strength theory. The nonlinear unified failure function, as mentioned in Hershcy (1954), Ehlers (1995), and Krenk (1996), is also not a truly unified model. In order to systematically solve the influence of middle principal stress and the unity of strength theory, the study presented by Yu (2011) has established a new and systematic unified strength theory system with more than 40 years of energy, since Yu first put forward the concept of the Twin Shear Strength Theory in 1961. It has been highly praised and widely used by many well-known scholars.

However, the Twin Shear Strength Theory system is a new theoretical system. There are still some loopholes and deficiencies which has no deterministic explanation, such as the structural form of the strength model and the selection of test points, as well as the adaptability of different parameters in existing models, and the characteristics of the existing models. Although the multiaxial strength test can be carried out in the true triaxial testing machine, the multiaxial compressive test is still difficult for brittle (or quasi-brittle) materials such as concrete. Meanwhile, it is difficult to gain on the general testing machine for the high-strength materials, due to the high triaxial isobaric compressive strength.

For the biaxial tensile strength of concrete, the study presented by Guo (1997) concluded that the biaxial tensile strength is close to its uniaxial tensile strength of concrete under any stress ratio. For the triaxial tensile strength of concrete, the study presented by Che (1979) concluded that the triaxial tension strength is equal to the uniaxial tensile strength, while the study presented by Song (1988) held that the triaxial tension strength is only 0.7 times of the uniaxial tensile strength. The study (Guo, 1997) held that the triaxial tensile strength under any stress ratio is slightly lower than the uniaxial tensile strength. In the concrete structures design standard in Chinese (GB50010-2002), the value of multiaxial strength is generally checked by the specific empirical formula. For triaxial compressive strength, not only the influence of middle principal stress (σ_2) and hydrostatic stress (σ_m) is not considered, but also the triaxial isobaric strength selects the value which is 5 times than that of uniaxial compressive strength, and the tension isobaric strength selects the value which is 0.9 times that of uniaxial compressive strength. Therefore, the unsolved problem is the method to determine the multiaxial strength theoretically through the uniaxial tension or compressive test parameters of materials. At the same time, the mathematical model of multiaxial compressive strength is generally through test fitting method or complex mathematical expression (Yao et al., 2004), due to the complexity of multiaxial compressive strength, which brings inconvenience to engineering application. It is necessary to propose a simple model to express the multiaxial strength of concrete.

As mentioned above, this chapter will start with the discussion about the equation form of the Twin Shear Strength Theory and the selection of test stress points (boundary conditions). Then the multiaxial compressive strength of concrete is taken as an example to discuss the construction law of the double shear strength model. Based on the revealed internal law of the Twin Shear Strength Theory model, the following key contents are studied: the relationship of the multiaxial strength parameters, the simple expression method of the multiaxial strength model, the effect of middle principle stress and hydrostatic stress on multiaxial compressive strength of concrete, and its embodiment in Twin Shear Strength model.

3.3.2 The Twin Shear Strength Theory

The reason why the Twin Shear Strength Theory system is a new theoretical system is that uses the new unit body as the foundation, and thus the framework of the Twin Shear Strength Theory is originally set up. The unit body of the new theory is different from that of the traditional regular hexahedron, which is mainly orthogonal octahedron (there are three groups and each group can be divided into horizontal and vertical parts), and can be evolved into 5-hedron, 4-pyramid, 12-hedron, and 26-hedron in turn. The first three units are paired units (also divided into horizontal and vertical parts). The differences between the element body of the Twin Shear Strength Theory and the traditional regular hexahedron are that the different sections are used to intercept the regular hexahedron, so as to obtain a series of new element bodies, and the relationship between different sections of unit body is expressed in the study presented by Huang (2008).

For any inclined section on a hexahedron, the maximum shear stress and principal stress on the inclined section can be deduced by using the mathematical methods of material mechanics or elasticity mechanics, that is, the section is 45° to the two principal stress axes, or the cosine in both directions is $\pm\sqrt{2}/2$. If it is parallel to the other principal stress axis, the direction cosine will be 0, the corresponding stress will be recorded as twin shear principal shear stress and twin shear normal stress respectively (i.e., twin shear stress defined by Yu (2011) which is the maximum normal stress and maximum shear stress in M–C theory). The specific expressions of the three types of principal stresses and principal shear stresses are as follows:

$$\sigma_{13}/\tau_{13} = (\sigma_1 \pm \sigma_3)/2 \tag{3.1a}$$

$$\sigma_{12}/\tau_{12} = (\sigma_1 \pm \sigma_2)/2 \tag{3.1b}$$

$$\sigma_{23}/\tau_{23} = (\sigma_2 \pm \sigma_3)/2 \tag{3.1c}$$

where, the σ_m is the hydrostatic stress, $\sigma_m = (\sigma_1 + \sigma_2 + \sigma_3)\,/\,3$.

Therefore, the Twin Shear Strength Theory considers the middle principal stress (σ_2) and the hydrostatic stress (σ_m) (assuming the twin shear stress as an independent physical quantity). Furthermore, based on transforming the Lode parameter of the stress, the following expressions are gained:

$$\mu_\tau = \frac{\tau_{12}}{\tau_{13}} = \frac{\sigma_1 - \sigma_2}{\sigma_1 - \sigma_3} = \frac{3\sigma_m - 2\sigma_{23} - \sigma_2}{3\sigma_m - 4\sigma_{23} + \sigma_2} \tag{3.2a}$$

$$\mu'_\tau = \frac{\tau_{23}}{\tau_{13}} = \frac{\sigma_2 - \sigma_3}{\sigma_1 - \sigma_3} = \frac{3\sigma_m - 4\sigma_{13} + \sigma_{12} - \sigma_2}{3\sigma_m - 4\sigma_{23} + \sigma_2} \tag{3.2b}$$

The transformation shows that the stress parameter in the Twin Shear Strength Theory is the ratio of two principal shear stresses and the ratio of the radius (diameter) in two stress circles. It can be a parameter to reflect the middle principal stress (σ_2), or a parameter of the stress state which is related to the hydrostatic stress (σ_m), or a characterization parameter of the deviation principal stress condition. Based on the relationship and the selected values of μ_τ and μ'_τ, the stress condition types can be clearly defined by Yu (2002). It can be found that the Twin Shear Strength Theory has clear physical significance and rich connotation, which also provides a theoretical foundation for a reasonable explanation of the results to be obtained in the following section.

As shown in Section 3.2, the principal stress, the principal shear stress, the middle principal stress (σ_2), and the hydrostatic stress (σ_m) are the key parameters for the failure criterion model of concrete under multiaxial compressive strength. In order to consider the four main factors, the Twin Shear Strength Theory can be the foundation to describe the various materials under complex stress condition (Yu, 2011). As shown above, it is well known that this theory considers the influence of the middle principal stress and different tensile-compressive properties of materials in detail. The theory assumes that both the intermediate principal shear stress and the corresponding principal stress are important parameters. If the combination of these stresses reaches the maximal value, the materials will be destroyed. The mathematical model of the Twin Shear Strength Theory is shown in Eq. (3.3). In the theory, $\tau_{13} + \tau_{12}$ and $\tau_{13} + \tau_{23}$ can reflect the variation in the principal shear stress, and $\sigma_{13} + \sigma_{12}$ and $\sigma_{12} + \sigma_{23}$ can reflect the variation in the principal stresses. Lastly, β is the effective coefficient of the principal stresses.

$$F = \tau_{13} + \tau_{12} + \beta \times (\sigma_{13} + \sigma_{12}) = CF \geq F' \tag{3.3a}$$

$$F' = \tau_{13} + \tau_{23} + \beta \times (\sigma_{13} + \sigma_{23}) = CF < F' \tag{3.3b}$$

where τ_{12}, τ_{23} and τ_{13} are the three principal shear stresses, and the calculation equations are shown in Eq. (3.1); σ_{12}, σ_{23} and σ_{13} are the corresponding principal stresses, and the calculation equations are shown in Eq. (3.1).

With the above parameters, the Twin Shear Strength Theory model can fully balance the influence of the principal shear stresses, the principal stresses, and the variations in the middle principal stress (σ_2). The model is applied to determine the experimental characteristic point at the plane stress, the uniaxial tensile stress (f_t) and the biaxial compressive stress (f_{cc}) can be computed using Eq. (3.3a), and the uniaxial compressive stress (f_{co}) and the biaxial tensile stress (f_{tt}) can be computed using Eq. (3.3b).

According to the M–C theory and the Twin Shear Strength Theory, the twin shear theory system is used for isotropic materials which means that the three elastic constants (E, G and γ) are the same in three stress directions. In addition, because the isotropic materials have symmetry for their elastic constants in any direction (i.e., the elastic constants of the materials remain unchanged after rotating 180° around this direction), the principal stress only causes the linear strain, while the shear stress only causes the shear strain in the same stress plane along the coordinate axis (or stress vector) under the conditions of online elastic range and small deformation. The relationship between the linear strain and the principal stress can be obtained by the superposition principle, which means when the three principal stresses exist, respectively, the linear strain in a certain direction is superimposed. When the three principal stresses exist at the same time, the linear strain in the corresponding direction is superimposed, respectively, which is named the generalized Hooke's law.

Based on the above analysis, it can be found that the Twin Shear Strength Theory model is not applied for any uncharacteristic point in the plane stress state and space stress state of concrete. It is necessary to modify the Twin Shear Strength Theory to establish the reasonable failure criterion models for the concrete under multiaxial compressive pressure.

3.3.3 The model establishment process

Based on the study presented by Huang (2008), the Twin Shear Strength Theory model has low applicability. If the corresponding model is used to verify the actual test results, the prediction results will be seriously distorted. Partial models can only predict the characteristic test points, but most models will not make correct calculation and inversion for the characteristic test points. On the other hand, the suitable corresponding model should be judged first according to the judgment conditions, even for the same batch of test data. Therefore, the deciding method of the adopted model will lead to inconvenience in the model application. Meanwhile, the selection of test stress points is highly consistent with the adaptability of the twin shear strength model. That is, if the plane test stress points are used as the boundary condition, the corresponding deterministic model can only adapt to the plane stress case. Although some models have poor prediction effect on noncharacteristic stress test, the predicted values of characteristic test stress points are accurate, which provides convenient conditions for using specific models to invert specific characteristic test points, such as f_t, f_c, and f_{cc} (or f_{tt}) values are used to invert f_{tt} (or f_{cc}) values. It can be used to gain some characteristic values which are not available in the same batch of tests or cannot be obtained, due to limited test conditions. For example, f_{ccc} is generally more than ten times or higher than f_c. For the materials with high strength, the value of f_{ccc} can be tested in the general testing machine.

In order to propose a theoretical foundation which is suitable for concrete under multiaxial compressive pressure by improving the Twin Shear Strength Theory model, the structural characteristics in the Twin Shear Strength Theory need to be analyzed. The following two aspects cause uncertainty in the model. One aspect is that the model is a polynomial, each term is twin shear stress (twin shear principal shear stress and twin shear normal stress) or other forms of stress (i.e., hydrostatic stress, octahedral stress, etc.). However, there are quite a lot of structural possibilities in terms of mathematical theory, such as the algebraic formula of a single stress or multiple stresses, the corresponding coefficients, power exponent, etc. On the other hand, according to the selection of boundary conditions (test characteristic stress points, some characteristic points in the stress space or deviator stress plane, such as the highest hydrostatic stress point on the tensile and compressive meridian, etc.), there will be a combination of multiple characteristic test points for a kind of theory model. Especially for the fewer parameters in the model (e.g. three parameters), more boundary conditions should be chosen and there are more stress items in the model, which brings great difficulties to the model establishment and verification. However, the Twin Shear Strength Theory model is deterministic. For the various configurations of the Twin Shear Strength Theory, the number of models is in a limited minority, rather than many possible configurations in the mathematical theory. The main reasons can be divided into the following aspects: the symmetry of the Twin Shear Strength Theory model, the two sub-models of the same type model are relatively symmetrical for the same test point; the characteristics of the multiaxial characteristic stress (e.g., σ_1, σ_2 and σ_3 are partially equal to 0, or equal, or the sign is opposite); the characteristics of the solution of the equation system, whether the rank of the coefficient matrix is full rank; the actual test data, whether it conforms to the test results. Therefore, the Twin Shear Strength Theory brings great convenience to studies and the application of failure criterion model. It is effective to establish the failure criterion of concrete under multiaxial compression pressure by improving the Twin Shear Strength Theory.

Based on the study results concluded by Huang (2008), the three-parameter Twin Shear Strength Theory model has many forms, such as considering different stress expression terms, different powers, and so on. For boundary conditions, if only multiaxial characteristic stress test points are considered, there will be the combinations of (f_t, f_c, f_{tt}), (f_t, f_c, f_{cc}), (f_t, f_c, f_{ttt}), (f_t, f_c, f_{ccc}), etc. If the characteristic stress points in the shear test, the stress space, or the deviator stress plane are considered, there will be more combinations. In general, the uniaxial characteristic stress point should be used as the basis, because the test is easy and the result is reliable. And the shear test results should not be recommended as the characteristic test point, due to the nonuniqueness of the shear test method. Therefore, this chapter only studies the structural forms to explore their laws. Based on the above discussion, the multiparameter Twin Shear Strength Theory model will have more mathematical possibilities. There are more possible choices for each term in the polynomial with the increase of parameter, then it may also cause the rank of the coefficient matrix in the equation system to be non-full rank. Therefore, it is more difficult to distinguish the four-parameter Twin Shear Strength Theory model. However, according to the above conclusions, the adaptability of the Twin Shear Strength Theory model is highly consistent with the boundary conditions. Therefore, the multiparameter Twin Shear Strength Theory model will be more suitable for the plane stress situation. Furthermore, in order to consider the boundary conditions of practical problems (i.e., initial conditions, stress ratio, and constant lateral pressure conditions), the model is at least quadratic polynomial for the plane stress case. Therefore, this chapter will discuss the general quadratic polynomial models to explore the general construction law of the multiparameter Twin Shear Strength Theory model, based on the stress analysis of concrete under multiaxial compression pressure in Section 3.2.

In conclusion, the Twin Shear Strength Theory model has good adaptability to the characteristic stresses, but it cannot uniquely determine a certain characteristic stress value, due to the lack of interdependencies between multiaxial characteristic strengths, which is named as confining condition. Against this background, the extreme strength model, which is suitable for general concrete under complex stress conditions, is established by the following steps

(1) the influence factors and the parameters of the target model are determined
(2) the physical meaning of the non-full rank of the equations is identified
(3) the dependence of the multiaxial characteristic strength is established, based on the indefinite solution of the corresponding Twin Shear Strength Theory model
(4) the deterministic multiparameter Twin Shear Strength Theory model is obtained at last.

For the stress boundary condition in solving process, the test values of f_t, f_c, f_{cc}, and f_{ccc} can be accurately obtained on the true triaxial testing machine for low-strength materials, and the test values of f_t, f_c, f_{tt}, and f_{ttt} can be obtained more accurately for the concrete with higher strength grade. In extreme cases, even if there are only test values of f_t and f_c, the proposed strength model can be used to calculate the corresponding missing characteristic test values and all possible ultimate strength values under complex stress conditions.

Based on the previous discussion, there are four main factors for the failure criterion of concrete under multiaxial compressive pressure: the principal stress, the principal shear stress, the intermediate principal stress (σ_2), and the hydrostatic stress (σ_m). Furthermore, the principal shear stress and hydrostatic stress are the most important factors. In the following section, the Twin Shear Strength Theory is introduced and modified for concrete under multiaxial compressive pressure. And three kinds of multi-parameters Twin Shear Strength Theory models are established for the failure criterion of concrete under multiaxial compression, where the principal shear stress and hydrostatic stress are respectively as the main influence factors. The model establishment process is shown in Fig. 3.4.

3.4 Failure criterion models of concrete under multiaxial compressive pressure

In order to study the effect of the four influence factors (principal stress, the shear stress, the intermediate principal stress, and the hydrostatic stress), the Twin Shear

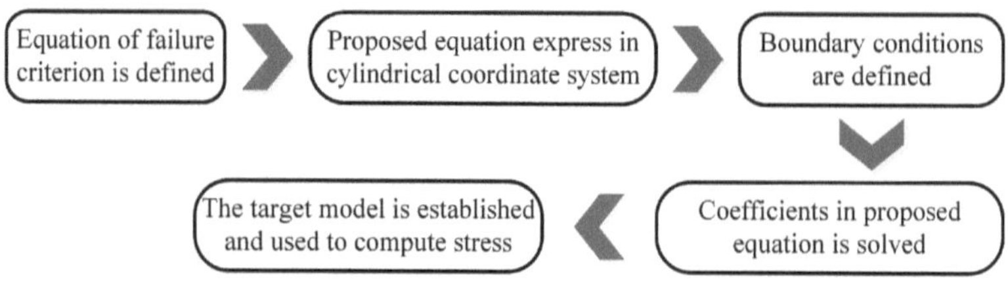

Figure 3.4 The model establishment process.

Strength Theory (Yu, 2011) is introduced and modified in this section. The theory is suitable for the various materials which have different tensile-compressive properties, such as concrete. And the multi-parameters in the equation is not only deeply reflected by the influence of middle principal stress, but also suitable for the materials under complex stress condition. Based on the theoretical foundation, the stress failure surface is described as a kind of stress condition, which means the material will be destroyed when the combination of these stresses reaches the extremum condition. The mathematical equation of the Twin Shear Strength Theory is shown in Eq. (3.1), the two expressions are used to reflect the effect of middle principal stress by given determining condition. In addition, the principal shear stress parameters (i.e., $\tau_{13} + \tau_{12}$ or $\tau_{13} + \tau_{23}$) are used to reflect the effect of three kinds of principal shear stresses. The principal stress parameters (i.e., $\sigma_{13} + \sigma_{12}$ and $\sigma_{12} + \sigma_{23}$) are used to reflect the effect of the hydrostatic stress and the minimum/maximum principal stress. And the coefficient β is also used to further reflect the effect of the hydrostatic stress and the minimum/maximum principal stress. Based on the stress parameters and the determining condition, the effect of various stresses can be reflected concurrently in Twin Shear Strength Theory, including the hydrostatic stress, the minimum or maximum principal stress, the principal shear stress, and the middle principal stress.

Based on the stress parameters in the model, the Twin Shear Strength Theory model can fully balance variations of the principal shear stress and principal stress. Especially for variations of the middle principal stress and the minimum or maximum principal stress. The model is applied to determine the characteristic point at the plane stress condition, where the uniaxial tensile stress (f_t) and the biaxial compressive stress (f_{cc}) can be computed by using Eq. (3.1a), the uniaxial compressive stress and the biaxial tensile stress can be computed by using Eq. (3.1b). And the parameters of the Twin Shear Strength Theory model have a linear effect on the stress failure surface of the concrete. However, the linear increasing trend cannot be used to reflect the increase of axial compression strength under lateral compression pressure, according to Chapter 2. The Twin Shear Strength Theory model cannot be applied for any uncharacteristic point in the plane stress state and space stress condition of concrete under multiaxial compressive stress. In order to further consider the effect of the shear stress and the hydrostatic stress, the parameters are introduced as quadratic variation, including $(\tau_{13} + \tau_{23})$, $(\tau_{13} + \tau_{12})$, and σ_m. By various combination methods, the three kinds of modified models are proposed, namely five-parameter failure criterion A, five-parameter failure criterion B, and six-parameter failure criterion.

3.4.1 Five-parameter failure criterion A (principal shear stresses are as main influence factors)

It has been known that hydrostatic stress and the principal shear stress are the two influence factors for the stress failure surface of concrete under multicompression pressure. Therefore, the two factors are introduced in the Twin Shear Strength Theory model, and the following assumptions are set: the principal shear stress is the main influence factor and shows a quadratic trend; the hydrostatic stress is the secondary influence factor and shows a linear trend. Based on the above combination method, the five-parameter failure criterion A can be established, and the mathematical expression is given as follows:

$$F = \tau_{13} + \tau_{12} + \beta(\sigma_{13} + \sigma_{12}) + A\sigma_m + B_1(\tau_{13} + \tau_{12})^2 = CF \geq F' \tag{3.4a}$$

$$F' = \tau_{13} + \tau_{23} + \beta(\sigma_{13} + \sigma_{23}) + A\sigma_m + B_2(\tau_{13} + \tau_{23})^2 = CF < F' \tag{3.4b}$$

where, σ_m is the hydrostatic stress, $\sigma_m = (\sigma_1 + \sigma_2 + \sigma_3) / 3$; A is the coefficient to reflect the effect of hydrostatic stress; and B_1 and B_2 are the coefficients to reflect effect of the shear stresses.

In this proposed model, the hydrostatic stress is introduced and controlled by the same coefficient in the two equations. Meanwhile, the two kinds of principal shear stress parameters are introduced, so as to set two coefficients. Therefore, the model is a kind of univariate quadratic equation for the principal shear stresses, and it is a kind of linear equation for the principal stresses and the hydrostatic stress. It means that the principal shear stresses are the main influence factors in the proposed five-parameter failure criterion A.

As shown in Eq. (3.4), five unknown coefficients (i.e., β, A, B_1, B_2, and C) are needed to be solved. However, the parameters are dependent between each other and each characteristic stress point should be judged before putting it into the equation, which makes the solving method become complex. In order to simplify the solving method, the cylindrical coordinate system is introduced. Then the stress parameters should be transformed into the new coordinate parameters.

A cylindrical coordinate system refers to a coordinate system that uses plane polar coordinates and Z-direction distance to define the spatial coordinates of an object. If the spatial rectangular coordinate system is the same, there will be a variable value in the cylindrical coordinate system. The three coordinate variables in the cylindrical coordinate system are r, φ, and z. It is similar to the space rectangular coordinate system, there is also a z variable in the cylindrical coordinate system. Where r is the distance between the origin point (O) and the projection point (M') of point M on the plane (xoy), $r \in [0, +\infty)$. The φ is the angle turned counterclockwise from the x-axis to OM′ in the positive z-axis, $\varphi \in [0, 2\pi)$. The z is the height of the cylinder, $z \in R$. However, the ordinary cylindrical coordinate system is also not suitable for the stress coordinate transformation. Then stress cylindrical coordinate system is defined by the following basic condition: the hydrostatic stress axis as the primary axis, and the π plane as the cross-section through the origin. The point coordinate can be expressed as P (ξ, ρ, φ) in the cylindrical coordinate system, where ρ is the projection distance in π plane, ξ is the projection height in hydrostatic stress axis, the φ is the angle turned counterclockwise from the $\sigma 1$-axis. Therefore, a primary stress point P (σ_1, σ_2, σ_3) can be expressed as the coordinate P (ξ, ρ, φ) in the cylindrical coordinate system. And the transformation relationship between the point in the primary stress system (σ_1, σ_2, σ_3) and the point in the cylindrical coordinate system (ξ, ρ, φ) can be expressed as follows:

$$\xi = (\sigma_1 + \sigma_2 + \sigma_3)/\sqrt{3} \tag{3.5a}$$

$$\rho = \sqrt{(\sigma_1 - \sigma_2)^2 + (\sigma_2 - \sigma_3)^2 + (\sigma_1 - \sigma_3)^2}/\sqrt{3} \tag{3.5b}$$

$$\cos(\varphi) = \frac{2\sigma_1 - \sigma_2 - \sigma_3}{\sqrt{2(\sigma_1 - \sigma_2)^2 + 2(\sigma_2 - \sigma_3)^2 + 2(\sigma_1 - \sigma_3)^2}} \tag{3.5c}$$

Based on the above transformation relationship, the principal stress can be expressed in the cylindrical coordinate system by the following equations:

$$\sigma_1 = \left(\xi + \sqrt{2}\rho\cos(\varphi)\right)/\sqrt{3} \tag{3.6a}$$

$$\sigma_2 = \left(\xi + \sqrt{2}\rho\cos\left(\varphi - 2\pi/3\right)\right)/\sqrt{3} = \left[2\xi + \sqrt{2}\rho\left(\sqrt{3}\sin(\varphi) - \cos(\varphi)\right)\right]/\sqrt{3} \tag{3.6b}$$

$$\sigma_3 = \left(\xi + \sqrt{2}\rho\cos\left(\varphi + 2\pi/3\right)\right)/\sqrt{3} = \left[2\xi - \sqrt{2}\rho\left(\sqrt{3}\sin(\varphi) + \cos(\varphi)\right)\right]/\sqrt{3} \tag{3.6c}$$

Based on the combination of Eq. (3.1) and Eq. (3.2), the principal shear stress in the multiparameter failure criterion model can be transformed into the expressions in the cylindrical coordinate system, and the expressions are as follows:

$$\tau_{12} + \tau_{13} = \sqrt{6}\rho\cos(\varphi)/2 \tag{3.7a}$$

$$\tau_{13} + \tau_{23} = \sqrt{6}\rho\cos\left(\varphi - \pi/3\right)/2 \tag{3.7b}$$

$$\sigma_{12} + \sigma_{13} = \left(4\sqrt{3}\xi + \sqrt{6}\rho\cos(\varphi)\right)/6 \tag{3.7c}$$

$$\sigma_{13} + \sigma_{23} = \left(4\sqrt{3}\xi - \sqrt{6}\rho\cos(\varphi - \pi/3)\right)/6 \tag{3.7d}$$

$$\sigma_m = (\sigma_1 + \sigma_2 + \sigma_3)/3 = \sqrt{3}\xi/3 \tag{3.7e}$$

Therefore, Eq. (3.4) can be expressed in the cylindrical coordinate system by the above transformation equations (i.e., Eq. 3.7), as follows:

$$\xi = \frac{\sqrt{3}C}{A+2\beta} - \frac{\sqrt{2}\beta + 3\sqrt{2}}{2A+4\beta}\rho\cos(\varphi) - \frac{3\sqrt{3}B_1}{2A+4\beta}(\rho\cos(\varphi))^2 0 \le \varphi \le \varphi_b \tag{3.8a}$$

$$\xi = \frac{\sqrt{3}C}{A+2\beta} - \frac{3\sqrt{2} - \sqrt{2}\beta}{2A+4\beta}\rho\cos(\varphi - \pi/3) - \frac{3\sqrt{3}B_2}{2A+4\beta}\left(\rho\cos(\varphi - \pi/3)\right)^2 \varphi_b \le \varphi \le 60^\circ \tag{3.8b}$$

In order to simplify the solution process, the new unknown coefficients (i.e., a_i and b_i, i is the coefficient number) should be defined, and Eq. (3.8) can be expressed as follows:

$$\xi = a_0 - a_1\rho\cos(\varphi) - a_2(\rho\cos(\varphi))^2 0 \le \varphi \le \varphi_b \tag{3.9a}$$

$$\xi = b_0 - b_1\rho\cos(\varphi - \pi/3) - b_2\left(\rho\cos(\varphi - \pi/3)\right)^2 \varphi_b \le \varphi \le 60^\circ \tag{3.9b}$$

It is easy found that the following coefficient relationships: $a_0 = b_0 = \sqrt{3}C / (A + 2\beta)$, $a_1 = -(3\sqrt{2} + \sqrt{2}\beta) / (2A + 4\beta)$, $a_2 = -3\sqrt{3}B_1 / (2A + 4\beta)$, $b_1 = (3\sqrt{2} - \sqrt{2}\beta) / (2A + 4\beta)$, and $b_2 = -3\sqrt{3}B_2 / (2A + 4\beta)$. Then the unknown quantities (β, A, B_1, B_2 and C) can be expressed as follows: $A = (6b_1 - 6a_1 - 3\sqrt{2}) / (a_1 + b_1)$, $B_1 = 2\sqrt{6}a_2 / (3a_1 + 3b_1)$, $B_2 = 2\sqrt{6}b_2 / (3a_1 + 3b_1)$, $C = -\sqrt{6}a_0 / (a_1 + b_1)$, and $\beta = 3(b_1 - a_1) / (a_1 + b_1)$.

To solve the five unknown parameters (i.e., a_0, a_1, a_2, b_1, and b_2), five characteristic stress points should be used as the boundary conditions, and these stress points are selected in test results. By the selection of stress points in the particular stress plane, Eq. (3.9) can be simplified further. Therefore, the stress points in the tensile meridian ($\sigma_1 \ge \sigma_2 = \sigma_3$, $\varphi = 0$) and the compressive meridian ($\sigma_1 = \sigma_2 \ge \sigma_3$, $\varphi = 60°$) are selected to solve the unknown coefficients for convenient calculation. When $\varphi = 0$, the $\cos(\varphi) = 1$. When $\varphi = 0$, the $\cos(\varphi - \pi/3) = 1$. Based on the reasonable selection of stress points, Eq. (3.9) can be simplified into the following equations:

$$\xi = a_0 - a_1\rho - a_2\rho^2 \varphi = 0 \tag{3.10a}$$

$$\xi = b_0 - b_1\rho - b_2\rho^2 \varphi = 60^\circ \tag{3.10b}$$

3.4.1.1 The ordinary solution of five-parameter failure criterion A

As mentioned in the above discussion, five stress points should be selected from the tensile meridian and the compressive meridian to be the boundary conditions. According to the convenient test results, the following stress points can be selected in the ordinary stress coordinate system: for $\varphi = 0$, the uniaxial tensile strength (f_t, 0, 0), the biaxial compressive strength (0, $-f_{bc}$, $-f_{bc}$) and a stress point in the tensile meridian for $\varphi = 60°$, the uniaxial compressive strength (0, 0, $-f_c$), and a stress point in compressive meridian. After normalization processing by compressive strength of unconfined concrete, the boundary conditions can be expressed as follows: for $\varphi = 0$, the corresponding points are ($\sqrt{3}\alpha/3$, $\sqrt{3}\alpha/3$), ($-2\sqrt{3}\alpha_{bc}/3$, $\sqrt{6}\alpha_{bc}/3$) and ($-\xi_1$, ρ_1); for $\varphi = 60°$, the corresponding points are ($-\sqrt{3}/3$, $\sqrt{6}/3$) and ($-\xi_2$, ρ_2). Therefore, the following equations can be gained by substituting the boundary conditions into Eq. (3.10).

For the uniaxial compressive strength:

$$-\sqrt{3} = 3a_0 + \sqrt{6}b_1 - 2b_2 \tag{3.11a}$$

For the uniaxial tensile strength:

$$\sqrt{3}\alpha = 3a_0 + \sqrt{6}a_1\alpha - 2a_2\alpha^2 \tag{3.11b}$$

For the biaxial compressive strength:

$$-2\sqrt{3}\alpha_{bc} = 3a_0 + \sqrt{6}a_1\alpha_{bc} - 2a_2\alpha_{bc}^2 \tag{3.11c}$$

For a point in the tensile meridian:

$$\xi_1 = a_0 - a_1\rho_1 - a_2\rho_1^2 \tag{3.11d}$$

For a point in the compressive meridian:

$$\xi_2 = b_0 - b_1\rho_2 - b_2\rho_2^2 \tag{3.11e}$$

Based on the boundary condition training, the calculation process for the above coefficients in Eq. (3.10) is as follows:

for Eq. (3.11b) – Eq. (3.11c):

$$\sqrt{3}(\alpha + 2\alpha_{bc}) = a_0 + \sqrt{6}a_1(\alpha - \alpha_{bc}) - 2a_2\left(\alpha^2 - \alpha_{bc}^2\right) \tag{3.12a}$$

for Eq. (3.11b) – Eq. (3.11d):

$$\sqrt{3}\alpha - 3\xi_1 = a_1\left(\sqrt{6}\alpha - 3\rho_1\right) - a_2\left(2\alpha^2 - 3\rho_1^2\right) \tag{3.12b}$$

Then, the calculation equation of a_2 can be gained as follows:
for Eq. (3.12b) × $\sqrt{6}(\alpha-\alpha_{bc})$ / $(\sqrt{6}\alpha-3\rho_1)$ – Eq. (3.12a):

$$a_2 = \frac{\sqrt{3}\alpha\left(\sqrt{2}\xi_1 + \rho_1 - \sqrt{6}\alpha_{bc}\right) + \alpha_{bc}\left(2\sqrt{3}\rho_1 - \sqrt{6}\xi_1\right)}{\left(\sqrt{6}\alpha/3 - \rho_1\right)\left(\sqrt{6}\alpha\rho_1 + 2\alpha_{bc}^2 - 2\alpha\alpha_{bc} - \sqrt{6\alpha_{bc}\rho_1}\right)} \tag{3.13a}$$

then, for Eq. (3.11b) – Eq. (3.11c) × α/α_{bc}:

$$a_0 = b_0 = \frac{3\sqrt{3}\alpha\alpha_{bc} + 2\alpha_2\alpha\alpha_{bc}(\alpha - \alpha_{bc})}{3\alpha_{bc} - 3\alpha} \tag{3.13b}$$

then, for Eq. (3.11b) – Eq. (3.11c):

$$a_1 = \frac{\alpha(\alpha + 2\alpha_{bc})}{\sqrt{2}(\alpha_{bc} - \alpha)} - \sqrt{6}a_2(\alpha + \alpha_{bc})/3 \tag{3.13c}$$

then, for Eq. (3.11a) – Eq. (3.11e) × 2 / $3{\rho_2}^2$:

$$b_1 = \frac{2\xi_2 - \sqrt{3}\rho_2^2 - a_0\left(3\rho_2^2 - 2\right)}{\sqrt{6}\rho_2^2 - 2\sqrt{3}\rho_2} \tag{3.13d}$$

then, for Eq. (3.11a) – Eq. (3.11e) × $\sqrt{6}$/ $3\rho_2$:

$$b_2 = \frac{\sqrt{6}\xi_2 - \sqrt{3}\rho^2 - a_0\left(3\rho_2 - \sqrt{6}\right)}{2\rho_2 - \sqrt{6}\rho_2^2} \tag{3.13e}$$

3.4.1.2 Simplification solution of five-parameter failure criterion A

For stress points in the tensile meridian, there are same values of the two larger principle stress (i.e., σ_2 and σ_3) in multiaxial compressive condition. However, these stress points are difficult to gain in multiaxial compressive test. Therefore, the triaxial tensile strength (f_{ttt}) is tried to replace the stress point in the tensile meridian under the compressive condition. It can be assumed that the triaxial tensile stress is close to the uniaxial tensile strength (i.e., $f_t = f_{ttt}$), due to the tensile stress of concrete being very low. Based on the above simplification method and normalization processing by the compressive strength, the new boundary conditions can be set in the cylindrical coordinate system (ξ, ρ, φ), as follows: the new boundary conditions are uniaxial compressive strength ($-\sqrt{3}$ / 3, $\sqrt{6}$ / 3, 60°), the uniaxial tensile strength ($\sqrt{3}$ α / 3, $\sqrt{6}\alpha$ / 3, 0), the biaxial compressive strength ($-2\sqrt{3}$ α_{bc} / 3, $\sqrt{6}\alpha_{bc}$ / 3, 0), the triaxial tensile strength ($\sqrt{3}\alpha$, 0, 0) and a point in the compressive meridian ($-\xi_2$, ρ_2, 60°). Therefore, the following equations can be gained by substituting the boundary conditions into Eq. (3.10). For the uniaxial compressive strength:

$$-\sqrt{3} = 3a_0 + \sqrt{6}b_1 - 2b_2 \tag{3.14a}$$

For the uniaxial tensile strength:

$$\sqrt{3}\alpha = 3a_0 + \sqrt{6}a_1\alpha - 2a_2\alpha^2 \tag{3.14b}$$

For the biaxial compressive strength:

$$-2\sqrt{3}\alpha_{bc} = 3a_0 + \sqrt{6}a_1\alpha_{bc} - 2a_2\alpha_{bc}^2 \tag{3.14c}$$

For the triaxial tensile strenngth:

$$\sqrt{3}\alpha = a_0 \tag{3.14d}$$

For a point in the compressive meridian:

$$\xi_2 = b_0 - b_1\rho_2 - b_2\rho_2^2 \tag{3.14e}$$

Then, the calculation equation of a_2 can be gained as follows:

$$a_0 = b_0 = \sqrt{3}\alpha \tag{3.15a}$$

for Eq. (3.14b)–Eq. (3.14c) $\times$ α / α_{bc}:

$$a_2 = \frac{3\sqrt{3}\alpha}{2\left(\alpha\alpha_{bc} - \alpha_{bc}^2\right)} \tag{3.15b}$$

then, for Eq. (3.14b) $\times$ $\alpha_{bc}{}^2$ – Eq. (3.14c) $\times$ α_2:

$$a_1 = \frac{3\alpha^2 + 2\alpha_{bc}\alpha - 2\alpha_{bc}^2}{\sqrt{2}\left(\alpha_{bc}^2 - \alpha\alpha_{bc}\right)} \tag{3.15c}$$

then, for Eq. (3.14e) × 2–Eq. (3.14a) × $\rho_2{}^2$:

$$b_1 = \frac{2\xi_2 - \sqrt{3}\rho_2^2 - \sqrt{3}\alpha\left(3\rho_2^2 - 2\right)}{\sqrt{6}\rho_2^2 - 2\rho_2} \tag{3.15d}$$

then, based on Eq. (3.14e), it can be gained as:

$$b_2 = \frac{\sqrt{6}\xi_2 - \sqrt{3}\rho_2 - \sqrt{3}\alpha\left(3\rho_2 - \sqrt{6}\right)}{2\rho_2 - \sqrt{6}\rho_2^2} \tag{3.15e}$$

3.4.2 Five-parameter failure criterion B (hydrostatic stress is main influence factor)

Based on stress analysis mentioned in Section 3.2, the hydrostatic stress is also an important influence factor for the stress failure surface of concrete under multiaxial compression. Therefore, the following assumptions are set in this section: the hydrostatic stress is the main influence factor and shows a quadratic trend; the principal shear stress is the secondary influence factor and shows a linear trend. Based on the above combination method, the five-parameter failure criterion B is proposed, and the mathematical expression is given as follows:

$$F = \tau_{13} + \tau_{12} + \beta(\sigma_{13} + \sigma_{12}) + A\sigma_m + B_1\sigma_m^2 = CF \geq F' \tag{3.16a}$$

$$F' = \tau_{13} + \tau_{23} + \beta(\sigma_{13} + \sigma_{23}) + A\sigma_m + B_2\sigma_m^2 = CF < F' \tag{3.16b}$$

Based on the transformation method which is shown in Eq. (3.7), Eq. (3.16) can be expressed in the cylindrical coordinate system, the mathematical expression is given as follows:

$$\rho\cos(\varphi) = \frac{\sqrt{6}C}{3+\beta} - \frac{2\sqrt{2}\beta + \sqrt{2}A_1}{3+\beta}\xi - \frac{\sqrt{6}B_1}{9+3\beta}\xi^2 0 \leq \varphi \leq \varphi_b \tag{3.17a}$$

$$\rho\cos\left(\varphi - \pi/3\right) = \frac{\sqrt{6}C}{3-\beta} - \frac{2\sqrt{2}\beta + \sqrt{2}A_2}{3-\beta}\xi - \frac{\sqrt{6}B_2}{9-3\beta}\xi^2 \varphi_b \leq \varphi \leq 60^\circ \tag{3.17b}$$

In order to simplify the solving process, the stress points in the tensile meridian ($\sigma_1 \geq \sigma_2 = \sigma_3$, $\varphi = 0$) and the compressive meridian ($\sigma_1 = \sigma_2 \geq \sigma$, $\varphi = 60^\circ$) are selected to solve the unknown coefficients for convenient calculation. When $\varphi = 0$, the $\cos(\varphi) = 1$. When $\varphi = \pi/3$, the $\cos(\varphi - \pi/3) = 1$. Based on the reasonable selection of stress points, Eq. (3.17) can be simplified into the following equations:

$$\rho = a_0 - a_1\xi - a_2\xi^2 \varphi = 0 \tag{3.18a}$$

$$\rho = b_0 - b_1\xi - b_2\xi^2 \varphi = 60° \tag{3.18b}$$

where $a_0 = \sqrt{6}C / (3 + \beta)$, $a_1 = -\sqrt{2}(A_1 + 2\beta) / (3 + \beta)$, $a_2 = -\sqrt{6}B_1 / (9 + 3\beta)$, $b_0 = \sqrt{6}C / (3 - \beta)$, $b_1 = -(\sqrt{2}A_2 + 2\beta) / (3 - \beta)$, and $b2 = \sqrt{6}B_2 / (9 - 3\beta)$. Then the unknown quantities (β, A_1, $A2$, B_1, B_2, and C) can be expressed as follows: $A_1 = (6a_0 - 6b_0 - 3\sqrt{2}a_1b_0) / (a_0 + b_0)$, $A_2 = (6a_0 - 6b_0 - 3\sqrt{2}a_0b_1) / (a_0 + b_0)$, $B_1 = 3\sqrt{6}a_2b_0 / (a_0 + b_0)$, $B_2 = 3\sqrt{6}a_0b_2 / (a_0 + b_0)$, $C = \sqrt{6}a_0b_0 / (a_0 + b_0)$, and $\beta = 3(b_0 - a_0) / (a_0 + b_0)$.

Based on the discussion in the above section, the five stress points should be selected from the tensile meridian and the compressive meridian to be the boundary conditions. In order to simplify the solving process, the following stress points can be selected in the ordinary stress coordinate system: for $\varphi = 0$, the uniaxial tensile strength (f_t, 0, 0), the biaxial compressive strength (0, $-f_{bc}$, $-f_{bc}$) and the triaxial tensile strength (f_t, f_t, f_t); for $\varphi = 60°$, the uniaxial compression strength (0, 0, $-f_c$), and a stress point in compressive meridian. After normalization processing by the compressive strength of unconfined concrete, the boundary conditions can be expressed as follows: for $\varphi = 0$, the corresponding points are ($\sqrt{3}\alpha/3$, $\sqrt{6}\alpha/3$), ($-2\sqrt{3}\alpha_{bc}/3$, $\sqrt{6}\alpha_{bc}/3$) and ($\sqrt{3}\alpha$, 0); for $\varphi = 60°$, the corresponding points are ($-\sqrt{3}/3$, $\sqrt{6}/3$) and ($-\xi_2$, ρ_2). Based on the boundary conditions, the following equations can be gained by substituting the boundary conditions into Eq. (3.18).

For the uniaxial tensile strength:

$$\sqrt{3}\alpha = 3a_0 + \sqrt{6}a_1\alpha - 2a_2\alpha^2 \tag{3.19a}$$

For the biaxial compressive strength:

$$-2\sqrt{3}\alpha_{bc} = 3a_0 + \sqrt{6}a_1\alpha_{bc} - 2a_2\alpha_{bc}^2 \tag{3.19b}$$

For the triaxial tensile strength in Eq. (3.18a):

$$0 = a_0 + \sqrt{3}\alpha a_1 - 2\alpha^2 a_2 \tag{3.19c}$$

For the uniaxial compressive strength:

$$-\sqrt{3} = 3a_0 + \sqrt{6}b_1 - 2b_2 \tag{3.19d}$$

For a point in the compressive meridian:

$$\rho_2 = b_0 - b_1\xi_2 - b_2\xi_2^2 \tag{3.19e}$$

For triaxial tensile strength in Eq. (3.18b):

$$0 = b_0 + \sqrt{3}\alpha b_1 - 2\alpha^2 b_2 \tag{3.19f}$$

Based on the boundary condition training, the calculation process for the above coefficients in Eq. (3.18) is given as follows:
for Eq. (3.19c) – Eq. (3.19a):

$$-\sqrt{6}=2\sqrt{3}a_1+8a_2\alpha \tag{3.20a}$$

then, for Eq. (3.19c) – Eq. (3.19b):

$$-\sqrt{3}\alpha_{bc}=\left(\sqrt{3}\alpha+2\sqrt{3}\alpha_{bc}/3\right)a_1+\left(3\alpha^2-4\alpha_{bc}^2/3\right)a_2 \tag{3.20b}$$

then, for Eq. (3.19d) × 3 – Eq. (3.19f):

$$\sqrt{6}/3=-\left(\sqrt{3}/3+\sqrt{3}\alpha\right)b_1+\left(1/3-3\alpha^2\right)b_2 \tag{3.20c}$$

then, for Eq. (3.19e) × 3 – Eq. (3.19f):

$$\rho_2=\left(\xi_2+\sqrt{3}\alpha\right)b_1+\left(\xi_2^2-3\alpha^2\right)b_2 \tag{3.20d}$$

Therefore, the calculation equation of a_2 can be gained as follows:
for Eq. (3.20a) × $(\sqrt{3}\alpha+2\alpha_{bc}/\sqrt{3})$ – Eq. (3.20b) × $2/\sqrt{3}$:

$$a_2=\frac{-3\sqrt{6}\alpha}{2(3\alpha+2\alpha_{bc})(\alpha+2\alpha_{bc})} \tag{3.21a}$$

then based on Eq. (3.20a), it can be gained that:

$$a_1=\frac{\sqrt{6}+8\alpha a_2}{2\sqrt{3}} \tag{3.21b}$$

then, for Eq. (3.19a) × 3 – Eq. (3.19c):

$$a_0=\frac{\sqrt{6}\alpha+2\alpha^2a_2}{2} \tag{3.21c}$$

Meanwhile, the calculation equation of b_2 can be gained as follows:
for Eq. (3.20c) × $(\xi_0+\sqrt{3}\alpha)$ – Eq. (3.20d) × $(\sqrt{3}/3+\sqrt{3}\alpha)$:

$$b_2=\frac{\sqrt{6}(\xi_2+\sqrt{3}\alpha)-\sqrt{3}\rho_2(1+3\alpha)}{\left(\xi_2+\sqrt{3}\alpha\right)\left(1+3\alpha-3\sqrt{3}\xi_2\alpha-\sqrt{3}\xi_2\right)} \tag{3.21d}$$

then based on Eq. (3.20c), it can be gained that:

$$b_1 = \frac{(1-9\alpha^2)b_2 - \sqrt{6}}{\sqrt{3}(1+3\alpha)} \tag{3.21e}$$

then, Eq. (3.20e) × $\sqrt{3}\alpha$ − Eq. (3.20f) × ξ_0:

$$b_0 = \frac{\sqrt{3}\rho_2 - (\sqrt{3}\xi_2^2 + 3\alpha^2\xi_2)b_2}{\xi_2 + \sqrt{3}\alpha} \tag{3.21f}$$

3.4.3 Six-parameter failure criterion (both principal shear stress and hydrostatic stress are main influence factors)

It has been discussed that both the principal shear stress and the hydrostatic stress are important factors for the stress failure surface of concrete under multiaxial compressive pressure. However, only one factor has been considered by quadratic variation in the two kinds of six-parameter failure criterion models. In order to fully consider both the hydrostatic stress and principal shear stress as the two important factors, the six-parameter failure criterion model is proposed by the Twin Shear Strength Theory. In the proposed model, both the two stresses show a quadratic trend, and the mathematical expression is given as follows:

$$F = \tau_{13} + \tau_{12} + \beta(\sigma_{13} + \sigma_{12}) + A\sigma_m + B\sigma_m^2 + C_1(\tau_{13}+\tau_{12})^2 = DF \geq F' \tag{3.22a}$$

$$F' = \tau_{13} + \tau_{23} + \beta(\sigma_{13} + \sigma_{23}) + A\sigma_m + B\sigma_m^2 + C_2(\tau_{13}+\tau_{23})^2 = DF < F' \tag{3.22b}$$

Based on the transformation method which is shown in Eq. (3.7), Eq. (3.16) can be expressed in the cylindrical coordinate system, the mathematical expression is given as follows:

$$6C_1(\rho\cos(\varphi))^2 + \left(\frac{\sqrt{6}}{2} + \frac{\sqrt{6}\beta}{6}\right)\rho\cos(\varphi) = D - \left(\frac{\sqrt{3}A}{3} + \frac{2\sqrt{3}\beta}{3}\right)\xi - \frac{B}{3}\xi^2 0 \leq \varphi \leq \varphi_b \tag{3.23a}$$

$$6C_1\left(\rho\cos\left(\varphi - \pi/3\right)\right)^2 + \left(\frac{\sqrt{6}}{2} - \frac{\sqrt{6}\beta}{6}\right)\rho\cos\left(\varphi - \pi/3\right) = D - \left(\frac{\sqrt{3}A}{3} + \frac{2\sqrt{3}\beta}{3}\right)\xi - \frac{B}{3}\xi^2 \varphi_b \leq \varphi \leq 60^\circ \tag{3.23b}$$

To solve for the unknown quantities, six experimental characteristic stresses should be used as boundary conditions. In order to simplify the solving process, the stress points in the tensile meridian ($\sigma_1 \geq \sigma_2 = \sigma_3$, $\varphi = 0$) and the compressive meridian ($\sigma_1 = \sigma_2 \geq \sigma_3$, $\varphi = 60°$) are selected to solve the unknown coefficients for convenient calculation. When $\varphi = 0$, the $\cos(\varphi) = 1$. When $\varphi = 60°$, the $\cos(\varphi - \pi/3) = 1$. Based on the reasonable selection of stress points, the Eq. (3.23) can be simplified into the following equations:

$$a_1\rho^2 + \rho = a_2 - a_3\xi - a_4\xi^2 \varphi = 0 \tag{3.24a}$$

$$b_1\rho^2 + \rho = b_2 - b_3\xi - b_4\xi^2 \varphi = 60° \tag{3.24b}$$

where $a_1 = 6\sqrt{6}C_1 / (3 + \beta)$, $a_2 = \sqrt{6}D / (3 + \beta)$, $a_3 = -\sqrt{2}(A + 2\beta) / (3 + \beta)$, $a_4 = -\sqrt{6}B / (9 + 3\beta)$, $b_1 = 6\sqrt{6}C_2 / (3 - \beta)$, $b_2 = \sqrt{6}D / (3 - \beta)$, $b_3 = -\sqrt{2}(A + 2\beta) / (3 - \beta)$, and $b_4 = -\sqrt{6}B / (9 - 3\beta)$. Then the unknown quantities (β, A, B, C_1, C_2 and D) can be expressed as $A = (6a_2 - 6b_2 - 3\sqrt{2}a_3b_2) / (a_2 + b_2)$, $B = -3\sqrt{6}a_4b_2 / (a_2 + b_2)$, $C_1 = \sqrt{6}a_1b_2 / (6a_2 + 6b_2)$, $C_2 = \sqrt{6}a_2b_1 / (6a_2 + 6b_2)$, $D = \sqrt{6}a_2b_3 / (a_2 + b_2)$ and $\beta = 3(b_2 - a_2) / (a_2 + b_2)$. In addition, the following relationship can be found: $a_3 / a_2 = b_3 / b_2$ and $a_4 / a_2 = b_4 / b_2$.

3.4.3.1 The boundary conditions with triaxial tensile stress

Based on the above discussions, six stress points should be selected from the tensile meridian and the compressive meridian to be the boundary conditions. Therefore, the following stress points can be selected in the ordinary stress coordinate system: for $\varphi = 0$, the uniaxial tensile strength (f_t, 0, 0), the biaxial compressive strength (0, $-f_{bc}$, $-f_{bc}$), the triaxial tensile strength (f_t, f_t, f_t) and a stress point in tensile meridian; for $\varphi = 60°$, the uniaxial compressive strength (0, 0, $-f_c$), and a stress point in compressive meridian. After normalization processing by the compressive strength, the boundary conditions can be expressed as follows: for $\varphi = 0$, the corresponding points are ($\sqrt{3}\alpha/3$, $\sqrt{6}\alpha/3$), ($-2\sqrt{3}\alpha_{bc}/3$, $\sqrt{6}\alpha_{bc}/3$), ($\sqrt{3}\alpha$, 0) and ($-\xi_1$, ρ_1); for $\varphi = 60°$, the corresponding points are ($-\sqrt{3}/3$, $\sqrt{6}/3$) and ($-\xi_2$, ρ_2). Based on the boundary conditions, the following equations can be gained by substituting the boundary conditions into Eq. (3.24).

For the uniaxial tensile strength:

$$2a_1\alpha^2 + \sqrt{6}\alpha = 3a_2 + \sqrt{3}a_3\alpha - 2a_4\alpha^2 \tag{3.25a}$$

For the biaxial compressive strength:

$$2\alpha_1\alpha_{bc} + \sqrt{6}\alpha_{bc} = 3a_2 - 2\sqrt{3}a_3\alpha_{bc} + 4a_4\alpha_{bc}^2 \tag{3.25b}$$

For a point in the compressive meridian:

$$a_1\rho_1^2 + \rho_1 = a_2 - \alpha_3\xi_1 - a_4\xi_1^2 \tag{3.25c}$$

For the triaxial tensile strength in Eq. (3.24a):

$$0 = a_0 + \sqrt{3}\alpha a_3 + 2\alpha^2 a_4 \tag{3.25d}$$

For the uniaxial compressive strength:

$$2b_1 + \sqrt{6} = 3b_2 + \sqrt{3}b_3 + b_4 \tag{3.25e}$$

For a point in the tensile meridian:

$$b_1\rho_2^2 + \rho_2 = b_2 - b_3\xi_2 - b_4\xi_2^2 \tag{3.25f}$$

For the triaxial tensile strength in Eq. (3.24b):

$$0 = b_2 + \sqrt{3}\alpha b_3 - 3\alpha^2 b_4 \tag{3.25g}$$

Based on the boundary condition training, the calculation process for the coefficients in Eq. (3.25) is given as follows:
for Eq. (3.25a) – Eq. (3.25d):

$$2a_1\alpha^2 + \sqrt{6}\alpha = -2\sqrt{3}\alpha a_3 - 8\alpha^2 a_4$$

for Eq. (3.25b) – Eq. (3.25d):

$$2a_1\alpha_{bc}^2 + \sqrt{6}\alpha_{bc} = -\sqrt{3}a_3(2\alpha_{bc} + 3\alpha) + a_4\left(4\alpha_{bc}^2 - 9\alpha^2\right)$$

for Eq. (3.25c) – Eq. (3.25d):

$$a_1\rho_1^2 + \rho_1 = -\alpha_3\left(\xi_1 + \sqrt{3}\alpha\right) + a_4\left(\xi_1^2 - 3\alpha^2\right)$$

These equations can be equivalent to the following forms:

$$a_1x_1 + \sqrt{6}\alpha/3 = a_3x_2 + a_4x_3 \tag{3.25h}$$

$$a_1y_1 + \sqrt{6}\alpha_{bc}/3 = a_3y_2 + a_4y_3 \tag{3.25i}$$

$$a_1z_1 + \rho_1 = \alpha_3z_2 + a_4z_3 \tag{3.25j}$$

where $x_1 = 2\alpha^2 / 3$, $x_2 = -2\alpha / \sqrt{3}$, $x_3 = -8\alpha^2 / 3$, $y_1 = 2\alpha_{bc}{}^2 / 3$, $y_2 = -\sqrt{3}(3\alpha + 2\alpha_{bc}) / 3$, $y_3 = (4\alpha_{bc}{}^2 - 9\alpha^2) / 3$, $z_1 = \rho_1{}^2$, $z_2 = -\sqrt{3}\alpha - \xi_1$, and $z_3 = \xi_1{}^2 - 3\alpha^2$.

Furthermore, the following equations can be gained:
for Eq. (3.25h) × y_1 − Eq. (3.25i) × x_1:

$$\sqrt{6}(\alpha y_1 + \alpha_{bc} x_1)/3 = a_3(x_2 y_1 - x_1 y_2) + a_4(x_3 y_1 - x_1 y_3) \tag{3.25k}$$

for Eq. (3.25h) × y_1 − Eq. (3.25j) × x_1:

$$\sqrt{6}(\alpha y_1 + \alpha_{bc} x_1)/3 = a_3(x_2 y_1 - x_1 y_2) + a_4(x_3 y_1 - x_1 y_3) \tag{3.25l}$$

Therefore, the calculation equation of a_4 can be obtained as follows: Eq. (3.25k) × $(x_2 z_1 - x_1 z_2)$ − Eq. (3.25l) × $(x_2 y_1 - x_1 y_2)$:

$$a_4 = \frac{\sqrt{6}\alpha(x_1 y_2 z_1 - x_1 y_1 z_1) + \sqrt{6}\alpha_{bc}\left(x_1^2 z_2 - x_1 x_2 z_1\right) + 3\rho_1\left(x_1 x_2 y_2 - x_1^2 y_2\right)}{3\left(x_1^2 y_3 z_2 - x_1 x_2 y_1 z_3 + x_1 x_3 y_2 z_1 - x_1^2 y_2 z_3 - x_1 x_2 y_3 z_1 - x_1 x_3 y_1 z_2\right)} \tag{3.26a}$$

Then the calculation equation of a_3 can be obtained from Eq. (3.25k) and Eq. (3.26a):

$$a_3 = \frac{\sqrt{6}(\alpha y_1 - \alpha_{bc} x_1) - 3a_4(x_3 y_1 - x_1 y_3)}{3(x_2 y_1 - x_1 y_2)} \tag{3.26b}$$

Then, for Eq. (3.25e) × y_2 − Eq. (3.25f) × x_2:

$$a_1 = \frac{\sqrt{6}(\alpha y_2 - \alpha_{bc} x_2) - 3a_4(x_3 y_2 - x_2 y_3)}{x_2 y_1 - x_1 y_2} \tag{3.26c}$$

Then the calculation equation of a_3 can be gained from Eq. (3.25d):

$$a_2 = -\left(\sqrt{3}a_3\alpha + 3a_4\alpha^2\right) \tag{3.26d}$$

Since there is the relationship that: $a_3 / a_2 = b_3 / b_2$ and $a_4 / a_2 = b_4 / b_2$, The Eqs. (3.25h) and (3.25i) can be equivalent to the following forms:

$$2b_1 + \sqrt{6} = b_2\left(3 - \sqrt{3}a_3/a_2 + a_4/a_2\right) \tag{3.27a}$$

$$b_1\rho_2^2 + \rho_2 = b_2\left(1 - \xi_2 a_3/a_2 - \xi_2^2 a_4/a_2\right) \tag{3.27b}$$

Then, for Eq. (3.27a) × $\rho_2{}^2$ − Eq. (3.27b) × 2/3:

$$b_2 = \frac{\sqrt{6}\rho_2^2 - 2\rho_2}{3\rho_2^2 - 2 - a_3\left(\sqrt{3}\rho_2^2 - 2\xi_2\right)/a_2 + a_4\left(\rho_2^2 - 2\xi_2^2\right)/a_2} \tag{3.28a}$$

Based on Eq. (3.27a), it can be found that:

$$b_1 = b_2\left(3 - \sqrt{3}a_3/a_2 + a_4/a_2\right)/2 - \sqrt{6}/2 \tag{3.28b}$$

And it can be found in Eq. (3.24) that:

$$b_3 = a_3 b_2/a_2 \tag{3.28c}$$

$$b_4 = a_4 b_2/a_2 \tag{3.28d}$$

3.4.3.2 The boundary conditions with triaxial compressive stress

In the cylindrical coordinate system (ξ, ρ, φ), the new boundary conditions are the uniaxial compressive strength ($-\sqrt{3}/3$, $\sqrt{6}/3$, 60°), the uniaxial tensile strength ($\sqrt{3}\alpha/3$, $\sqrt{6}\alpha/3$, 0), the biaxial compressive strength ($-2\sqrt{3}\ \alpha_{bc}/3$, $\sqrt{6}\ \alpha_{bc}$ / 3, 0), the triaxial tensile strength ($-\sqrt{3}\alpha$, 0, 0), a point in the tensile meridian ($-\xi_1$, ρ_1, 0), and a point in the compressive meridian ($-\xi_2$, ρ_2, 60°). All the characteristic stresses are from the tensile meridian ($\sigma_1 \geq \sigma_2 = \sigma_3$, $\varphi = 0$) and the compressive meridian ($\sigma_1 = \sigma_2 \geq \sigma_3$, $\varphi = 60°$). According to the above section, the corresponding coefficients can be expressed as follows:

$$a_4 = \frac{\sqrt{6}\alpha(x_1y_2z_1 - x_1y_1z_1) + \sqrt{6}\alpha_{bc}\left(x_1^2z_2 - x_1x_2z_1\right) + 3\rho_1\left(x_1x_2y_2 - x_1^2y_2\right)}{3\left(x_1^2y_3z_2 - x_1x_2y_1z_3 + x_1x_3y_2z_1 - x_1^2y_2z_3 - x_1x_2y_3z_1 - x_1x_3y_1z_2\right)} \tag{3.29a}$$

$$a_3 = \frac{\sqrt{6}(\alpha y_1 - \alpha_{bc}x_1) - 3a_4(x_3y_1 - x_1y_3)}{3(x_2y_1 - x_1y_2)} \tag{3.29b}$$

$$a_1 = \frac{\sqrt{6}(\alpha y_2 - \alpha_{bc}x_2) - 3a_4(x_3y_2 - x_2y_3)}{x_2y_1 - x_1y_2} \tag{3.29c}$$

$$a_2 = -\left(\sqrt{3}a_3\alpha_{cc} + 3a_4\alpha_{cc}^2\right) \tag{3.29d}$$

$$b_2 = \frac{\sqrt{6}\rho_2^2 - 2\rho_2}{3\rho_2^2 - 2 - a_3\left(\sqrt{3}\rho_2^2 - 2\xi_2\right)/a_2 + a_4\left(\rho_2^2 - 2\xi_2^2\right)/a_2} \tag{3.29e}$$

$$b_1 = b_2\left(3 - \sqrt{3}a_3/a_2 + a_4/a_2\right)/2 - \sqrt{6}/2 \tag{3.29f}$$

$$b_3 = a_3 b_2/a_2 \tag{3.29g}$$

$$b_4 = a_4 b_2 / a_2 \tag{3.29h}$$

where $x_1 = 2\alpha^2 / 3$, $x_2 = \sqrt{3}(\alpha + 3\alpha_{cc}) / 3$, $x_3 = (\alpha_2 + 9\alpha_{cc}^2) / 3$, $y_1 = 2\alpha_{bc}^2 / 3$, $y^2 = \sqrt{3}(3\alpha_{cc} - 2\alpha_{bc}) / 3$, $y_3 = (4\alpha_{bc}^2 - 9\alpha_{cc}^2) / 3$, $z_1 = \rho_1^2$, $z_2 = \sqrt{3}\alpha_{cc} - \xi_1$, and $z_3 = \xi_1^2 - 3\alpha_{cc}^2$

3.5 Failure criterion model validation

In order to verify the accuracy and the reliability of the proposed failure criterion models, comparisons between the maximum principal stress (σ_3) of the model computation results and the test results are set. And the model results are computed by the minimum principal stress (σ_1) and middle principal stress (σ_2). In addition, the relative error (E_i) is introduced to reflect the difference between the computation results and the experimental results, and the mathematical expression is given as follows:

$$E_i = \frac{\left(\sigma'_{3i} - \sigma_3^{exp}\right)}{\sigma_3^{exp}} (i = 1, 2, 3...) \tag{3.30}$$

where σ_3^{exp} is the test result; σ'_{3i} is the computation result by the proposed model, and i is the model number.

3.5.1 Five-parameter failure criterion model A

For the ordinary solution of the proposed model, the unknown coefficients can be computed by Eqs. (3.9) and (3.13). For the simplification solution of the proposed model, the unknown coefficients can be computed by Eqs. (3.9) and (3.15). After the determination of each coefficient, the maximum principal stress can be calculated in the following method.

At first, Eq. (3.9) can be equivalent to a monadic quadratic equation:

$$3B_1\sigma_3^2 + \sigma_3[6\beta + 4A - 6 - 6B_1(2\sigma_1 - \sigma_2)] + 6(2\sigma_1 - \sigma_2) + 6\beta(2\sigma_1 + \sigma_2) + 4A(\sigma_1 + \sigma_2) + 3B_1(2\sigma_1 - \sigma_2)^2 - 12C = 0$$

$$12B_2\sigma_3^2 + \sigma_3[12\beta + 4A - 12 - 12B_2(\sigma_1 + \sigma_2)] + 3B_2(\sigma_1 + \sigma_2)^2 + (\sigma_1 + \sigma_2)(6 + 6\beta + 4A) - 12C = 0$$

To simplify the above equations, the substitute coefficients are introduced, then the simplified equations can be expressed as follows:

$$a_1\sigma_3^2 + b_1\sigma_3 + c_1 = 0 \tag{3.31a}$$

$$a_2\sigma_3^2 + b_2\sigma_3 + c_2 = 0 \tag{3.31b}$$

where $a_1 = 3B_1$, $a_2 = 12B_2$, $b_1 = 6\beta + 4A - 6 - 6B_1(2\sigma_1 - \sigma_2)$, $b_2 = 12\beta + 4A - 12 - 12B_2(\sigma_1 + \sigma_2)$, $c_1 = 6(2\sigma_1 - \sigma_2) + 6\beta(2\sigma_1 + \sigma_2) + 4A(\sigma_1 + \sigma_2) + 3B_1(2\sigma_1 - \sigma_2)^2 - 12C$ and $c_2 = 3B_2(\sigma_1 + \sigma_2)^2 + (6 + 6\beta + 4A)(\sigma_1 + \sigma_2) - 12C$.

Based on the solution formula of the univariate quadratic equation, the maximum principal stress (σ_3) can be computed as follows:

$$\sigma_3{}' = \frac{-b_1 - \sqrt{b_1^2 - 4a_1c_1}}{2a_1} \tag{3.32a}$$

$$\sigma_3{}'' = \frac{-b_2 - \sqrt{b_2^2 - 4a_2c_2}}{2a_2} \tag{3.32b}$$

It should be noted that σ_3 should select the larger one in $\sigma_3{}'$ and $\sigma_3{}''$, due to the judgment condition in Eq. (3.9), which is given as follows:

$$\sigma_3 = \max(\sigma_3{}', \sigma_3{}'') \tag{3.33}$$

The comparison results are shown in Table 3.3, where $\sigma_3{}^{exp}$ is the test result, $\sigma_{31}{}'$ is computed by Eqs. (3.9), (3.13), (3.31), and (3.32), and σ_{32}' is computed by Eqs. (3.9), (3.15), (3.31), and (3.32).

It is easily found that E_1 and E_2 for specimen TCS10 and E_1 for specimen TCS13 are 0 in Table 3.3, because the data for these specimens are used to compute the model coefficients. There is little difference between σ'_{31} and σ'_{32}, which means that the simplified solution can be used to solve the unknown coefficients in the

Table 3.3 The validation of the five-parameter failure criterion model A.

Concrete	$\sigma_3{}^{exp}$ (MPa)	σ'_{31} (MPa)	E_1 (%)	σ'_{32} (MPa)	E_2 (%)
TCS3	325.34	725.67	123.05	655.06	101.35
TCS4	52.83	51.81	−1.93	51.89	−1.78
TCS5	72.87	77.91	6.92	77.92	6.93
TCS6	71.77	85.46	19.07	85.47	19.09
TCS7	69.82	89.30	27.90	88.30	26.46
TCS8	61.65	73.58	20.37	72.67	18.88
TCS9	57.23	59.40	3.79	58.52	2.25
TCS10	150.36	150.36	0.00	150.36	0.00
TCS11	188.01	180.25	4.12	180.24	4.13
TCS12	187.28	217.35	16.06	213.76	14.14
TCS13	166.69	166.70	0.00	158.71	4.79
TCS14	83.96	161.15	101.73	137.00	71.49

proposed failure criterion model. However, the comparisons in Table 3.3 also show that E_1 and E_2 for several specimens (i.e., TCS3, TCS6, TCS7, TCS8, and TCS14) are large (exceeding 15%), which means that the five-parameter failure criterion model A cannot provide accurate prediction value for concrete in the stress condition on which the hydrostatic stress has dominant influence. In addition, if the stress condition is $\sigma_1 \neq \sigma_2 \neq \sigma_3$ or approximately $\sigma_1 = \sigma_2 = \sigma_3$, the hydrostatic stress becomes more important.

3.5.2 Five-parameter failure criterion model B

For the solution of the proposed model, the unknown coefficients can be computed by Eqs. (3.16) and (3.21). After the determination of each coefficient, the maximum principal stress can be calculated in the following method.

In order to simplify the equations, the substitute coefficients are introduced, then Eq. (3.16) can be equivalent to a monadic quadratic equation:

$$a_1\sigma_3^2 + b_1\sigma_3 + c_1 = 0 \tag{3.34a}$$

$$a_2\sigma_3^2 + b_2\sigma_3 + c_2 = 0 \tag{3.34b}$$

where $a_1 = B_1/9$, $a_2 = B_2/9$, $b_1 = \beta/2 + A_1/3 - 1/2 + 2B_1(\sigma_1 + \sigma_2)$, $b_2 = \beta + A_2/3 - 1 - 2B_2(\sigma_1 + \sigma_2)/9$, $c_1 = \sigma_1(1 + \beta - A_1/3) + \sigma_2(A_1/3 + \beta/2 - 1/2) + 3B_1(2\sigma_1 + \sigma_2)^2/9 - C$ and $c_2 = B_2(2\sigma_1 + \sigma_2)^2/9 + (1/2 + \beta/2 - A_2/3)(\sigma_1 + \sigma_2) - C$.

Based on the solution formula of the univariate quadratic equation, the maximum principal stress (σ_3) can be computed as follows:

$$\sigma_3{}' = \frac{-b_1 - \sqrt{b_1^2 - 4a_1c_1}}{2a_1} \tag{3.35a}$$

$$\sigma_3{}'' = \frac{-b_2 - \sqrt{b_2^2 - 4a_2c_2}}{2a_2} \tag{3.35b}$$

It should be noted that σ_3 should select the larger one in $\sigma_3{}'$ and $\sigma_3{}''$, due to the judgment condition in Eq. (3.16), which is given as follows:

$$\sigma_3 = \max(\sigma_3{}', \sigma_3{}'') \tag{3.36}$$

To verify the accuracy and reliability of the five-parameter failure criterion model B, the σ_1 and σ_2 are used as the known data to compute $\sigma 3$, based on Eqs. (3.16), (3.21), (3.35), and (3.36). The comparison results are shown in Table 3.4.

It is easily found that E_3 of specimen TCS10 is 0 in Table 3.4 because the specimen data is used to compute the coefficients for the model. If the three principal

Table 3.4 The validation of the five-parameter failure criterion model B.

Concrete	σ_3^{exp} (MPa)	σ'_{33} (MPa)	E_3 (%)
TCS3	325.34	83.15	−74.44
TCS4	52.83	58.93	11.55
TCS5	72.87	90.91	24.76
TCS6	71.77	95.03	32.41
TCS7	69.82	74.55	6.77
TCS8	61.65	60.57	−1.75
TCS9	57.23	49.91	−12.79
TCS10	150.36	150.36	0.00
TCS11	188.01	166.45	−11.46
TCS12	187.28	98.05	−47.65
TCS13	166.69	60.06	−63.95
TCS14	83.96	90.91	8.28

stresses are different from each other ($\sigma_1 \neq \sigma_2 \neq \sigma_3$) or if they become similar ($\sigma_1 = \sigma_2 = \sigma_3$), the proposed model can provide the accurate prediction result. The result comparisons in Table 3.5 show that E_3 of some specimens (i.e., TCS3, TCS5, TCS6, TCS12, and TCS13) are large (exceeding 15%), which means that the five-parameter failure criterion model B cannot provide an accurate prediction value for concrete in the stress condition which shear stress has a dominant influence. Furthermore, based on the validation of σ_{33}, models A and B show complementary prediction results.

3.5.3 Six-parameter failure criterion model

For the boundary conditions with triaxial tensile stress, the unknown coefficients can be computed by Eqs. (3.22) and (3.25). For the boundary conditions with triaxial compressive stress, the unknown coefficients can be computed by Eqs. (3.22) and (3.28). After the determination of each coefficient, the maximum principal stress can be calculated in the following method.

In order to simplify the equations, the substitute coefficients are introduced, then the Eq. (3.16) can be equivalent to a monadic quadratic equation:

$$a_1\sigma_3^2 + b_1\sigma_3 + c_1 = 0 \tag{3.37a}$$

$$a_2\sigma_3^2 + b_2\sigma_3 + c_2 = 0 \tag{3.37b}$$

where $a_1 = B/9 + C_1$, $a_2 = B/9 + 4C_2$, $b_1 = \beta/2 + A/3 - 1/2 + 2B(\sigma_1 + \sigma_2) - 2C_1(2\sigma_1 - \sigma_2)$, $b_2 = \beta + A/3 + 2B(\sigma_1 + \sigma_2)/9 + 4C_2(\sigma_1 + \sigma_2) - 1$, $c_1 = (2\sigma_1 - \sigma_2)/2 + \beta(2\sigma_1 + \sigma_2)/2 + A(\sigma_1 + \sigma_2)/3 + B(\sigma_1 + \sigma_2)^2/9 + C_2(2\sigma_1 - \sigma_2)^2 - D$ and $c_2 = (B/9 + C_2)(2\sigma_1 + \sigma_2)^2/9 + (\beta/2 + A/3 + 1/2)(\sigma_1 + \sigma_2) - D$.

Table 3.5 The validation of the six-parameter failure criterion model.

Concrete	σ_3^{exp} (MPa)	σ'_{34} (MPa)	E_4 (%)	σ'_{35} (MPa)	E_5 (%)
TCS3	325.34	325.34	0.00	325.34	0.00
TCS4	52.83	55.61	5.26	56.26	6.49
TCS5	72.87	80.11	14.05	79.94	9.70
TCS6	71.77	90.74	26.43	79.93	11.37
TCS7	69.82	78.69	12.70	59.66	−14.55
TCS8	61.65	65.98	7.02	56.41	−8.50
TCS9	57.23	52.27	−8.67	43.43	−24.11
TCS10	150.36	150.36	0.00	149.45	−0.60
TCS11	188.01	174.56	−7.15	172.80	−8.09
TCS12	187.28	165.56	−11.89	96.00	−48.74
TCS13	166.69	124.83	−25.07	166.67	−0.05
TCS14	83.96	92.68	10.39	83.67	−0.01

Based on the solution formula of the univariate quadratic equation, the maximum principal stress (σ_3) can be computed as follows:

$$\sigma_3' = \frac{-b_1 - \sqrt{b_1^2 - 4a_1c_1}}{2a_1} \tag{3.38a}$$

$$\sigma_3'' = \frac{-b_2 - \sqrt{b_2^2 - 4a_2c_2}}{2a_2} \tag{3.38b}$$

It should be noted that σ_3 should select the larger one in σ_3' and σ_3'', due to the judgment condition in Eq. (3.22), which is given as follows:

$$\sigma_3 = \max(\sigma_3', \sigma_3'') \tag{3.39a}$$

To verify the accuracy and reliability of the six-parameter failure criterion model, the σ_1 and σ_2 are used as the known data to compute σ_3. The comparison results are shown in Table 3.5, where σ_3^{exp} is the test result, σ'_{34} is computed by Eqs. (3.22), (3.25), (3.38), and (3.39), and σ'_{35} is computed by Eqs. (3.22), (3.28), (3.38), and (3.39).

It is easily found that the computation result of most specimens can be fitted with the experiment results. However, some computation result of the proposed model under different boundary conditions are very different, which means that the prediction results of the six-parameter failure criterion model mostly depend on the boundary conditions.

3.6 Conclusions

The failure criterion model is important for the stress analysis of the concrete in multiaxial stress conditions. To establish the theoretical and accurate constitutive failure criterion, the biaxial compressive strength and triaxial compressive strength of concrete are studied respectively to find the main influence factors. Then, the reasonable theory foundation is found, based on the reviews of strength theory models. Finally, based on the Twin Shear Strength Theory, three kinds of failure criterion models are established and verified, which can be the foundation for nonlinear analysis of concrete under multiaxial compression. As shown above, the main conclusions can be drawn:

(1) Based on biaxial and triaxial compressive experiments on concrete specimens, six mechanical effects are examined, which show the intermediate principal stress, the hydrostatic pressure, and the principal shear stress are the most important factors.

(2) The Twin Shear Strength Theory can be modified by introducing stress parameters for the concrete under multiaxial compression pressure. Based on the design idea, three kinds of multifailure criterion models with clear physical significance are established, solved, and verified.

(3) The five-parameter failure criterion model A can be applied to examine concrete structures under low confining effect. However, this proposed model is not applied to the stress condition which is $\sigma_1 \neq \sigma_2 \neq \sigma_3$ or approximate to $\sigma_1 = \sigma_2 = \sigma_3$.

(4) The five-parameter failure criterion model B can be applied to complicated stress conditions. Furthermore, the five-parameter failure criterion model A and model B show complementary prediction results.

(5) The six-parameter failure criterion model can predict the accurate prediction results for the concrete under multiaxial compression. However, the proposed model has multiple parameters and complex forms, while the prediction results are likely to depend on the boundary conditions.

References

Asteris, PG, & Plevris, V (2016). Anisotropic masonry failure criterion using artificial neural networks. *Neural Computing and Applications*, *28*, 2207–2229.

Che, AA (1979). Triaxial uniform tensile property of concrete. *Concrete and Electron*, *6*, 33–34.

Guo, ZH (1997). *Strength and deformation of concrete: Test basis and constitutive relationship*. Tsinghua University Press.

Huang, H (2008). *Revision of several problems on double shear strength theory*. Wuhan Polytechnic University.

Kaya, A, Dawood, M, & Gencturk, B (2015). Repair of corroded and buckled short steel columns using concrete-filled GFRP jackets. *Construction and Building Materials*, *94*, 20–27.

Menetrey, P, & Willam, KJ (1995). Triaxial failure criterion for concrete and its generalization. *ACI Materials Journal*, *92*(3), 311–318.

Mogi, K (1967). Effect of intermediate principal stress on rock failure. *Journal of Geophysical Research*, *72*, 5117–5131.

Øverli, J (2016). A density-dependent failure criterion for concrete. *Construction and Building Materials*, *124*, 566–574.

Ozbakkaloglu, T (2013). Compressive behavior of concrete-filled FRP tube columns: Assessment of critical column parameters. *Engineering Structures*, *51*, 188–199.

Piscesa, B, Attard, MM, & Samani, AK (2016). A lateral strain plasticity model for FRP confined concrete. *Composite Structures*, *158*, 160–174.

Plevris, V, & Asteris, PG (2014). Modeling of masonry failure surface under biaxial compressive stress using neural networks. *Construction and Building Materials*, *55*, 447–461.

Rong, C, Shi, QX, Zhang, T, & Zhao, HC (2018). New failure criterion models for concrete under multiaxial stress in compression. *Construction and Building Materials*, *161*, 432–441.

Rossi, P, Tailhan, JL, & Maou, FL (2013). Creep strain versus residual strain of a concrete loaded under various levels of compressive stress. *Cement and Concrete Research*, *51*, 32–37.

Shang, S, & Song, Y (2013). Dynamic biaxial tensile-compressive strength and failure criterion of plain concrete. *Construction and Building Materials*, *40*, 322–329.

Shkolnik, IE (2008). Influence of high strain rates on stress-strain relationship, strength and elastic modulus of concrete. *Cement and Concrete Composites*, *30*(10), 1000–1012.

Song, YP (1988). *Study on mechanical model in finite element analysis of reinforced concrete*. Dalian University of Technology.

Syrmakezis, CA, & Asteris, PG (2001). Masonry failure criterion under biaxial stress state. *Cement and Concrete Research*, *13*(1), 58–64.

Tang, XW, & Zhang, CH (2009). Study on concrete in macro-and meso-scale mechanical properties based on homogenization theory. *Chinese Journal of Computational Mechanics*, *26*, 876–881.

Wang, HW, Wu, CQ, Zhang, FR, Fang, Q, Xiang, H, Li, P, Li, Z, Zhou, Y, Zhang, Y, & Li, J (2017). Experimental study of large-sized concrete filled steel tube columns under blast load. *Construction and Building Materials*, *134*, 131–141.

Wang, YB, Richard, JY, Lee, SC, & Xiong, DX (2016). Experimental study of ultra-high-strength concrete under triaxial compression. *ACI Materials Journal*, *113*, 105–112.

Xiao, J, Li, W, & Liu, Q (2011). Meso-level numerical simulation on mechanical properties of modeled recycled concrete under uniaxial compression. *Journal of Tongji University*, *39*, 791–797.

Yan, D, & Lin, G (2008). Influence of initial static stress on the dynamic properties of concrete. *Cement and Concrete Composites*, *30*(4), 327–333.

Yao, YP, Lu, DC, Zhou, AN, & Zou, B (2004). Generalized nonlinear strength theory and its transformed stress space. *Science China (E)*, *11*, 1283–1299.

Yu, MH (2002). *Concrete strength theory and application*. Higher Education Press.

Yu, MH (2011). *New system of strength theory: Theory, development and applications*. Xi'an Jiaotong University Press.

Zhang, B (1998). Relationship between pore structure and mechanical properties of ordinary concrete under bending fatigue. *Cement and Concrete Research*, *28*(5), 699–711.

Zheng, D, Li, Q, & Wang, L (2007). Rate effect of concrete strength under initial static loading. *Engineering Fracture Mechanics*, *74*, 2311–2319.

Zhou, J, Chen, X, Wu, L, & Kan, X (2011). Influence of free water content on the compressive mechanical behaviour of cement mortar under high strain rate. *Sadhana*, *36*, 357–369.

Zhu, WC, Tang, CA, & Wang, SY (2005). Numerical study on the influence of mesomechanical properties on macroscopic fracture of concrete. *Structural Engineering and Mechanics*, *19*(5), 519–533.

Analysis-oriented constitutive model

4

4.1 Introduction

A series of concrete composite columns have been studied in existing researches and applied in structural engineering for a long time, including the hoops confined concrete columns in early years, the concrete filled in steel tube columns in later years, and the fiber reinforced polymers (FRP) confined concrete columns in recent years. The outer materials can provide a confining effect on the core concrete, which can improve the mechanical properties of the concrete. Therefore, confined concrete plays an important role in these high-performance composite columns. To understand mechanical behavior of the confined concrete in detail, many experimental and theoretical studies (Li et al., 2009; Lim & Ozbakkaloglu, 2015; Liu et al., 2020; Saberi et al., 2020) have been presented over the last two decades. Based on these studies, the following conclusions can be drawn:

(1) all types of confined concrete can be summarized into the actively confined concrete (i.e., the outer confining pressure is constant) and the passively confined concrete (i.e., the outer confining pressure increases with the increase of lateral strain)
(2) the stress behaviors of the actively and passively confined concrete are approximately equal, which means the stress is independent of the loading method
(3) the strain behaviors of the actively and passively confined concretes show many differences, which means the strain depends on the loading method.

In addition, many studies have established various constitutive models. However, most studies just focused on the stress−strain curve prediction of the particular confined concrete, few studies focused on the unified constitutive model. It is necessary to find a unified establishment method for the constitutive model to satisfy the design and application of the composite concrete columns.

For the concrete in a steel tube or hoops cage, the lateral confining pressure tends to be a constant value in the yield state of steel. We can simply classify the concrete into actively confined concrete. Meanwhile, the FRP has been commonly applied as a confining material in both the reinforcement of existing concrete structures, and in confining the core concrete of new construction, because of its high strength-to-weight ratio and high corrosion resistance. Under the confining effect provided by FRP, the lateral confining pressure increases with the increase of axial loading, due to the linear property of FRP. Therefore, the FRP confined concrete belongs to the passively confined concrete and has a considerable complex state of stress and strain. It is necessary to present accurate and theoretical studies on stress and strain calculation methods for the analysis-oriented constitutive models

Concrete Composite Columns. DOI: https://doi.org/10.1016/B978-0-323-85171-8.00004-5

of confined concrete. However, most of the strength models and the strain models have less theoretical and inaccuracy. Especially for the strain models, the study by Bisby et al. (2015) showed that the existing models poorly correlated with the experimental results of the confined concrete strain, and the mean absolute error of the strain prediction varies from 35% to 250%. It is necessary to study the relationship between the actively and passively confined concrete and propose the accurate stress and strain prediction method, then propose a unified establishment method for constitutive models of the two kinds of confined concrete.

In order to establish the accurate and theoretical constitutive models for the confined concrete, the theoretical analysis should be presented on the stress and strain models. For the stress models, the accuracy failure criterion is the key foundation. The proposed failure criterion in Chapter 3 is discussed and selected for the actively and passively confined concrete at first. Then a new axial-strength model for the two confined concrete circular columns is established, based on the improved failure criterion. And the proposed models are verified by comparing them with a large extensive database. For the strain models, the energy-balance approach is selected to be the foundation. This chapter studies the relationship between the axial strain energy of the core concrete and the absorbed energy of the confining materials around the core concrete, including the actively and passively confined concrete. Based on the energy analysis, we analyze deeply the relationship between the actively confined concrete and the passively confined concrete. Then we propose a new strain calculation method for the confined concrete, including the strain characteristics, the discrimination method of the softening stage, the stress−strain curve trend of the softening concrete, and the hardening concrete. Finally, we propose a unified calculation method of the constitutive models for the confined concrete, based on the stress models and the strain models. In addition, we verify these proposed models by the experiment results of the steel tube confined concrete (represent actively confined concrete) and the FRP confined concrete (represent passively confined concrete).

4.2 Behavior of the confined concrete

The confining pressure can improve the mechanical properties of the core concrete. And the increased mode of the lateral confining pressure is the key to distinguishing the two types of confined concrete. The lateral compress pressure of the actively confined concrete is a constant value, whereas that of passively confined concrete increases with the expansion of the lateral strain. However, the two types of confined concrete show different mechanical behaviors and different stress−strain curve trends. The comparisons of stress−strain curves are shown in Fig. 4.1 (Lam & Teng, 2003a).

In order to compare the stress−strain curves in Fig. 4.1, we adopt the normalized stress and the normalized strain for the two types of confined concrete. In these normalized stress−strain curves, the axial stress (σ_c) is normalized by the axial peak

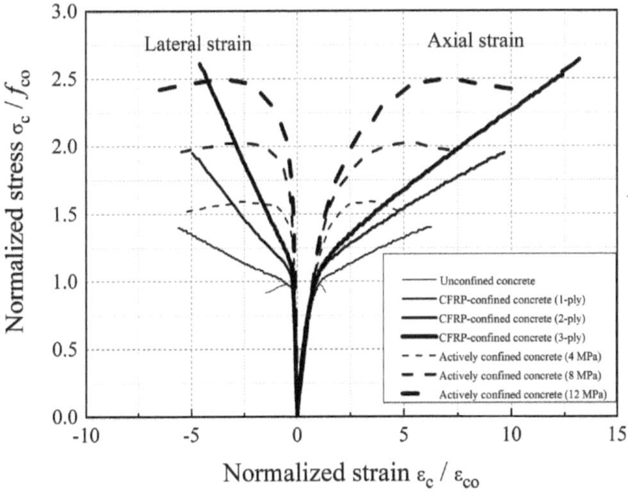

Figure 4.1 Preoperative X-rays and computed tomography scans. Axial normalized stress-axial and lateral normalized strains of FRP confined and actively confined concrete.

stress (f_{co}) of the unconfined concrete, and the axial strain (ε_c) is normalized by the axial peak strain (ε_{co}) of the unconfined concrete, and the lateral strain (ε_l) is normalized by the lateral peak strain ($\varepsilon_{l,co}$) of the unconfined concrete. Meanwhile, the experimental data of unconfined concrete, actively confined concrete, and the carbon FRP (CFRP) confined concrete are selected in the existing experimental studies (Candappa et al., 2001; Lam et al., 2004; Lam & Teng, 2003b, 2004). The normalized stress−strain curve of unconfined concrete shown in Fig. 4.1 is gained from a concrete cylinder (152 mm × 305 mm) in the study (Lam et al., 2004), which has an axial peak strength of 41 MPa and an axial peak strain of 0.00258. The normalized stress−strain curves of the FRP confined concrete with one and two plies of CFRP are gained from the studies (Lam et al., 2004; Lam & Teng, 2003b), the concrete cylinders have the size of 152 mm × 305 mm, an axial peak stress of 35.9 MPa, and an axial peak strain of 0.00203. The normalized stress−strain curves of the FRP confined concrete with three plies of CFRP are gained from the study (Lam & Teng, 2004), the concrete cylinder has the size of 152 mm × 305 mm, an axial peak stress of 34.3 MPa, and an axial peak strain of 0.00188. In addition, the CFRP tube has a nominal thickness of 0.165 mm per ply and an elastic modulus of 250 GPa (Lam et al., 2004; Lam & Teng, 2003b, 2004). The normalized stress−strain curves of the actively confined concrete are obtained from the study (Candappa et al., 2001), the concrete cylinders have the size of 100 mm × 200 mm, an axial peak stress of 41.9 MPa, and an axial peak strain of 0.0024, the actively confining effect provided by three different lateral pressures (i.e., 4, 8, and 12 MPa).

In Fig. 4.1, it is easily found that the stress−strain curves of actively confined concrete and passively confined concretes show different developing trends. We can divide all the stress–strain curves of the confined concrete into two parts, which can reflect the different stress states of the confined concrete. For all confined concrete, the first part of the stress–strain curve is similar, which can be described as a second-degree parabola. This stage reflects that the concrete doesn't reach the damage plane. The second part of the stress–strain curve of actively confined concrete differs from that of the passively confined concrete and the unconfined concrete. The second part of the passively confined concrete or the unconfined concrete shows a softening branch, which means the damage plane of concrete reduces with the increase of axial strain. In contrast, the second part curve of CFRP confined concrete is steadily increasing, which means the damage plane of concrete increases with the increase of axial strain. The major reason is that the confining pressure provided by the FRP jacket increases with the increase of the lateral strain of concrete.

In the above-mentioned studies (Lam et al., 2004; Lam & Teng, 2003b, 2004), the FRP confined concrete cylinders are under the large confining effect, which causes the stress–strain curve shows the bilinear trend. However, the stress state of the FRP confined concrete is more complex under the insufficient confining effect. Therefore, we should study the FRP confined concrete cylinder under the low confining effect. Based on the experimental study by the first author (Rong et al., 2020), the experimental axial stress-axial and lateral strains curves are shown in Fig. 4.2. The concrete cylinders have the size of 150 mm × 300 mm, an axial peak stress of 44.6 MPa, and an axial peak strain of 0.00354. In the experiment, the concrete cylinders are confined by one, three, and five plies of glass FRP (GFRP).

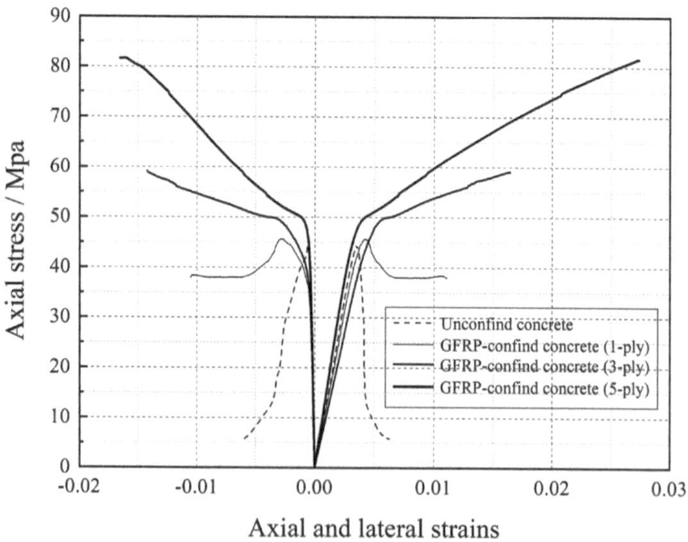

Figure 4.2 Axial stress-axial and lateral strains of FRP confined concrete. The special stress–strain relationship.

In addition, the GFRP tube has the nominal thickness of 0.165 mm per ply, an elastic modulus of 82.11 GPa, and a tensile strength of 1223.4 MPa.

The curve comparison in Fig. 4.2 shows that FRP confined concrete with insufficient confining effect has a trilinear trend. The third stage of the stress–strain curve closes to a horizontal line because the low consistent stiffness can prevent the decrease of stress. In addition, the FRP confined concrete with a low confining effect (i.e., the specimen with 3-ply FRP) shows a horizontal transition between the two increasing stages, which is similar to that of FRP confined high strength concrete (Ozbakkaloglu & Lim, 2013). We can conclude that FRP confined concrete with enough confining effect has a bilinear increasing curve, or a softening branch is possible in the curve. Therefore, there must be a threshold value of the confining effect between the two curves, which is key for determining the shape of the second part of the stress–strain curve of the FRP confined concrete. However, this threshold value has been studied in a few studies (Dong et al., 2015; Imran & Pantazopoulou, 1996; Samani & Attard, 2014; Teng et al., 2007a), and these studies just proposed an approximate criterion condition.

It should be noted that the steel tube confined concrete is not analyzed in this chapter. This is because there is no consensus has been reached in the existing studies on the changing trend of confining pressure in steel tube confined concrete. A part of the studies believes that the confining pressure remains constant after the yielding of steel tube, and concludes that the steel tube confined concrete can be assumed as a type of actively confined concrete. However, another part of studies has proposed that the lateral confining pressure changes until the steel is yielding and there is a continuous increase of lateral confining pressure after the yielding of steel tube, due to the hardening stage. And they divide the steel tube confined concrete into the passively confined concrete. Therefore, the steel tube confined concrete is not studied in this chapter, due to the complex stress state.

Based on the above analysis, the peak stress and peak strain are the keys to establishing the first stage of the constitutive model of all confined concrete. The threshold value is the key to distinguishing the developing trend of the later stage of the FRP confined concrete. Both the stress model and the strain model are important for the secondary increasing stage of the FRP confined concrete with a large confining effect. Therefore, we study the stress model and strain model and established them in the following sections, respectively.

4.3 Stress model of the confined concrete

4.3.1 Lateral confining pressures

The lateral confining pressure is the key to both improving the mechanical properties of core concrete and distinguishing the type of confined concrete. The action mode of the confining pressure is shown in Fig. 4.3A. For the actively confined concrete, the lateral confining pressure (f_l) is a constant value, and the confining action is shown in Fig. 4.3B.

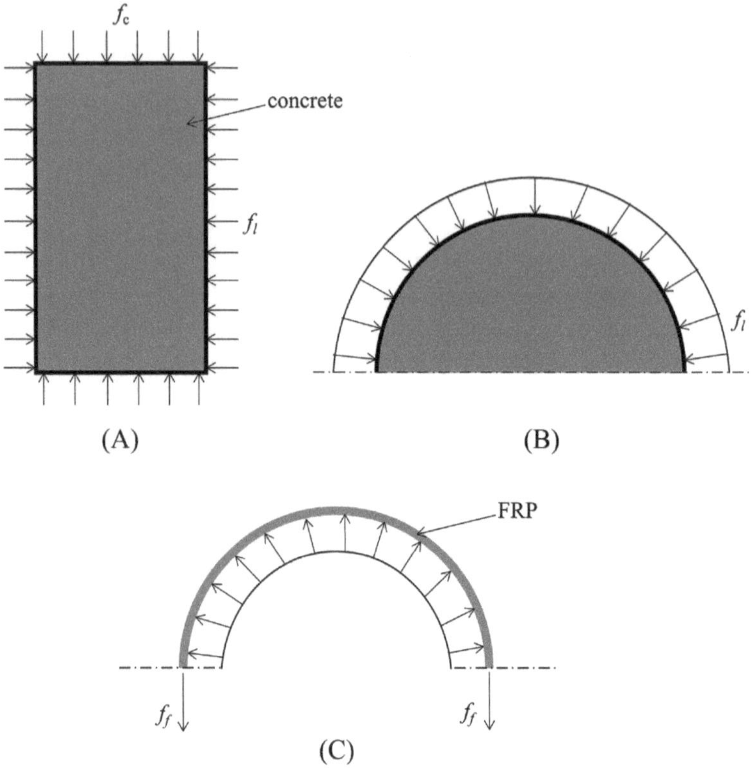

Figure 4.3 The confining pressure around core concrete. (A) Confining effect on the concrete column; (B) actively confined concrete; (C) FRP confined concrete.

For the passively confined concrete, the lateral confining pressure (f_l) is subject to the axial compressive strain, which causes the concrete lateral expansion. Therefore, the FRP confined concrete is a typical passively confined concrete, because of the linear elastic property of the FRP tube. The FRP jacket can provide the hoop tension force to confine the lateral expansion of concrete, and the confining pressure increases with the increase of lateral strain. In the FRP confined concrete cylinders, the lateral confining pressure can be assumed to uniformly distribute around the core concrete, as shown in Fig. 4.3C. The lateral confining pressure f_l can be computed as follows:

$$f_l = (2f_f t)/D \tag{4.1}$$

where f_f is the hoop tensile strength of the FRP jacket, t is the thickness of the FRP jacket, and D is the diameter of the core concrete. Because of the linear elastic property of the FRP jacket, the hoop tensile strength (f_f) relates to the lateral strain of concrete (ε_l), and can be computed as follows:

$$f_f = E_f rp \varepsilon_l \tag{4.2}$$

where E_{frp} is the elastic modulus of the FRP jacket. The lateral confining pressure reaches the maximum value (f_{lc}) at the rupture strain of the FRP jacket ($\varepsilon_{l,rup}$):

$$f_{lc} = (2f_{f,rup}t)/D = (2E_f rp \varepsilon_{l,rup})t)/D \tag{4.3}$$

As we discussed above, the lateral confining pressure (f_l) provided by the FRP tube can greatly improve the mechanical properties of core concrete in the FRP confined concrete column. Therefore, the accuracy of the lateral strain of the FRP tube is the key parameter to establish the axial-strength model. We should study the material properties of the FRP tube in detail, because of the difference between the FRP tube and the FRP jacket. Then the stress model for FRP confined concrete can be established.

It is well known that the actual hoop rupture strain ($\varepsilon_{l,rup}$) of the FRP tube is lower than the rupture tensile strain of the fiber (ε_f) and FRP cloth (ε_{frp}) (Lam & Teng, 2003a; Pessiki et al., 2001; Wu & Jiang, 2013; Yu, 2004). The study presented by Pessiki et al. (2001) introduced the strain reduction coefficient (k_ε) to reflect the relationship between the actual hoop rupture strain of the FRP tube and the rupture tensile strain of the fiber and FRP jacket, the relationship can be expressed as follows:

$$\varepsilon_{h,rup} = k_{\varepsilon,f}\varepsilon_f = k_{\varepsilon,frp}\varepsilon_{frp} \tag{4.4}$$

where $k_{\varepsilon,f}$ and $k_{\varepsilon,frp}$ are the strain reduction coefficients, corresponding to the rupture tensile strain of the fiber (ε_f) and the FRP jacket (ε_{frp}), respectively.

Then, the study (Lam & Teng, 2003a) proposed the equation for the lateral confining pressure ($f_{lu,a}$) based on the strain reduction coefficients, and it can be computed as follows:

$$f_{lu,a} = (2E_f t_f \varepsilon_{h,rup})/D = (2E_{frp}t_{frp}\varepsilon_{h,rup})/D \tag{4.5}$$

where E_f and E_{frp} are the elastic modulus of fiber and FRP cloth respectively, t_f and t_{frp} are the thickness of fiber and FRP cloth, respectively.

Based on the existing researches (Youssef et al., 2007; Yu, 2004), the compressive strength of unconfined concrete (f'_{co}) and the elastic modulus of confining materials (i.e., E_f or E_{frp}) is the key to establishing the model of strain reduction coefficient. In the studies by Ozbakkaloglu and Akin (2011) and Ozbakkaloglu and Lim (2013), a series of FRP confined concrete columns are studied, including the CFRP, the glass FRP, the high-performance FRP, or the ultrahigh performance

CFRP. The studies show that the FRP types have less effect on the strain reduction coefficient, and the coefficient can be calculated as follows:

$$k_{\varepsilon,f} = 0.9 - 2.3f'_{co} \times 10^{-3} - 0.75E_f \times 10^{-3} \quad (4.6)$$

4.3.2 Existing stress models

In the early studies on axial compressive behavior of confined concrete, many confinement models for FRP confined concrete were proposed, such as the Mandal model (Mandal et al., 2015), Cui and Sheikh model (Cui & Sheikh, 2010), Berthet model (Berthet et al., 2006), and so on. For most of the existing models of confined concrete, the popular expression of the peak strength (f'_{cc}) is based on the following equation proposed by Richart (Richart et al., 1928):

$$f'_{cc}/f'_{co} = 1 + kf_l/f'_{co} \quad (4.7)$$

where f'_{co} (MPa) is the peak strength of the unconfined concrete; f'_{cc} (MPa) is the peak strength of the confined concrete strength; and k is the lateral confining coefficient. In practice, k is gained by fitting a large database, the value is 4.1 based on the actively confined concrete tests (Mander et al., 1988), the value is 3.3 as proposed by Teng et al. (2007b), and the value is 3.5 as proposed by Teng et al. (2009).

In general, most of the existing studies focused on the value of k than on other variables. Except for the constant value, some studies suggested that k should be expressed by a mathematical equation and computed by using the confining ratio (f_l/f'_{co}). Shehata et al. (2002) and Li et al. (2003) proposed the equation of the lateral confining coefficient (k), based on Mohr-Columb failure theory, which can be computed as follows:

$$k = tan^2(45^\circ + \varphi/2) \quad (4.8)$$

where φ is the internal friction angle of concrete.

Meanwhile, Shi et al. (2017) presented a further study on the lateral confining coefficient (k), based on the twin shear unified strength theory (Yu, 2004), which is suitable for various materials under the complex stress state. The equation of k is proposed by the mathematical derivation, based on the internal friction angle (φ) of concrete, which is given as follows:

$$k = (1 + \sin\varphi)/(1 - \sin\varphi) \quad (4.9)$$

However, the internal friction (φ) angle of concrete is difficult to test for Eqs. (4.5) and (4.6) because it varies from the lateral confining pressure and the unconfined compressive strength. Therefore, the coefficient (k) should be determined by the fitting method based on a test database (Li et al., 2003; Shehata et al., 2002). The fitting solving process reflects that most models just showed a trend of the confined concrete strength, and the accuracy of proposed models is mainly influenced by the size and reliability of the database. It is necessary to establish the model by a high theoretical method, which is not depend on the database.

Except for the fitting method, some studies introduced a series analysis of methods to establish the model in different expression forms. Pham and Hadi (2014) introduced the diameter–thickness ratio (t / D) into the study of FRP confined concrete strength to establish the stress model, the proposed model is expressed as follows:

$$f'_{cc} = 0.91f'_{co} + 1.88f_l + 7.6t/D \tag{4.10}$$

The study presented by Mander et al. (1988) adopt the damage plane of concrete under multiaxial stress to estimate the compressive strength of confined concrete. The form of the proposed model differs from the previous models, and can be computed as follows:

$$f'_{cc} = f'_{co}(-1.254 + 2.254\sqrt{1 + 7.94f_l/f'_{co}} - 2f_l/f'_{co}) \tag{4.11}$$

The Hoek–Brown failure criterion (Hoek & Brown, 1980) is adopted to establish the strength model for confined concrete by Wu and Zhou (2010), the mathematical expression is as follows:

$$f'_{cc}/f'_{co} = f_l/f'_{co} + \sqrt{(16.7/f'^{0.42}_{co} - f'^{0.42}_{co}/16.7)f_l/f'_{co} + 1} \tag{4.12}$$

As aforementioned, most existing models are got by the fitting method in the test database. However, the proposed fitting model is only suitable for the particular confined concrete, which is influenced by the types, size, and accuracy of the database. And the physical meaning of the relationship is unclear. Against the fitting method, some failure criteria are as the foundation to establish the constitutive model of confined concrete in a few researchers. However, some models are not suitable for most types of confined concrete, because these proposed models compromise the extensive database or have an unsuitable theoretical foundation; however, the models proposed by Mander et al. (1988) and Wu and Zhou (2010) proposed the effective theoretical analysis method. It is clearly demonstrated that the constitutive models of the confined concrete can be established by an accuracy failure criterion, then the proposed model will not only provide accurate prediction results but also have physically meaningful and experimentally determinable parameters.

4.3.3 Improved failure criterion

As we discussed in Chapter 3, the second principal stress and the hydrostatic stress are two key factors in the failure criterion. If the former is focused and as the major factor, we can establish the five-parameter failure criterion model A. The model verification shows that this proposed model applies to examine concrete structures under the low confining effect. And the mechanical expression of the five-parameter failure criterion model A is given as follows:

$$F = \tau_{13} + \tau_{12} + \beta(\sigma_{13} + \sigma_{12}) + A\sigma_m + B_1(\tau_{13} + \tau_{12})^2 = C, F \geq F' \quad (4.13a)$$

$$F' = \tau_{13} + \tau_{23} + \beta(\sigma_{13} + \sigma_{23}) + A\sigma_m + B_2(\tau_{13} + \tau_{23})^2 = C, F \leq F' \quad (4.13b)$$

where σ_m is the hydrostatic stress, $\sigma_m = (\sigma_1 + \sigma_2 + \sigma_3) / 3$; A is the influence coefficient of the hydrostatic stress; and B_1 and B_2 are the influence coefficients of shear stresses.

In order to solve the unknown quantities (β, A, B_1, B_2, and C), we adopt the coordinate transformation between the cylindrical coordinate system and the stress coordinate system and introduce some new quantities. The transformation method has been discussed in Chapter 3, and the relationship is expressed as follows: $A = (6b_1 - 6a_1 - 3\sqrt{2})/(a_1 + b_1)$, $B_1 = 2\sqrt{6}a_2/(3a_1 + 3b_1)$, $B_2 = 2\sqrt{6}b_2/(3a_1 + 3b_1)$, $C = -\sqrt{6}a_0/(3a_1 + 3b_1)$ and $\beta = 3(a_1 - b_1)/(a_1 + b_1)$.

In the cylindrical coordinate system (ξ, ρ, φ), the five characteristic stress points are adopted as the boundary conditions, including the uniaxial compressive stress ($-3^{1/2}/3$, $6^{1/2}/3$, 60°), the uniaxial tensile stress ($3^{1/2}\alpha / 3$, $6^{1/2}\alpha / 3$, 0), the biaxial compressive stress ($(-2) \times 3^{1/2}\alpha_{bc} / 3$, $6^{1/2}\alpha_{bc} / 3$, 0), a point in the tensile meridian ($-\xi_1$, ρ_1, 0), and a point in compressive meridian ($-\xi_2$, ρ_2, 60°). Through the boundary conditions training, the calculation for the new coefficients are in the following forms:

$$a_2 = \frac{\sqrt{3}\alpha(\sqrt{2}\xi_1 + \rho_1 - \sqrt{6}\alpha_b c) + \alpha_b c(2\sqrt{3}\rho_1 - \sqrt{6}\xi_1)}{(\sqrt{6}\alpha/3 - \rho_1)(\sqrt{6}\alpha\rho_1 + 2\alpha_b c^2 - 2\alpha\alpha_b c - \sqrt{6}\alpha_b c\rho_1)}$$

$$a_0 = b_0 = \frac{3\sqrt{3}\alpha_b c\alpha + 2\alpha_2\alpha\alpha_b c(\alpha - \alpha_b c)}{3\alpha_b c - 3\alpha}$$

$$a_1 = \sqrt{2}\alpha(\alpha + 2\alpha_b c)/(2\alpha_b c - 2\alpha) - \sqrt{6}(\alpha + \alpha_b c)a_2/3$$

$$b_1 = (2\xi_2 - \sqrt{3}\rho_2^2 - (3\rho_2^2 - 2)a_0)/(\sqrt{6}\rho_2^2 - 2\rho_2)$$

and $b_2 = ((\sqrt{6}\xi_2 - \sqrt{3}\rho_2) - (3\rho_2 - \sqrt{6})a_0)/(2\rho_2 - \sqrt{6}\rho_2^2)$.

where $\alpha = f_t / f'_{co}$, $\alpha_{bc} = f_{cc} / f'_{co}$, $\xi_i = 3^{1/2}\sigma_m / f'_{co} = (\sigma_1 + \sigma_2 + \sigma_3) / 3^{1/2}$ and $\rho_i = 2\tau_m / f'_{co} = [(\sigma_1 - \sigma_2)^2 + (\sigma_2 - \sigma_3)^2 + (\sigma_1 - \sigma_3)^2]^{1/2} / f'_{co}$ ($i = 1, 2, 3$).

Based on the comparison between model results and test results (see Table 3.3 in Chapter 3), the following conclusion can be drawn: the proposed failure criterion can well apply to the characteristic stresses, which closes to the tensile meridian and the compressive meridian under the low first principal stress ratio ($\sigma_1 / f'_{co} <$ 3.5). Based on the stress analysis of the confined concrete, including the steel tube confined concrete and the FRP confined concrete, the stress state is at the compressive meridian (i.e., $\sigma_1 = \sigma_2 = f_l$, $\sigma_3 = f'_{cc}$) and the lateral pressure ratio (f_l/f'_{co}) is less than 2.3. The stress state means that the proposed failure can agree well with the FRP confined concrete. Therefore, the five-parameter failure criterion model A can be the foundation to establish the stress model for the confined concrete.

4.3.4 Stress models

The stress state of the confined concrete includes one axial stress and two same lateral stress, which belong to the compressive meridian. Therefore, Eq. (4.13b) can establish the stress model of confined concrete. For the stress state (i.e., $\sigma_1 = \sigma_2 = f_l$, $\sigma_3 = f'_{cc}$) in the confined concrete, the lateral confining pressure ($\sigma_1 = \sigma_2 = f_l$) is as the known data to calculate the axial stress ($\sigma_3 = f'_{cc}$). The stress model can be expressed in the following form:

$$f'_{cc} = \frac{2B_2 f_l + \beta + A/3 - 1 + \sqrt{B_2 f_l (8\beta + 4A) + (\beta + A/3)(\beta + A/3 - 2) - 4B_2 C + 1}}{2B_2} \tag{4.14}$$

In order to simplify the calculation method, we can translate Eq. (4.14) into the following equation:

$$f'_{cc} = f_l + a + \sqrt{b f_l + c} \tag{4.15}$$

In order to solve the three unknown coefficients (i.e., a, b, and c), we should select three stress points to be the boundary conditions. The three stress points are the peak stress of the unconfined concrete ($0, 0, f'_{co}$), two stress points of the confined concrete (f_{l1}, f_{l1}, f'_{cc1}), and (f_{l2}, f_{l2}, f'_{cc2}). The two stress points (f_{l1}, f_{l1}, f'_{cc1}) and (f_{l2}, f_{l2}, f'_{cc2}) are selected from the experimental results. In addition, we can select the two stress points in the second increasing stage of the stress–strain curves of FRP confined concrete with a high confining effect, because of the load-path independence of stress. Based on the boundary conditions training, the three coefficients can be calculated as follows:

$$a = \frac{(f'_{cc2} - f_{l2})^2 f_{l1} - (f'_{cc1} - f_{l1})^2 f_{l2} - (f_{l1} - f_{l2}) f'^2_{co}}{2(f'_{cc2} - f'_{co})^2 f_{l1} - 2(f'_{cc1} - f'_{co})^2 f_{l2}} \tag{4.16a}$$

$$b = \frac{(f'_{cc1} - f_{l1} - a)^2 - (f'_{cc2} - f_{l2} - a)^2}{f_{l1} - f_{l2}} \tag{4.16b}$$

$$c = \frac{(f'_{cc2} - f_{l2} - a)^2 f_{l1} - (f'_{cc1} - f_{l1} - a)^2 f_{l2}}{f_{l1} - f_{l2}} \tag{4.16c}$$

4.3.5 Value method of coefficients in the proposed model

Because the calculation is complex in Eq. (4.16), it is necessary to gain specific values for these coefficients for the confined concrete. In order to study the value method of coefficients, we collected a large experimental database of various confined concrete in the monotonic uniaxial compressive test, including the hoops confined concrete, the steel tube confined concrete, and the FRP confined concrete (including five types of FRP tube). In addition, the peak compressive strength of the unconfined concrete ranges from 20 to 56 MPa. Based on the data analysis and existing studies, the types of FRP tube have little effect on the stress model. Whereas the peak compressive strength of the unconfined concrete strength is the key influence factor. It means that the coefficients in the stress model of the confined concrete change with the unconfined concrete strength. Therefore, we can assign the value of these coefficients with the unconfined concrete strength. In order to ensure both the convenience of calculation and the accuracy of the proposed stress model, we can divide the database into 11 ranges, based on the unconfined concrete strength. And the three quantities of the proposed model have the known values in each strength range. The values of coefficients in each range are solved and summarized in Table 4.1. Furthermore, the stress state of the confined concrete belongs to the triaxial compressive stress; thus, this chapter assumes that the compressive stress is a positive value, and the stress state will be: $|\sigma_3| > |\sigma_2| \geq |\sigma_1|$.

In order to identify the changing trend of the coefficients in the proposed model (i.e., a, b, and c), we analyze the relationship between the coefficient values and the unconfined concrete strength and then draw the relationship curve in Fig. 4.4. The relationship between a and the unconfined concrete strength shows a logarithmic developing trend. The relationship between b and the unconfined concrete strength shows a reciprocal developing trend. The relationship between c and the unconfined concrete strength shows an exponentially developing trend. These relationships are unexpected and surprising discoveries, which will be studied in our future researches. For the convenience of the stress model and the strain analysis in the following sections, the changing rules of the parameters (a, b, c) can be expressed in Eq. (4.17) by the fitting method.

$$\left\{ \begin{array}{l} a = 22\ln f_{co} + 67.9 b = 110/(f_{co} - 20.8) + 143 \\ c = 33.4e^{0.06 f_{co}} + 350 \end{array} \right\} \tag{4.17}$$

4.3.6 Verification of the stress model

The proposed stress models are verified by a large experimental database of confined concrete cylinders from the existing research. The database comprises 654 data points which are selected from experimental studies conducted from 1982 to 2014, where the stress data of the actively confined concrete is selected in the study presented by Shi et al. (2017) and the stress data of the FRP confined concrete in

Table 4.1 Calculated values of the quantities for the proposed stress model.

The range of strength (MPa)	**>20 ≤ 23**	**>23 ≤ 26**	**>26 ≤ 30**	**>30 ≤ 33**	**>33 ≤ 36**	**>36 ≤ 40**	**>40 ≤ 43**	**>43 ≤ 46**	**>46 ≤ 50**	**>50 ≤ 53**	**>53 ≤ 56**
a	−0.63	1.91	5.03	8.6	10.2	11.5	14.4	15	17.9	18.1	19.61
b	268	211	179	146	145	149	141	146	143	154	152.7
c	482	510	528	538	596	700	734	873	908	1117	1217

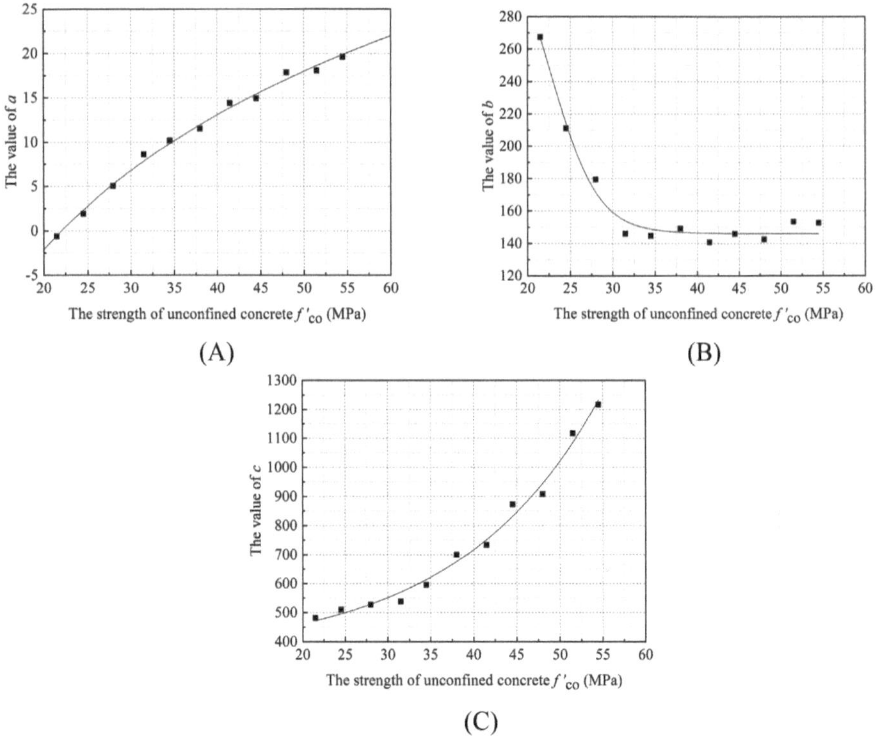

Figure 4.4 The relationship between the coefficient values and the unconfined concrete strength. (A) The relationship between a and the unconfined concrete strength; (B) the relationship between b and the unconfined concrete strength; (C) the relationship between c and the unconfined concrete strength.

Table 4.2 Parameters of the verification database.

Item	Min	Max
f_{co} (MPa)	20.4	55.2
f_l (MPa)	0.7	108.7
f'_{cc} (MPa)	35.6	192.2
$f_{lu,a}/f_{co}$	0.02	2.3
f'_{cc}/f_{co}	1.02	6.8

the study presented by Ozbakkaloglu and Lim (2013) included five FRP types (CFRP, GFRP, AFRP, HM CFRP, and UHM CFRP). Therefore, the verification database is comprehensive, and the distribution of some parameters in the database is summarized in Table 4.2.

Note: f'_{co} is the peak strength of the unconfined concrete, f'_{cc} is the ultimate strength of the confined concrete, and f_l is the lateral confining pressure.

In order to verify the proposed models, the strength enhancement (f'_{cc} / f'_{co}) is used to compare. The stress comparisons between the model results and test results are summarized in Fig. 4.5, based on the experimental database. In addition, statistical indicators are used to quantify the accuracy of the proposed models. The statistical indicators are as follows: the average absolute error (E), which can be computed by Eq. (4.18a) and directly reflect the absolute error; the standard deviation (SD), which can be computed by Eq. (4.18b) and reflect the magnitude of the associated dispersion of the model results; the correlation coefficient (R^2), which can reflect the similarity of the data results; the slope (S) of the linear fitting curve (in a format of $y = Sx$) to reflect the average fitting degree of the proposed model.

$$E = \left(\sum_{1}^{n} \left| (m_i - e_i)/e_i \right| \right) / n \tag{4.18a}$$

$$SD = \sqrt{\left(\sum_{1}^{n} \left(m_i/e_i - m_{avg}/e_{avg}{}^2 \right)/(n-1) \right)} \tag{4.18b}$$

where m_i is the prediction result by the proposed model, e_i is the experimental results, m_{avg} is the average value of the model results, and e_{avg} is the average value of the test result.

It has been concluded that the strain reduction coefficient (k_ε) is important for the stress model of the FRP confined concrete and can compute the actual ultimate strain of the FRP tube. In order to verify the proposed stress model and the strain reduction coefficient equation, the two types of stress comparisons are set:

1. the comparisons of $Model_T$ which is calculated by the test data, based on the ultimate lateral strain measured by strain gauges in the experiment
2. the comparisons of $Model_K$ which is calculated by the strain reduction coefficient (Eq. 4.6) and the stress equation (Eq. 4.15).

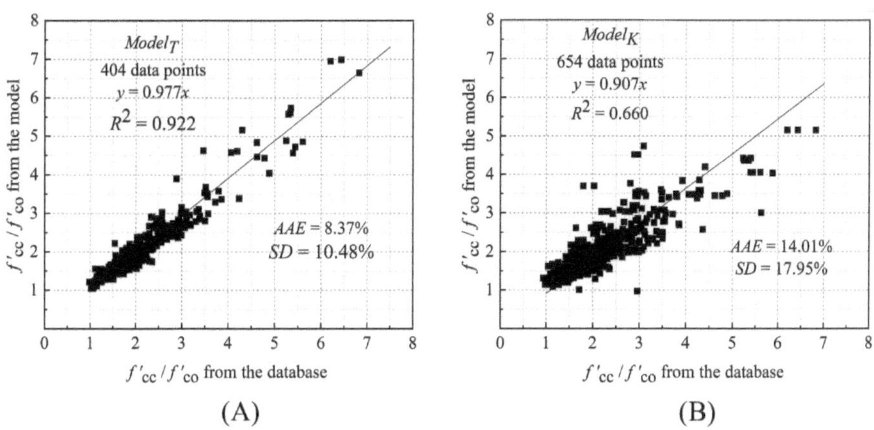

Figure 4.5 Comparison between model results and test results. (A) Strength enhancement ratios of $Model_K$; (B) strength enhancement ratios of $Model_T$.

The stress comparisons are summarized in Fig. 4.5. It is easily known that not all the data points of FRP confined concrete (a total of 550 data points are test results of the FRP confined concrete) in the database can provide actual rupture strain of the FRP tube. Therefore, the verification of $Model_T$includes 404 data points, where 300 data points are test results of the FRP confined concrete and 104 data points are test results of other confined concrete.

Based on the stress comparisons in Fig. 4.5, both $Model_T$ and $Model_K$ can provide an accurate prediction of axial stress for all types of confined concrete. However, the comparisons of the statistical indicators in Fig. 4.5 show that the prediction results of $Model_T$ are much more accurate than that of $Model_K$. Therefore, further analysis of the strain reduction coefficient (k_ε) should be presented in future studies.

In order to deeply analyze the accuracy of the proposed model, we select and verify five existing models (Keshtegar et al., 2017; Ozbakkaloglu & Lim, 2013; Pham & Hadi, 2014; Realfonzo & Napoli, 2011; Wu & Zhou, 2010) in the selected database, and we use all models as comparisons to the proposed model. To compare the validity of these models, we assemble the 404 data points in the verification of $Model_T$ as the database. The normalized prediction results of the five existing models and the proposed model are shown in Fig. 4.6. In addition, the statistical indicators of these models are presented in Fig. 4.7. Based on the parameter comparisons, the proposed model can provide accurate prediction results.

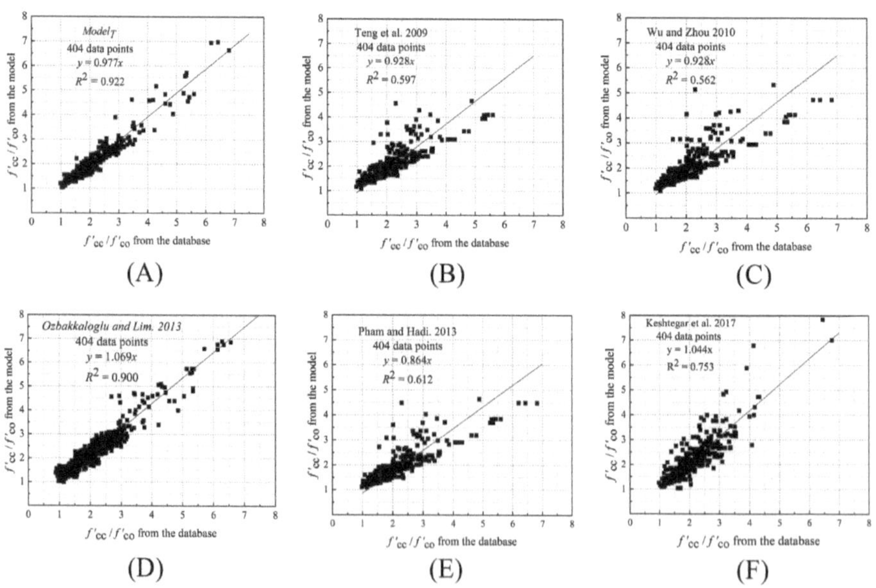

Figure 4.6 Comparison of the existing models and the proposed model. One proposed model and five existing models.

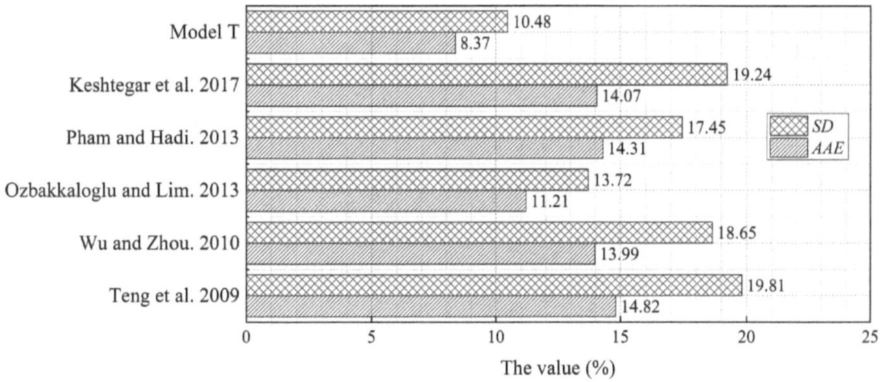

Figure 4.7 Accuracy of the existing models and the proposed model. The other comparisons.

4.4 Strain analysis of the confined concrete by the energy method

There are few studies focused on strain prediction of confined concrete. Especially, the study presented by Bisby et al. (2015) concluded that it poorly correlated the existing models with the experimental results of the strain of the confined concrete. It is necessary to find a reasonable strain analysis for the confined concrete. Therefore, we revised the existing models at first. Afterward, the section follows with the analysis of the strain state of the confined concrete. Finally, we establish the strain models and verify the proposed model by the confined concrete, which can establish the analysis-oriented constitutive model.

4.4.1 Existing strain models

Based on existing studies on the constitutive models of the confined concrete, the stress models have been extensively studied and can provide accurate prediction results, whereas the strain models (e.g., the axial strain prediction, the relationship between axial and lateral strains, the relationship between stress and strain) should be gained further attention and careful analysis.

Based on the study on actively confined concrete proposed by Richart et al. (1928), the relationship between the axial peak strain and the lateral confining pressure shows a linear trend, which has been the foundation for many theoretical studies. Based on the above assumption, some researchers (Chen et al., 2018; Li et al., 2009) present studies on the strain models of the confined concrete by a fitting method. In addition, the volume strain and dilation behavior of the confined concrete are analyzed by Lam and Teng (2003a) and Shehata et al. (2002). They introduced the corresponding parameters to establish the strain model for the FRP confined concrete, based on the polynomial fitting method. Most of the earlier existing studies focused on the behavior mechanism of confined concrete, and then proposed a strain estimation method by fitting experimental results.

Except for the above method, the energy-balance method is also used to analyze the strain state of confined concrete. For the energy-balance condition of the confined concrete column, it can assume that the applied load can produce the additional axial strain energy of the confined concrete column, which is equal to the addition lateral strain energy of the confining device produced by the concrete extension. Mander et al. (1988) first introduces the energy-balance method into the strain analysis and establishes the peak strain model of the hoops confined concrete. However, there will be an energy loss in the transfer process from the axial strain energy to the lateral strain energy, based on the study by Saadatmanesh et al. (1994). Based on the assumption, the study by Pham and Hadi (2014) proposes a strain estimation for the FRP confined concrete.

It is clearly known that the strain energy can be computed by multiplying the corresponding stress and strain, which can reflect the relationship between stress and strain. Therefore, the energy-balance method is not only used to establish the strain models, but also applied in characteristic strain analysis and constitutive model establishment for confined concrete. We will study the strain state analysis for the actively and passively confined concrete in detail by the energy-balance method in the following sections.

4.4.2 Strain state analysis by the energy-balance method

Before the strain energy analysis, the basic analysis parameters should be defined, including the axial strain energy and the lateral strain energy. The axial strain energy of a concrete member is the energy absorbed from the axial external load. And we can compute the axial strain energy by the load–displacement curve of the concrete column, which is equal to the curve coverage area. Because the load–displacement curve of the concrete column shows a nonlinear developing trend, the axial strain energy can be calculated by the curve integral, the mathematical expression as follows:

$$Q = W = \int_0^L F dL \tag{4.19}$$

where Q is the axial strain energy; W is the working energy of the applied load; F is the applied load; L is the displacement of the concrete column; and dL is an increment of the column displacement.

With the confining effect, the confined concrete column has a good load-bearing capacity and good ductility than the unconfined concrete column. It means the lateral confining pressure can increase the axial strain energy of the confined concrete column. At the same time, the lateral confining component can absorb the strain energy by the lateral expansion of the confined concrete column. The energy-balance method proposed by Mander et al. (1988) assumed that the axial strain energy was equal to the lateral strain energy. However, some studies (Lu & Yu, 2004; Xiao et al., 2010) confirm that the direct balance relationship is too sketchy to establish the strain model for the confined concrete, because of two reasons: the energy transformation is a complex process with the energy dissipation; the lateral component is not the unique material to store energy and the energy stored in core concrete can be not ignored. Therefore, only if the strain energy in the confined concrete column has been identified and analyzed clearly, the reasonable energy-balance analysis method can be proposed for the strain state analysis and corresponding model establishment.

It can be understood that we can translate the working energy of the applied load into three parts:

(1) the first part is the axial strain energy stored in the core concrete, which leads to the lateral expansion and complex mechanical state
(2) the second part is the lateral strain energy stored in the lateral confining component, which is caused by the lateral expansion of the core concrete
(3) the third part is the loss energy, including the plastic strain energy and the energy dissipation in two transformation processes (i.e., one is from the working energy to the axial strain energy, another is from the axial strain energy to the lateral strain energy).

However, we just should analyze the energy dissipation from the axial strain energy to the lateral strain energy, because other kinds of loss energy have been covered in the integral process of the load–displacement curve. Based on the above analysis, we would study the strain energy of the confined concrete column in the following section.

4.4.3 Strain energy in the confined concrete column

In order to study deeply the strain energy, we should study deeply the stress–strain curves of all types of concrete, including the unconfined concrete, the actively confined concrete, and the FRP confined concrete under the different confining effects. The experimental data can be found in Fig. 4.1. In order to analyze the strain energy easily, we should define the characteristic points first. For the stress–strain curve of the unconfined concrete shown in Fig. 4.1, the characteristic points are given as follows: f_{co} is the axial peak stress of the unconfined concrete, ε_{co} is the axial peak strain corresponding to the axial peak stress, $f_{co,res}$ is the residual stress in the softening branch of the unconfined concrete, and $\varepsilon_{co,res}$ is the residual strain corresponding to the residual stress. For the stress–strain curves of the actively confined concrete shown in Fig. 4.1, the characteristic points are given as follows: f_{cc} is the axial peak stress of the actively confined concrete, ε_{cc} is the axial peak strain corresponding to the axial peak stress, $f_{cc,res}$ is the residual stress in the softening branch of actively confined concrete, and $\varepsilon_{cc,res}$ is the residual strain corresponding to the residual stress. In addition, we will discuss the residual stress and the residual strain of the unconfined or actively confined concrete in the section about the strain energy analysis. For the stress–strain curves of the FRP confined concrete without a softening branch shown in Fig. 4.1, the second half stage is a linear increasing trend which is recommended by ACI 440.2R-08 (2008). For this type of FRP confined concrete, the characteristic points are mentioned as follows: f_{cc} is the axial ultimate stress when the FRP tube ruptures, ε_{cc} is the axial ultimate strain corresponding to the ultimate stress, f_{ce} is the demarcation stress at the point between first and second increasing stages, and ε_{ce} is the demarcation strain corresponding to the demarcation stress.

4.4.3.1 Axial strain energy

Based on the above-defined characteristic points in the stress−strain curves, the strain energy can be computed and analyzed. For the unconfined concrete and the actively confined concrete, the peak strain energy, which corresponds to the energy of the first half part in the stress−strain curve, can be computed in the following equation:

$$Q_{co,fir}/Q_{cc,fir} = A_c H \int_0^{\varepsilon_{co}/\varepsilon_{cc}} f_c d\varepsilon_c \tag{4.20}$$

where $Q_{co,fir}/\ Q_{cc,fir}$ is the strain energy of the concrete column in the increasing stage, A_c is the section area of the concrete column, H is the height of the concrete column, f_c is the stress of the concrete column in the increasing stage, and $d\varepsilon_c$ is the micro-increment of strain in the concrete column.

Then the residual strain energy is computed from the peak point to the residual point. It can be computed as follows:

$$Q_{c,res} = A_c H \int_{\varepsilon_{cc}}^{\varepsilon_{co,res}/\varepsilon_{cc,res}} f_c d\varepsilon_c \tag{4.21}$$

where $Q_{c,res}/\ Q_{co,res}$ is the residual strain energy of unconfined concrete in the softening stage.

In addition, the selection of the residual point is too complex to determine in this section, we can select it after strain energy analysis.

For the FRP confined concrete, the mechanical properties of the core concrete are improved by confining effect. It can be found that the increase of the strength and ductility in the initial increasing stage of the stress−strain curve depends on the unconfined concrete properties and the lateral confining effect, which is similar to the actively confined concrete. For the second increasing stage of the stress−strain curve, the strength increase and the curve trend mainly depend on the lateral confining pressure provided by the FRP tube. If the lateral confining pressure provided by the FRP tube is low, the demarcation stress (f_{ce}) is approximately equal to the peak stress (f_{co}) of the unconfined concrete, and the second half part of the stress−strain curve shows a decreasing trend (namely softening stage), as shown in the stress−-strain curve of the GFRP confined concrete (1-ply) in Fig. 4.2. If the lateral confining pressure provided by the FRP tube is high, the demarcation stress (f_{ce}) is higher than the peak stress (f_{co}) of the unconfined concrete, and the second half part of the stress–strain curve shows a linear increasing trend (namely hardening stage), as shown in the stress–strain curve of the GFRP confined concrete (3-ply, 5-ply) in Fig. 4.2 and the CFRP confined concrete in Fig. 4.1. Therefore, it is important to determine the demarcation point between the first increasing stage and the second increasing stage of the stress–strain curve of the FRP confined concrete with the high confining pressure.

Based on the above analysis, the strain energy of the first increasing stage of the stress–strain curve can be computed as follows:

$$Q_{ce} = A_c H \int_0^{\varepsilon_{ce}} f_c d\varepsilon_c \tag{4.22}$$

where Q_{ce} is the strain energy of the first increasing stage in the FRP confined concrete column, ε_{ce} is the demarcation strain corresponding to the demarcation stress (f_{ce}).

For FRP confined concrete with high confined pressure, we can modify the second increasing stage of the stress–strain curve to get a simple integration, due to the linear increasing trend, then the strain energy can be computed as follows:

$$Q_c s = A_c H \int_{\varepsilon_{ce}}^{\varepsilon_{cc}} f_c d\varepsilon_c = (A_c H(\varepsilon_{cc} - \varepsilon_{ce})(f_{cc} + f_{ce}))/2 \tag{4.23}$$

where Q_{cs} is the strain energy of the first increasing stage of the FRP confined concrete column.

4.4.3.2 Lateral strain energy

For the actively confined concrete, the lateral confining pressure (f_l) is a constant value. Therefore, the strain energy is controlled by the lateral expansion strain of the core concrete. Then the strain energy absorbed by the lateral confining component can be computed directly as follows:

$$Q_{la} = F_l L_l = RHf_l \times 2R\pi\varepsilon_l = 2A_c H f_l \varepsilon_l \tag{4.24}$$

where Q_{la} is the strain energy absorbed by the lateral confining component, F_l is the lateral force provided by the confining component, L_l is the lateral displacement around the core concrete, R is the radius of the core concrete section, H is the height of the FRP confined concrete column, f_l is the confining pressure around the core concrete and ε_l is the lateral strain of the core concrete, and A_c is the section of the core concrete.

For the FRP confined concrete, the lateral confining pressure (f_l) increases with the increase of the lateral expansion of the core concrete, because of the linear elastic behaviors of the FRP tube. Therefore, the strain energy is also controlled by the lateral expansion strain of the core concrete. Then the strain energy absorbed by the FRP tube can be computed as follows:

$$Q_{lfrp} = (RHf_l \times D\pi\varepsilon_l)/2 = E_{frp}HtR\pi\varepsilon_l^2 \quad (4.25)$$

where $Q_{l,frp}$ is the strain energy absorbed by FRP tube and E_{frp} is the elastic modulus of FRP tube.

In summary, the above equations (from Eqs. 4.20 to 4.25) can compute the energy absorbed by the concrete column and the lateral confining component. Then we would discuss the relationship between these strain energy parameters in the following section.

4.4.4 Strain energy analysis of the confined concrete

As we discussed in the above section, the direct energy-balance method (i.e., using the relationship between axial strain energy and lateral strain energy) is not suitable for confined concrete, because of the following reasons:

(1) a part of strain energy absorbed by concrete column
(2) a part of strain energy will lose in the transformation process from the axial strain energy to the lateral strain energy. Therefore, we introduce the energy transformation coefficient (k) to reflect the relationship between these energy parameters.

4.4.4.1 Strain energy analysis for the actively confined concrete

For the actively confined concrete, the three types of strain energy (i.e., $Q_{co,fir}$, $Q_{co,res}$, and Q_{la}) should be studied in detail. In order to gain the actual axial strain, the strain energy of the core concrete should be eliminated from the total axial strain energy, including $Q_{co,fir}$ and $Q_{co,res}$. Then the energy transformation coefficient (k) in the increasing stage of the actively confined concrete can be computed as follows:

$$k_{a,fir} = (Q_{cc,fir} - Q_{co,fir})/Q_{la,cc} \quad (4.26)$$

where $k_{a,fir}$ is the energy transformation coefficient in the increasing stage, $Q_{cc,fir}$ is the strain energy in the increasing stage of the actively confined concrete, $Q_{co,fir}$ is the strain energy in the increasing stage of unconfined concrete, and $Q_{la,cc}$ is the lateral strain energy before the peak stress of actively confined concrete.

Referring to the axial strain energy in the increasing stage, the axial strain energy in the softening stage should remove the strain energy stored in the core concrete. Then the energy transformation coefficient (k) in the softening stage of the actively confined concrete can be computed as follows:

$$k_{a,res} = (Q_{cc,res} - Q_{co,res})/Q_{la,res} \quad (4.27)$$

where $k_{a,res}$ is the energy transformation coefficient in the softening stage, $Q_{cc,res}$ is the residual strain energy in the softening stage of the actively confined concrete, $Q_{co,res}$ is the residual strain energy in the softening stage of the unconfined concrete, and $Q_{la,res}$ is the strain energy after the peak stress of the actively confined concrete.

The stress–strain curves of the actively confined concrete in Fig. 4.1 are used to analyze the energy transformation coefficient. For three types of lateral pressure (i.e., 4, 8, and 12 MPa), we can compute the values of two types of energy transformation coefficients (i.e., $k_{a,fir}$ and $k_{a,res}$) by Eqs. (4.26) and (4.27), then summarize the results in Table 4.3.

The comparisons in Table 4.3 clearly show that $k_{a,fir}$ is greatly larger than $k_{a,res}$, which means that the energy loss is large before the peak stress, and the transformation efficiency is improved once the concrete is in the softening stage.

For unconfined concrete and actively confined concrete, the axial stress decreases after the peak point, which means that the strain energy is dissipated with the damage to the core concrete. In order to study the strain energy of concrete in the softening stage, the dissipative energy is determined for the softening stage of stress-strain curves. For the energy comparisons between the unconfined concrete and the actively confined concrete, we can calculate the dissipative energy from the peak point to the appointed point. The appointed strain is 1.3 times larger than the peak point. The calculation assumption is that the axial stress is still constant, then the extra energy can be defined as the dissipative energy. Therefore, the numerical expression is given as follows:

$$Q_{dis,co} = A_c H f_{co}(\varepsilon_{1.3co} - \varepsilon_{co}) - Q_{co,res} \tag{4.28a}$$

$$Q_{dis,cc} = A_c H f_{cc}(\varepsilon_{1.3cc} - \varepsilon_{cc}) - Q_{cc,res} \tag{4.28b}$$

where $Q_{dis,co}$ is the dissipative energy of the unconfined concrete, and $Q_{dis,cc}$ is the dissipative energy of the actively confined concrete, $\varepsilon_{1.3co}$ is 1.3 times greater than ε_{co}, and $\varepsilon_{1.3cc}$ is 1.3 times greater than ε_{cc}.

Based on the computation results, the dissipative energies of the unconfined concrete and the actively confined concrete are all similar. The particular collusion shows that the dissipative energy can be used to calculate the trend of the stress—-strain curve in the softening stage. Moreover, the dissipative energy can study the softening stage of the FRP confined concrete with small confining pressure.

Table 4.3 The values of energy transformation coefficient of actively confined concrete.

Lateral stress	**4 MPa**	**8 MPa**	**12 MPa**
$k_{a,fir}$	7.9	8.4	7.5
$k_{a,res}$	4.4	3.6	4.5

4.4.4.2 Strain energy analysis for the FRP confined concrete

For the FRP confined concrete, the three types of strain energy (i.e., Q_{ce}, Q_{cs}, and Q_{lfrp}) should be studied in detail. Similar to the above section, the strain energy of the core concrete should also be eliminated. Then the energy transformation coefficient (k) in the increasing stage of the FRP confined concrete can be computed as follows:

$$k_{ce} = (Q_{ce} - Q_{co,fir})/Q_{lfrp,cc} \tag{4.29}$$

where k_{ce} is the transformation factor of the first stage, Q_{ce} is the strain energy of the first stage of the FRP confined concrete, $Q_{co,fir}$ is the strain energy of the first stage of the unconfined concrete, and $Q_{lfrp,cc}$ is the lateral strain energy before the demarcation point of the FRP confined concrete.

For the FRP confined concrete with large confining pressure, the stress-strain curve shows a bilinear increasing trend without a softening stage. Therefore, we can assume that the axial strain energy is totally transferred into the lateral strain energy in the FRP tube. Then the energy transformation coefficient (k) in the hardening stage can be computed as follows:

$$k_{cs} = Q_{cs}/Q_{lfrp,s} \tag{4.30}$$

where k_{cs} is the energy transformation coefficient in the hardening stage, Q_{cs} is the strain energy in the hardening stage of the FRP confined concrete, and $Q_{lfrp,s}$ is the strain energy after the demarcation point of the FRP confined concrete.

The stress–strain curves of the FRP confined concrete in Fig. 4.1 are used to analyze the energy transformation coefficient. For the three types of CFRP tube (i.e., 1-ply, 2-ply, and 3-ply), we can compute the values of two types of the energy transformation coefficients (i.e., k_{ce} and k_{cs}) by Eqs. (4.29) and (4.30), then summarize the results in Table 4.4.

The comparisons in Table 4.4 clearly show that $k_{a,fir}$ is almost equal to $k_{a,res}$, which means that the characteristics of the stress and strain state also remain unchanged in the hardening stage, the energy transformation coefficient is similar to that of the initial increasing stage. Furthermore, the conclusion can prove the load-path independence of the stress development in concrete without a softening stage.

Table 4.4 The values of energy transformation coefficient of FRP confined concrete.

CFRP tube type	1-ply	2-ply	3-ply
K_{ce}	8	8.5	9.2
K_{cs}	7.8	8.3	8.4

Based on the coefficient comparisons between the actively confined concrete (see Table 4.3) and the FRP confined concrete (see Table 4.4), the following conclusions can be drawn:

(1) the strain energy continuously increases in the hardening stage, due to the linear increasing curve
(2) except for the softening stage, the energy transformation coefficients of all concrete (i.e., the unconfined concrete, the two kinds of confined concrete) are similar, which means that the relationship between the strain energy absorbed by column totally transfer into the FRP tube and the transformation efficiency is unchanged
(3) the energy transformation coefficient can be averaged by Tables 4.3 and 4.4, then it can establish the strain model of the peak strain of the actively confined concrete and the demarcation strain and the hardening strain of the FRP confined concrete with large confining pressure
(4) if the FRP confined concrete with small confining pressure, the stress–strain curve has a softening stage, the dissipative energy of the actively confined concrete can study the softening stage in the following section.

4.5 Strain model for the confined concrete

There are many characteristics between the actively confined concrete and the passively confined concrete. Therefore, we will establish the strain models for the two types of the confined concrete, separately. For the actively confined concrete, the peak strain is the unique parameter. For the passively confined concrete (i.e., the FRP confined concrete), we should consider the following characteristic points:

(1) the demarcation point between the two linear increasing stages
(2) the ultimate strain at the FRP tube rupture
(3) the discriminative state of the confining pressure, which decides the softening stage of the stress–strain curves of FRP confined concrete.

In addition, the conclusions in Section 4.4 show that there is a linear relationship between the axial strain energy absorbed by the concrete column and the lateral strain energy absorbed by the lateral confining component. Furthermore, the value of the energy transformation coefficient can be gained by regression analysis, based on the test results in this chapter and the studies (Ozbakkaloglu & Lim, 2013; Pham & Hadi, 2014), which find that $k = 8.2$.

4.5.1 Strain model of the actively confined concrete

The actively confined concrete is much simpler than passively confined concrete, which just has a peak point as the characteristic point. To establish the strain model at the peak point by strain energy analysis, we should define the constitutive model of unconfined concrete first. Based on many existing studies (Lam & Teng, 2003a) the stress–strain curves can be divided into two parts: the increasing stage and the decreasing stage, and the constitutive model can be expressed into two equations. The first part of the constitutive model can be expressed in the following equation:

$$f_c = f_{cc}[\alpha\varepsilon_c/\varepsilon_{cc} - (3-2\alpha)(\varepsilon_c/\varepsilon_{cc})^2 + (\alpha-2)(\varepsilon_c/\varepsilon_{cc})^3] \tag{4.31}$$

where α is the parameter of increasing stage, $\alpha = 2.4 - 0.0125f_{cc}$.

In order to simplify the solving method, the parameter (α) value is 2, based on the study by Teng et al. (2009). Therefore, the strain energy can be computed as follows:

$$Q_{fir} = A_c H \int_0^{\varepsilon_{cc}} f_c d\varepsilon_c = 2A_c H f_{cc}\varepsilon_{cc}/3 \tag{4.32}$$

Then the peak strain can be gained by Eq. (4.25), the corresponding parameters can be computed by Eqs. (4.19), (4.24), and (4.31), then the equation can be expressed as follows:

$$\varepsilon_{cc} = (3kf_l\varepsilon_{l,cc} + f_{co}\varepsilon_{co})/f_{cc} \tag{4.33}$$

where k is the energy transformation coefficient, the value is 8.2, $\varepsilon_{l,cc}$ is the lateral strain of the confined concrete at the peak stress.

It is clearly shown that the lateral strain is an unknown parameter in Eq. (4.32), but it can be calculated by the peak stress of the confined concrete. The studies by Samani and Attard (2014) and Piscesa et al. (2016) have been presented on actively confined concrete. The results show that the relationship between the axial strain and the lateral strain relates to the lateral confining pressure and the strength of the unconfined concrete. Based on the theoretical analysis, the relationship can be expressed as follows:

$$\varepsilon_{l,cc}/\varepsilon_{cc} = n = \left\{2.5 - 2\tanh[2.5(f_l/f_{co})^{0.0915\sqrt{f_{co}}}]\right\}\alpha \tag{4.34}$$

where n is the parameter to reflect the ratio of the lateral strain to the axial strain in the confined concrete, and α is the ratio of the plastic strain to the peak strain.

In the studies by Samani and Attard (2014) and Piscesa et al. (2016), the lateral strain is caused by the plastic strain, so it can be calculated by the equation: $\alpha = 1 - E_c/E_{co}$, where E_c is the secant modulus and E_{co} is the elastic modulus. In addition, the value of α generally varies between 0.25 and 0.4. Based on the above analysis, the peak strain of the actively confined concrete can be expressed as follows:

$$\varepsilon_{cc} = (f_{co}\varepsilon_{co})/(f_{cc} - 3knf_l) \tag{4.35}$$

4.5.2 Strain model of the FRP confined concrete

In order to establish the constitutive model of the FRP confined concrete, the following characteristic strain state should be studied: the demarcation point between the two linear increasing stages of the stress–strain curve, which determines the critical strain state at end of the initial stage; the demarcation confining condition, which determines the developing trend of the curve after the demarcation point; the ultimate point, which determines the ultimate strain state at rupture when the FRP tube break occurs. Based on the above strain states, the second part of the constitutive model of the FRP confined concrete can be further established.

4.5.2.1 Demarcation strain of the FRP confined concrete

The demarcation point is the starting point in the second stage of the stress–strain curve. Based on the analysis in Section 4.4, the stress increase in the second increasing stage of the FRP confined concrete mainly depends on the lateral confining pressure. Therefore, we can analyze the moment strain energy transformation between the axial strain energy and the lateral strain energy by Eq. (4.30). Then we can calculate the energy balance between the composite column and the FRP tube at the demarcation point in the following equation:

$$A_c H f_{ce} d\varepsilon_3 = kRHf_{le}D\pi d\varepsilon_1 \tag{4.36}$$

where f_{ce} is the axial stress of core concrete at the demarcation point, f_{le} is the lateral pressure provided by the FRP tube at the demarcation point, $d\varepsilon_3$ is the moment addition strain in the axial direction, and $d\varepsilon_1$ is the moment addition strain in the lateral direction. In addition, the moment addition stress of f_{ce} and f_{le} is too small to ignore in the energy-balance analysis.

Based on the proposed stress model (Eq. 4.15), Eq. (4.36) can be transformed into the following equation:

$$(d\varepsilon_1)/(d\varepsilon_3) = (f_{le} + a + \sqrt{bf_{le} + c})/(2kf_{le}) = \beta \tag{4.37}$$

where β is the moment strain ratio of $d\varepsilon_1$ to $d\varepsilon_3$ of the FRP confined concrete.

We can calculate the moment strain ratio (β) by the model proposed by Piscesa et al. (2016), which differs from Eq. (4.34). Then, Eq. (4.37) can be expressed in another form, which is as follows:

$$\beta = 2.5 - (2 + 0.0275\phi)tanh[q(f_{le}/f_{co})^p] \tag{4.38}$$

$$q = 2.5 + 0.25\sqrt{E_L/f_{co}}p = 0.0915\sqrt{f_{co}}exp[-0.00192\phi(f_{co} - 73)]$$

$$\phi = \sqrt{f_{le}E_L/f_{co}E_L} = t_{frp}E_{frp}/R$$

The lateral strain at the demarcation point of the FRP tube can be calculated by Eq. (4.38). In order to gain a clear mathematical equation of the lateral strain, the corresponding equations should be combined, including Eqs. (4.15), (4.22), (4.25),

(4.29), (4.32), and (4.38). Then, the equation of the lateral strain can be expressed as follows:

$$\varepsilon_{ce} = (3kE_{frp}t\varepsilon_l e^2/R + 2f_{co}\varepsilon_{co})/(2f_{co} + 7f_{le}) \tag{4.39}$$

4.5.2.2 The discriminative confining state

For FRP confined concrete, it is well known that the developing trend of the second part of the stress–strain curve depends on the value of strain energy absorbed by the FRP tube. Thus, the discriminative confining condition of the FRP tube is the key to the judgment of the curve trend. Based on the strain state analysis in Section 4.4, the axial strain energy completely transfers to the lateral strain energy in the second increasing stage. Therefore, it can be assumed that if the FRP tube doesn't absorb enough lateral strain energy, there will be a softening curve shape. The balanced relationship could be the discriminative confining state. Meanwhile, we have proved that the demarcation point is equal to the peak point of the unconfined concrete, under low confining pressure. Based on the above analysis, the discriminative confining state can be expressed as follows:

$$f_{co}/(2kf_{ld}) = \beta \tag{4.40}$$

where f_{ld} is the lateral confining pressure at the demarcation point, it can be calculated by Eqs. (4.3) and (4.34).

Based on further analysis, Eqs. (4.33) and (4.40) can be combined to gain the strain state. Then we introduce the confining coefficient (ρ_d) to represent the judgment. If the stress–strain curve of the FRP confined concrete doesn't have a softening stage, the confined coefficient ρ_d should satisfy the following condition:

$$\rho_d = E_{frp}t/(E_cR) \geq 0.03 \tag{4.41}$$

where ρ_d is the confining coefficient of the FRP confined concrete, E_c is the secant modulus at the compressive point of the unconfined concrete, $E_c = f_{co}/\varepsilon_{co}$.

4.5.2.3 Ultimate strain of the FRP confined concrete

In order to simplify the computation method, we select the demarcation point as the ultimate point of the confined concrete, based on the strain state analysis in Section 4.4. Therefore, the strain energies absorbed by the core concrete and the FRP tube should be removed from the total strain energies. Therefore, the corresponding equations should be combined to gain the ultimate strain of the FRP confined concrete, including Eqs. (4.3), (4.23), (4.25), (4.30), and (4.32). Then the equation of the ultimate strain can be expressed as follows:

$$\varepsilon_{cc} = \varepsilon_{co} + (6kE_{frp}t\varepsilon_{l,rup})^2/R + 4f_{co}\varepsilon_{co})/(3f_c) \tag{4.42}$$

where, $\varepsilon_{l,rup}$ is the rupture strain of the FRP tube.

4.5.2.4 Verification of the proposed models

In order to verify the proposed models, we compare the test results with the computation results by the proposed models. Furthermore, the statistical indicators are used to quantify the accuracy of the models, including the average absolute error (E) which can be computed by Eq. (4.18a), standard deviation (SD) which can be computed by Eq. (4.18b), and the correlation coefficient (R^2) which is used to reflect the similarity of the results. The corresponding strain state verifications will be shown in the following sections.

Verification of the actively confined concrete

It can be found that both Eqs. (4.33) and (4.35) can provide the prediction results for the peak strain of actively confined concrete. The differences between the two equations are the value method of the lateral strain. Therefore, in order to analyze the accuracy of the proposed models, the two strain comparisons are presented:

(1) the peak strain calculated using Eq. (4.33) and the lateral strain is gained from the experimental data

(2) the peak strain calculated directly using Eq. (4.35). In addition, the experimental database was collected by Ozbakkaloglu and Lim (2013).

The verifications of the proposed models in Eqs. (4.33) and (4.35) are shown in Fig. 4.8. The verification of the proposed model in Eq. (4.35) shows 61 data points of the normalized predicted strain versus the experimental results. And the verification of the proposed model in Eq. (4.35) shows 80 data points. It is easily found that the data points in Fig. 4.8A are less than that in Fig. 4.8B because not all the experimental data can provide the lateral strain.

As shown by the statistical indicators in Fig. 4.8, the prediction results provided by Eq. (4.33) are more accurate than those provided by Eq. (4.35). However, the prediction results provided by Eq. (4.35) also have enough accuracy. It suggests

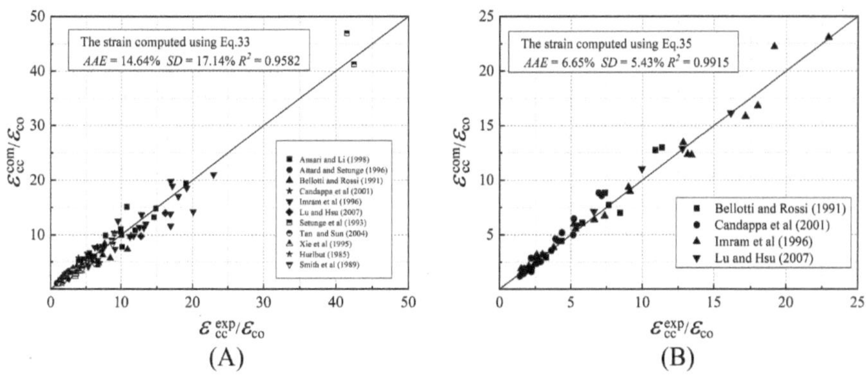

Figure 4.8 Verification of the actively confined concretes. (A) Prediction results provided by Eq. (4.33); (B) prediction results provided by Eq. (4.35).

that the strain estimation can be calculated directly without the lateral strain by Eq. (4.35).

Verification of the FRP confined concrete

We establish three models to estimate the strain state of the FRP confined concrete in the above section. Each model will be verified by a respective method in the following sections.

(1) The demarcation strain between two increasing stages

Based on the above strain state analysis, the lateral strain, axial stress, and axial strain at the demarcation point can be calculated by using Eqs. (4.38) and (4.39). And the experiment results can be gained from the stress–strain curves reported in the studies presented by Xiao and Wu (2003). Then, 12 data points of the normalized prediction strain versus the experiment results are shown in Table 4.5. Based on the data comparisons, we can find that the proposed models can provide accurate prediction results for the FRP confined concrete.

(2) The discriminative confining condition

In order to verify the discriminative confining state, the experiment results can be gained from the studies presented by Xiao and Wu (2003) and Shin and Andrawes (2010). Then, 18 data points of the prediction results versus the experiment results are shown in Table 4.6. Based on the data comparisons, we can find that the proposed models can provide accurate predictions for the developing trends of the second stage of the FRP confined concrete.

Table 4.5 Verification for the demarcation point of the FRP confined concrete.

Source of data	f_{ce}^{exp}/f_{co}	f_{ce}^{com}/f_{co}	$\varepsilon_{ce}^{exp}/\varepsilon_{co}$	$\varepsilon_{ce}^{com}/\varepsilon_{co}$	$\varepsilon_{le}^{exp}/\varepsilon_{co}$	$\varepsilon_{le}^{com}/\varepsilon_{co}$
Teng	1.043	1.127	1.401	1.354	1.192	1.18
Teng	1.114	1.136	1.27	1.175	0.64	0.699
Teng	1.211	1.167	1.191	1.127	0.484	0.536
Xiao	1.14	1.139	1.277	1.321	0.766	1.035
Xiao	1.215	1.151	1.191	1.128	0.511	0.562
Xiao	1.218	1.161	1.064	1.056	0.426	0.399
Xiao	1.204	1.111	1.226	1.233	0.821	0.946
Xiao	1.261	1.116	1.222	1.076	0.43	0.494
Xiao	1.273	1.12	1.168	1.021	0.312	0.34
Xiao	1.137	1.098	1.322	1.224	0.894	0.998
Xiao	1.21	1.102	1.186	1.076	0.569	0.518
Xiao	1.284	1.103	1.153	1.022	0.356	0.35
AAE	6.78%		6.57%		11.15%	
SD	6.36%		5.14%		11.79%	

Source: From Teng, J. G., & Lam, L. (2004). Behavior and modeling of fiber reinforced polymer-confined concrete. *Journal of Structural Engineering*, *130*(11), 1713–1723; Xiao, Y., & Wu, H. (2003). Compressive behavior of concrete confined by various types of FRP composite jackets. *Journal of Reinforced Plastics and Composites*, *22* (13), 1187–1201.

Table 4.6 Verification for the discriminative confining condition of FRP confined concrete.

Source of data	E_c (MPa)	$E_{frp}t/R$ (MPa)	ρ	Softening branch
Xiao and Wu (2003)	13,732	526	0.038	No
Xiao and Wu (2003)	13,732	1053	0.077	No
Xiao and Wu (2003)	13,732	1579	0.115	No
Xiao and Wu (2003)	15,688	526	0.034	No
Xiao and Wu (2003)	15,688	1053	0.067	No
Xiao and Wu (2003)	15,688	1579	0.101	No
Xiao and Wu (2003)	18,715	526	0.028	Yes
Xiao and Wu (2003)	18,715	1053	0.056	No
Xiao and Wu (2003)	18,715	1579	0.084	No
Shin and Andrawes (2010)	24,324	55	0.0023	Yes
Shin and Andrawes (2010)	24,324	110	0.0045	Yes
Shin and Andrawes (2010)	24,324	165	0.0068	Yes
Shin and Andrawes (2010)	24,324	220	0.009	Yes
Shin and Andrawes (2010)	24,324	275	0.0113	Yes
Shin and Andrawes (2010)	26,797	55	0.0021	Yes
Shin and Andrawes (2010)	26,797	110	0.0041	Yes
Shin and Andrawes (2010)	26,797	165	0.0062	Yes

Source: From Xiao, Y., & Wu, H. (2003). Compressive behavior of concrete confined by various types of FRP composite jackets. *Journal of Reinforced Plastics and Composites*, *22*(13), 1187–1201; Shin, M., & Andrawes, B. (2010). Experimental investigation of actively confined concrete using shape memory alloys. *Engineering Structures*, *32*(3), 656–664.

(3) The ultimate strain

Based on the model establishment, the ultimate strain of the FRP confined concrete can be calculated by using Eq. (4.42). In addition, a large experimental database collated by Ozbakkaloglu and Lim (2013) are used to verify the proposed model. The database has 214 data points and five types of FRP (i.e., CFRP, GFRP, AFRP, HM CFRP, and UHM CFRP). The comparison between the experimental results and the calculation results provided by the proposed model is shown in Fig. 4.9. The comparison results show that the proposed model can provide accurate prediction results for the ultimate strain of the FRP confined concrete.

4.6 Analysis-oriented constitutive model

Based on the behavior analysis of the confined concrete in Section 4.2, the stress–strain curves of the FRP confined concrete under a large confining effect show a bilinear increasing trend. And the stress–strain curves of other types of confined concrete have the softening stage, including the passively confined concrete under small confining effect and the actively confined concrete. Furthermore, the former confined concrete shows the load-path independence of the stress, the constitutive model could be established directly by the energy-balance method. The latter confined concrete can be predicted by the dissipative energy. Then the constitutive model can be established by the mechanical property of the unconfined concrete, the lateral confining pressure, and the ratio of lateral strain to axial strain.

4.6.1 Analysis-oriented constitutive model of the FRP confined concrete under large confining effect

For the FRP confined concrete under large confining effect, there is less damage in the core concrete and the axial stress continuously increases with the increase of the axial loading. In the loading process, the axial strain energy absorbed by the concrete column totally transforms into the lateral energy absorbed by the FRP tube. Therefore, the direct energy-balance method can establish the constitutive model. In addition, the strain prediction method is the key to the establishment of the constitutive model. Based on a given small increment in the lateral strain, we can calculate the axial strain by a single iteration, and the equation is given as follows:

$$\varepsilon_2 = \varepsilon_1 + 8\pi k E_{frp} t({\varepsilon_{l2}}^2 - {\varepsilon_{l1}}^2)/(f_1 + f_2) \tag{4.43}$$

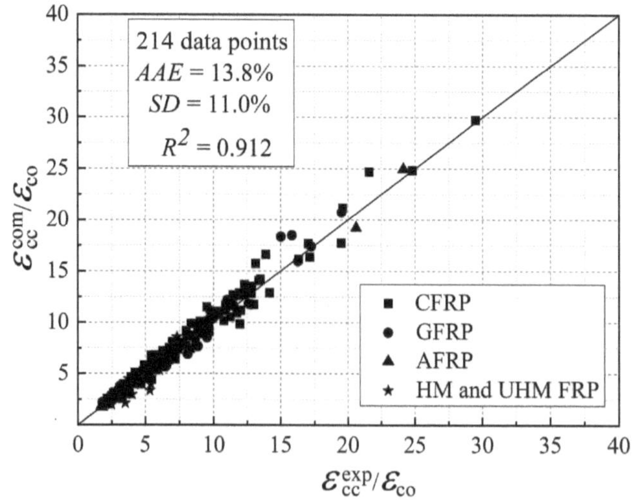

Figure 4.9 Verification of the ultimate strain of FRP confined concretes. The model verification.

where ε_1 is the axial strain at the previous step; ε_2 is the axial strain to be solved in the current step; f_1 and f_2 are axial stresses at previous and current steps, respectively, and they can be computed by the lateral strain; ε_{11} and ε_{12} are the lateral strains at previous and current steps respectively, we should set the small strain increment between the two lateral strains at first.

Base on above computation method, the analysis-oriented constitutive model for the FRP confined concrete under large confining effect can be expressed in following steps:

(1) based on the demarcation strain model, the demarcation stress and the demarcation strain can be computed by Eqs. (4.15) and (4.39), and the stress–strain curve before the demarcation point can be computed by Eqs. (4.15), (4.17), and (4.31), then stress–strain curve after the demarcation point can be computed by following simple iteration method
(2) a small increment of lateral strain is set, which can be 0.0001 after trialing, and the lateral pressure can be computed by the lateral strain increment and the properties of FRP tube
(3) the axial stress could be computed by the lateral pressure, based on Eqs. (4.15) and (4.16)
(4) the axial strain can be computed by Eq. (4.43)
(5) based on the computation axial strain, the next axial strain can be computed by the lateral strain increment and the corresponding equations, until the lateral strain leads to the rupture strain of the FRP tube.

In order to have a clear description of the circulated computation process, we draw each step of the circulated computation process in Fig. 4.10. It should be noted that the smoothing method should be used between the first stage and the second stage to make a smooth curve at the intersection point in stress–strain curves of the FRP confined concrete.

4.6.2 Analysis-oriented constitutive model of the confined concrete with a softening stage

It has been known that two types of confined concrete have the softening stage, namely the actively confined concrete and the passively confined concrete under a small confining effect. The damage to the core concrete is exacerbated after the demarcation point, which leads to a decrease in stress. Based on the strain state analysis in Section 4.4, the dissipative energies of different confined concrete in the softening stage are similar at the point whose strain is 1.3 times larger than the demarcation point. Therefore, the damage concrete can be regarded as a new lower strength concrete, and the softening

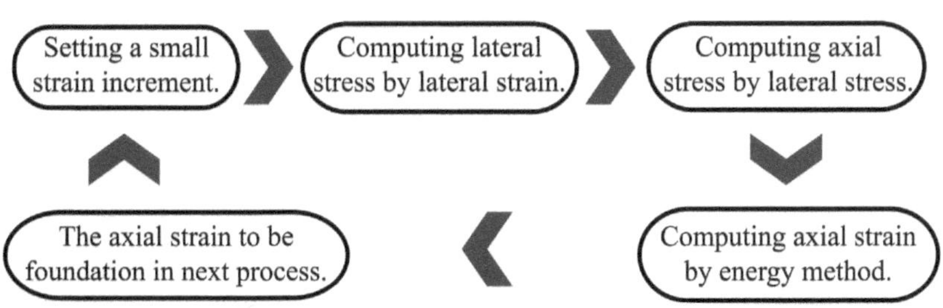

Figure 4.10 The circulated process of FRP confined concrete under a large confining effect. A flow chart.

curve is a process of the reduction of the failure surface. In this assumption, the concrete is changed into lower-strength concrete in the softening process, and the strain ratio of the strain to the peak strain of the assumption concrete is equal to that of the targeted confined concrete. Then the discriminative confining state can be reassessed by the properties of the assumption concrete and the properties of the FRP tube.

For the analysis-oriented constitutive model of the confined concrete, the stress–strain curves before the peak point can be computed by Eqs. (4.15), (4.17), and (4.31). After the demarcation point, the stress–strain curve shows a softening trend, it can be calculated in following steps:

(1) based on strain model in Section 4.5, the peak stress and peak strain can be computed by Eqs. (4.15) and (4.33), the stress–strain curves after the demarcation point can be computed by the next iteration method
(2) a small increment of lateral strain is set, which can be 0.0001 after trialing
(3) based on the study by Jian and Ozbakkaloglu (2015), the relationship between axial strain and lateral strain is gained (see Eq. 4.44), then the axial strain can be computed
(4) the strain ratio of the axial strain to the peak strain of the confined concrete is computed
(5) based on the stress–strain curves of the unconfined concrete (Keshtegar et al., 2017), above strain ratio is used to compute the peak stress of the assumption unconfined concrete, as shown in Eq. (4.45)
(6) based on the lateral confining pressure and the peak stress of the unconfined concrete, the axial stress can be computed by Eq. (4.15)
(7) based on the computation axial strain, a new strain increment can be reset for the next circulated process.

$$\begin{cases} \varepsilon_c = \dfrac{\varepsilon_l}{\upsilon_i\left[1+\left(\dfrac{\varepsilon_l}{\upsilon_i\varepsilon_{co}}\right)^n\right]^{1/n}} + 0.04\varepsilon_l^{0.7}\left[1+21\left(\dfrac{f_l}{f_{co}}\right)^{0.8}\right] \\ \upsilon_i = 8\times10^{-6}f_{co}^2 + 0.0002f_{co} + 0.138 \\ n = 1 + 0.03f_{co} \end{cases} \tag{4.44}$$

$$y = x/(\alpha_d(x-1)^2 + x) \tag{4.45}$$

where $y = \sigma_c/\sigma_{co}$, $x = \varepsilon_c/\varepsilon_{co}$.

In order to have a clear description of the circulated computation process, each step of the circulated process is drawn in Fig. 4.11. In addition, if the confining effect of passively confined concrete reaches equilibrium condition, based on Eq. (4.41), the increase of stress in the later curve will be too small to ignore. Then the later stress–strain curve can be assumed as a leveling line.

4.6.3 Verification of the proposed models

In order to verify the proposed models, the comparisons between computation results by the proposed models and experiment results of the actively confined

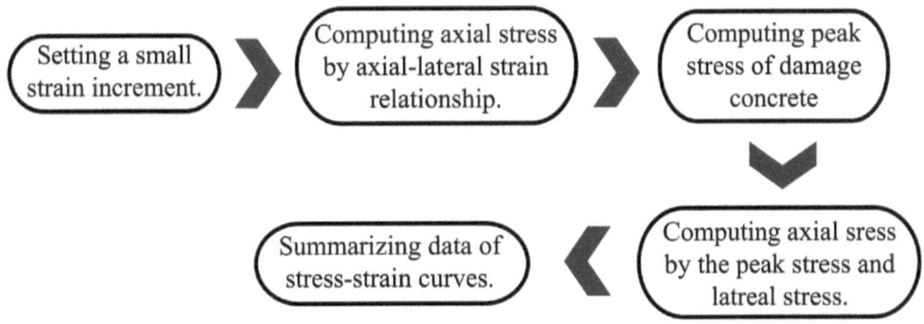

Figure 4.11 The circulated process of confined concrete with a softening stage. A flow chart.

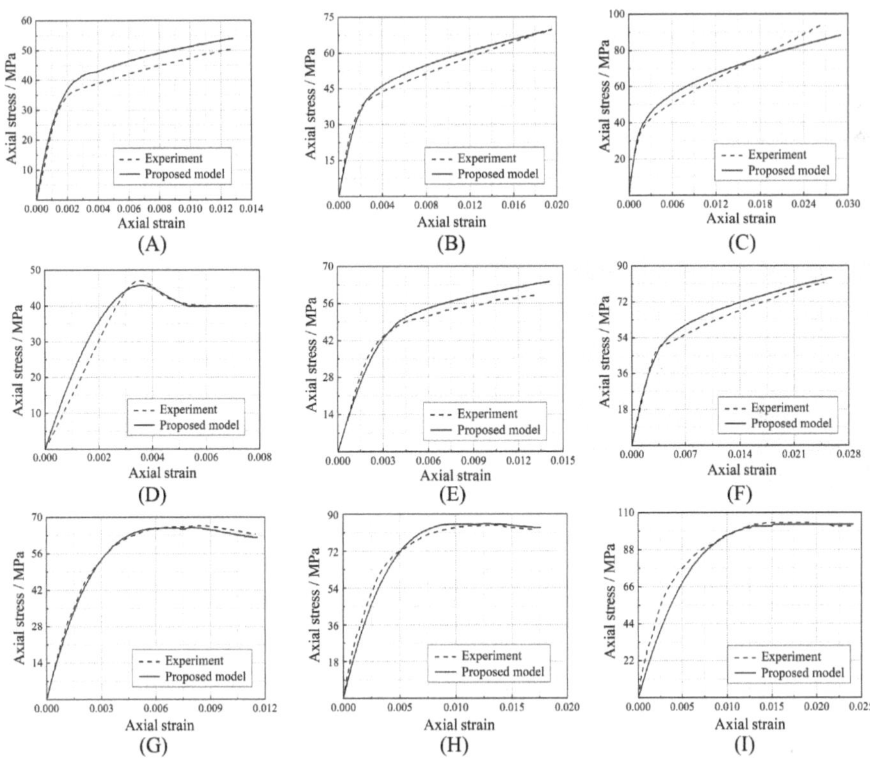

Figure 4.12 The comparison between computational results and experimental results. (A) CFRP-confined concrete (1-ply) in Fig. 4.1; (B) CFRP-confined concrete (2-ply) in Fig. 4.1; (C) CFRP-confined concrete (3-ply) in Fig. 4.1; (D) GFRP-confined concrete (1-ply) in Fig. 4.2; (E) GFRP-confined concrete (3-ply) in Fig. 4.2; (F) GFRP-confined concrete (5-ply) in Fig. 4.2; (G) actively confined concrete (4 MPa) in Fig. 4.1; (H) actively confined concrete (8 MPa) in Fig. 4.1; (I) actively confined concrete (12 MPa) in Fig. 4.1.

concrete and the FRP confined concrete are shown in Fig. 4.12. The curve comparison results show that the computation curves by proposed models are similar to the experiment curves. Therefore, their proposed analysis-oriented constitutive model can provide accurate prediction results for both the actively confined concrete and the passively confined concrete.

4.7 Conclusions

In this chapter, we study the behaviors of confined concrete in detail, and analyze the stress state and strain state of the actively confined concrete and the FRP confined concrete. Based on the existing experimental data of the actively confined concrete, the suitable failure criterion of the confined concrete is highlighted. Based on the theoretical analysis foundation and the mathematical derivation method, the stress model of confined concrete is established. Except for the stress analysis, a new strain energy-balance method is further developed for the strain state analysis. Based on the theoretical analysis, several models for the actively and passively confined concrete (i.e., FRP-wrapped) are proposed. Finally, we establish the analysis-oriented constitutive models by the combination between the stress model and the strain model. These models can provide accurate prediction results for both the actively confined concrete and the passively confined concrete. As above, the following conclusions can be drawn:

1. The stress properties between the active confined concrete and the passive confined concrete are similar. The hydrostatic stress is the key factor for the failure criterion of confined concrete, because of the low lateral pressure ratio ($f_l\,/\,f'_{co} < 2.3$).
2. The stress model is established by the proposed failure criterion and mathematical derivation, which have a high theoretical significance. Three coefficients in the model show a significant relationship with the unconfined concrete strength, which should be further studied.
3. The hoop rupture strain is the key part of the stress model. Therefore, the strain reduction coefficient (k_ε) is introduced to establish the stress model ($Model_K$), whereas the predictions of $Model_T$ based on the test hoop rupture strain show much accuracy. Therefore, a more accurate strain reduction equation needs to be studied.
4. Based on the new energy-balance method, the energy transformation coefficient (k) is introduced and analyzed in detail, the value is 8.2.
5. For actively confined concrete, the dissipative energy of the confined concrete relates to the unconfined concrete, which can be the foundation to study the softening stage of the confined concrete.
6. The demarcation point, the peak strain, and the discriminative confining condition of confined concrete are studied, and the corresponding models for these characteristic strain states are established and verified.
7. The analysis-oriented constitutive models are established and verified for both actively confined concrete and passively confined concrete. In the models, the stress model is based on the failure criterion of concrete under multiaxial compression, the strain model is based on the energy-balance method. Therefore, the constitutive models have the high theoretical establishment process.

Acknowledgments

This work was supported by the National Natural Science Foundation of China [grant numbers 52108171; 51878540]. The authors gratefully acknowledge two Master's students (Mr. Qv Yunsong and Mr. Tian Wenkai) of Xi'an University of Architecture and Technology for the drawing works.

References

ACI 440.2R-08. (2008). *Guide for the design and construction of externally bonded FRP systems for strengthening concrete structures.* Farmington Hills: American Concrete Institute (ACI).

Berthet, JF, Ferrier, E, & Hamelin, P (2006). Compressive behavior of concrete externally confined by composite jackets: Part B: Modeling. *Construction and Building Materials, 20*(5), 338–347.

Bisby, LA, Dent, AJS, & Green, MF (2015). Comparison of confinement models for fiber-reinforced polymer-wrapped concrete. *ACI Structural Journal, 102*(1), 62–72.

Candappa, DC, Sanjayan, JG, & Setunge, S (2001). Complete triaxial stress-strain curves of high-strength concrete. *Journal of Materials in Civil Engineering, 13*(3), 209–215.

Chen, P, Wang, Y, & Liu, C (2018). Confinement path-dependent analytical model for FRP-confined concrete and concrete-filled steel tube subjected to axial compression. *Composite Structures, 201*, 234–247.

Cui, C, & Sheikh, SA (2010). Analytical model for circular normal- and high-strength concrete columns confined with FRP. *Journal of Composites for Construction, 14*(5), 562–572.

Dong, CX, Kwan, AKH, & Ho, JCM (2015). A constitutive model for predicting the lateral strain of confined concrete. *Engineering Structures, 91*, 155–166.

Hoek, E, & Brown, ET (1980). Empirical strength criterion for rock masses. *Journal of Geotechnical and Geoenvironmental Engineering, 106*, 1013–1035.

Imran, I, & Pantazopoulou, S (1996). Experimental study of plain concrete under triaxial stress. *ACI Materials Journal, 93*(6), 589–601.

Jian, CL, & Ozbakkaloglu, T (2015). Lateral strain-to-axial strain relationship of confined concrete. *Journal of Structural Engineering, 141*(5), 130–138.

Keshtegar, B, Sadeghian, P, Gholampour, A, & Ozbakkaloglu, T (2017). Nonlinear modeling of ultimate strength and strain of FRP-confined concrete using chaos control method. *Composite Structures, 163*, 423–431.

Lam, L, & Teng, JG (2003a). Design-oriented stress-strain model for FRP-confined concrete. *Construction and Building Materials, 17*(6), 471–489.

Lam, L, & Teng, JG (2003b). *Hoop rupture strains of FRP jackets in FRP-confined concrete* (Vol. 1, pp. 601–612). World Scientific Publishing.

Lam, L, & Teng, JG (2004). Ultimate condition of FRP-confined concrete. *Journal of Composites for Construction, 8*(6).

Lam, L, Teng, JG, Cheung, CH, & Xiao, Y (2004). *Behavior of FRP-confined concrete under cyclic axial compression* (pp. 12–14). .

Li, P, Wu, Y, Zhou, Y, & Xing, F (2009). Stress-strain model for FRP-confined concrete subject to arbitrary load path. *Composite Part B: Engineering, 163*, 9–25.

Li, YF, Lin, CT, & Sung, YY (2003). A constitutive model for concrete confined with carbon fiber reinforced plastics. *Mechanics of Materials*, *35*(3), 603–619.

Lim, JC, & Ozbakkaloglu, T (2015). Unified stress-strain model for FRP and actively confined normal-strength and high-strength concrete. *Journal of Composites for Construction*, *19*(4), 1–14.

Liu, X, Tao, W, Chen, H, & Liu, Y (2020). Compressive stress-strain behavior of CFRP-confined lightweight aggregate concrete reinforced with hybrid fibers. *Composite Structures*, *224*, 112288.

Lu, G, & Yu, TX (2004). *Energy absorption of structures and materials*. Woodhead Publishing.

Mandal, S, Hoskin, A, & Fam, A (2015). Influence of concrete strength on confinement effectiveness of fiber-reinforced polymer circular jackets. *ACI Structural Journal*, *102* (3), 383–392.

Mander, JB, Park, R, & Priestley, MJN (1988). Theoretical stress-strain model for confined concrete. *Journal of Structural Engineering*, *114*(8), 1804–1826.

Ozbakkaloglu, T, & Akin, E (2011). Behavior of FRP-confined normal- and high-strength concrete under cyclic axial compression. *Journal of Composites for Construction*, *16*(4), 451–463.

Ozbakkaloglu, T, & Lim, JC (2013). Axial compressive behavior of FRP-confined concrete: Experimental test database and a new design-oriented model. *Composites Part B: Engineering*, *55*(12), 607–634.

Pessiki, S, Harries, KA, Kestner, J, Sause, R, & Ricles, JM (2001). The axial behavior of concrete confined with fiber reinforced composite jackets. *Journal of Composites for Construction*, *5*(4), 237–245.

Pham, TM, & Hadi, MNS (2014). Confinement model for FRP confined normal- and high-strength concrete circular columns. *Construction and Building Materials*, *69*(69), 83–90.

Piscesa, B, Attard, MN, & Samani, AK (2016). A lateral strain plasticity model for FRP confined concrete. *Composite Structures*, *158*, 160–174.

Realfonzo, R, & Napoli, A (2011). Concrete confined by FRP systems: Confinement efficiency and design strength models. *Composites Part B:Engineering*, *42*(4), 736–755.

Richart, FE, Brandtzaeg, A, & Brown, RL (1928). A study of the failure of concrete under combined compressive stress. *Univ. of Illinois Engineering Experimental Station*, *26* (12), 2971509.

Rong, C, Shi, QX, & Zhao, HC (2020). Behaviour of FRP composite columns: Review and analysis of the section forms. *Advances in Concrete Construction*, *9*(2), 125–137.

Saadatmanesh, H, Ehsani, MR, & Li, MW (1994). Strength and ductility of concrete columns externally reinforced with fiber-composite straps. *ACI Structural Journal*, *91*(4), 434–447.

Saberi, H, Bui, TQ, Furukawa, A, Rahai, A, & Hirosea, S (2020). FRP-confined concrete model based on damage-plasticity and phase-field approaches. *Composite Structures*, *244*, 112263.

Samani, AK, & Attard, MM (2014). Lateral strain model for concrete under compression. *ACI Structural Journal*, *111*, 441–451.

Shehata, IAEM, Carneiro, LAV, & Shehata, LCD (2002). Strength of short concrete columns confined with CFRP sheets. *Materials and Structures*, *35*(1), 50–58.

Shi, QX, Rong, C, Zhang, T, & Wang, QW (2017). A practical stress-strain model for confined concrete. *Steel and Composite Structures*, *20*, 49–54.

Shin, M, & Andrawes, B (2010). Experimental investigation of actively confined concrete using shape memory alloys. *Engineering Structures*, *32*(3), 656–664.

Teng, JG, Huang, YL, Lam, L, & Ye, LP (2007a). Theoretical model for fiber-reinforced polymer-confined concrete. *Journal of Composites for Construction*, *11*(2), 201–210.

Teng, JG, Huang, YL, Lam, L, & Ye, LP (2007b). Theoretical model for fiber-reinforced polymer-confined concrete. *Journal of Composites for Construction*, *11*(2), 201–210.

Teng, JG, Jiang, T, Lam, L, & Luo, YZ (2009). Refinement of a design-oriented stress-strain model for FRP-confined concrete. *Journal of Composites for Construction*, *13*, 269–278.

Wu, YF, & Jiang, JF (2013). Effective strain of FRP for confined circular concrete columns. *Composite Structures*, *95*, 479–491.

Wu, YF, & Zhou, YW (2010). Unified strength model based on Hoek-Brown failure criterion for circular and square concrete columns confined by FRP. *Journal of Composites for Construction*, *14*(2), 175–184.

Xiao, QG, Teng, JG, & Yu, T (2010). Behavior and modeling of confined high-strength concrete. *Journal of Composites for Construction*, *14*(3), 249–259.

Xiao, Y, & Wu, H (2003). Compressive behavior of concrete confined by various types of FRP composite jackets. *Journal of Reinforced Plastics and Composites*, *22*(13), 1187–1201.

Youssef, MN, Feng, MQ, & Mosallam, AS (2007). Stress-strain model for concrete confined by FRP composites. *Composites Part B: Engineering*, *38*(5), 614–628.

Yu, MH (2004). *Unified strength theory and its applications*. Springer-Verlag.

Steel frame confined concrete column

5

5.1 Introduction

In recent years, the confined concrete columns have been wildly studied in existing researches and applied in practical engineering, because of the great axial mechanical properties of concrete improve by confining materials. As we discussed in previous sections, the value and the developing trend of confining pressure provided by outer materials influence the properties and the types of confined concrete (Li et al., 2018; Montuori et al., 2012; Rong et al., 2018). The confined concrete columns include the stirrup confined concrete columns and the concrete-filled steel tubes (CFTs) which are similar to the actively confined concrete, FRP confined concrete columns which belong to passively confined concrete, and FRP−steel−concrete composite columns which belong to composite confined concrete. Meanwhile, the section shapes can determine the stress distribution around core concrete and have an obvious effect on the properties (Ahmed, 2018; Ozbakkaloglu & Jian, 2013; Rong & Shi, 2018; Wang et al., 2019). For the circle section, the core concrete is confined by the uniform lateral pressure; however, for the noncircular section, the core concrete is confined by the nonuniform lateral pressure. And the noncircular section can be divided into the effective area and the ineffective area. Therefore, the confining materials provide a low effective confining effect on the core concrete in the noncircular section. It is necessary to design a reasonable section form to increase the material utilization for the noncircular confined concrete before the application.

It is well known that the confining materials provide the nonuniform confining effect on the core concrete in the noncircular section, and the effective confining areas are distributed on the four corners. Therefore, we can design a new section form to increase the material utilization. We just set the angle steels in the four corners and connect these angle steels with steel bars or steel battens. The combination between the angle steel and the steel battens can form a steel frame to provide a well-confining effect on the core concrete. Fig. 5.1 shows the sketch of the steel frame. The emerging form column can be named steel frame confined concrete column (SFC), which is improved by the steel-encased RC columns. In SFCs, the setting method, which is the angle steel in the effective area, can ensure the reasonable utilization rate of angle steel.

There are a series of existing studies have been presented on the axial mechanical behaviors of SFCs. The experiment studies by Dritsos et al. (1993) and Dritsos and Pilakoutas (1992) on the axial mechanical behavior of the steel-encased RC columns show that the angle steel frame can be considered as a steel component in the new construction. Based on the comparisons of the steel reinforced concrete

Concrete Composite Columns. DOI: https://doi.org/10.1016/B978-0-323-85171-8.00005-7

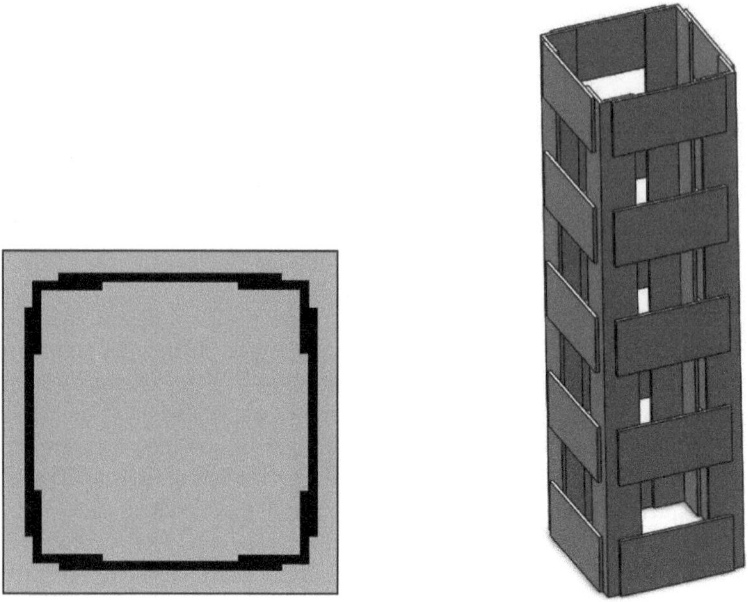

(A) (B)

Figure 5.1 The sketch of SFCs.
(A) the cross-section; (B) steel frame.

and two types of cross-section of SFCs, the study by Hwang et al. (2016) shows that the section form shown in Fig. 5.1 has an effective confining effect, and the columns have the great mechanic and seismic properties. The theoretical study presented by Montuori and Piluso (2009) proposed a fiber model and they improved the model to simulate SFCs through a finite element software. Julio et al. (2011) presented an experimental and theoretical study on the axial and eccentrically mechanical properties of SFCs. However, the test specimens were too small to reflect the accurate stress and strain state of the practical member. Ji and Zheng (2008), Zhang et al. (2012), and Zheng and Ji (2008a, 2008b) presented a series of deep studies on the mechanical properties of SFCs, and then they established the mechanical model of the steel frame confined concrete. However, these studies failed to consider the confining effect provided by the steel battens. The existing stress models of SFCs are summarized by Kim and Hwang (2018), and the theoretical analysis showed that the steel batten is the most effective connection between the angle steels to provide the lateral confining effect. In conclusion, the steel frame can greatly confine the core concrete and improve the mechanical behaviors of the composite columns. However, the existing studies cannot illustrate the influence of each parameter, so these proposed models may not provide accurate prediction results. It is necessary to present further experimental and theoretical studies on the mechanical behaviors of SFCs.

Based on the above analysis, it can be found that the steel frame is embedded in the concrete column, and the steel bars or the steel battens are used to connect the four-angle steels. However, the analysis results suggest we should remove the cover concrete around the steel frame and use more steel battens to connect the angle steels in the experimental studies on axial mechanical behaviors of SFCs, because of the following reasons:

(1) the cover concrete is not confined by the steel frame, so it would affect the study of the confining effect on the core concrete
(2) the steel battens can provide the main confining effect
(3) the steel battens can restrict the local buckling of the angle steels because the incoherence arrangement makes the steel battens bear less axial load and shows less deformation.

In addition, the existing studies have also been limited to the experimental analysis of SFCs. Few studies focused on the theoretical analysis, such as the confining mechanism of the steel frame and the constitutive model of the steel frame confined concrete. Against the background, this paper presents a systematic experimental study on the axial mechanical behaviors of SFCs. The influence factors include the size and the arrangement of angle steel and steel battens, the unconfined concrete strength, the confining effect, and so on. Further, we establish the finite element analysis models for SFCs to study the influence factors in detail. Finally, we analyze the mechanical mechanism of SFCs and establish the constitutive model of the steel frame confined concrete.

5.2 Experimental program

The main purposes of the experiment are:

(1) the study needs to be conducted on the axial mechanical properties of SFCs to prove the feasibility of using the steel frame to confine the core concrete and the effectiveness of the new composite form to improve the axial mechanical behaviors of the composite columns
(2) the analysis of the specimens under the axial loading to explain the failure mechanism of SFCs, including the stress characteristics and the deformation characteristics of SFCs, the strain distribution characteristics of the angle steels and the steel battens, and the influence of various factors
(3) the parameter analysis of the axial mechanical behaviors of STCs to discuss the effect of various factors, including the ductility and the strength of SFCs.

5.2.1 Specimen design

A total of 12 square column specimens were prepared and tested under axial concentric compression, including 11 SFCs without the cover concrete around the steel frame and one steel frame. All the specimens had a side length of 300 mm and a height of 900 mm. It is clearly found that the outer concrete cover can only have the axial bearing capacity under the axial loading. In order to study directly the confining effect of the angle steel form, we removed the concrete cover of all specimens in the experiment, so we only poured the concrete in the steel form. In addition, the bottom of the angle steel is welded on a steel plate (the thickness is 20 mm) to ensure the levelness and the location of the specimen. In order to prevent the local buckling to occur on angle steel, the length of the steel batten should be increased to make the steel batten fully in contact with the angle steel. Meanwhile, the steel batten and the angle steel should be welded at the three sides to prevent weld damage. In the loading process, the angle steels directly confine the core concrete in the SFC specimen, and the steel battens resist the angle steels. Therefore, the core concrete is confined by the combination of angle steel and steel batten in fact. Based on the above analysis, the main variation parameters of the specimen are as follows: the concrete strength grade; the size of angle steel; the arrangement of steel batten; the width of steel batten; the thickness of steel batten; and the effect of the inner spiral hoops. In summary, the details of all the specimens are summarized in Table 5.1.

In Table 5.1, we give a name to each specimen, which starts with some letters ("SFC" represents the composite column and "SF" represents the steel frame without the core concrete), followed by one number to identify each specimen. The steel frame in all specimens has five steel battens and four blank spacings between the adjacent battens, which leads to the different heights of specimens, due to the different sizes of the steel battens, that is, SFC-7 has a height of 870 mm, SFC-1 ~ 5 have a height of 900 mm, SFC-9 ~ 11 have a height of 900 mm, SFC-8 has a height of 935 mm, and SFC-6 has a height of 950 mm.

5.2.2 Specimen preparation

The preparation process of the SFC specimens should be followed in several steps:

(1) the steel frame is assembled and the steel battens are welded on the angle steels
(2) the bottom of the steel frame is welded on a thick steel plate to ensure the bottom keep in level and the location in the concrete casting, as shown in Fig. 5.2A
(3) the adhesive tape is wrapped on the blank layer between the two battens to prevent concrete mixture overflow
(4) the wood formwork is supported close to the steel frame
(5) casting the concrete in the formwork
(6) removing the formwork after five days, as shown in Fig. 5.2B
(7) curing the specimens about one month by watering.

5.2.3 Basic properties of the material

In this experiment, the angle steels and the steel battens were Q235 grade, and the three types of thicknesses (i.e., 4 mm, 6 mm, and 8 mm) were used in the specimens. The inner spiral hoops equipped in the column adopt HRB400 hot-rolled ribbed reinforcement with a diameter of 8 mm. Therefore, the standard tensile tests (JGJ101-96, 1997) of the flat coupons, which were cut from different shapes

Table 5.1 Details of specimens.

Item	Angle steel bar (mm)		Steel batten (mm)			Remarks
	L × t	Steel content (%)	L × B × t	Spacing	Steel content (%)	(Factor comparison)
SFC-1	L70 × 8	4.69	260 × 100 × 6	100	3.47	
SFC-2	L70 × 8	4.69	260 × 100 × 6	100	3.47	Lower concrete strength
SFC-3	L70 × 8	4.69	260 × 100 × 6	100	3.47	Higher concrete strength
SFC-4	L70 × 5	3	260 × 100 × 6	100	3.47	Thin-angle steel bars
SFC-5	L100 × 8	6.83	260 × 100 × 6	100	3.47	Wide angle steel
SFC-6	L70 × 8	4.69	260 × 50 × 6	50	3.47	Different arrangements of steel battens
SFC-7	L70 × 8	4.69	260 × 70 × 6	130	2.43	Narrow steel battens
SFC-8	L70 × 8	4.69	260 × 135 × 6	65	4.68	Wide steel battens
SFC-9	L70 × 8	4.69	260 × 100 × 4	100	2.31	Thin steel battens
SFC-10	L70 × 8	4.69	260 × 100 × 8	100	4.62	Thick steel battens
SFC-11	L70 × 8	4.69	260 × 100 × 6	100	3.47	Adding spiral hoops
SF	L70 × 8	—	260 × 100 × 6	100	3.47	Without concrete

(A) (B)

Figure 5.2 Specimen preparation.
(A) angle steel frame; (B) the specimen without formwork.

Table 5.2 Properties of steel component.

Shape	Thickness (mm)	Elastic modulus (MPa)	Yield stress (MPa)	Tensile strength (MPa)
Steel batten	4	2.1×10^5	340	502
Steel batten	6	2.1×10^5	319	434
Angle steel	5	2.1×10^5	343	505
Angle steel or steel batten	8	2.1×10^5	254	376
Spiral hoops	8	2.1×10^5	417	620

were conducted to determine the material properties of the steel, including the different sizes of the angle steels, and the different sizes of the steel battens and the steel bars. In addition, we prepared three samples for the standard tensile test. The average values of the elastic modulus, the yield stress, and the tensile strength of the flat coupons are summarized in Table 5.2.

For the SFC-2 and SFC-3 specimens, we adopted self-mixed concrete. And we poured three plain concrete prisms (150 mm × 300 mm) of each type of concrete for the material test. For most of the SFC specimens, we poured a batch of the commercial concrete at one time and poured 12 plain concrete prisms (150 mm × 300 mm) for the material test. All SFC specimens and the plain concrete prisms are cured by watering in the outdoor natural environment. Then we tested all the plain concrete prisms for each batch to identify the compressive strength. The average concrete strengths tested from these concrete prism tests are 16.7 MPa (SFC-2), 27.0 MPa (SFC-3), and 28.4 MPa (most SFC specimens) with a standard deviation of 1.1 MPa, 1.6 MPa, and 1.4 MPa.

5.2.4 Test set-up and instrumentation

We tested all specimens at the Xi'an University of Architecture and Technology using a Denison Compression Testing Machine with a load capacity of 20,000 kN, as shown in Fig. 5.3. In addition, the top section of the specimens should be leveled to ensure the uniformly loading, the leveling process is adopted the following steps:

(1) the super-strength gypsum paste is evenly poured over the top section
(2) a steel plate (the thickness is 20 mm) is covered in the top section
(3) pre-load (about 30 kN) is set for 30 minutes to ensure the gypsum has hardened.

Figure 5.3 The specimen in axial compression test.
Test process.

We applied the load with the displacement controlling, and the loading process adopted the following steps:

(1) the loading rate is 0.5 mm/min at the loading stage before the peak strength of each specimen
(2) the loading rate is increased to 0.7 mm/min at the loading stage after the peak strength of each specimen
(3) the loading rate is increased to 1.0 mm/min once the strength begins to stabilize.

In order analysis the axial behaviors of SFCs, we should record the following contents in the test:

(1) the failure process and the final failure mode of the SFC specimen
(2) the bearing capacity of the STC specimen
(3) the displacement of the whole SFC specimen
(4) the displacement in the middle part of the SFC specimen
(5) the longitudinal and transverse strain of the angle steels
(6) the transverse strain of the steel battens. For the steel section in each SFC specimen, six strain gauges by a length of 5 mm were installed at one vertical section, as shown in Fig. 5.4A.

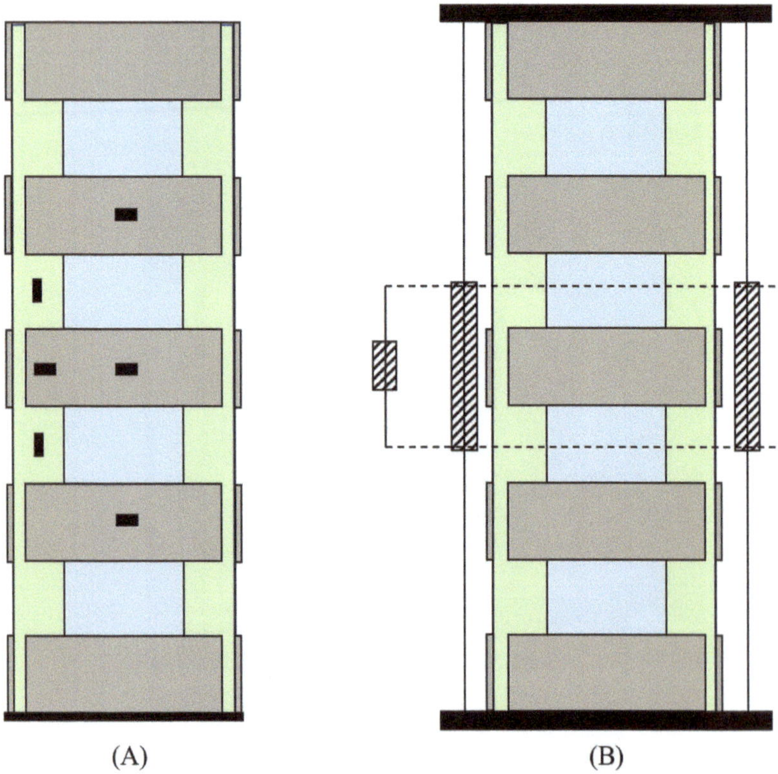

(A) (B)

Figure 5.4 Layout of strain gauges and LVDTs.
(A) strain gauges at one vertical section; (B) two pairs of LVDTs.

In order to compare the inside strain and the outside strain of the steel battens, two lateral strain gauges were installed at the steel battens in the mid-height part. In order to measure enough data, the above strain gauges were arranged in the same way on the three faces of the test piece. The total axial shortening of the SFC specimen was measured by two linear variable displacement transducers (LVDTs) placed 180° apart from each other. The central axial shortening (height was 200 mm) of the SFC specimens was measured by two LVDTs at two steel bars through the opposing concrete sides, and the layout of the two LVDTs was shown in Fig. 5.4B. All data during the test are collected automatically by the data acquisition instrument.

5.3 Experimental results

5.3.1 The failure process and failure mode

In the experiment, we applied axial monotonic compression load to study the axial mechanical behaviors of all SFC specimens. The failure processes of all the SFC specimens are relatively simple and similar. Therefore, we describe uniformly the failure processes of all SFC specimens in this section. During the loading, the failure process includes the following steps:

(1) before the peak load, the surface of the specimen has no significantly crack
(2) after the peak load, the concrete between the three steel battens in the middle is flaky exfoliate at first, which is mainly because there is no connection on the vertical of the concrete part between the adjacent steel battens, and the spalling occurs after being squeezed by the steel battens, meanwhile the steel frame can confine the core concrete
(3) with the continue loading, the angle steels between the steel battens gradually shows the obvious local buckling, the buckling position of the angle steel mainly occurs at the upper and lower sides of the middle steel batten, accompanied by the external bulge of the steel batten
(4) with the continue loading, many parts on the angle steel show the obvious local buckling
(5) in later loading stage, the internal concrete partially crushed, but the crushing degree of the internal concrete is not serious.

Finally, when the load tends to be stable, the loading can be stopped, and the specimen still maintains good integrity without the obvious brittle failure.

The failure modes of all SFC specimens are shown in Fig. 5.5. We can find in Fig. 5.5 that except for the spalling of the concrete surface, the core concrete of each specimen has no obvious brittle failure, which shows that the steel frame can provide an obvious confining effect on the core concrete. And the failure model of each specimen has its respective characteristics. We can describe the detailed failure modes as follows:

(1) for the SFC-1 specimen, the local buckling position of the angle steel is at the upper three layers, the upper second and third battens show both the lateral bulging expansion and the bias deformation, the bias deformation is mainly caused by the local buckling of the angle steel

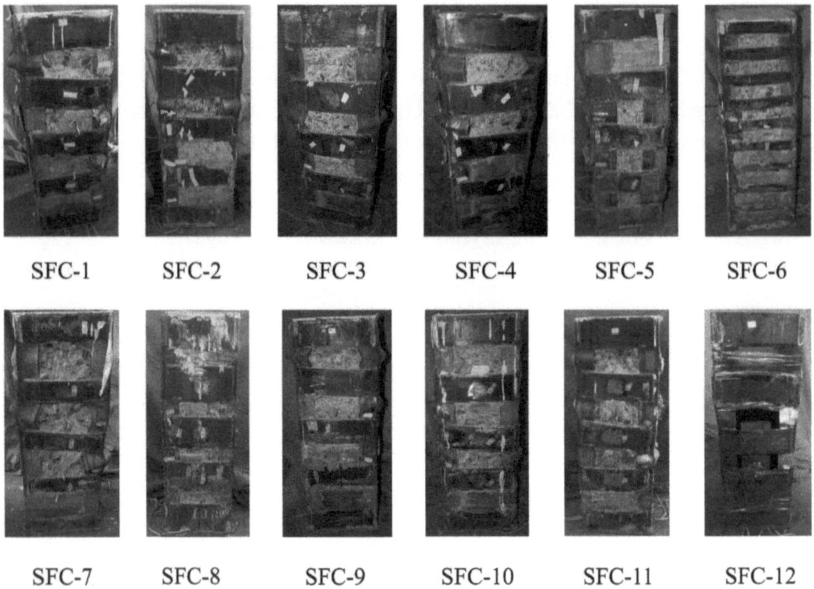

Figure 5.5 The failure modes of SFC specimens.
Specimens after the test.

(2) for the SFC-2 specimen, the local buckling of the angle steels mainly occur on the angle steels of the upper two layers, and the steel battens bulges outward
(3) for the SFC-3 specimen, the angle steel has local buckling on the upper three layers, and the obvious bulging expansion occurs on the middle steel batten
(4) for the SFC-4 specimen, the obvious local buckling occurs on the angle steel of the upper two layers, the steel batten of the second layer shows both the lateral bulging expansion and the bias deformation, because of the local buckling
(5) for the SFC-5 specimen, the angle steel of the lower three layers shows the serious local buckling, which also causes the bias deformation of the lower steel batten
(6) for the SFC-6 specimen, the angle steel has no obvious local buckling, whereas the middle steel batten has the obvious lateral bulging expansion
(7) all parts of the angle steel showed obvious local buckling, the core concrete shows the obvious exfoliation
(8) for the SFC-8 specimen, only the angle steel of upper two layers has the obvious local buckling
(9) for the SFC-9 and the SFC-10 specimens, the angle steel of upper three layers has the obvious local buckling, accompanied by the lateral bulging expansion of the steel battens
(10) for the STC-11 specimen, the angle steel of upper three-layer buckles has the obvious local buckling, and the steel battens of the upper second and third layers show both the lateral bulging expansion and the bias deformation, because of the local buckling
(11) for the empty steel frame (i.e., SF specimen), the local buckling mainly occurs on the middle part of the angle steel.

Based on the above analysis of the failure modes of all specimens, the following conclusions can be drawn:

(1) steel battens can greatly restrict the local buckling of angle steel
(2) the simple reduction of the batten spacing has less effect on the buckling restriction of the angle steel, but the dense distribution of the narrow small battens can effectively restrict the local buckling of the angle steel
(3) the steel frame with less steel battens shows the low confining effect, which means that the steel battens provide the major lateral confining pressure
(4) with the addition of the inner spiral hoops, the failure mode has no obvious difference.

5.3.2 The confining effect of the angle steel frame

In order to analyze the confining effect provided by the steel frame on the core concrete, we take the SFC-1 and SF specimens as examples to study, and the load-- strain curves are shown in Fig. 5.6. For these curves, the axial strain is the overall longitudinal strain of the specimen, which is the ratio of the total displacement to the column height. In addition, the curve of "SF + unconfined concrete" is the superposition curve combined with the load−strain curve of the steel frame specimen and the stress−strain curve of the unconfined concrete. Based on curve comparisons in Fig. 5.6, the following conclusions can be drawn:

(1) the peak load of the SFC-1 specimen is much larger than that of "SF + unconfined concrete" (approximately 1.6 times), the confined concrete causes the addition load, it reflects that steel frame can provide a great confining effect on the core concrete
(2) the confining effect increases the peak strain of the specimens, which leads to the good ductility of all specimens

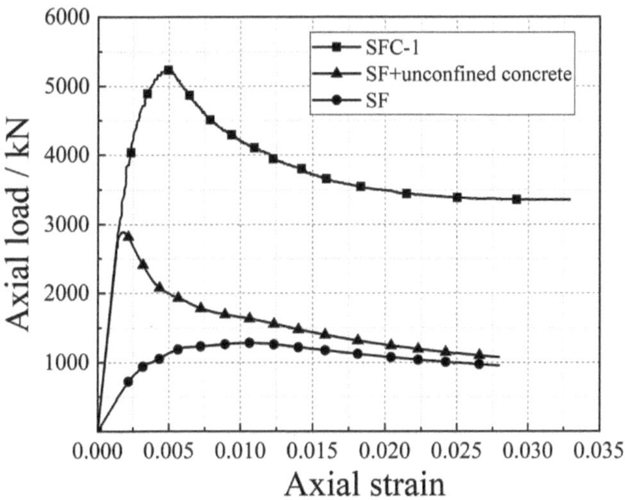

Figure 5.6 The load−strain curve of SFC-1, SF, and superposition results. The load−strain relationship.

(3) the later stage of the load–strain curve of the SFC-1 specimen tends to be horizontal gradually, because of the ductile failure of the core concrete under the confining effect.

5.3.3 The load–strain curves of all specimens

As mentioned in Chapter 2, the SFC specimens have different heights. In order to study reasonably the mechanical behaviors of all specimens, the deformation analysis adopts the axial strain instead of the axial displacement. Therefore, the axial strain gained by the different methods should be discussed. The load–strain curve of each SFC specimen is shown in Fig. 5.7, where the strain of the solid line is the ratio of the total axial displacement to the column height, and the strain of the dotted line is the ratio of the central axial displacement to the central height (i.e., 200 mm).

Based on the curve observation of each specimen, the following conclusions can be drawn: all specimens have no large stress suddenly drop stage, which shows good ductility; all curves show a trend of descending first and then entering the gentle stage, which is similar to the developing trend of the stress–strain curve of the actively confined concrete; before the peak stress, the developing trend of each specimen is similar, and the difference is mainly concentrated in the softening stage; under the large confining effect (i.e., the steel batten has a large width or a large thickness, the angle steel has a large size), the specimen shows the following performance: the high bearing capacity; the stress of the curve decreases quickly after the peak stress; the curve enters the gentle declining stage earlier.

There are some differences between the two types of curves reflected by the total strain and the central strain, which are as follows:

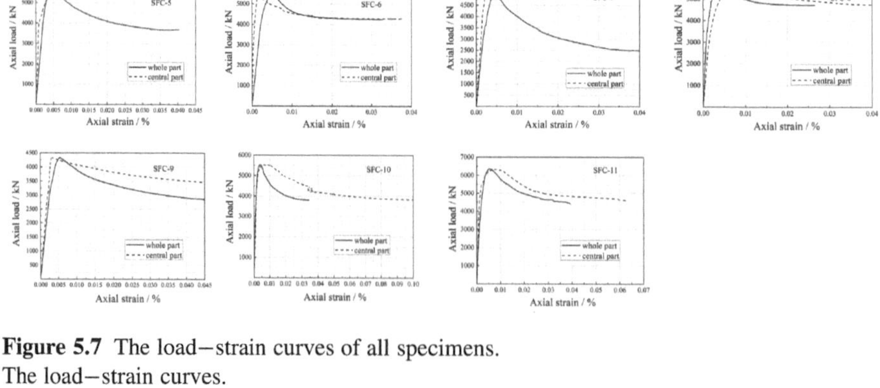

Figure 5.7 The load–strain curves of all specimens.
The load–strain curves.

(1) before the peak load, the central strain of most specimens is less than the overall strain, only a few specimens show that the central strain is equal to the overall strain, and even exceptional specimens show that the central strain is larger than the overall strain
(2) after the peak load, the central strain grows faster, then gradually increases to the larger value than total strain
(3) the load-central strain curves have the earlier and larger horizontal stage, and it shows that the central plastic deformation is larger than the overall plastic deformation, because of the end of columns are confined by the loading plants and the local buckling on the central part of the angle steels.

As mentioned above, it is known that the central strain differs greatly from the overall strain. Although the central strain is less affected by the confining effect provided by the loading machine. However, the central strain has many errors in the experiment results, which are affected by many influence factors, including the deformation of the steel bar, the axial nonuniform deformation of the core concrete, and the nonuniform expansion of the core concrete in the square section—such as the load−strain curves of few specimens show different trends. Meanwhile, the overall strain can reflect the overall mechanical properties of all specimens. Therefore, we will use the overall strain in the analysis of load−strain curves in the following sections.

5.3.4 Analysis of the influence factors

The core concrete in SFCs is confined by the combination of the angle steels and the steel battens, namely the steel frame confined concrete. In order to analyze the influence factors of the mechanical properties of SFCs, it is necessary to study the confining effect provided by the angle steel or the steel battens. Therefore, we compared and analyzed the load−strain curves of SFCs and the stress−strain curves of the core concrete in this section. In addition, the stress of the core concrete is calculated by removing the load of the steel frame from the total load.

5.3.4.1 Strength of the unconfined concrete

SFC-1, SFC-2, and SFC-3 specimens have the same inner steel frame, which can provide the same confining pressure. And the changed parameter is the concrete strength, among the SFC-2 and SFC-3 specimens adopting the self-mixture concrete. The load−strain curves are shown in Fig. 5.8A. Based on the curve comparisons, the following conclusions can be drawn:

(1) the concrete strength is the major factor for the bearing capacity
(2) the specimen with the lower concrete strength shows a well-confining effect under the same confining pressure, such as the larger peak strength increase, the gentler declining stage
(3) the curve trends of the specimens with larger concrete strength are similar.

In order to study the properties of the core concrete, the load provided by the steel frame should be removed at first, then the remaining load is divided by the section area

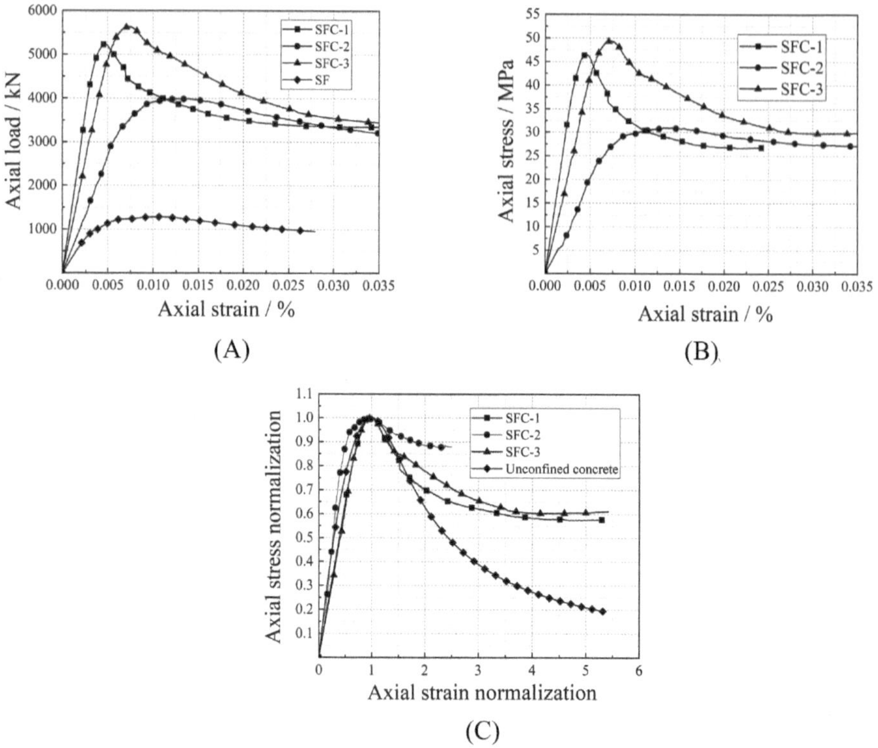

Figure 5.8 The influence of concrete strength on specimens.
(A) load–strain curves; (B) stress–strain curves of confined concrete; (C) normalized stress–strain curve.

of the core concrete. Based on the process, the stress–strain curves of three types of core concrete can be gained, as shown in Fig. 5.8B. Based on further comparisons, the specimen with the lower concrete strength has a gentle softening stage under the same confining effect, while the specimen with the high concrete strength has an obvious softening stage. Meanwhile, the residual stresses of all specimens in the final stage are similar, which means that the controlled parameter is the confining pressure.

As we knew, the mechanical properties of the unconfined concrete in SFC-2 and SFC-3 differ from that in SFC-1, because of the self-mixed method. Therefore, we should use the normalized stress–strain curves to analyze the mechanical properties of the core concrete, which are based on the peak stress and the peak strain. Then the normalized curves are summarized in Fig. 5.8C. We can find in Fig. 5.8 that all curves show a similar trend, including the following characteristics:

(1) before the peak stress, the curves show an increasing trend
(2) after the peak stress, the curves show the steep softening trend
(3) finally, all curves show the horizontal trend. In addition, the larger confining effect leads to the gentle softening trend, and the curves will reach the horizontal stage earlier.

5.3.4.2 Angle steel size

The comparisons of the load–strain curves of the specimens with the different angle steel sizes are shown in Fig. 5.9, where SFC-4 is the specimen with the thin angle steel and SFC-5 is the specimen with the wide angle steel. The curves show that the specimens with the higher angle steel content can provide a higher axial bearing capacity and the higher residual bearing capacity after the loading, because the angle steel can bear the main axial load.

The purpose of the design method (the angle steel distributed in the effective confining areas) is to improve the utilization ratio of the angle steel. Therefore, we should study the angle steel which provides the obvious confining effect in detail. In order to analyze the properties of the confined concrete in all specimens, we can calculate the stress–strain curves by removing the bearing capacity of the angle steel. It should be noted that the steel ratio of the angle steel leads to the unique bearing capacity. Thus, the removing method of all specimens (i.e., SFC-4 and SFC-5) should be combined with the section area of the angle steel and the bearing capacity of the steel frame. The stress–strain curves of SFC-1, SFC-4, and SFC-5 are shown in Fig. 5.9. Based on curves comparisons, it can be concluded that: (A) the curves show similar trends; (B) the angle steel size can influence obviously the mechanical properties of the confined concrete; (C) the angle steel has an obvious confining effect on the core concrete and have the great utilization rate. It is necessary to consider the confining effect of the angle steel in the constitutive model of the core concrete in SFCs.

5.3.4.3 Assembly of the steel battens

The comparisons of the load–strain curves of the specimens with a different assembly of the steel battens are shown in Fig. 5.10, where SFC-6 is the specimen with the narrow steel battens and the narrow spacing, SFC-7 is the specimen with the narrow steel battens and the wide spacing, SFC-8 is the specimen with the wide

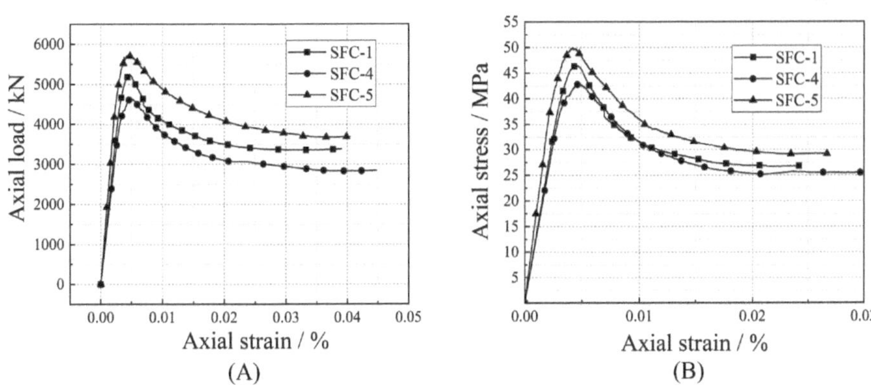

Figure 5.9 The influence of angle steel size.
(A) load–strain curves of specimens; (B) stress–strain curves of SF confined concrete.

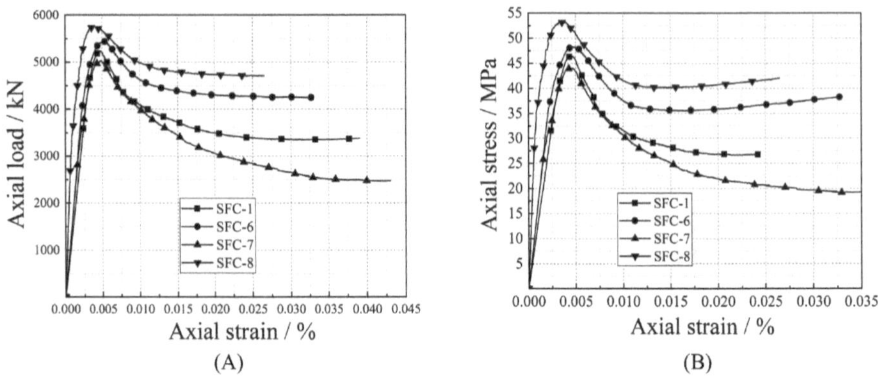

Figure 5.10 The influence of assembly of steel battens.
(A) load–strain curves; (B) stress–strain curves.

steel battens and the narrow spacing. Based on the curve comparisons of SFC specimens, we can find that: the specimen with the wide steel batten has the high bearing capacity and the high residual bearing capacity; the assembly of the narrow steel battens and the narrow spacing can provide a higher confining effect under the same steel ratio of the steel battens.

As shown above, the size and assembly of the steel battens can provide obvious different confining effects on the core concrete. Meanwhile, the confining pressure provided by the steel batten should be reduced, because of the spacing between two steel battens, namely the confining effect reduction. In order to analyze the properties of the confined concrete in the specimen, we should calculate the stress–strain curves by removing the bearing capacity of the angle steel, as shown in Fig. 5.10. Based on comparisons of stress–strain curves, the following conclusions can be drawn:

(1) the small spacing leads to the high confining effect, due to the low confining effect reduction
(2) the wider steel battens also lead to the low confining effect reduction
(3) the confining effect can improve the peak strength
(4) under the large confining effect, the stress–strain curve has the gentle softening stage, even the later curve shows a rising trend.

As above, the assembly of the steel battens is an important factor for SFCs, because of the following reasons:

(1) the steel battens can provide obvious confining pressure, a larger angle steel ratio leads to a larger confining pressure
(2) the spacing between two steel battens would reduce the confining effect
(3) the confining effect reduction relates to the width and spacing of the steel battens.

5.3.4.4 Thickness of the steel batten

The comparisons of the load–strain curves of the specimens with different thicknesses of the steel batten are shown in Fig. 5.11, where SFC-9 is the specimen with

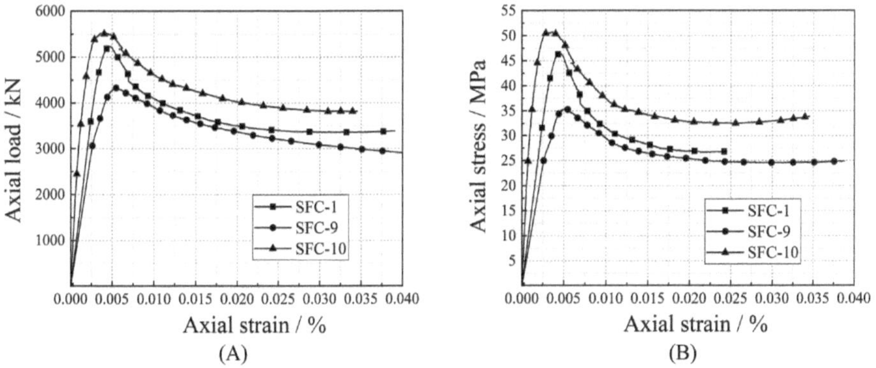

Figure 5.11 The influence of the thickness of steel batten.
(A) load−strain curves; (B) stress−strain curves.

the thin steel batten, and SFC-10 is the specimen with the thick steel battens. The curves of the two types of SFC specimens show similar trends. Meanwhile, the thicker steel batten can provide a larger confining effect, which leads to the higher bearing capacity and the higher residual bearing capacity, but the faster-decreasing trend of the softening stage. Furthermore, the stress−strain curves of the core concrete in the SFC specimens are shown in Fig. 5.11, all curves show a similar trend. However, the curves of specimens with the thin steel battens have less peak stress but the gentle softening stage.

Based on the analysis in the above sections, the confining effect provided by the steel batten is complex, the size and assembly of the steel battens can provide the combination confining effect. Therefore, we summarize all stress−strain curves of the corresponding specimens in Fig. 5.12 to analyze the combination confining effect. It should be noted that SFC-7 and SFC-9 specimens have the similar steel volume ratio of the steel battens, and ASFC-8 and ASFC-10 have the similar steel volume ratio of the steel battens. Based on the comparisons of these curves, the peak stresses are in the descending order: ASFC-8 > ASFC-10 > ASFC-6 > ASFC-1 > ASFC-7 > ASFC-9, the residual stresses are in the descending order: ASFC-8 > ASFC-6 > ASFC-10 > ASFC -1 > ASFC-9 > ASFC-7. Furthermore, the following conclusions can be drawn:

(1) the thickness of the steel battens has an obvious effect on peak stress, but the effect reduces with the increase in the thickness
(2) the wide steel batten can contribute to the peak stress of the steel frame confined concrete, and the batten widthhas a great effect on the core concrete in SFCs with a high steel ratio
(3) the spacing has a great effect on the residual stress of the steel frame confined concrete.

In summary, the thickness of the steel batten should be considered first under the low steel ratio, whereas the width of steel battens should be considered first under the high steel ratio.

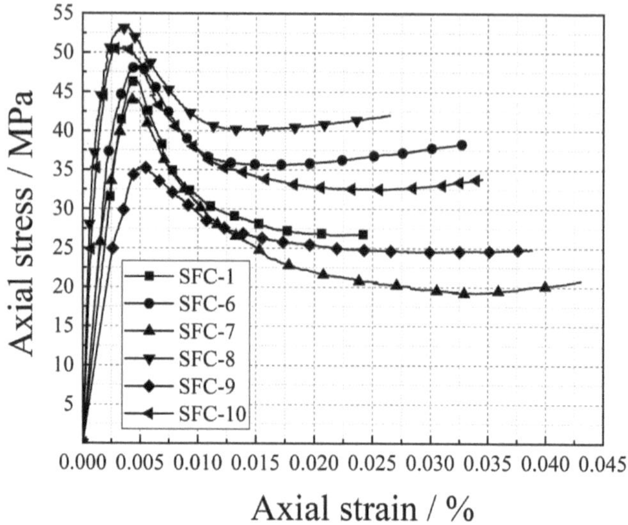

Figure 5.12 The influence of steel battens.
The stress–strain relationship.

5.3.4.5 Layout of the spiral hoops

The comparisons of the load–strain curves and the stress–strain curves of the specimens with the addition inner confining effect are shown in Fig. 5.13, where SFC-11 is the specimen with the inner spiral hoops. The curve comparisons show that the steel frame and the spiral hoops can provide the double confining pressure to improve significantly the mechanical behaviors of both SFC specimens and the core concrete. Meanwhile, the two types of curves show a similar trend. However, the double confining effect is complex, and we will analyze the confining effect in the following chapters.

5.3.5 Strains of the steel frame

5.3.5.1 Axial strain of the angle steel

As we knew, the angle steel can provide both the axial stress to bear the axial load, and the lateral pressure to confine the core concrete in the loading process. In order to analyze the strain state of the angle steel, we summarize the relationships between the axial strain of the angle steel and the axial load of SFC specimens in Fig. 5.14, where the axial strains are averaged from the measured values of the axial strain gauges. It has been known that the yield strain of the angle steel is about 0.002 by the material property test. Fig. 5.14 shows that the axial strains of all specimens reach the yield strain at the peak load, and the load rapidly declines with the increase in the axial strain. It means that the angle steel reaches the yield

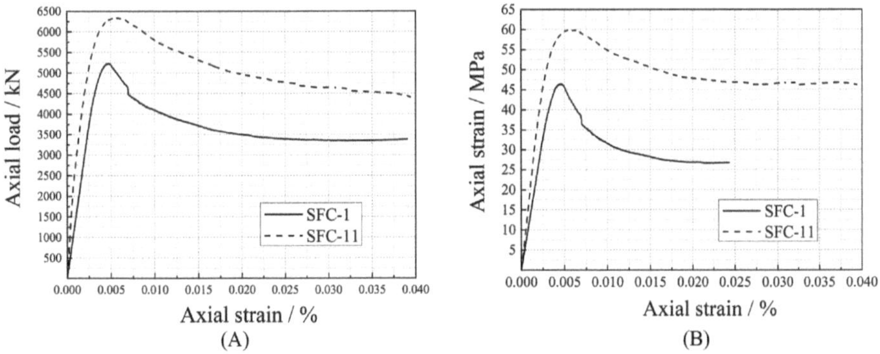

Figure 5.13 The influence of layout of spiral hooping.
(A) load–strain curves; (B) stress–strain curves.

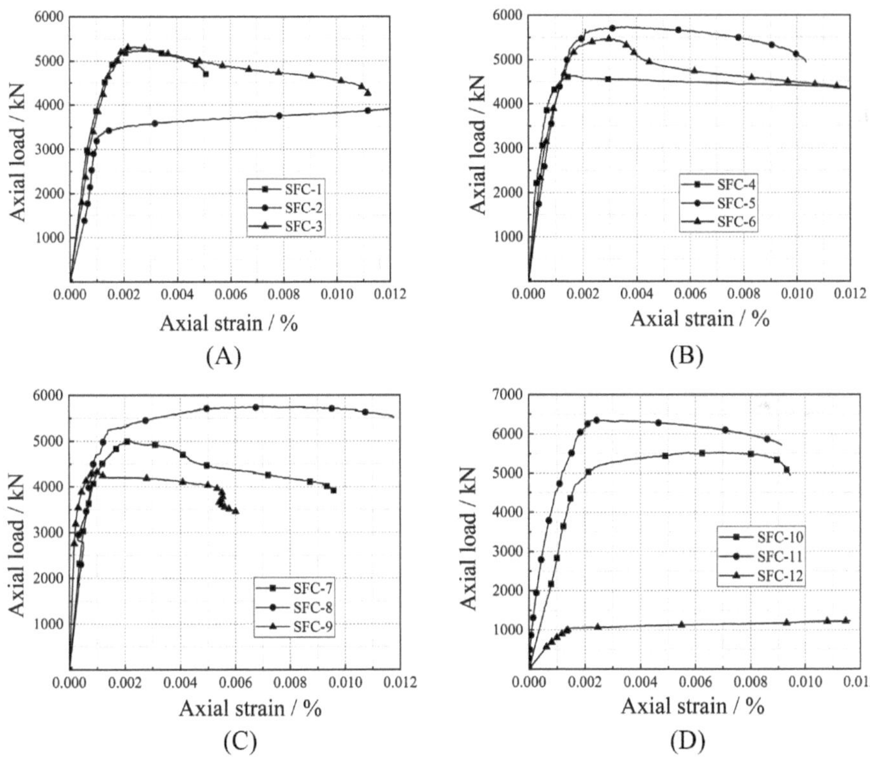

Figure 5.14 The load-axial strain curves of angle steel.
(A) SFC-1 ~ SFC-3; (B) SFC-4 ~ SFC-6; (C) SFC-7 ~ SFC-9; (D) SFC-10 ~ SFC-12.

state before reaching the peak load, then the strains rapidly grow into the buckling state, which leads to the obvious local buckling on the angle steels.

Furthermore, the strains of the angle steel in the SFC specimens are earlier reaching the yield state than that in the SF specimen, because of the bidirectional stress effect on the angle steel in the SFC specimen, including the axial compressive stress and the later tensile stress. In summary, the angle steels have been in the yield state and completely provide the axial bearing capacity at the peak load.

5.3.5.2 Lateral strain of the steel batten

It is clearly found that steel batten can provide the lateral confining pressure to confine the core concrete in the loading process. We summarize the axial strain-lateral strain curves of the steel battens in all specimens in Fig. 5.15, where the lateral strains were averaged from the measured value of the lateral strain gauges. It can be found that the curves in Fig. 5.15 represent the lateral strain at different locations of the steel battens, respectively. The "higher part" represents the lateral strain gauges of the higher steel batten, "middle part" represents the lateral strain of the middle steel battens, "lower part" represents the lateral strain of the lower steel batten, "inner part" represents the lateral strain of the inside steel batten, "spiral hoops" represents the lateral strain of the inner spiral hooping, and the "side part" represents the lateral strain of the overlap between the angle steel and the steel batten.

Based on the curve comparisons in Fig. 5.15, the following conclusions can be drawn:

(1) with the increasing axial strain, the increasing rate of the lateral strain shows slowly at first and fast afterward, which has a close relationship with the lateral expansion of the core concrete

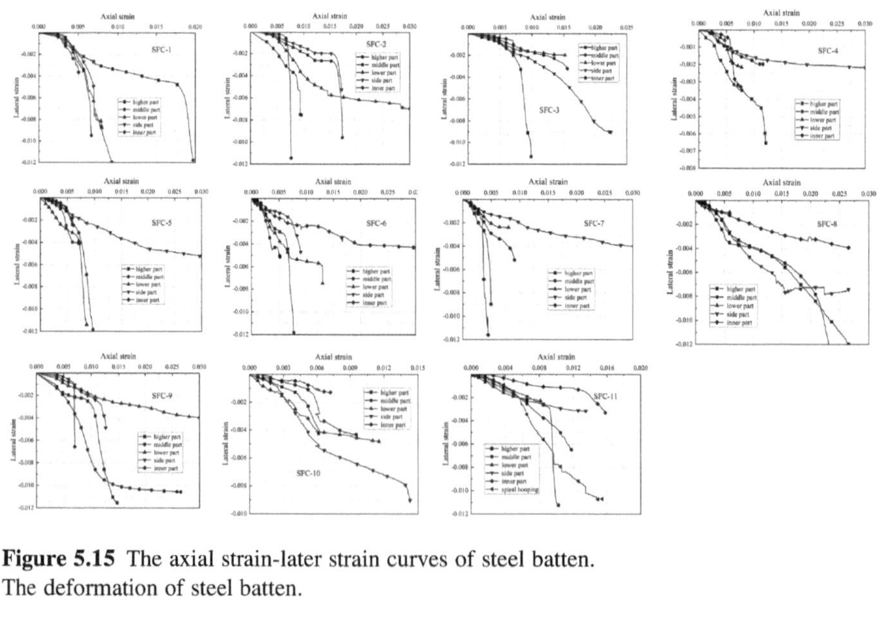

Figure 5.15 The axial strain-later strain curves of steel batten.
The deformation of steel batten.

(2) the strain of the middle three battens is not uniform, but all lateral strains reach the yield state before the axial peak load, it means that the steel batten can fully provide the confining effect
(3) the strains of "side part" are less than that of the steel battens, but they can also reach the yield state before the axial peak load
(4) the strains of the inside parts are less than that of the outside parts because the uneven deformation occurred by the thickness of the steel batten.

In summary, all the lateral strains have been in a yield state before the peak load of the SFC specimens. It means that the steel frame would play a full confining effect in the proposed section of SFCs. We should adopt the yield stress for the angle steel and the steel batten in both the bearing capacity computation of SFCs and the mechanical property analysis of the core confined concrete.

5.4 The influence factor analysis by the finite element model

Based on the experimental study on the axial compression behaviors of the SFC specimens, we analyze a series of test results, including the failure process, the failure modes, the stress and strain state, and the weak parts of the SFC specimens. In order to study the influencing factors in detail, we can use ABAQUS software to establish the finite element analysis models of SFC and verify the proposed model by the experimental results. Then we study the influence of the different parameters on the mechanical behaviors of SFCs, including the steel volume ratio of the angle steel, the width of the steel batten, the thickness of steel batten, the spacing and arrangement of the steel batten. Finally, we study the stress mechanism in the entire process of the steel frame confined concrete composite columns under the axial load.

5.4.1 Establishment of the finite element model

5.4.1.1 Constitutive models

In the studies on the steel frame confined concrete composite column structure, we select the bilinear model of the stress−strain relationship curve without the strengthening stage of the steel in the analysis and the simulation. When the short column bears the axial load, the lateral extrusion force of the concrete in the core area will cause the large deformation of the angle steel, which easily leads to the local buckling, especially when the thickness of the angle steel in the SFCs is thin. Even if the angle steel is strengthened, it cannot bring the addition strength into full play. The elastic modulus of the steel component is selected by the material property test in the experiment, and Poisson's ratio is 0.3.

For the constitutive model of the core concrete, we adopt the plastic damage model provided by the software, which can show significant advantages in simulating concrete as a solid element. Because the model introduces the damage factor of the core concrete to reduce the elastic matrix of concrete, we can simulate the characteristics of the unloading stiffness of the concrete, which decreases with the increase in the concrete internal damage. Meanwhile, we introduce the non-associated hardening parameter into the elastic-plastic constitutive model. Therefore, the plastic damage model can effectively and accurately simulate the nonlinear behavior of a concrete structure under axial force. In establishing the finite element model, we can calculate the stress–strain curve of the confined concrete according to the axial compression constitutive model of the confined concrete established in Chapter 4, and the Poisson's ratio is 0.2.

5.4.1.2 Model element and model mesh

In order to simulate the AFC specimens, both the steel frame and the core concrete use solid elements (C3D8R). The C3D8R element is a three-dimensional solid element in the eight nodes reduced integral format, which can not only meet the calculation accuracy but also reduce the number of degrees of freedom, so as to save the calculation time and meet the calculation efficiency. For the core concrete, we adopted the C3D8R element to facilitate the control element shape to be hexahedron during the grid division, and the structural grid division technology is implemented. For the steel frame welded by the angle steels and the steel battens, we used the C3D8R element to define it. There are two reasons for the above element selection, which are as follows: the component model in the simulation analysis is not complex and the operation time is not limited by the element; in the analysis and calculation by using three-dimensional solid elements, the local coordinate system is consistent with the overall coordinate system, the relevant stress and strain do not need to be converted during subsequent treatment, which is more convenient.

In ABAQUS, the size of the mesh is very important for finite element analysis. This paper uses the sweep (omitted) technology to divide the different mesh densities in different areas, so as to increase the possibility of the model convergence, and meet the requirements of the calculation accuracy and calculation time. After several trial calculations, the reasonable size is determined that the main unit size of the core concrete is 30 mm and the main unit size of the steel frame is 15 mm, which can not only meet the calculation efficiency but also meet the corresponding accuracy requirements. The grid division of the steel frame confined concrete composite short column is shown in Fig. 5.16.

5.4.1.3 Interaction, constraint, and boundary conditions

For the finite element analysis of the ordinary concrete filled steel tube, the bond-slip effect between the contact surface of the steel and the concrete has little effect on the analysis results. Therefore, we can assume that the steel tube and the core concrete are completely bonded and the bond-slip effect is ignored. Based on the study results of the concrete-filled steel tube, the bond-slip effect between the angle

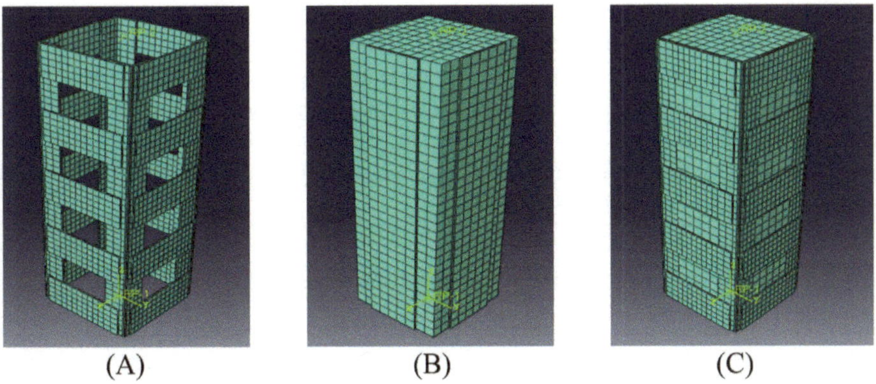

Figure 5.16 Mesh division of SFC.
(A) steel frame; (B) concrete column; (C) the SFC.

steel surface and the concrete surface is ignored in the finite element nonlinear analysis of the SFCs. At the same time, the influence of the concrete shrinkage and creep on the structure is ignored. Therefore, we set the normal direction of the contact surface between the core concrete and the steel frame as the correlation attribute of "hard contact", set the tangential direction as "penalty function". And the friction coefficient is taken as 0.5. At the same time, we ignored the influence of the concrete shrinkage and creep on the structure. Therefore, we set the normal direction of the contact surface between the core concrete and the steel frame as the correlation attribute of "hard contact", while setting the tangential direction as "penalty function". The friction coefficient is taken as 0.5.

It should be noted that the establishment of the SFC model mainly focuses on the whole deformation. Therefore, the welding area should not be accurately modeled, so as to save calculation time and increase the possibility of model convergence. Then the "merge" command is used to combine the angle steels and the battens into a whole to form the steel frame.

In order to make the established finite analysis model consistent with the test specimen, we set the reference points (i.e., RP1, RP2) and coupled the two points with the bottom and top of the column, respectively. Then we can define the boundary conditions of the column bottom in the initial step as follows: the RP1 point to impose all constraints on the displacement (U1, U2, U3) and rotation angle (UR1, UR2, UR3). Finally, in the subsequent analysis step (Step 1), we can define the boundary conditions of the column top surface as follows: the displacement (U1, U2) and the rotation angles (UR1, UR2, UR3) are constrained by using the reference point RP2, then the point RP2 apply the displacement load in the Z direction.

5.4.2 The verification of the finite element model

In order to verify the proposed mode, we used the above finite element model establishment methods to simulate the axial compression behaviors of the SFC specimens. The models were carried out based on four specimens in Section 5.2, that is, SFC-1, SFC-3, SFC-5, and SFC-6. The parameters of each model are exactly the same as the size of the test specimen, and the specific specimen parameters are shown in Table 5.1. Based on the comparison and verification of the test results, this paper aims to prove the accuracy of the finite element model and provide a reliable guarantee for the following analysis of the influence factors based on the finite element software simulation.

5.4.2.1 Verification of the load–displacement curve

Based on the comparison between the calculation curves by the proposed models and test curves in Section 5.2, as shown in Fig. 5.17, the following conclusions can be drawn:

(1) the finite element simulation results of SFCs are basically consistent with the variation trend of the load–displacement curve of the test result

(2) for the peak bearing capacity of the specimens, the finite element simulation results are slightly higher, but the error is within the acceptable range.

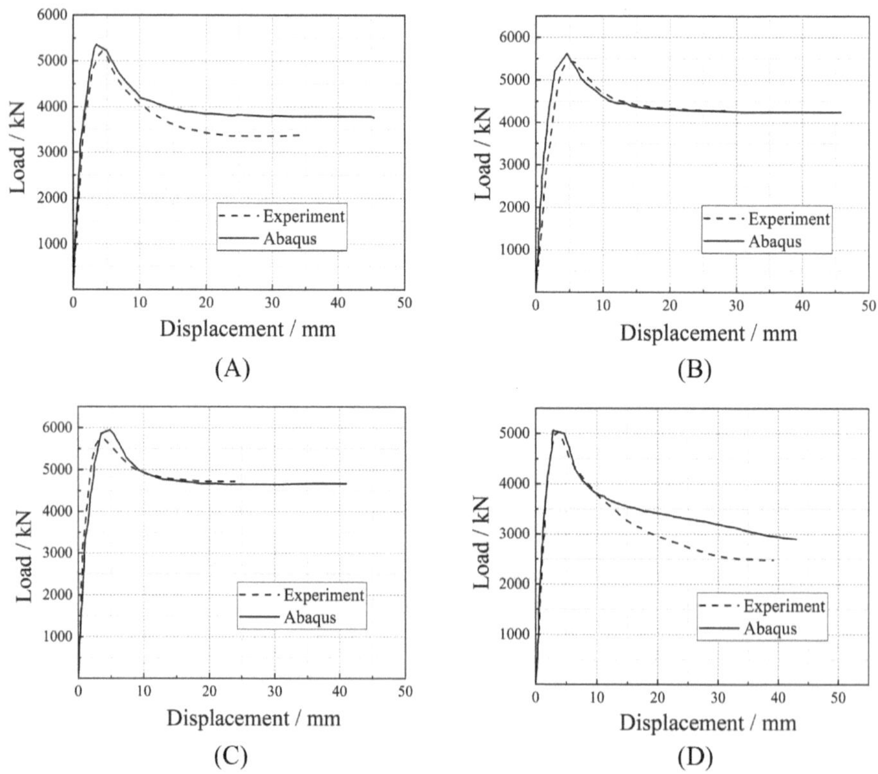

Figure 5.17 The comparison between calculation results and test results. (A) SFC-1; (B) SFC-3; (C) SFC-5; (D) SFC-6.

The reason may be that the specimen has some initial defects because of the improper operation in the processing process.

For the load–displacement curve of the specimen, the finite element simulation curves basically coincide with the test curves in the elastic stage. Compared with test curves, the curves show a trend of first falling and then entering the gentle horizontal stage after the peak load. However, the decreasing stage of the finite element simulation results is small and enters the gentle section earlier, while the decreasing stage of the test curve is large and enters the gentle section late. The reason for the above inconsistency is that the actual test specimen cannot reach the ideal condition because of the influence of external conditions. Further, the ABAQUS finite element calculation results are too complete and ideal to lead to certain errors with the model parameters established, including the material properties, the interaction relationship, and the boundary conditions.

In general, the finite element simulation results of ABAQUS are in good agreement with the basic trend of the load–displacement curve measured in the test. Therefore, the establishment method of the finite element model can be used to further study and analyze the mechanical behaviors and the influence factors of the SFC specimens.

5.4.2.2 Verification of the failure modes

After the model calculation, we also compared the stress nephogram in the visualization with the failure modes of the test specimen in Section 5.2, as shown in Fig. 5.18 to verify the proposed model. Based on the comparisons, the following conclusions can be drawn:

(1) For all specimens, the numerical simulation results by ABAQUS are consistent with the overall failure modes of the test specimens.

(2) In the loading process, the core concrete at the middle part of the specimen cracks and peels first. In the later loading stage, the core concrete at the middle part crushes seriously. And the central angle steels have the obvious local buckling and the batten plates bulge out. Moreover, the stress nephogram shows that the stress in the middle of the specimen is the largest, which is in good agreement with the test failure modes.

In order to further verify the model, we also compared the local failure mode. For example, when the SFC-6 specimen reaches the ultimate failure, the whole layer of the angle steel at the third layer shows the fully local buckling, accompanied by the bulging and the inclination of the steel batten, and the inner core concrete seriously crushes. We set the comparison between the local stress nephogram of the numerical simulation results by ABAQUS and the test failure results in Fig. 5.18. We can find that the local failure mode and the location of local buckling in the test specimens are similar to the numerical simulation results.

Based on the comparisons of the load–displacement curve and the failure mode characteristics, the numerical simulation results can greatly coincide with the test results. Therefore, the finite element models established in this paper are accurate

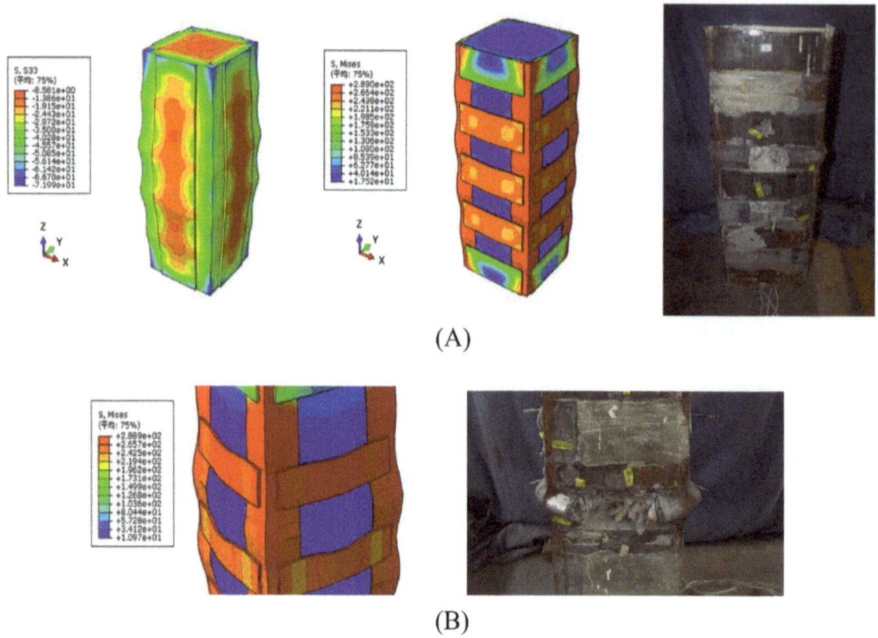

Figure 5.18 Comparison of failure modes and numerical simulation stress diagram. (A) whole failure modes; (B) the local buckling of SFC-6.

and effective, and the model establishment method can make sufficient preparations for the further analysis of the influencing factors for SFCs.

5.4.3 Analysis of influence factors by the numerical simulation results

In order to analyze the mechanical mechanism of SFCs, we deeply study the following influence factors in this section: the concrete strength grade, the steel ratio of the angle steel, the spacing between two steel battens, and the width and thickness of the steel battens.

5.4.3.1 Concrete strength grade

In order to study the influence of concrete strength on the mechanical properties of SFCs, we designed and simulated three specimens with the concrete strength grade, and the concrete strength is the only variable parameter. It shows the specific parameters of specimen size in Table 5.3. And these proposed models were given the names (i.e., SFC-A1, SFC-A14, and SFC-A15).

Based on the numerical simulation analysis and calculation, the load–displacement curves of the three specimens are shown in Fig. 5.19. Based on the comparison results of the bearing capacity and the peak displacement, it can be found that

Table 5.3 The sizes of specimens with different concrete strength.

Item	Angle steel (mm)		Steel batten (mm)			Spacing (mm)	Concrete strength
	Length	Thickness	Length	Width	Thickness		
SFC-A3	70	8	260	100	6	100	C40
SFC-A14	70	8	260	100	6	100	C30
SFC-A15	70	8	260	100	6	100	C50

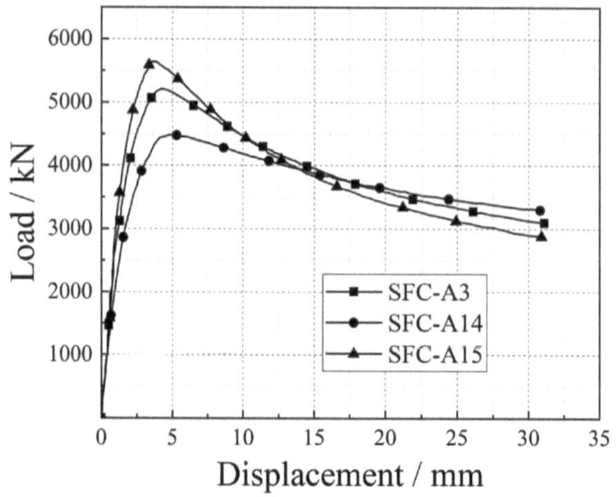

Figure 5.19 The load–displacement curves of specimens with different concrete strengths. Analysis results.

the peak bearing capacity of the specimens increases with the increase in the concrete strength grade. However, after the stopping load, the final bearing capacity of the specimen decreases slightly, so the confining effect is more obvious on the lower-strength concrete.

Based on the load–displacement curves of SFC with the three types of concrete strength, these curves show the same developing trend. After the peak load, the curve goes to the softening stage first and then to the gentle stage. When the concrete strength is lower, the confining effect can show more significant strength improvement. And the decreasing stage of the curve is shorter, which will lead to the gentle stage earlier and gentle. When the concrete strength is higher, there will be a more obvious falling and longer stage in the load–displacement curve.

5.4.3.2 Steel ratio of the angle steel

In order to study the influence of steel ratio of angle steel on the mechanical properties of SFCs, we designed and simulated five specimens with different sizes of the angle steels. And the size of the angle steel is the only variable parameter. It shows the specific parameters of specimen size in Table 5.4. And these proposed models were given the names (i.e., SFC-A1, SFC-A2, SFC-A3, SFC-A4, and SFC-A5).

Based on the numerical simulation analysis and calculation, the load–displacement curves of the five specimens are shown in Fig. 5.20. Based on the comparison results of the bearing capacity and the peak displacement, it can be found that the peak bearing capacity and the corresponding peak displacement of the specimens increase with the increase in the steel ratio of the angle steel. Meanwhile, the final bearing capacity of the specimen increases accordingly after the stopping load. Because the angle steel can bear the vertical load, the steel ratio of the angle steels

Table 5.4 The sizes of specimens with different steel ratios of angle steel.

Item	Angle steel (mm)		Steel batten (mm)			Spacing (mm)	Concrete strength
	Length	Thickness	Length (%)	Width	Thickness		
SFC-A1	70	4	2.48	260	100	6	100
SFC-A2	70	6	3.63	260	100	6	100
SFC-A3	70	8	4.76	260	100	6	100
SFC-A4	56	8	3.72	260	100	6	100
SFC-A5	100	8	6.93	260	100	6	100

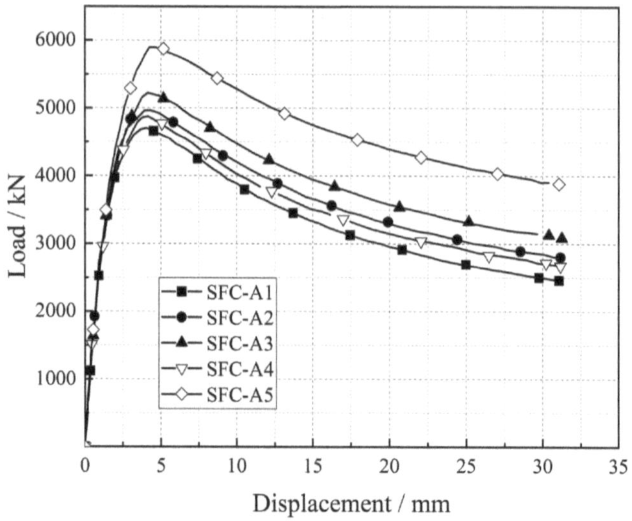

Figure 5.20 The load–displacement curves of specimens with different steel ratios of angle steel. Analysis results.

has a significant effect on the peak bearing capacity of SCFs. However, we cannot ignore the confining effect of the increasing steel ratio of SCFs in the stress analysis.

It is easily found that the steel ratio is influenced by the length and thickness of the angle steel. Therefore, the two influence factors will be further analyzed, respectively. Based on the comparisons between SFC-A2 with SFC-A4, the steel ratios of the two simulated angle steels are similar, but the SFC-A2 has a higher bearing capacity and better ductility. Therefore, the method of increasing the leg length of the angle steel can more effectively improve the axial mechanical behaviors of SFC.

5.4.3.3 Spacing between the steel battens

In order to study the influence of the spacing between two steel battens on the mechanical properties of SFCs, we designed and simulated four specimens with different spacing distances. And the spacing distance is the only variable parameter. It shows the specific parameters of specimen size in Table 5.5. And these proposed models were given the names, that is, SFC-A3, SFC-A6, SFC-A7, and SFC-A8.

Based on the numerical simulation analysis and calculation, the load–displacement curves of four specimens are shown in Fig. 5.21. Based on the comparison results of the bearing capacity and the peak displacement, it can be found that the peak bearing capacity and the corresponding peak displacement of the specimens increase with the decrease in the spacing between two steel battens, due to the enhancing confining effect. It should be noted that if the spacing is larger than 150

Table 5.5 The sizes of specimens with different spacing distance.

Item	Angle steel (mm)		Steel batten (mm)			Spacing (mm)	Concrete strength (%)
	Length	Thickness	Length	Width	Thickness		
SFC-A3	70	8	260	100	6	100	3.47
SFC-A6	70	8	260	100	6	50	4.62
SFC-A7	70	8	260	100	6	150	2.77
SFC-A8	70	8	260	100	6	200	2.31

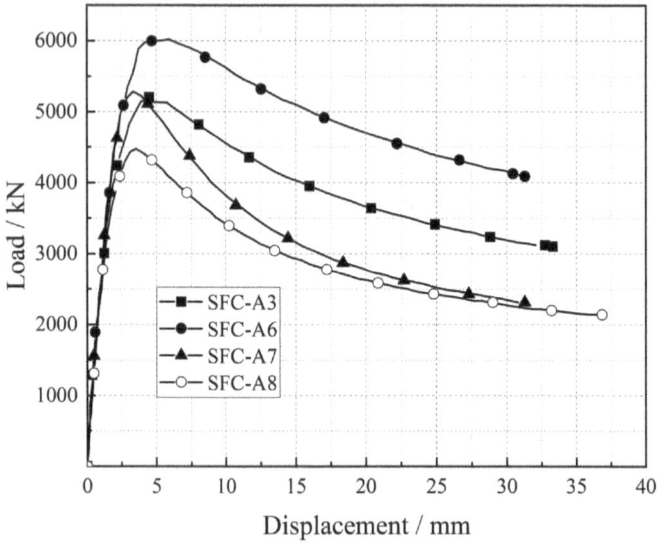

Figure 5.21 The load–displacement curves of specimens with different spacing distances. Analysis results.

mm, the load of the specimen will decrease significantly after the peak load, and the confining effect will be weakened obviously. Therefore, we suggest that the spacing between two steel battens should not exceed 150 mm in the specimen design.

Because the steel batten is intermittently configured, the confining effect along the column height direction is uneven and will be reduced with the spacing. If the spacing is smaller, the reduction effect will be weaker, then the confining effect provided by the steel frame is better, which leads to the well axial mechanical behaviors of SFCs. In general, the steel volume ratio of the specimen reflects the spacing distance in the existing studies. However, we find that the steel volume ratio is also related to the width and the thickness of the steel battens, so we will study the two influence factors in the following sections.

5.4.3.4 *Width of the steel batten*

In order to study the influence of the width of steel batten on the mechanical properties of SFCs, we designed and simulated three specimens with different widths, and the width is the only variable parameter. It shows the specific parameters of specimen size in Table 5.6. And these proposed models were given the names, that is, SFC-A3, SFC-A9, SFC-A10, and SFC-A11.

Based on the numerical simulation analysis and calculation, the load–displacement curves of four specimens are shown in Fig. 5.22. Based on the comparison results of the bearing capacity and the peak displacement, it can be found that the peak bearing capacity and the corresponding peak displacement of the specimens

Table 5.6 The sizes of specimens with different widths of steel batten.

Item	Angle steel (mm)		Steel batten (mm)			Spacing (mm)	Concrete strength (%)
	Length	Thickness	Length	Width	Thickness		
SFC-A3	70	8	260	100	6	100	3.47
SFC-A9	70	8	260	50	6	100	2.31
SFC-A10	70	8	260	150	6	100	4.16
SFC-A11	70	8	260	200	6	100	4.62

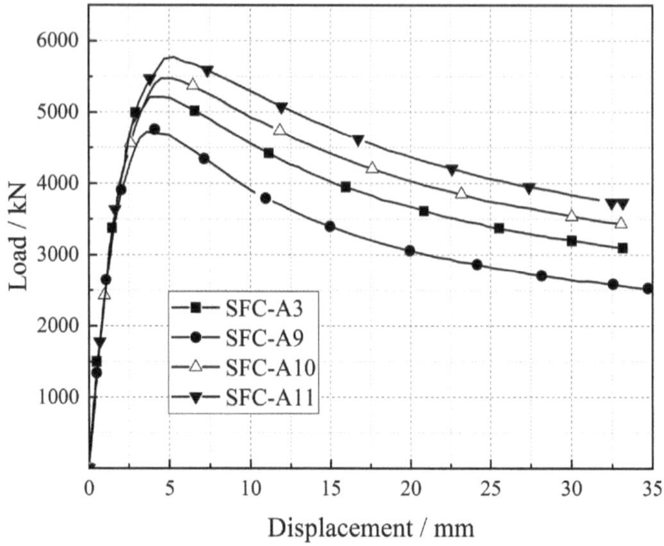

Figure 5.22 The load–displacement curves of specimens with different widths of steel batten.
Analysis results.

increase with the increase in the width of the steel battens, due to the enhancing confining effect. And the final bearing capacity of the specimen increases accordingly after the stopping load.

5.4.3.5 Thickness of the steel batten

In order to study the influence of the thickness of steel batten on the mechanical properties of SFCs, we designed and simulated three specimens with different widths, and the width is the only variable parameter. It shows the specific parameters of specimen size in Table 5.7. And these proposed models were given the names, that is, SFC-A3, SFC-A12, and SFC-A13.

Based on the numerical simulation analysis and calculation, the load–displacement curves of the three specimens are shown in Fig. 5.23. Based on the comparison results of the bearing capacity and the peak displacement, it can be found that the peak bearing capacity and the corresponding peak displacement of the specimens increase with the increase in the thickness of the steel battens, due to the enhancing confining effect. And the final bearing capacity of the specimen increases accordingly after the stopping load.

5.4.3.6 The design of the steel batten

It is easily found that the steel volume ratio of the steel batten is determined by the following factors: the spacing, the width, and the thickness. And the inner confining

Table 5.7 The sizes of specimens with different thicknesses of steel batten.

Item	Angle steel (mm)		Steel batten (mm)			Spacing (mm)	Concrete strength (%)
	Length	Thickness	Length	Width	Thickness		
SFC-A3	70	8	260	100	6	100	3.47
SFC-A12	70	8	260	100	4	100	2.31
SFC-A13	70	8	260	100	8	100	4.62

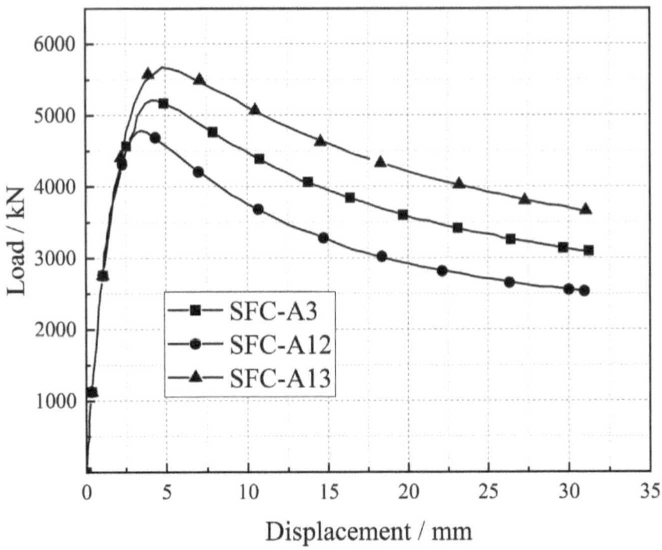

Figure 5.23 The load–displacement curves of specimens with different thicknesses of steel batten.
Analysis results.

effect and the mechanical properties of the specimens increased with the increase in these parameters. Therefore, the influence of the steel batten on the confining effect is complex. Then, we selected the SFC-A6, SFC-A11, and SFC-A13 to analyze the influence of each factor. In these models, the steel battens of the three specimens have the same volume steel ratios, but they have different assembly forms. The load–displacement curves of the three specimens are summarized in Fig. 5.24, and it can be found that with the same volume steel ratio, the confining effect provided by the different batten forms on the core concrete is in the following order: the reduction in the batten spacing, the increase in the batten width, and the increase in the batten thickness. After stopping load, the confining effect provided by the increase in the width or the thickness of the steel batten is not as good as that of the reduction in the spacing between the steel battens.

As above, we can suggest that the ratio of the spacing to the width is 1:2, and the spacing between steel battens is less than 150 mm. Based on the design, the confining effect of the steel frame is better, which can improve the axial mechanical behaviors of SCFs.

5.4.4 *Mechanism analysis of SCFs*

Based on the numerical simulation results, we can observe the changes in the stress nephogram in each stage of SFC under the axial load. And we take SCF-A3 as an example to analyze the stress mechanism of the composite column in the entire loading process, including the elastic stage, the plastic stage, and the failure stage.

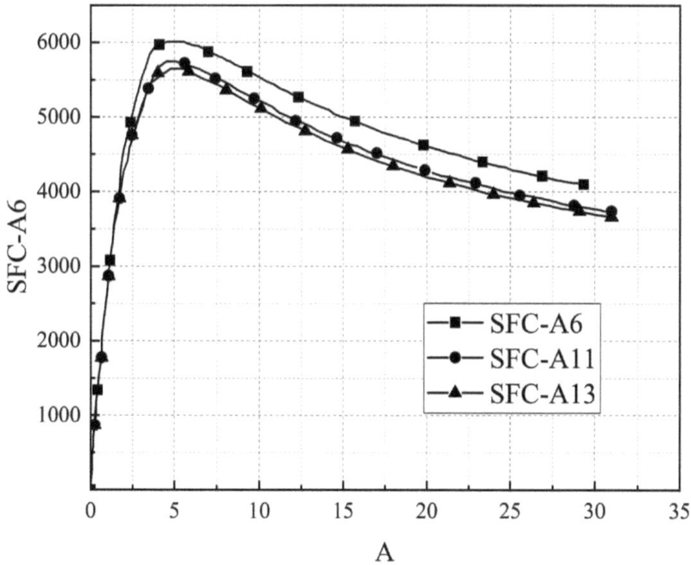

Figure 5.24 The load–displacement curves of specimens with different arrangements of steel batten.
Analysis results.

5.4.4.1 Elastic stage in the loading process

When the load is less than 80% of the peak load of the specimen, the whole specimen is basically still in the elastic stage, and the relationship between load and displacement is still in linear growth in this stage. We summarize the corresponding nephograms in Fig. 5.25, including the axial stress nephogram and the cumulative plastic strain nephogram of the core concrete, the Mises stress nephogram, and the cumulative plastic strain nephogram of the steel frame, respectively. It can be found that there is basically no plastic strain in both the core concrete and the steel frame. Therefore, the entire steel frame has no confining effect on the core concrete in the elastic stage.

5.4.4.2 Plastic stage in the loading process

When the specimen reaches the peak load, the whole specimen has been in the plastic stage, and the displacement of the core concrete in X and Y directions is shown in Fig. 5.26. It can be found that the deformation in the middle part of the specimen is the largest, expanding outward in the transverse direction to squeeze the steel frame. Fig. 5.26 shows the axial stress nephogram and the Mises stress nephogram of the core concrete. It can be found that the core concrete basically reaches the ultimate stress state at the peak bearing capacity. We show the cross-section in the middle part of the specimen in the view section of Fig. 5.26E, it can be clearly seen that the area with the largest stress concentrates in the four corners which are close

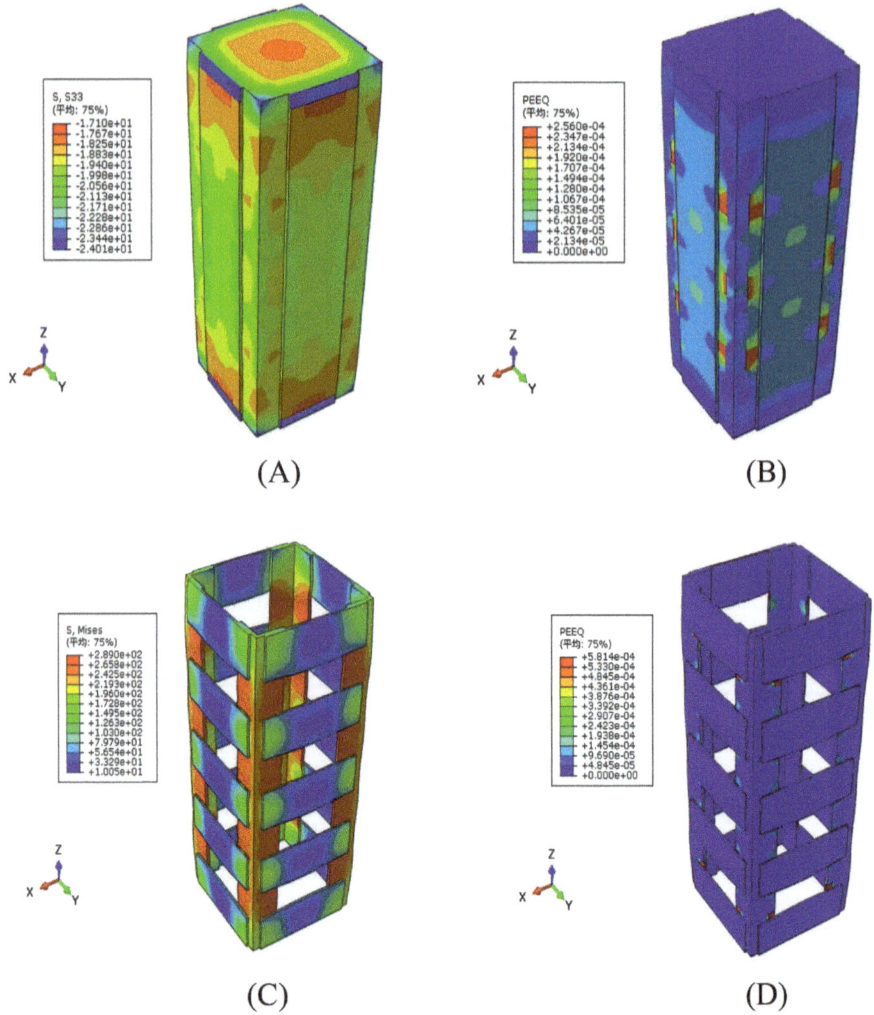

Figure 5.25 The stress and strain nephogram in elastic stage.
(A) axial stress nephogram of core concrete; (B) plastic strain nephogram of core concrete; (C) Mises stress nephogram of steel frame; (D) plastic strain nephogram of the steel frame.

to the angle steel. The core concrete in the center also shows the obvious stress distribution. And the area with the four edges close to the batten plate has the smallest stress. The stress states show that the steel frame has an obvious confining effect on the concrete at the four corners, which is similar to the effective confining area and the invalid confining area of the square section in the SFC specimens. The cumulative compressive plastic strain of the core concrete (see Fig. 5.26F) shows that the plastic development in the middle of the column is the most sufficient, so the weak

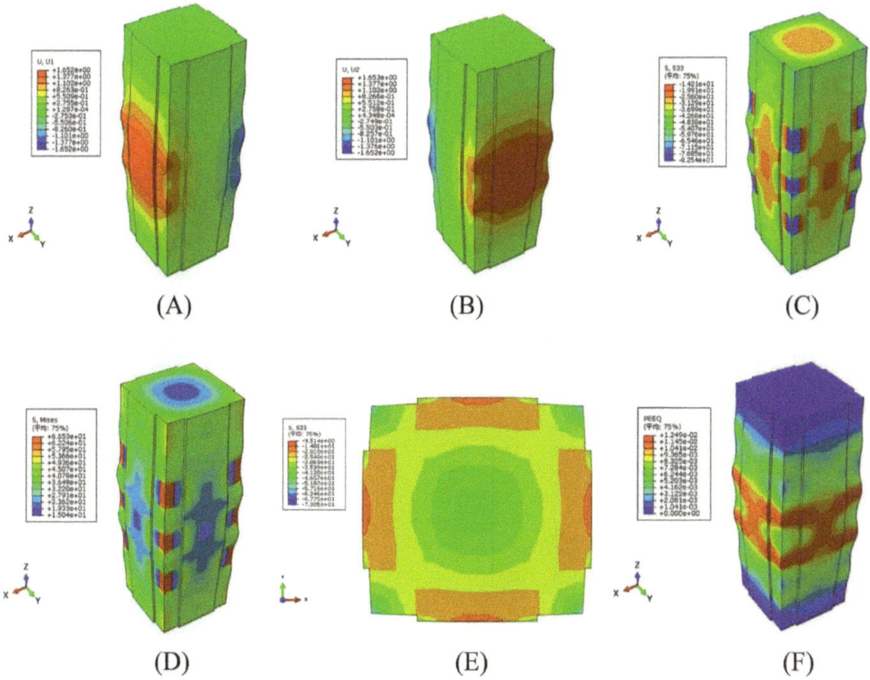

Figure 5.26 The stress and strain nephogram of core concrete in plastic stage. (A) X-direction displacement nephogram; (B) Y-direction displacement nephogram; (C) axial stress nephogram; (D) Mises stress nephogram; (E) view section; (F) cumulative plastic strain.

area is formed in this area, where the concrete is most likely to be crushed and peeled off after the continuous loading.

The stress nephogram and the strain nephogram in Fig. 5.27 show the displacement of the steel frame in the X-direction and the Y-direction. It can be found that the deformation is the largest in the area near the middle part of the specimen, and the area most likely occurs the failure. Similarly, the angle steel near the middle height of the specimen shows the local buckling at first, and the steel batten bulges outward in the experiment. The distribution of the Mises stress nephogram of the steel frame (see Fig. 5.27) shows that when the specimen reaches the peak load, the angle steel has reached the yield state, and the steel battens at the middle height of the specimen reach the yield state. The stress states also confirm the assumption that the angle steel and the steel batten just reach the yielding state once the specimen reaches the peak load. The accumulated plastic strain nephogram of the steel frame, as shown in Fig. 5.27, shows that the plastic strain is the largest near the middle height of the specimen, so the steel frame in this area yields first. At this time, the transverse steel batten connects the four-angle steels to form a closed section, and the entire steel frame can provide an obvious confining effect on the core concrete.

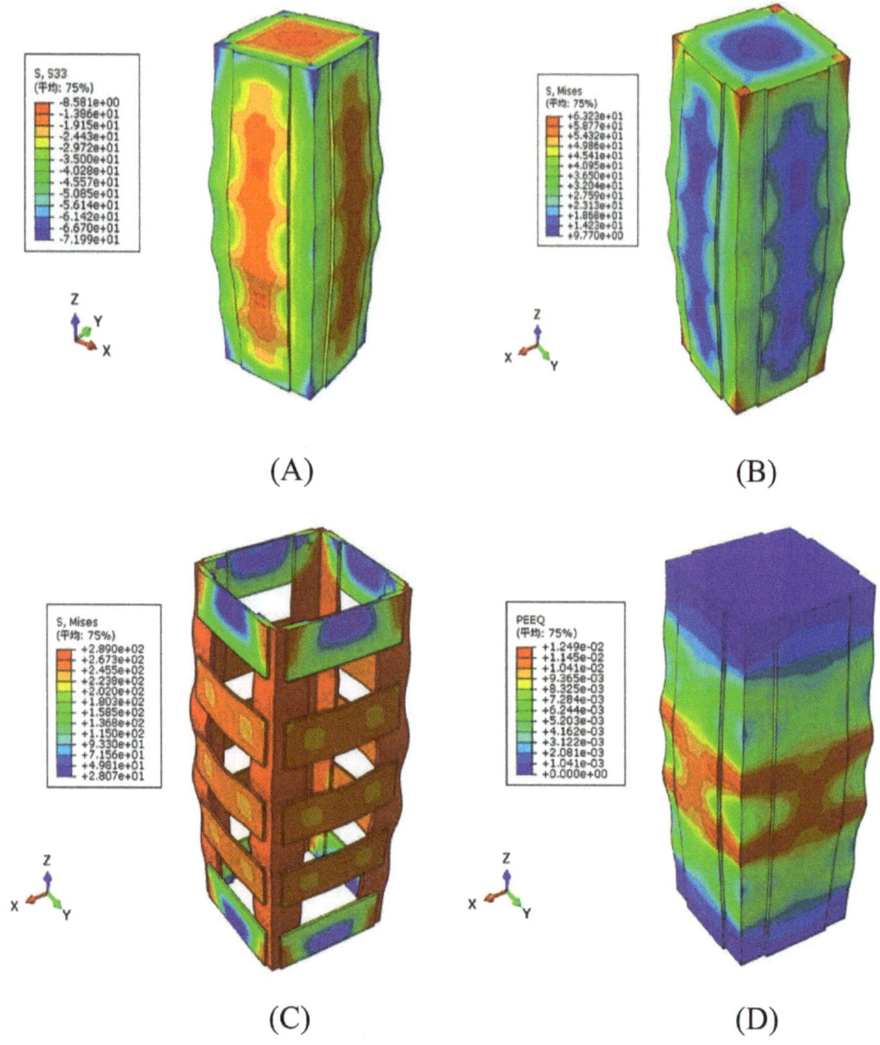

Figure 5.27 The stress and strain nephogram of steel frame in plastic stage. (A) X-direction displacement nephogram; (B) Y-direction displacement nephogram; (C) Mises stress nephogram; (D) cumulative plastic strain.

5.4.4.3 Failure stage in the loading process

After the peak load, the load–displacement curve enters the failure stage. Based on the stress nephogram and the strain nephogram of the core concrete in the failure stage, as shown in Fig. 5.28, we can find that the weak area of the core concrete gradually develops from the middle part of the specimen to the upper and lower sides. The strain nephogram shows that the compression plasticity deformation of the core concrete can develop more extensively so that the core concrete will crush

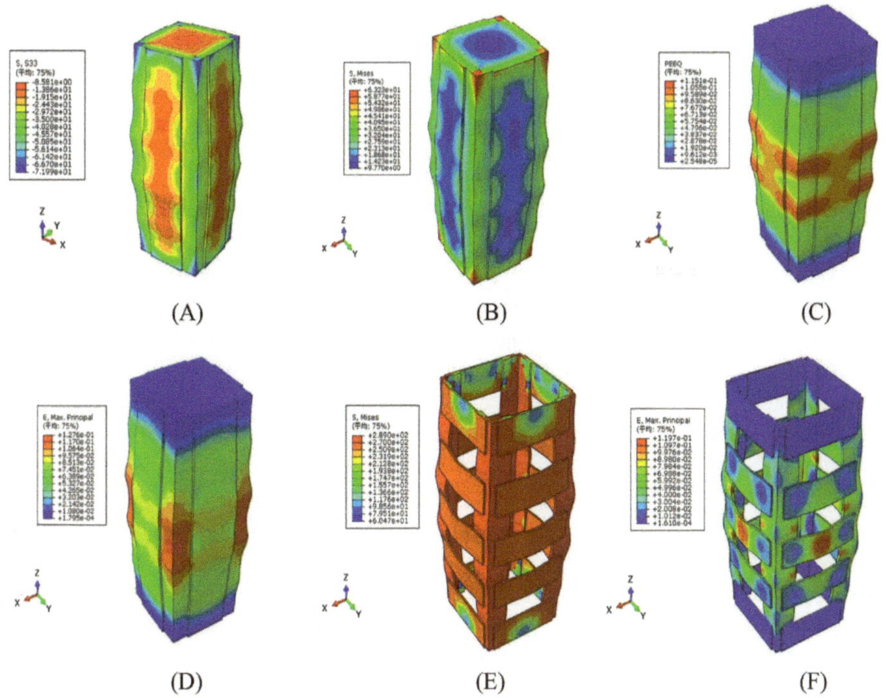

Figure 5.28 The stress and strain nephogram of SFC specimen in failure stage. (A) axial stress nephogram of core concrete; (B) Mises stress nephogram of core concrete; (C) cumulative plastic strain of core concrete; (D) total strain nephogram of core concrete; (E) Mises stress nephogram of steel frame; (F) total strain nephogram of the steel frame.

and flake along the column height. Based on the stress and strain nephogram of the failure of the core concrete in the failure stage, as shown in Fig. 5.28, the angle steel shows the obvious local buckling. Therefore, we can find that the angle steel not only completely yields, but also has serious bending deformation. Meanwhile, the yield of the steel batten gradually develops from the middle part to the upper and lower sides of the specimen, and the steel batten fully yields and seriously bulges outward. Therefore, the steel frame has a good confining effect on the core concrete, which not only improves the bearing capacity of the SFC specimen but also improves the ductility of the composite column.

5.5 Design-oriented constitutive model for the steel frame confined concrete

It is well known that the design-oriented constitutive model is important for the application of SFC in engineering structures. Therefore, we should identify the confining mechanism of the steel frame first. Then we should propose the equation of the confining pressure. Finally, we can establish the design-oriented constitutive model for SF confined concrete.

5.5.1 Confining mechanism of the steel frame

As we discussed in the experimental and simulation results, both the angle steel and the steel batten will provide the confining effect on the core concrete in SFCs under the axial load. And the confining effect can be influenced by the following factors: the specimen size, the angle steel size, the steel batten size, and the assembly of the steel battens. However, the confining effect provided by the steel batten is complex, due to the axial incoherence layout. In order to identify the confining mechanism, we can make the angle steel and the steel batten as the two independent components to analyze the confining effect provided by the steel frame in the following sections. Therefore, the total confining pressure can be assumed as follows:

$$f_l = f_{la} + f_{lb} \tag{5.1}$$

where f_l is the total lateral confining pressure; f_{la} is the lateral confining pressure provided by the angle steels; f_{lb} is the lateral confining pressure provided by the steel battens.

5.5.1.1 Confining pressure provided by the angle steel

In the square section, the angle steel can provide axial evenly stress to bear the axial load, and the lateral uneven pressure to confine the core concrete. The lateral uneven confining pressure will divide the core concrete section into the effective confining area and the ineffective confining area, as shown in Fig. 5.29, where the shadow area is the effective confining area.

As we discussed, the steel battens can restrict the axial and lateral deformation of the angle steels, which can prevent the local buckling of the angle steel and make the angle steel provide fully the confining effect on the core concrete. And the stress state of the angle steels is similar to that of four corner parts in the square steel tube. Therefore, the confining pressure distribution of the angle steel can refer to as the square concrete filled in the steel tube. According to Chapter 2, the distribution equation is $q(x) = 4qx^2/L^2$ along the length of one side, where x is the distance between the target point and the middle point; q is the max lateral confining pressure, $q = 2tf_y/L$. Then, the mathematical equation is derived by the finite integration method, as follows:

$$f_{la} = 2/L \int_{L/2-L_s}^{L/2} q(x)dx = (2t_1 f_{y1}[L^3 - (L-2L_S)^3])/(3L^4) \tag{5.2}$$

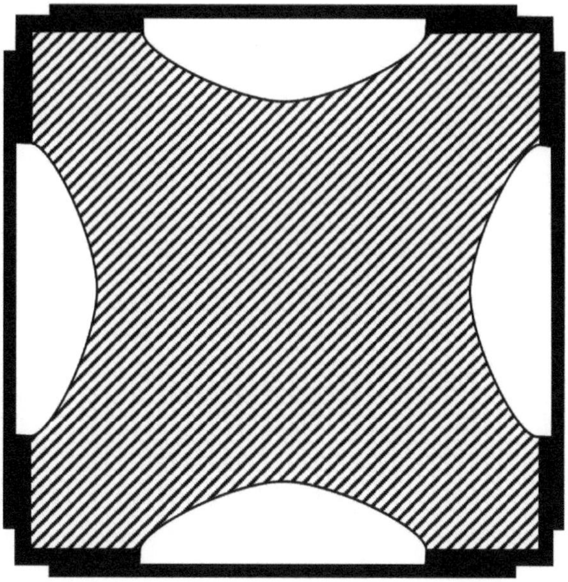

Figure 5.29 Confining distribution in the cross-section of core concrete. Diagrammatic sketch.

where t_1 is the thickness of the angle steel; f_{y1} is the yield strength of the angle steel; L_s is the side length of the angle steel; L is the side length of the core concrete section.

5.5.1.2 Confining pressure provided by the steel batten

It has been known that steel battens just provide the lateral pressure to confine the core concrete under the axial loading, due to the longitudinal separation. However, the longitudinal separation makes the uneven distribution of the confining pressure along the column height, as shown in Fig. 5.30, which reduces the confining pressure provided by the steel battens. Then the confining pressure can be computed by the following equation:

$$f_{lb} = 2at_2f_{y2}/L \tag{5.3}$$

where t_2 is the thickness of the steel battens; f_{y2} is the yield strength of the steel battens; L is the side length of the core concrete section; a is the reduction ratio.

Based on the axial mechanical mechanism of the steel confined concrete, the following layout of steel battens can lead to less reduction ratio:

(1) increasing width of the steel batten
(2) increasing thickness of the steel batten
(3) reducing the spacing between the steel battens.

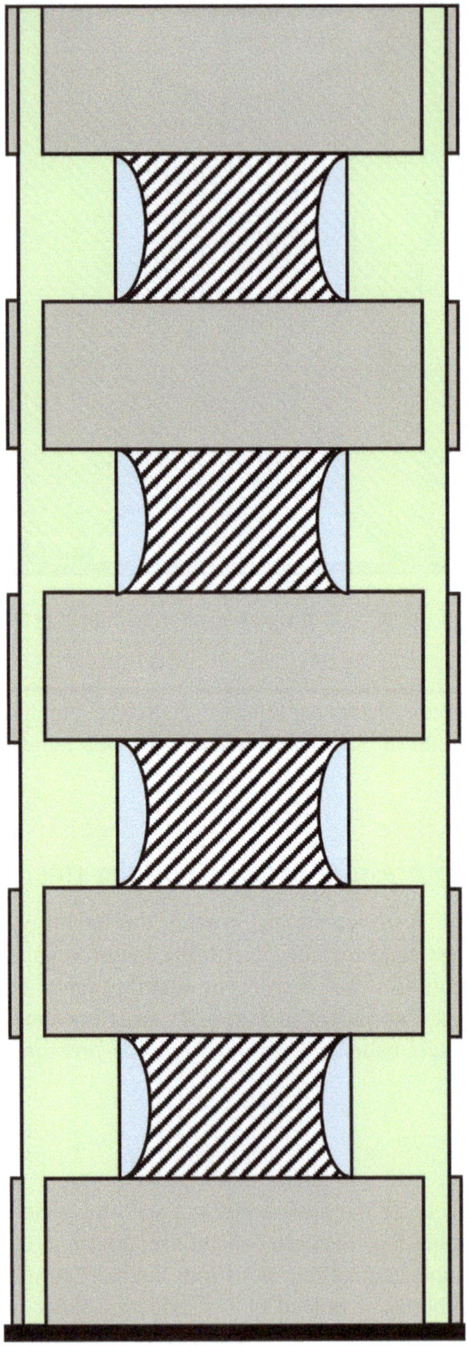

Figure 5.30 Confining distribution in the longitudinal direction.
Diagrammatic sketch.

Therefore, we should propose the reduction ratio by the thickness of the steel battens, the spacing between the steel battens, and the area ratio of the spacing. Based on experimental results and the existing research (Campione, 2012), the equation of reduction ratio is proposed as follows:

$$a = 1.31exp(1.4t_2/S - 2.05V_b) \leq 1 \tag{5.4}$$

where t_2 is the thickness of steel batten; S is the spacing between steel battens; V_b is the area ratio of the spacing, $V_b = S\ (L - 2L_s)\ /\ L\ /\ (S + B)$, B is the width of steel batten, L is the side length of the concrete section.

5.5.1.3 Bearing capacity model of SFC

To verify the proposed model of the steel frame confined concrete, we establish the bearing capacity model of SFC by referring to the concrete filled in steel tube. The bearing capacity model is added directly by two parts including the steel frame and the steel frame confined concrete, which is given as follows:

$$N_{st} = N_s + N_{cc} = f_y A_s + f_{cc} A_{cc} \tag{5.5}$$

where f_y is the yield stress of the angle steels; f_{cc} is the peak stress of the steel frame confined concrete; A_s is the section area of the angle steel; A_{cc} is the section area of the steel frame confined concrete.

We can find that the steel frame confined concrete can be simplified to the actively confined concrete. So the peak stress of the steel frame confined concrete can refer to previous research by Shi et al. (2017), based on the twin shear strength theory (Yu, 2004). The equation is expressed as follows:

$$f_{cc} = \gamma_u f_{co} + k f_l \tag{5.6}$$

where γ_u is the strength reduction parameter (Sakino et al., 2004), $\gamma_u = 1.67L^{-0.112}$, L is the side length of concrete section; f_{co} is the compressive strength of unconfined concrete; f_l is the total confining pressure; k is the confining parameter and can be computed as follows:

$$k = 1 + 1/(3.5\omega^{0.6973}) \tag{5.7}$$

where ω is the relative confining coefficient, $\omega = f_l/\ f_{co}$, f_l is the lateral confining pressure, f_{co} is the compressive strength of unconfined concrete.

Finally, the comparisons between the computation results and the experiment results are summarized in Table 5.8. The comparisons of the bearing capacity of SFC specimens show the computation results are in good agreement with the experiment results. Therefore, the proposed models are accurate.

5.5.2 The Design-oriented constitutive model

As a type of actively confined concrete, we can establish the design-oriented model of the steel fame confined concrete by modifying the existing constitutive model proposed by Guo et al. (1982), which has been discussed in Chapter 2. It should be noted that the proposed model is based on the characteristics of SF confined concrete in dimensionless coordinates. Therefore, the parameters in the increasing stage and the softening stage can be identified by fitting the various confined concrete. It is well known that the increasing stage of the stress−strain curves of the core concrete show a similar trend and the lateral confining pressure has a significant effect on the peak stress, the peak strain, and the softening stage curve.

Based on the above discussion, the parameters in the proposed model can be modified by the following attention:

(1) the parameters in the increasing stage should consider the peak stress of the confined concrete
(2) the parameter in the softening stage should consider the peak stress and the peak strain
(3) the slope parameter should be introduced in the equation of the softening stage, because of the complex strain state of the steel frame confined concrete
(4) the uneven parameter should be introduced in the proposed model to reflect the uneven confining effect of the steel battens.

In summary, the design-oriented model can be expressed as follows:

$$y=\left\{\begin{array}{l}\alpha_a x+(3-2\alpha_a)x^2+(\alpha_a-2)x^3\\ \dfrac{x}{\psi\alpha_d(x-1)^n+x}\end{array}\right\} \tag{5.8}$$

where $x = \varepsilon_c/\ \varepsilon_{cc}$, ε is the strain in the stress−strain curve; $y = \sigma\ /\ f_{cc}$, σ is the axial stress; α_a is the increasing parameter, $\alpha_a = 2.4 - 0.0124f_{cc}$; Ψ is the uneven coefficient, $\Psi = 2.3\ /\ \omega$; n is the slope parameter, $n = 1 + \exp(-3\omega)$; α_d is the decreasing coefficient, it can be computed by the following equation:

$$\left\{\begin{array}{l}\alpha_d=\dfrac{3d}{17(d-1)^n}\\ d=0.0033/\varepsilon_{cc}\times[1+2.18\times10^5(2\omega f_c\varepsilon_{cc}/f_y)^{1.2}]\end{array}\right\} \tag{5.9}$$

In the proposed model, ε_{cc} is the peak strain, which is influenced by the stress state of the steel frame confined concrete. Based on the experimental and theoretical studies on confined concrete (Shi et al., 2017), the stress state of the steel frame confined concrete can be divided into two parts: the hydrostatic stress and the self-stress removed from the hydrostatic stress. It can be found that the latter is higher than the peak stress of the unconfined concrete, due to the confining effect.

Table 5.8 The verification of the bearing capacity of SFCs.

Item	Confining stress (MPa)	Experimental results (kN)	Computational results (kN)	Relative error (%)
SFC-1	14.04	5257.9	5179.8	−1.49
SFC-2	14.04	3862.9	3904.6	1.08
SFC-3	14.04	5518.9	5329.7	−3.43
SFC-4	13.04	4662	4813.5	3.25
SFC-5	17.11	6051.6	5980.2	−1.18
SFC-6	15.31	5419.9	5310.4	−2.02
SFC-7	12.76	5041.9	5048.2	0.12
SFC-8	16.59	5860.9	5440.1	−7.18
SFC-9	8.63	4492.9	4698.1	4.57
SFC-10	15.35	5572.9	5314	−4.64
SFC-11	28.04	6472.9	6577.2	1.61

Correspondingly, the peak strain can be divided into two parts: the strain ε_p generated by the hydrostatic pressure; the strain ε_c generated by the self-stress removed the hydrostatic stress. Based on the existing studies on confined concrete (Chung et al., 2002; Hong et al., 2006), the peak strain (ε_{cc}) can be computed as follows:

$$\varepsilon_{cc} = \varepsilon_c + \varepsilon_p \tag{5.10}$$

where $\varepsilon_c = (700 + 172f_c'^2) \times 10^{-6}$, $f_c' = f_c + (k-1)\,p$, $\varepsilon_p = 2.4 \times (2\omega)^2 \times 10^{-3}$

5.5.3 Verification of constitutive model

In order to verify the proposed constitutive model, we show the comparisons between the computation results by the proposed model and the experiment results in Fig. 5.31. The curve comparisons show that the proposed constitutive model can provide an accurate stress–strain curve for the steel frame confined concrete.

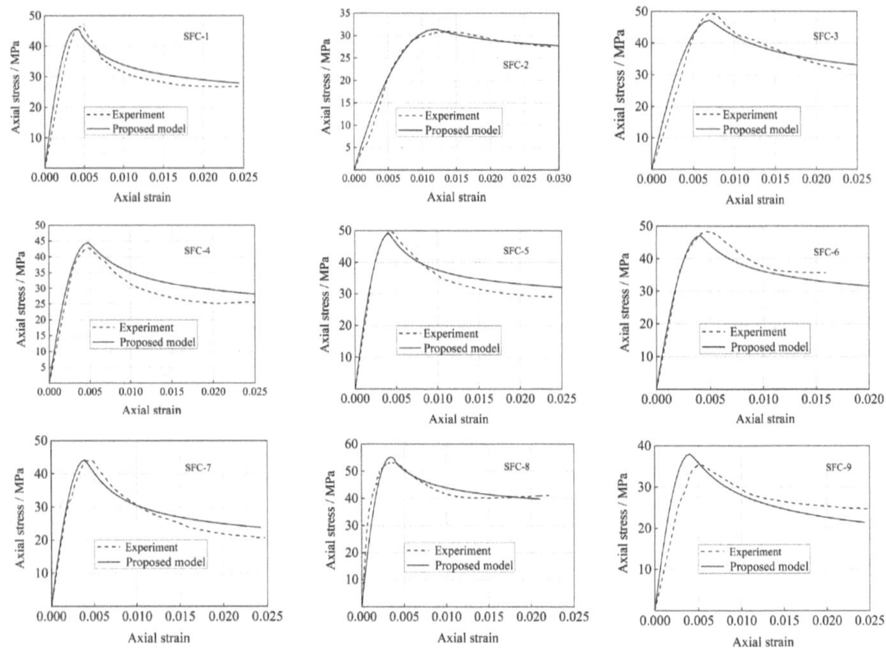

Figure 5.31 The verification of the proposed model.
The results comparisons.

5.6 Conclusions

This paper presents a series of studies on the section form and the compressive behaviors of SFCs. Based on the analysis in the experiment and simulation, we study the confining effect provided by the steel frame in detail. Based on the confining mechanism of the steel frame, we establish the constitutive model of the steel frame confined concrete. Based on the study results and discussions presented in the paper, the following conclusions can be drawn:

(1) Steel frame can provide an effectively confining effect on the core concrete in noncircular section columns. The confining effect can improve the mechanical properties of the core concrete.
(2) The same confining pressure has a stronger confining effect on the concrete with a lower strength grade, which leads to a higher increase in the stress and strain, and the gentler softening stage of the stress–strain curve.
(3) The higher steel ratio of the angle steel can improve the strength property of the core concrete but has little effect on the developing trend of the stress–strain curves.
(4) The steel batten is the key to the confining effect of the steel frame, and meanwhile, the thickness has an obvious effect under the lower steel ratio, the width has an obvious effect under the higher steel ratio, and smaller spacing between the steel battens leads to the higher residual stress.

(5) The inner additional confining materials (i.e., the inner steel spiral hoops) have a significant effect on the confined concrete and less effect on the curve trend.

(6) The influence factors for strength increase are in descending order: the inner additional confining material, the dense assembly of the steel battens, the wide the steel battens, the thick steel batten, and the big size of the angle steels.

(7) Based on the strain analysis of the steel frame in specimens, all the strains (i.e., the lateral strains of steel batten, and the axial and lateral strains of the angle steels) reach the yield state before the peak load of SFCs, which show that the steel frame can play the full steel properties to confine the core concrete in the section form.

Acknowledgments

This work was supported by the National Natural Science Foundation of China [grant numbers 52108171; 51878540]. The authors gratefully acknowledge two Master's students (Mr. Qv Yunsong and Mr. Tian Wenkai) of Xi'an University of Architecture and Technology for the drawing works.

References

Ahmed, AEF (2018). New axial stress-strain model of square concrete columns confined with lateral steel and FRP. *Composite Structures*, *202*, 731–751.

Campione, G (2012). Load carrying capacity of RC compressed columns strengthened with steel angles and strips. *Engineering Structures*, *40*, 457–465.

Chung, HS, Yang, KH, Lee, YH, & Eun, HC (2002). Stress-strain curve of laterally confined concrete. *Engineering Structures*, *24*, 1153–1163.

Dritsos, S, Georgopoulos, T, & Pilakoutas, K (1993). *Experimental study on a composite technique for repair/strengthening techniques of RC columns*, 189–193.

Dritsos, S, & Pilakoutas, K (1992). *Composite technique for repair/strengthening RC members*, 958–963.

Guo, ZH, Zhang, XQ, Zhang, DC, & Wang, RQ (1982). Experimental investigation of the complete stress-strain curve of concrete. *Journal of Building Structures*, *3*, 1–12.

Hong, KN, Han, SH, & Yi, ST (2006). High strength concrete columns confined by low volumetric ratio lateralties. *Engineering Structures*, *28*, 1346–1353.

Hwang, HJ, Eom, TS, Park, HG, & Lee, SH (2016). Axial load and cyclic lateral load tests for composite columns with steel angles. *Journal of Structural Engineering*, *142*(5), 04016001.

JGJ 101-96. (1997). *Specification of test methods for earthquake resistant building*. China Architecture and Building Press.

Ji, J, & Zheng, WZ (2008). Experimental research on flexural mechanical behavior of angle-steel concrete columns. *Industrial Construction*, *38*(8), 67–72.

Julio, GR, Jose, MA, & Pedro, AC (2011). Behaviour of RC columns strengthened by steel caging under combined bending and axial loads. *Construction and Building Materials*, *25*, 2402–2412.

Kim, CS, & Hwang, HJ (2018). Numerical investigation on load-carrying capacity of high-strength concrete-encased steel angle columns. *International Journal of Concrete Structures and Materials*, *12*(1), 1−17.

Li, P, Wu, Y, Zhou, Y, & Xing, F (2018). Cyclic stress-strain model for FRP-confined concrete considering post-peak softening. *Composite Structures*, *201*, 902−915.

Montuori, R, & Piluso, V (2009). Reinforced concrete columns strengthened with angles and battens subjected to eccentric load. *Engineering Structures*, *31*(2), 539−550.

Montuori, R, Piluso, V, & Tisi, A (2012). Comparative analysis and critical issues of the main constitutive laws for concrete elements confined with FRP. *Composites Part B: Engineering*, *43*(8), 3219−3230.

Ozbakkaloglu, T, & Jian, CL (2013). Axial compressive behavior of FRP-confined concrete: Experimental test database and a new design-oriented model. *Composites Part B Engineering*, *55*(12), 607−634.

Rong, C, & Shi, QX (2018). Axial-strength model for FRP-confined concrete based on the improved twin shear strength theory. *Composite Structures*, *202*, 102−110.

Rong, C, Shi, QX, Zhang, T, & Zhao, HC (2018). New failure criterion models for concrete under multiaxial stress in compression. *Construction and Building Materials*, *161*, 432−441.

Sakino, K, Nakahara, H, Morino, S, & Nishiyama, I (2004). Behavior of centrally loaded concrete-filled steel-tube short columns. *Journal of Structural Engineering*, *130*(2), 180−188.

Shi, QX, Rong, C, & Zhang, T (2017). A constitutive model for confined concrete in composite structures. *Steel and Composite Structures*, *24*(6), 689−695.

Wang, Y, Yang, L, Yang, H, & Liu, C (2019). Behaviour of concrete-filled corrugated steel tubes under axial compression. *Engineering Structures*, *183*, 475−495.

Yu, MH (2004). *Unified strength theory and its applications*. Springer-Verlag.

Zhang, FW, Ji, J, Lu, H, & Sui, H (2012). Behavior of angle reinforced concrete stub column under axial compression. *Journal of Building Structures*, *33*(4), 121−127.

Zheng, WZ, & Ji, J (2008a). Dynamic performance of angle-steel concrete columns under low cyclic loading-I: experimental study. *Earthquake Engineering and Engineering Vibration*, *1*, 67−75.

Zheng, WZ, & Ji, J (2008b). Dynamic performance of angle-steel concrete columns under low cyclic loading-II: Parametric study. *Earthquake Engineering and Engineering Vibration*, *2*, 137−146.

Confined recycled aggregate concrete column

6

6.1 Introduction

In order to make environment-friendly construction and keep the development of civil engineering sustainable, recycled aggregate concrete (RAC) has gained increasing deep studies and wide applications in the past two decades (Torgal & Jalali, 2011; Xiao et al., 2004). The RAC is mixed with the recycled aggregates which is used to replace the natural aggregate. We can gain the recycled aggregate from the waste construction materials (e.g., the waste concrete, the waste brick, and so on) after the crushing and screening processes. Therefore, the application of RAC can both protect the environment and reduce the cost of practice engineering. However, most of the recycled aggregate has weak and uncertain mechanical properties which lead to the following disadvantages of the mixed RAC which has the same mix proportion as ordinary concrete: the low strength, the high brittleness, the poor durability, and the great dispersion of mechanical properties.

Over the past two decades, there are many studies on a series of salient issues for the theoretical analysis and the practical application of RAC. In the early stage, the studies focused on the mix design methods and the compressive strength prediction models (Etxeberria et al., 2007; Meyer, 2008; Pacheco-Torgal et al., 2012). Over the past decades, the studies focused on the comprehensive mechanical properties and durability properties of different RAC (López-Uceda et al., 2016). In recent years, studies focused on microproperties, macroproperties, and the behaviors of RAC component (Torgal & Jalali, 2011; Yu et al., 2020). Based on existing studies, it is clearly found that the properties of RAC differ from the common concrete with the natural aggregate, the disadvantages can be summarized into the following aspects: (1) even for the same mix proportion, the RAC has a great dispersion of the properties, because of the different raw materials of the recycled aggregate; (2) the RAC has the weak mechanical properties, because of the weak properties of the recycled aggregate (Behera et al., 2014; Dimitriou et al., 2018). Therefore, it can conclude that recycled aggregate is the key to influencing the properties of RAC.

Some studies introduced various methods to improve the weak and dispersion properties of RAC. The study presented by Evangelista and De Brito (2007) crushed the waste concrete into the fine recycled aggregate to gain the RAC with substantial properties, but the replacement rate of the fine aggregate is low. Nam et al. (2016) added the fiber into the RAC to improve the properties, and they studied the fiber reinforcement effect on the mechanical properties, and the study results show that the fibers and the fine recycled aggregates can be used to mix the RAC with the great properties. Then, Gao et al. (2020) used fine recycled aggregates, fibers,

Concrete Composite Columns. DOI: https://doi.org/10.1016/B978-0-323-85171-8.00006-9

and high-activity binder materials to mix the RAC. The test results show that the proposed mix proportion can make the RAC have high strength and high ductility. The study presented by Meng et al. (2021) studied the mix proportion of RAC, and the strength-based mix design method is proposed by the analysis of a relatively extensive database. In summary, most studies adopted a complex pretreatment method for the recycled aggregate or used the high-property materials in the mix proportion design. However, we should note several issues as follows: the required concrete strength in most engineering structures is ranging from 30 to 60 MPa; the uncertainty properties of the recycled aggregates and the binder materials are the important factors, which lead to the different mix proportions for the same target strength; the proposed complex mix proportions may go against the economy and the environmental protection of the RAC. Therefore, most existing methods, which focused on the properties improvement of the RAC, are complex and unnecessary. For the wide application of RAC in practical engineering, it is necessary to propose a simple mix design method that can quickly gain the RAC with stable properties at first, and then study the structural measures to improve the properties of RAC.

Besides the property improvement of RAC, the properties of RAC can be simplified and improved by the reasonable section design method in the composite structure. We have known that the lateral confining pressure can improve the mechanical properties and the durability properties of the core concrete. And many high property materials can provide the lateral confining pressure, based on the combinations of different materials in the composite structure (Kaeseberg et al., 2020; Rong et al., 2020; Rong & Shi, 2021). In practical engineering, the steel tube and the FRP tube are widely used to provide the confining effect (Fam & Rizkalla, 2001; Li et al., 2017). There have been several existing studies on the RAC filled steel tube and the FRP confined RAC. The experimental study presented by He and Zhang (2014) focused on the strength characteristics and failure criterion of the RAC under triaxial compressive pressure. The study results show the lateral confining pressure can improve and stable the properties of RAC. The FRP confined concrete cylinders have been analyzed by Xu et al. (2017), the corresponding experiments are reviewed in this study, and the strength prediction model is proposed. Li et al. (2017) presented an experimental study on the RAC filled steel tube, which shows the steel tube can provide an obvious confining effect on the core RAC. All existing studies show that the properties of RAC can be improved under the confining effect provided by some high-property materials. The composite structure can expand the applications of the RAC members. However, the studies on the confining mechanism of the core RAC in steel tube and FRP tube are limited. It is necessary to have a theoretical study of the RAC under different lateral confining pressures.

Based on the previous chapters, we can divide the confined concrete into the actively confined concrete and the passively confined concrete. The steel tube confined concrete seemingly belongs to the actively confined concrete, the lateral compressive pressure around the core concrete almost tends to be constant after the yield stage of the steel tube. However, it also can be divided into the passively confined concrete in fine analysis, due to the hardening stage of steel. The FRP confined concrete absolutely belongs to the passively confined concrete, the lateral confining pressure increases with the increase in the lateral deformation of the core concrete, due to the linear elastic property of the FRP tube. Although the outer high-property materials can improve the properties of RAC, there are some disadvantages in the composite column by using single confining material: the steel tube has low corrosion resistance and will appear the local buckling under axial compression; the FRP tube will suddenly break by the concrete expansion. Therefore, it is necessary to propose the multiple composite forms for the RAC column.

Against the background, this chapter presents the experimental and theoretical studies on two objects: the simple mix design method of RAC and the axial compressive behavior of various confined RAC. Therefore, the contents of this chapter can be concluded into the following aspects: (1) the simple mix design method, including the orthogonal test method to identify the main influence factors of RAC, and the single-factor test to propose the strength prediction equation for each mix proportion of RAC; (2) the experimental study on axial compression behavior of various confined RAC, including the RAC filled steel tube, the GFRP confined RAC and the double confined RAC; (3) the confining mechanism of RAC under various confining effects, especially for the double confining effect; (4) the constitutive model for each confined RAC. The first study content focused on the required mix proportion by the target strength, the design target is to design an efficient and accurate mix proportion of RAC for the application in practical engineering. The last three contents are important for the theoretical studies of the confined RAC, including the constitutive models, the section design methods, and the applications of the composite RAC structures.

6.2 Simple mix design method of the recycled aggregate concrete

Based on the existing studies of RAC, the water–binder ratio is one of the main influencing factors for compressive strength. However, the optimal content of the fly ash and the bind-aggregate ratio are still controversial. In order to both increase the efficiency of the trial mixed method and reasonably consider the influence effect of the water–binder ratio, the bind-aggregate and the fly ash content in mix proportion of RAC, we can propose the simple mix design method by the compressive strength test of RAC. In the proposed method, we combine the orthogonal experiment and the single factor experiment to gain the mix proportion of RAC in this section. The orthogonal test method was first carried out to select the optimal combination of the influence factors, including the range analysis and the variance analysis of the orthogonal test results. The orthogonal test method can reduce the test times as much as possible and find the major factors accurately. After that, the single factor test method was carried out to gain the strength calculation equation of RAC, so as to determine the final mix proportion. The single factor test method is applied to propose the mix design equation by the major factors as a variation.

Based on the proposed mix design equation, the detailed mix proportion of RAC can be easily computed for the target compressive strength.

6.2.1 Raw material properties

In order to make mixed RAC have stable properties, we should test the basic properties of the material component first. Especially, the stable properties of the recycled aggregate should be the foundation to mix the RAC with stable properties.

6.2.1.1 Recycled coarse aggregate

The recycled coarse aggregate in this chapter is made by Yannan Mechanical Processing Factory in Xi'an, which is got from the waste concrete specimens after the crushing and screening process. The treated recycled coarse aggregate is shown in Fig. 6.1. The basic properties of the recycled coarse aggregate should meet requirements in the Chinese Standard (GBT 25177-2010, 2010) for the recycled coarse aggregate concrete. The Chinese Standard (GBT 14685-2011, 2011) suggests the test methods for the particle size gradation, the crushing index and apparent density of the pebbles, and the crushed stones for construction. Besides, the Chinese Standard (GBT 17431.2-2010, 2010) suggests the test methods of the water absorption ratio of the lightweight aggregate.

(1) Particle size gradation

The recycled coarse aggregate is screened by the shaker machine, as shown in Fig. 6.2. And the particle size gradation of the recycled coarse aggregate is

Figure 6.1 The treated recycled coarse aggregate a picture.

Figure 6.2 Particle gradation test of recycled coarse aggregate a picture.

Table 6.1 Particle gradation test results of recycled coarse aggregate.

Sieve Diameter (mm)	Sieve residue (g)	Each residue rate (%)	Accumulated residue rate (%)	Parameter in standard (%)
<2.36	16	0.3	100	—
2.36	7	0.1	100	95~100
4.75	1990	39.8	100	90~100
16	2958	59.2	60	30~70
26.5	27	0.5	1	0~5
31.5	0	0	0	0

summarized in Table 6.1. Based on the test results, the particle size of the recycled coarse aggregate used in this chapter is mainly in the range of 4.75–26.5 mm. Therefore, the particle grading meets the requirements in the Standard (GBT 14685-2011, 2011) for 5–25 mm continuous particle size.

(2) Crushing index

The crushing index is the capacity of the crushing resistance of the recycled coarse aggregate, the crushing index test is shown in Fig. 6.3. For both the crushing index test and the mix proportion of RAC, the aggregate particles should be

(A) (B)

Figure 6.3 Crushing index test of recycled coarse aggregate (A) Test process (B) Aggregate after crushing test.

Table 6.2 Crushing index test results of recycled coarse aggregate

Item	Sample mass (g)	Remaining mass (g)	Crushing index (%)		Standard value (%)
First group	3000	2618	12.7	Average value: 12	I type: <12;
Second group	3000	2674	10.9		II type: <20;
Third group	3000	2639	12		III type: <30

adopted as the particles without the needle and flake particles, and the particle size should be in the range from 9.50 to 19.0 mm. After the crushing index test (see Fig. 6.3), the fine particles with a size less than 2.36 mm should be screened out. And the percentage ratio of the remaining particle mass to the initial particle mass is the crushing index. The crushing index test results are shown in Table 6.2, where the arithmetic mean of the three test results is taken as the ultimate results. The test results show that the crushing index of the recycled coarse aggregate meets the requirements of the specification for the class I recycled coarse aggregate, which means it has a strong crushing resistance capacity.

(3) Apparent density

The apparent density test of the recycled coarse aggregate is carried in the jars. The test results are shown in Table 6.3. The arithmetic mean of the two test results is taken as the ultimate result. The test results show that the apparent density of the recycled coarse aggregate meets the requirements of the specification for the class I recycled coarse aggregate.

(4) Water absorption ratio

The test results of the water absorption ratio of recycled coarse aggregate are shown in Table 6.4. The average value of the three test results is taken as the

Table 6.3 Apparent density test results of recycled coarse aggregate.

Item	Mass after drying (g)	Total mass of test device without aggregate (g)	Total mass of test device with aggregate (g)	Apparent density (kg/m^3)		Standard value (kg/m^3)
First group	781	1873	2365	2696	Average value: 2690	I type: >2450; II type: >2350; III type: >2250
Second group	829	1876	2398	2691		

Table 6.4 Water absorption test results of recycled coarse aggregate.

Item	First group	Second group	Third group
Mass of immersed sample (g)	1582	1588	1585
Mass of drying sample (g)	1513	1518	1517
Water absorption ratio (%)	4.56	4.61	4.48
Standard value (%)	Average value: 4.6 I type: <3.0; II type: <5.0; III type: <8.0		

ultimate result. The test results show that the water absorption ratio of the recycled coarse aggregate meets the requirements of the specification for the class II recycled coarse aggregate.

The Chinese Standard (GBT 14685-2011, 2011) suggests the following points: the crushed stones for construction stipulate that the crushing indexes of class I, class II and class III crushed stones should be less than 10%, 20%, and 30%, respectively; the apparent density should not be less than 2600 kg/m^3; the water absorption ratio should be less than 1.0%, 2.0%, and 2.0%, respectively. Based on the comparison between the recycled coarse aggregate and the natural coarse aggregate, we can conclude that the recycled coarse aggregate has a smaller apparent density and larger water absorption. In addition, the crushing index of the recycled coarse aggregate in this test is small, which shows the strength is high, based on the above test. Meanwhile, the apparent density of the recycled aggregate is similar to the requirements of the natural gravel aggregate, but the water absorption is much greater than that of the natural gravel aggregate. Therefore, the overall properties of recycled coarse aggregate are relatively poor than that of natural gravel aggregate.

6.2.1.2 Other raw materials

(1) Cement

The cement used in the test is ordinary P42.5 Portland cement. The basic properties are shown in Table 6.5. All properties meet the requirements of the relevant standard.

(2) Fly ash

The fly ash used in the test is grade I, and the basic properties are shown in Table 6.6.

Table 6.5 Basic properties of cement.

	Loss on ignition	SO_3	MgO	Setting time (min)		Stability	Compressive strength (MPa)		Flexural strength (MPa)	
				Initial setting	Final setting		3d	28d	3d	28d
Measured value	2.14 ≤5.0	1.54 ≤3.5	1.82 ≤5.0	195 ≥45	260 ≤600	Good	27.9 ≥17.0	52.9 ≥42.5	5.9 ≥3.5	8.5 ≥6.5

Table 6.6 Basic properties of fly ash.

Fineness	Density	Loss on ignition	SiO_2	Al_2O_3	Fe_2O_3	CaO	MgO	SO_3
5.9	2.39	3.4	56.99	25.47	4.98	6.21	1.29	0.37

Table 6.7 Basic properties of coarse aggregates.

Particle gradation (mm)	Apparent density (kg/m^3)	Water absorption (%)	Crushing index (%)
5~25	2690	4.6	12

(3) Coarse aggregate

The coarse aggregates used in the test are recycled coarse aggregates. According to the test results in the above section, their basic properties are summarized in Table 6.7.

(4) Fine aggregate

The fine aggregate used in the test is the natural river sand, and the particle gradation test is carried out as shown in Fig. 6.4. The particle gradation of the fine aggregate meets the requirements of the natural sand in the grade Ⅱ. Based on the test results, the fineness modulus of the fine aggregate is 2.5, within the range from 2.3 to 3.0, which belongs to the medium sand. Finally, the particle gradation test results of the coarse aggregate and the fine aggregate are shown in Table 6.8.

(5) Admixture and water

The super-plasticizer used in the mixing process is the polycarboxylate high-performance superplasticizer of Subert. The mixed water and the curing water used in the experiment are ordinary tap water.

6.2.2 Orthogonal experiment

The orthogonal experimental design is an efficient, fast, and economical design method, which selects a part of the representative horizontal combination for the test. Based on tests with the multi factors and the multi-levels, we analyze the

Figure 6.4 Particle gradation test of the fine aggregate a picture.

Table 6.8 Particle gradation test results of the aggregate.

Item		Sieve Diameter (mm)					
Coarse aggregate		<2.36	2.36	4.75	16	26.5	31.5
	Sieve residue (g)	16	7	1990	2958	27	0
	Each residue rate (%)	0.3	0.1	39.8	59.2	0.5	0
	Accumulated residue rate (%)	100	100	100	60	1	0
	Parameter in standard (%)	—	95~100	90~100	30~70	0~5	0
Fine aggregate		<2.36	2.36	4.75	16	26.5	31.5
	Sieve residue (g)	16	7	1990	2958	27	0
	Each residue rate (%)	0.3	0.1	39.8	59.2	0.5	0
	Accumulated residue rate (%)	100	100	100	60	1	0
	Parameter in standard (%)	—	95~100	90~100	30~70	0~5	0

significance and the influence law of each factor of the proportion. Therefore, the important influence factors can be clearly and quickly determined in a limited number of tests.

6.2.2.1 Specimen design and fabrication

In this section, we selected the 7- and 28-day compressive strengths of RAC as the test indexes. And we selected the water–binder ratio (A), the binder–aggregate ratio (B), and the fly ash content (C) as the variable factors. Based on the existing study results, the three reasonable levels (i.e., A = 0.30, 0.35, and 0.40; B = 0.30, 0.33, and 0.36; C = 10%, 15%, and 20%) were selected in the orthogonal experimental design. The orthogonal experimental design scheme table selects $L_9(3^4)$ type in the experiment, the detailed parameters are shown in Table 6.9. The $L_9(3^4)$ type shows that we can arrange up to 4 variable factors, each factor contains 3 levels, so a total of 9 groups of specimens need to be tested. It can be found that there is a blank column (D) in Table 6.9 because we selected only 3 variable influence factors in this section.

It is assumed that the apparent density of the RAC mixture is 2300 kg/m^3, Based on the code for the mix proportion design in Standard (JGJ 55-2011, 2011) and combined with the orthogonal experimental design scheme, we designed the 9 groups of mix proportion of RAC as shown in Table 6.10. In addition, the dosage of the water reducer is 1% of the dosage of the binder material.

The pouring of the RAC cube test block should be carried out according to the test mix proportion in Table 6.10, and the size of the test block is 150 mm × 150 mm × 150 mm. Each group in the experiment has 9 test blocks, where 3 blocks

Table 6.9 Orthogonal experimental design scheme.

Test number	Factor			
	A	B	C	D
1	1(0.30)	1(0.30)	1(10%)	1
2	1(0.30)	2(0.33)	2(15%)	2
3	1(0.30)	3(0.36)	3(20%)	3
4	2(0.35)	1(0.30)	2(15%)	3
5	2(0.35)	2(0.33)	3(20%)	1
6	2(0.35)	3(0.36)	1(10%)	2
7	3(0.40)	1(0.30)	3(20%)	2
8	3(0.40)	2(0.33)	1(10%)	3
9	3(0.40)	3(0.36)	2(15%)	1

Table 6.10 Mix proportion in orthogonal experiment.

Test number	Water−binder ratio (A)	Binder−aggregate (B)	Fly ash ratio (C)	Blank factor (D)
1	0.3	0.3	10%	—
2	0.3	0.33	15%	—
3	0.3	0.36	20%	—
4	0.35	0.3	15%	—
5	0.35	0.33	20%	—
6	0.35	0.36	10%	—
7	0.4	0.3	20%	—
8	0.4	0.33	10%	—
9	0.4	0.36	15%	—

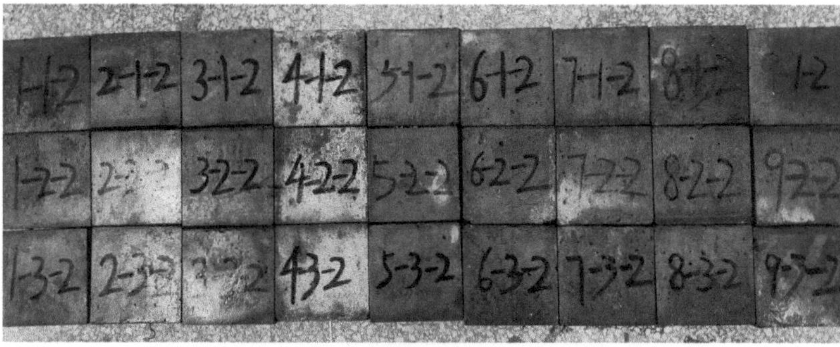

Figure 6.5 The test block in an orthogonal experiment. A picture.

are used to test the 7-day compressive strength and the other blocks are used to test the 28-day compressive strength. It should be noted that the formwork should be removed 24 hours after the pouring. All RAC blocks were watered for curing under normal temperature. Some test blocks are shown in Fig. 6.5.

6.2.2.2 Test process

Based on the standard for test methods of concrete block (GB/T 228.1-2010, 2010), the TYA-2000 electro-hydraulic pressure testing machine is used to test the compressive strength of the RAC cube, as shown in Fig. 6.6.

$$f = F/A \tag{6.1}$$

where F is the peak load, and A is the load-bearing area.

6.2.2.3 Analysis of test results

The range analysis method is usually used to analyze the orthogonal test results. The range analysis method can make an intuitive judgment on the influence law of various factors by a simple calculation. Therefore, the range analysis method is used to analyze preliminarily the influence law of the water–binder ratio (A), the binder–aggregate ratio (B), and the fly ash content (C) on the 7- and 28-day compressive strength of RAC, then the optimal level combination of each factor can be obtained. The cube compressive strength of RAC is computed by Eq. (6.1), and the test results of 7- and 28-day compressive strength of RAC are summarized in Table 6.11.

Based on the test results, the range analysis results are shown in Table 6.12. In the table of test results, K_i represents the sum of the corresponding test values when the factor level is i in the one-factor column (i.e., i = 1, 2, or 3), k_i represents the average of the corresponding test values when the factor level is i in one-factor column (i.e., i = 1, 2, or 3); T represents the sum of all test values under one factor. R is the range, which is the difference between the maximum and minimum test

Figure 6.6 The compressive test of RAC cube. A picture.

Table 6.11 Average compressive strength.

Item	1	2	3	4	5	6	7	8	9
7-day strength (MPa)	46.3	37.7	38.1	35.8	32.5	37.2	34.5	36.6	33
28-day strength (MPa)	59.9	52.4	55.7	49.7	46.8	54.1	47.9	47.2	42.8

values at each level of the one-factor column. The larger value of R means the greater effect of the factor on the test index.

Based on Table 6.12, the following conclusions can be drawn:

(1) For the 7-day compressive strength, the range of water–binder ratio is 6.0, the range of binder–aggregate ratio is 3.3, and the range of fly ash content is 5.0, that is, R(A) > R (C) > R(B).

(2) For the 28-day compressive strength, the range of water–binder ratio is 10.0, the range of binder–aggregate ratio is 3.7, and the range of fly ash content is 5.4, that is, R(A) > R(C) > R(B).

Table 6.12 The range analysis.

Item		Influence factors		
		A	B	C
7-day compressive strength	K1	122.1	116.6	120.1
	K2	105.5	106.8	106.5
	K3	104.1	108.3	105.1
	k1	40.7	38.9	40
	k2	35.2	35.6	35.5
	k3	34.7	36.1	35
	R	6	3.3	5
28-day compressive strength	K1	168	157.5	161.2
	K2	150.6	146.4	144.9
	K3	137.9	152.6	150.4
	k1	56	52.5	53.7
	k2	50.2	48.8	48.3
	k3	46	50.9	50.1
	R	10	3.7	5.4

(3) For the 7-day compressive strength, the binder–aggregate ratio shows that $k_1 > k_3 > k_2$, and the other two influence factors show that $k_1 > k_2 > k_3$, which means that the optimal level of water–binder ratio is A1, the optimal level of binder–aggregate ratio is B1 and the optimal level of fly ash content is C1.

(4) For the 28-day compressive strength, the water–binder ratio shows that $k_1 > k_2 > k_3$, and the other two influence factors show that $k_1 > k_3 > k_2$, that is, the optimal level of water–binder ratio is A1, the optimal level of glue bone ratio is B1 and the optimal level of fly ash content is C1.

Therefore, for both the 7- and 28-day compressive strength of RAC, the influence of the water–binder ratio is the largest, the influence of the fly ash content is the second, and the influence of the glue bone ratio is the smallest. Meanwhile, the 7- and 28-day compressive strength of RAC reaches the highest when the water–binder ratio is 0.30, the binder–aggregate ratio is 0.30, and the fly ash content is 10%.

In order to reflect the direct influence of the water–binder ratio, the binder–aggregate ratio, and the fly ash content, the relationship between the compressive strength of RAC and various factors is drawn in Figs. 6.7–6.9, where the level of each factor is the abscissa and the k_i value of each factor is the ordinate.

Based on the three figures, the following conclusions can be drawn:

(1) When the water–binder ratio increases from 0.30 to 0.40, the 7- and 28-day compressive strengths of RAC show a decreasing trend. And the 28-day compressive strength is closer to the linear change, which is similar to that of ordinary concrete.

(2) When the binder–aggregate ratio increases from 0.30 to 0.33, the 7- and 28-day compressive strengths of RAC show a decreasing trend. When the binder–aggregate ratio increases from 0.33 to 0.36, the changing trend of 7- and 28-day compressive strengths

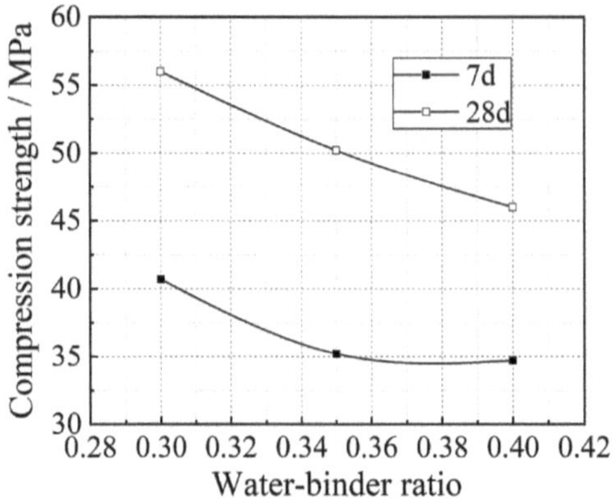

Figure 6.7 Influence of water–binder ratio. Test results analysis.

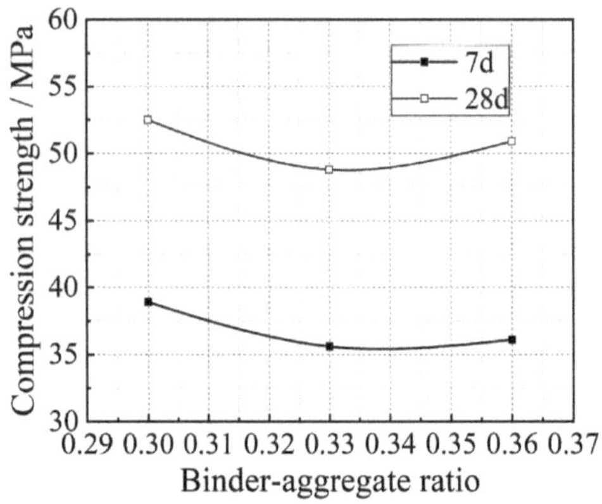

Figure 6.8 Influence of fly ash content. Test results analysis.

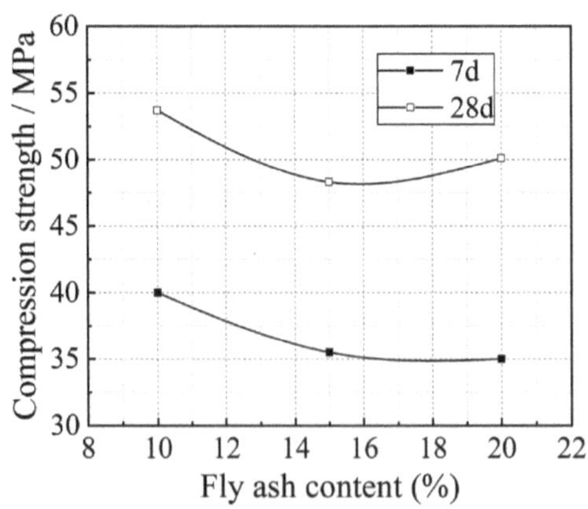

Figure 6.9 Influence of binder–aggregate ratio. Test result analysis.

of RAC is not obvious, and the influence of the binder–aggregate ratio on the compressive strength of RAC is relatively weak.

(3) When the fly ash content increases from 10% to 15%, the 7- and 28-day compressive strengths of RAC show a decreasing trend. When the fly ash content increases from 15% to 20%, the 7-day compressive strength of RAC still shows a slow decreasing trend, while the 28-day compressive strength shows a slowly increasing trend. The reason may be that the pozzolanic effect and the micro aggregate effect of the fly ash have not been fully stimulated in the early stage of the RAC mixture, and the effect of the fly ash will gradually play a role with the increase of the curing age. Meanwhile, the fly ash can improve the particle gradation of RAC, which leads to low porosity and high compactness, so as to improve the compressive strength of RAC.

6.2.2.4 Variance analysis

The variance analysis is also a common analysis method for the orthogonal test results. The variance analysis can divide the total variation into two parts: one is the variation caused by factors, and the other is the variation caused by errors. Based on F-test in variance analysis, we analyze the significant effect of the three influence factors. And the F-test is convenient to further reflect the influence effect of each factor on the test index in combination with the range analysis results.

The variance analysis of the 28-day compressive strength of RAC is shown in Table 6.13. Based on the comparisons between the F value and critical value F_a, the following four conditions can be drawn: (1) $F > F_{0.01}$, it shows that particularly significant impact, marked as "**"; (2) $F_{0.01} \geq F \geq F_{0.05}$, it shows the significant impact, marked with "*"; (3) $F_{0.05} \geq F \geq F_{0.1}$, it shows the certain impact, marked as "(*)"; (4) $F < F_{0.1}$, it shows the small impact. It can be seen from Table 6.13 that: the water–binder ratio has a significant effect on the 28-day compressive strength; the fly ash content has a certain influence on the 28-day compressive strength; the binder–aggregate ratio has little effect on the 28-day compressive strength.

6.2.3 Single-factor experiment

Based on the range analysis and the variance analysis of the orthogonal test results, among the three factors in the test, the water–binder ratio is the main influence

Table 6.13 Variance analysis of 28-day compressive strength.

Item	Quadratic sum	Free degree	Mean square difference	F	Critical value F_a	Significant degree
A	152.1	2	76.07	36.83	$F0.01(2,2) = 99$	significant impact, *
B	20.6	2	10.32	4.99	$F0.05(2,2) = 19$	small impact
C	45.8	2	22.9	11.08	$F0.1(2,2) = 9$	certain impact, (*)
Error	4.1	2	2.07			
Total	222.6	8				

factor on the 28-day compressive strength of RAC. When the binder–aggregate ratio is 0.30 and the fly ash content is 10%, the RAC can get higher compressive strength. Therefore, we can use the single-factor experiment to analyze the water–binder ratio for the determination of the final mix proportion. In this experiment, the binder–aggregate ratio is 0.30, the fly ash content is 10%, and the water-blinder ratios are 0.30, 0.34, 0.38, and 0.42. In addition, the RAC cube size, the fabrication process, and the curing process of the single-factor experiment are similar to the orthogonal experiment. The RAC test blocks are shown in Fig. 6.10. The test results are summarized in Table 6.14.

The relationship between the water–binder ratio and the 28-day compressive strength of RAC is shown in Fig. 6.11. It can be found that the compressive strength decreases linearly with the increase in the water–binder ratio. By the linear

Figure 6.10 The test block in single factor experiment a picture.

Table 6.14 Test results of single factor experiment.

Water–binder ratio	Material content (kg/m^3)					Compressive strength (MPa)
	Cement	Fly ash	Fine aggregate	Coarse aggregate	Water	
0.3	446	50	562	1090	149	59.8
0.34	442	49	556	1080	167	51.6
0.38	438	49	551	1070	185	45.4
0.42	434	48	546	1060	202	37.9

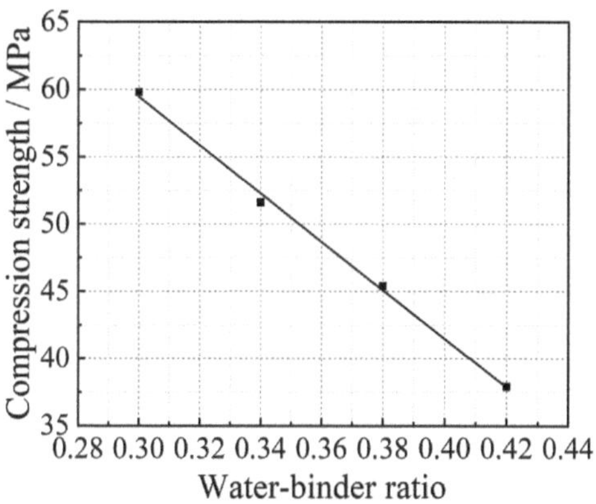

Figure 6.11 Relationship between water–binder ratio and compressive strength. Test result analysis.

regression, when the fly ash content is 10% and the binder–aggregate ratio is 0.30, the fitting equation of the 28-day compressive strength of RAC is as follows:

$$f = 113.4 - 179.8(W/B) \tag{6.2}$$

where, f is the compressive strength of the RAC cube, MPa; W/B is the water–binder ratio.

6.3 Axial compressive test of confined RAC cylinder

As we knew, the defects of recycled aggregate can affect the mechanical properties of RAC. Based on the reviews in Section 6.1, the axial compressive properties of RAC can be improved by the lateral confining pressure. When the steel tube is used as the confining material, the increase in the lateral confining pressure is small after the steel tube yields, so the steel tube confined RAC can be simply regarded as the actively confined concrete. When the GFRP tube is used as the confining material, the lateral confining pressure will increase with the increase in the lateral deformation of the core concrete, so the FRP confined RAC can be regarded as the passively confining concrete. In this section, the axial compression tests will be carried out on the three types of confined RAC columns, including the RAC filled steel tube, the GFRP confined RAC column, and the double confined recycled concrete column by the inner steel tube and the outer GFRP. Then, the axial compressive mechanical properties of the core RAC under different forms will be analyzed.

6.3.1 Experimental design

In order to study the confining mechanism of the confined RAC in detail, this section selects a circle section for the three types of composite short columns.

6.3.1.1 Specimen design

The design scheme of the confined RAC specimens is summarized in Table 6.15. The compressive test focuses on the confining effect of different forms on the same RAC, and a total of 25 specimens are categorized into 12 groups. The four RAC columns can be described as follows:

(1) A triplet of unconfined RAC columns (three specimens) has a height of 600 mm and a diameter of 200 mm, which is given the name of RCC.
(2) Three pairs of RAC filled steel tube (a total of six specimens), the size of each specimen is shown in Table 6.15, which are given names of SRCC.
(3) Four pairs of GFRP confined RAC cylinders (a total of eight specimens) have a height of 600 mm and an inner diameter of 200 mm, which are given the names of GRCC. Two pairs of specimens adopt prefabricated GFRP tubes with a thickness of 5 and 8 mm, respectively. The prefabricated GFRP tubes can directly serve as templates to simplify the manufacturing process. However, the manufacturer cannot provide prefabricated tubes with a smaller thickness. In order to study the RAC confined by the weak confining pressure, the other two pairs of specimens use the wet bonding method to wrap GFRP cloth on the RAC cylinder, which are wrapped by 2 layers (i.e., 1 mm thickness) or 4 layers (i.e., 2 mm thickness), respectively. This wet bonding method can get a similar confining effect to the prefabricated GFRP tube.
(4) Four groups of double confined RAC cylinders have a height of 600 mm and an inner diameter of 200 mm, which are given names of DRCC. In DRCC, the GFRP tube is used as the outer confining material and the steel tube is used as the inner confining material. The inner steel tube has a diameter of 102 or 140 mm and the GFRP tube has a thickness of 5 or 8 mm.

Table 6.15 Design scheme of confined RAC specimens.

Item	Specimen size (mm)		Steel tube size (mm)		GFRP tube size (mm)	
	Diameter	Height	Diameter	Thickness	Diameter	Thickness
RCC-1,2,3	200	600	—	—	—	—
SRCC102-1,2	102	300	102	6.5	—	—
SRCC127-1,2	127	400	127	5.5	—	—
SRCC140-1,2	140	400	140	6.5	—	—
GRCC(1)-1,2	202	600	—	—	202	1
GRCC(2)-1,2	204	600	—	—	204	2
GRCC(5)-1,2	210	600	—	—	210	5
GRCC(8)-1,2	216	600	—	—	216	8
DRCC102(5)-1,2	210	600	102	6.5	210	5
DRCC140(5)-1,2	216	600	140	6.5	216	8
DRCC102(8)-1,2	210	600	102	6.5	210	5
DRCC140(8)-1,2	216	600	140	6.5	216	8

Source: From Rong, C., Qv, Y. S., Shi, Q. X., Wang, P., & Li, C. F. (2022). Axial behaviors of recycled aggregate concrete cylinders under various confinement types: The experimental and theoretical analysis. *Construction and Building Materials*, *340*, 127821.

6.3.1.2 Preparation of specimens

(1) Preparation of formwork

The steel tubes were used as the formwork for the SRCC specimens, in which the steel tube with a diameter of 102 or 127 mm used the steel plates (150 mm × 150 mm × 10 mm) as the base plates, and the steel tubes with the diameter of 140 mm used the steel plates (250 mm × 250 mm × 10 mm) as the base plate.

For GRCC-5-1,2 and GRCC-8-1,2, the prefabricated GFRP tubes were used as the formwork, and the wood plates (250 mm × 250 mm × 10 mm) were used as the base plates and fixed with GFRP tubes by epoxy resin.

For RCC, GRCC-1-1,2, and GRCC-2-1,2, the GFRP tubes (the thickness is 5 mm) were used as formwork, the GFRP tube should be removed after 7-day curing. After that, GRCC-1-1,2 and GRCC-2-1,2 were got by wrapping GFRP cloth on RCC after 28-day curing. All of them used wood plate (250 mm × 250 mm × 10 mm) as the base plate and fixed with GFRP formwork by epoxy resin.

For DRCC specimens, the steel tubes were welded in the steel plates (250 mm × 250 mm × 10 mm) which were used as the base plate. Meanwhile, in order to lead out the connecting line of strain gauges, the GFRP tube opened a small hole at 15 mm from the top end. Then the strain gauge should be pasted on the outer surface of the steel tube. In addition, a layer of sealant was evenly applied on the surface of the strain gauge, then the gauze with soaked epoxy resin was evenly applied to prevent the damage to strain gauges during concrete pouring. At last, the centering and fixing of the GFRP tube and the steel tube was finished in the flowing steps: (1) the location of the GFRP tube should be marked on the bottom plate; (2) the GFRP tube should be placed at the mark; (3) the level laser should be used to align the top; (4) the bottom of the GFRP tube should be slightly raised and temporarily fixed by a wooden wedge; (5) the bottom should be fixed by epoxy resin; (6) after the epoxy resin is dry, the wooden wedge should be pulled out, and the slight gap can be ignored. The formwork before RAC pouring is shown in Fig. 6.12.

(2) Pouring and curing

The raw materials used for RAC are the same as those in Section 6.2. Based on the results of the orthogonal experiment and the single-factor experiment, it is determined that the water–binder ratio is 0.38, the binder–aggregate ratio is 0.30 and the fly ash content is 10%, and the detailed mix proportion is shown in Table 6.16.

The SRCC specimens were poured at one time. Each specimen was poured in three times and fully vibrated once per one-third height of pouring. After pouring, the concrete surface was leveled, watered for curing at normal temperature, and the formwork should be removed after 7 days. The DRCC after RAC pouring is shown in Fig. 6.13.

(3) Wrapping of GFRP cloth

The required length of GFRP cloths was calculated based on the design scheme, which included a lap length of 150 mm. In addition, the upper and lower ends of each specimen should be reinforced by the GFRP strips. The length can be consistent with the GFRP cloth, and the width is 100 mm. The wrapping processes are as

Figure 6.12 The formwork before RAC pouring. A picture.

Table 6.16 Mix proportion of RAC (kg/m^3).

Cement	Fly ash	Sand	Recycled aggregate	Water
438	49	551	1070	185

Source: From Rong, C., Qv, Y. S., Shi, Q. X., Wang, P., & Li, C. F. (2022). Axial behaviors of recycled aggregate concrete cylinders under various confinement types: The experimental and theoretical analysis. *Construction and Building Materials*, *340*, 127821.

follows: (1) the GFRP cloth and GFRP strip are cut according to the size for standby; (2) after the hardening of epoxy resin, the surface of the unconfined RAC cylinder is sanded by the angle grinder and the sandpaper, then the floating ash is removed to ensure that the bonding surface is flat; (3) the epoxy resin is taken by an appropriate amount in proportion, mixed evenly and applied on the bonding surface; (4) the GFRP cloth is pasted along the circumferential direction, it should

Figure 6.13 The DRCC specimen a picture.

ensure that the dense bonding between the cylinder surface and GFRP cloth, and between each subsequent two layers of GFRP cloth; (5) GFRP strips at both ends of the specimen are wrapped in the same way; (6) when the epoxy resin is slightly dry, the redundant fires on the surface of the specimen should be trimmed; (7) after the epoxy resin is dry and hard, the upper and lower ends of the specimen should be polished to make the surface flat.

6.3.1.3 Mechanical properties of materials

(1) Steel tube

The steel tubes use Q235 grade steel, three standard tensile coupons were prepared for each type of steel tube, based on the standard (GB/T 228.1-2010, 2010). The averaged material test results are summarized in Table 6.17.

In addition, the axial compression behaviors of the empty steel tubes are tested in this study, because it has an obvious axial load capacity. The load–strain curves

Table 6.17 Tensile properties of steel tube.

Steel tube size (mm)	Yield strength (MPa)	Elastic strength (MPa)	Elastic modulus (MPa)
102 × 6.5	405	548	2.23×10^5
127 × 5.5	340	507	2.10×10^5
140 × 6.5	380	520	2.11×10^5

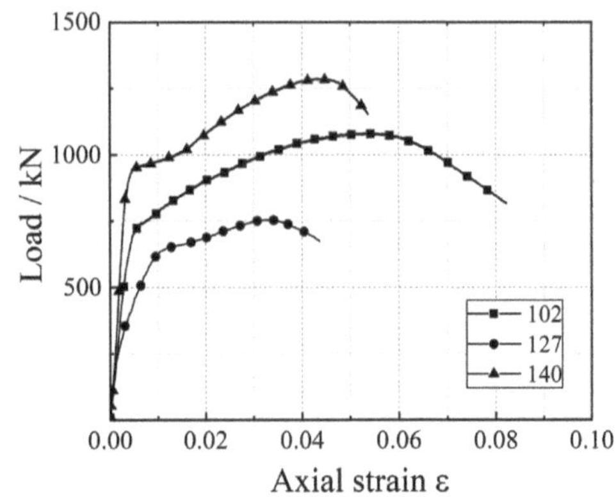

Figure 6.14 Load−strain curves of the steel tubes load-strain relationship.

of the steel tubes are summarized in Fig. 6.14, where the axial strain is the ratio of the total axial displacement to the column height. It can be concluded from Fig. 6.14 that we can divide the curves of a steel tube into three loading stages: in the elastic stage, the force characteristics of the steel tube are in the elastic state; in the hardening stage, the force characteristics of steel tube are in the yield and hardening state; in the softening stage, the force characteristics of steel tube are in the local buckling state.

(2) GFRP cloth

The GFRP cloths were directly wet-winded on RCC in GRCC-1 and GRCC-2. Three GFRP coupons were made for each thickness type of GFRP cloth, then tensile tests were conducted on these FRP coupons, based on the standard (Wu & Lv, 2003). The averaged material test results are summarized in Table 6.18.

(3) GFRP tube

The GFRP tubes were provided by Guangdong Fiber Glass Co., Ltd. The GFRP tubes were winded with a glass fiber angle of 80°, and the circumferential tensile test of the GFRP pipe was carried out by the separating disc method. The separating disc is shown in Fig. 6.15. The GFRP tube was cut into a 30 mm wide ring and installed on the outer circumference of the separation disc. The tensile force on the GFRP ring was exerted by applying a tensile force to the separation disc. In order to measure the circumferential tensile strain of the GFRP tube, two strain gauges were pasted on both sides of the GFRP ring and near the gap of the separation disc. The averaged material test results are summarized in Table 6.18.

Table 6.18 Tensile properties of GFRP.

Thickness (mm)	Tensile strength (MPa)	Elastic modulus (MPa)	Fracture strain (%)
1	181	11,786	1.24
2	190	13,972	1.13
5	465	42,608	1.38
8	473	53,062	1.29

Source: From Rong, C., Qv, Y. S., Shi, Q. X., Wang, P., & Li, C. F. (2022). Axial behaviors of recycled aggregate concrete cylinders under various confinement types: The experimental and theoretical analysis. *Construction and Building Materials, 340*, 127821.

(A) (B)

Figure 6.15 Tensile test of GFRP tube (A) test; (B) failure mode.

In addition, the axial compression behaviors of the empty GFRP tubes were tested in this study. It can be concluded that the GFRP tube is always in elastic condition under the loading process. However, the bearing capacity of the empty GFRP tube is too small to ignore, based on comparisons between the load of the GFRP tube and the RAC column.

6.3.1.4 Test instrumentation and procedure

The axial load data of the specimen was obtained by the force sensor embedded in the testing machine. Two linear variable displacement transducers (LVDTs) were diagonally installed beside the specimen to measure the axial displacement of the full height.

For the RCC specimens, three axial strain gauges were installed equidistantly along the circumferential direction on the middle surface. These strain gauges were used to measure the axial strain in the middle area of the specimen.

Figure 6.16 Axial compressive test of specimens. A picture.

For the SRCC specimens and the GRCC specimens, three pairs of strain gauges (including one axial strain gauge and one circumferential strain gauge) were installed equidistantly along the circumferential direction on the middle surface. For the DRCC specimens, three pairs of strain gauges were installed equidistantly along the circumferential direction on the middle surface of the outer FRP tube and the inner steel tube. These strain gauges were used to measure the axial and circumferential strain in the middle area of the specimen.

All specimens were loaded by the displacement controlling method on the YAS-5000 microcomputer-controlled electro-hydraulic servo pressure testing machine, as shown in Fig. 6.16. For RCC, GRCC-1, and GRCC-2 specimens, the displacement loading rate was 0.003 mm/s. For DRCC, GRCC-5, and GRCC-8 specimens, the displacement loading rate was 0.007 mm/s. For the SRCC specimens, the displacement loading rate before yielding was 0.005 mm/s, and the loading displacement rate after yielding was 0.012 mm/s.

6.3.2 Failure process and failure mode

For the RCC specimens, there was no obvious change in the appearance of the specimen in the initial stage of loading. With the increase in load, the deformation increased and there were more microcracks on the specimen. When the specimen was close to the peak load, the microcracks on the surface of the specimens appeared and increased gradually. When the specimen was damaged, there were vertical or oblique splitting cracks on the surface of the specimen, and parts of concrete were peeled. The specimen will quickly lose its bearing capacity and fail. The typical failure mode of the unconfined specimen is shown in Fig. 6.17.

For the SRCC specimens, the specimen was in an elastic state without an obvious change in appearance at the initial loading stage. With the increasing load, the rust on the steel tube surface peeled off when the load reached about 70% of the peak load, and the specimen deformation increases. After the peak load, the specimen deformation continued to increase, the local buckling appeared on the middle-height area of the steel tube. Meanwhile, the load of the specimen is almost unchanged. The typical failure modes of the SRCC specimens are shown in Fig. 6.18.

For the GRCC specimens, the specimens should be divided into two groups to discuss the failure process and the failure mode. And the failure modes of GRCC

Figure 6.17 Failure modes of RCC specimens. Three pictures.

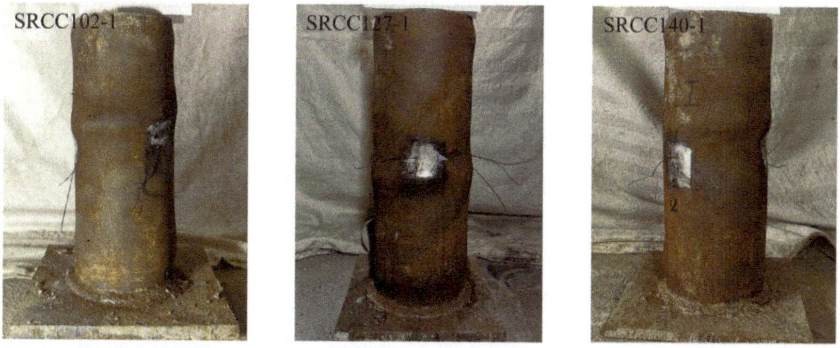

Figure 6.18 Failure modes of SRCC specimens. Three pictures.

Figure 6.19 Failure modes of GRCC specimens. Four pictures.

specimens are shown in Fig. 6.19. For the GRCC-1 and GRCC-2, the core concrete mainly bears the load at the initial loading stage. When the load reaches the compressive strength of the unconfined RAC cylinder, the GFRP cloth plays a confining role gradually. When the specimen was close to the ultimate load, the middle-height part of the GFRP cloth turned white and emitted a slight sound of fiber breakage. When the specimen was at the ultimate load, the GFRP cloth will crack and make a loud sound of the fiber breakage. It should be noted that the core RAC column shows vertical or oblique splitting cracks. It means that the weak confining effect has little effect on the failure modes of the RAC column. For GRCC-5 and GRCC-8 specimens, the load−displacement curves show a bilinear increasing trend. There was no obvious change on the surface of the specimens in the initial loading stage. When the specimen was at about 50% of the ultimate load, the surface of the GFRP tube appeared white texture along the fiber direction. When the specimen was close to failure, the sound of fiber breakage was gradually dense. Followed by a huge sound, the fiber completely broke, and the specimen failed. The GRPC tube was stripped from the broken part to both sides along the fiber direction, and the core concrete was crushed.

For the DRCC specimens, they did not occur the failure when the applied load was nearly 5000 kN. In order to ensure the safety of the compressive machine, the test stopped. In the initial stage of loading, there was no obvious change on the surface of the specimen. With the increase in load, the white texture along the filament winding direction appeared on the surface of the GFRP tube, and the sound of fiber breakage was emitted at the same time. In addition, the load−displacement curve of the DRCC specimen shows a bilinear increasing trend, which is similar to that of GRCC-5 and GRCC-8.

6.3.3 Load−displacement curve

The load−displacement curves of the RCC specimens are summarized in Fig. 6.20. Because of the brittleness splitting failure of the specimen, we did not measure the

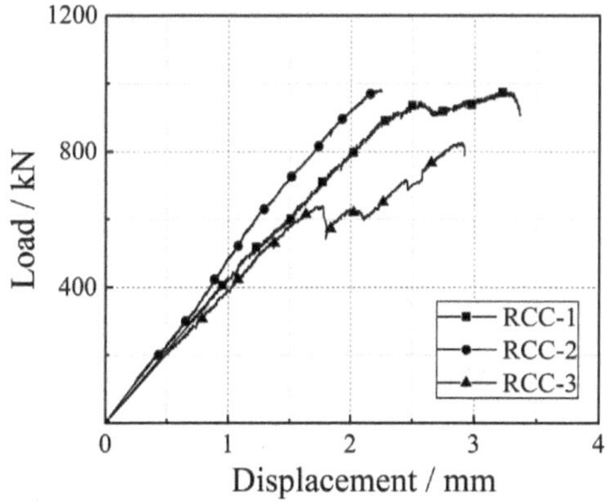

Figure 6.20 Load−displacement of RCC specimens load−displacement relationship.

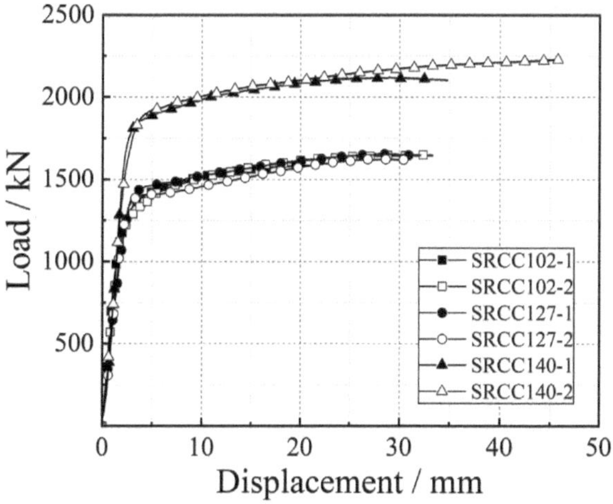

Figure 6.21 Load−displacement of SRCC specimens load−displacement relationship.

decreasing stage of the curve. It should be noted that the load decreases briefly and then increases slowly during the loading process for the RCC-1 and RCC-3 specimens. The reason may be that the recycled coarse aggregate is damaged under the compressive load (due to its own defects), and the internal cracks of the RAC cylinder increase with the following loading. But the damaged aggregate is squeezed tightly under the continuous load, which leads to the secondary increasing curve.

The load−displacement curves of the SRCC specimens are summarized in Fig. 6.21. In the initial elastic stage, the steel tube and the core recycled concrete can work together to bear jointly the axial load. Under the axial loading, the inner RAC occurs lateral expansion. When the core circumferential strain is larger than the circumferential strain of the steel tube, the core RAC will be confined by the

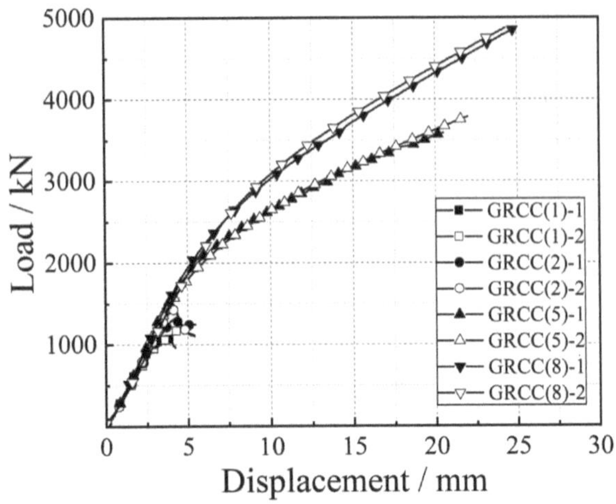

Figure 6.22 Load–displacement of GRCC specimens load–displacement relationship.

steel tube. The confining effect of the steel tube improves the strength and the ductility of the specimen. In the later stage of loading, the displacement of the specimen increases continuously, and the load remains stable.

The load–displacement curves of GRCC specimens are summarized in Fig. 6.22. In the initial stage, the curve slopes of GRCC-1 and GRCC-2 are roughly the same, and the curve slopes of GRCC-5 and GRCC-8 are slightly greater than the former. It means that the initial rigidity of the GRCC specimen increases with the increase in the lateral confining pressure. With the increase in displacement, the load increases slowly and the specimens show an elastic-plastic changing trend. For GRCC-1 and GRCC-2, the bearing capacity decreases after the peak load, the specimens show a softening changing trend. For GRCC-5 and GRCC-8, the load increases linearly with the increase in displacement, and the stiffness is much smaller that the initial rigidity. Therefore, the curves of GRCC-5 and GRCC-8 show an obvious bilinear increasing trend.

The load–displacement curves of DRCC specimens are summarized in Fig. 6.23. The developing trend of the curves is similar to that of specimens GRCC-5 and GRCC-8, which shows the obvious bilinear trend. However, compared with specimens GRCC-5 and GRCC-8, the bearing capacity of the DRCC specimen has been improved, due to the inner steel tube. In the elastic stage, the core RAC and the inner steel tube bear most of the load. In the strengthening stage, the GFRP tube has a great confining effect on the core RAC. For the DRCC specimens, all curves show a similar developing trend. When the same GFRP tube is used as the outer tube, the bearing capacity of the specimen with the 140 mm-diameter steel tube is slightly greater than that of the specimen with the 102 mm-diameter steel tube. When the same steel tube is used as the inner tube, the bearing capacity of the specimen with the 8 mm-thickness GRCC tube is slightly greater than that of the specimen with the 5 mm-thickness GRCC tube.

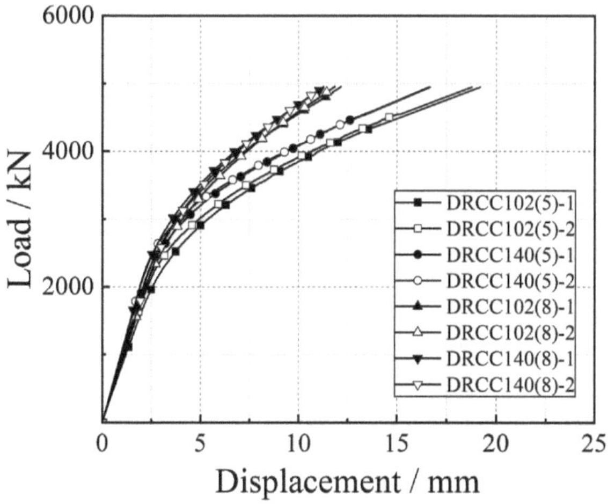

Figure 6.23 Load–displacement of DRCC specimens load–displacement relationship.

6.3.4 Load-axial strain curve

The load axial strain curves of all specimens with different confining modes are summarized in the following figures. In these figures, the overall strain of the specimen is the ratio of axial overall displacement to the height, and the central strain is the middle-height area of the specimen which is the average value of the strain values measured by the axial strain gauges on the specimen surface. Based on the curve comparisons, the following conclusions can be drawn:

(1) For the SRCC specimen (see Fig. 6.24), the axial strain increases with the increase of load in the initial loading stage. In the later loading stage, the axial strain continues to increase, but the load increases slowly and remains stable at last. Based on the comparisons between the central strain and the total strain, it can be found that: the central strain is less than the total strain in the early loading stage; the increasing rate of the central strain is larger than that of the total strain in the later loading stage; finally, the central strain is larger than that of the total strain. The latter two representations occurred by the local buckling of steel tube.

(2) For GRCC specimens (see Fig. 6.25), the axial strain increases linearly with the increase in the load in the initial loading stage. For GRCC-1 and GRCC-2, after the peak load, the axial strain continues to increase but the load gradually decreases. For GRCC-5 and GRCC-8, the axial strain continues to increase, and the load continues to increase in the later loading stage. Based on the comparisons between the central strain and the total strain, it can be found that the increasing rate of the central strain is similar to that of the total strain in the later loading stage, so the central strain is always less than the total strain.

(3) For GRCC specimens (see Fig. 6.26), the axial strain increases linearly with the increase of the load in the initial loading stage. The load continues to increase with the increase in the axial strain in the later loading stage. Based on the

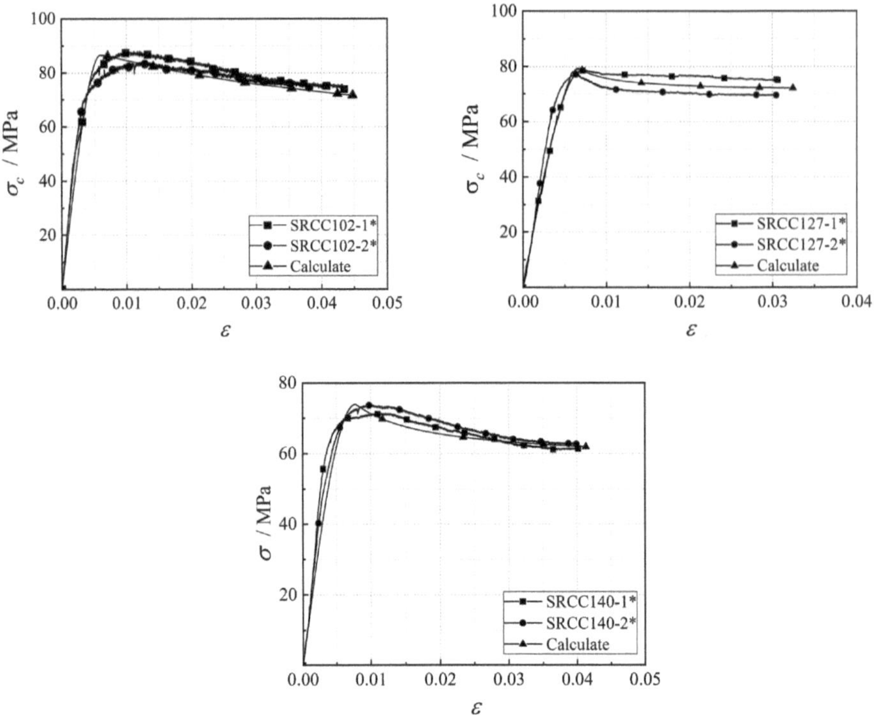

Figure 6.24 Load-axial strain of SRCC specimens load-axial strain relationship.

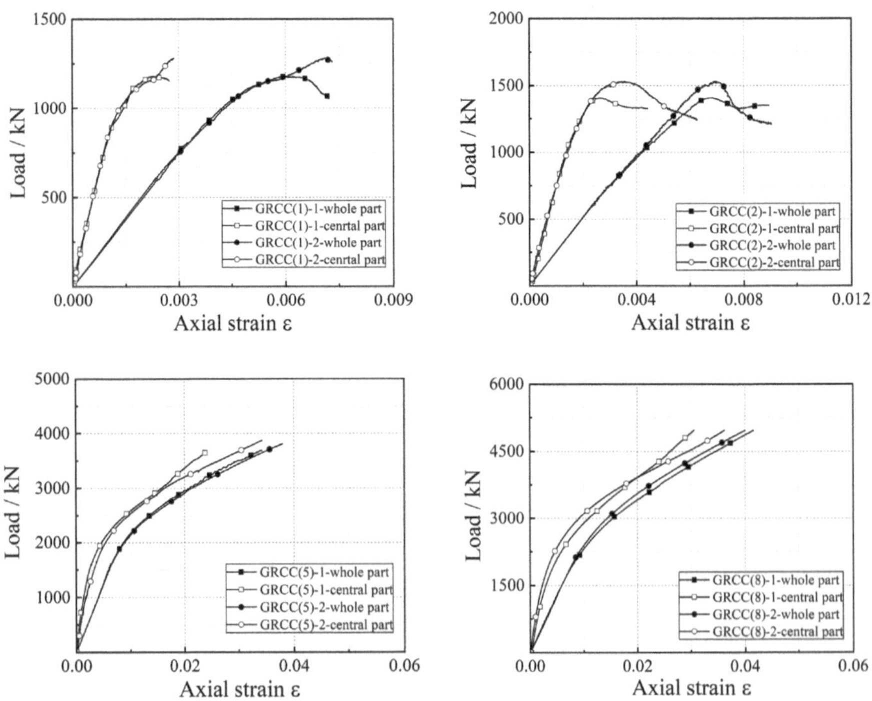

Figure 6.25 Load-axial strain of GRCC specimens load-axial strain relationship.

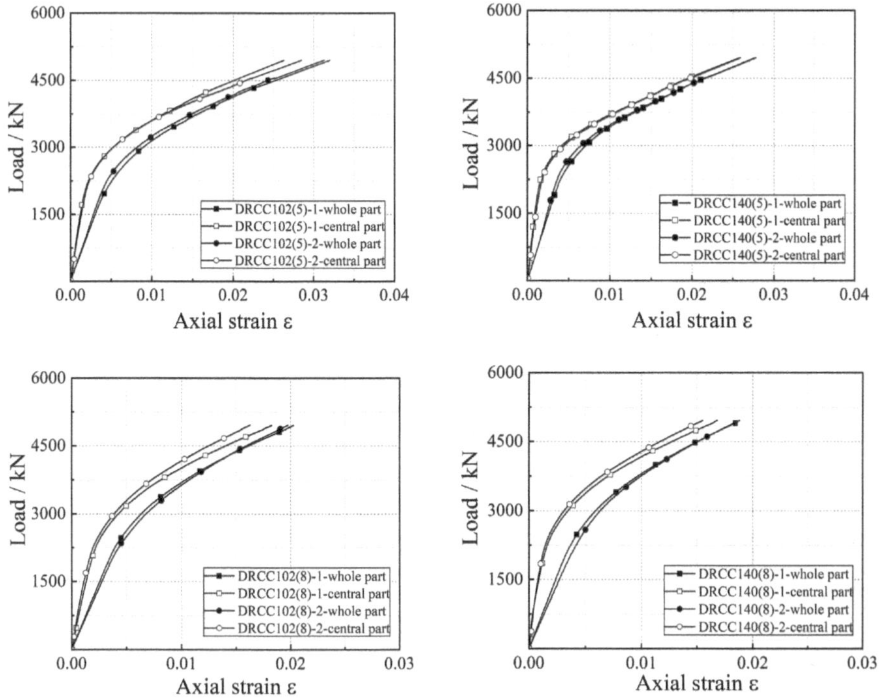

Figure 6.26 Load-axial strain of DRCC specimens load-axial strain relationship.

comparisons between the central strain and the total strain, it can be found that the increasing rate of the central strain is similar to that of the total strain in the later loading stage, so the central strain is always less than the total strain.

Although the central strain is larger than the total strain in the later loading stage, it is not representative. It is because the local buckling on the steel tube makes the central strain increase quickly. However, the axial strain gauge on the specimen surface is affected by many factors, such as the uneven expansion of concrete and confining materials, so it cannot measure accurately the axial strain of the specimen. However, the overall strain can reflect the stress state and the deformation of the total specimen. Therefore, the total strain is as the foundation to analyze the mechanical behavior of confined RAC in the following sections, unless otherwise specified.

6.4 Analysis of the confining effect

It has been known that the confining effect provided by the steel tube or the FRP tube is very different. Especially, the double confined concrete is tested and analyzed in this chapter. Therefore, the three types of confining effects will be analyzed in the section.

6.4.1 Steel confined RAC

(1) The load-axial strain curve

For the SRCC specimens, we can gain the axial load provided by the core RAC by deducting the axial load provided by the steel tube from the total load. The load-axial strain curves are shown in Fig. 6.27. The deducting process is shown in the following paragraph.

We take the SRCC102-1 as an example to illustrate the deducting process: (1) the strain of SRCC102-1 corresponds to the average load-axial strain curve of an empty steel tube (the diameter is 102), then we can gain the bearing capacity of the empty steel tube under the same strain as SRCC102-1; (2) the bearing capacity data of the empty steel tube can be subtracted from the bearing capacity data of SRCC102-1 in the same axial strain; (3) the bearing capacity of core RAC can be gained and drawn into the load-axial strain curve, as shown in Fig. 6.27. The black curve is the load-axial strain curve of SRCC102-1, the red curve is the load-axial strain curve of the empty steel tube, and the blue curve is the load-axial strain curve of core RAC, represented by "*". Meanwhile, the ultimate strain of core RAC is taken to be equal to the peak strain of the empty steel tube.

(2) The stress–strain curve

The stress-axial strain curves of steel confined RAC are summarized in Fig. 6.28. It can be found that the axial compressive bearing capacity and ductility of RAC are improved by the confining effect, which shows that the steel tube has a good confining effect on the core RAC. When the steel tube is thin (e.g., SRCC127-1,2), the development law of the curve before the peak stress of the steel tube confined RAC is almost the same as that of the unconfined RAC. After the peak stress, the curve decreases rapidly, and then enters the gentle trend. When the steel tube is thick (e.g., SRCC102-1,2), the curve near the peak stress increases

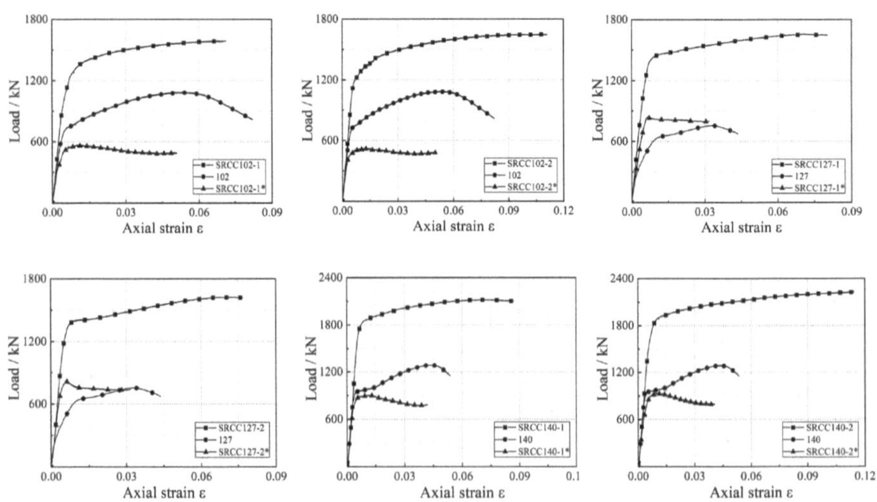

Figure 6.27 Load-axial strain of steel tube confined RAC Load-axial strain relationship.

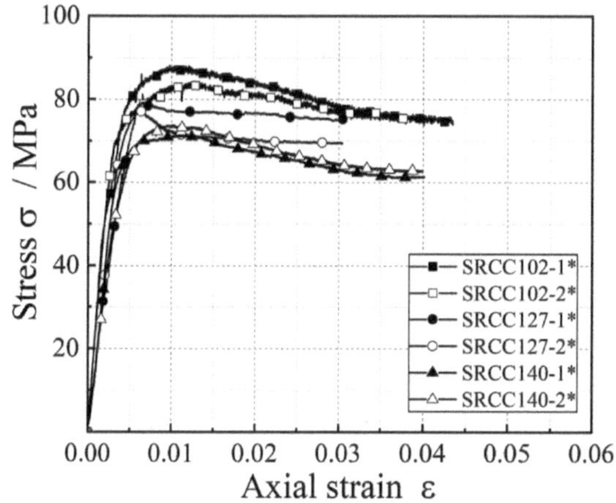

Figure 6.28 Stress−strain curve of the core RAC. Stress−strain relationship.

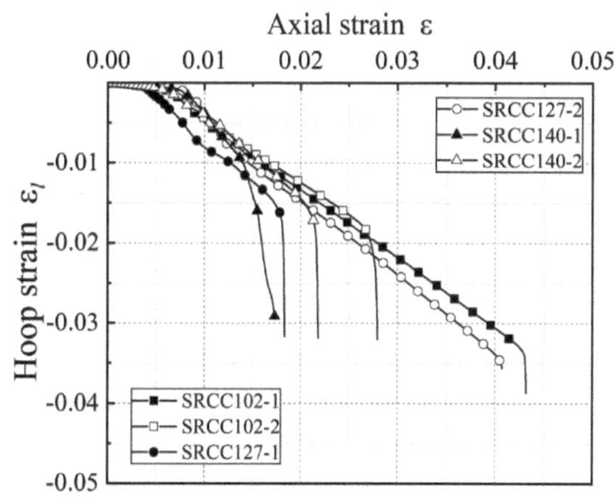

Figure 6.29 Axial strain-lateral strain curve of SRCC. Axial strain-lateral strain relationship.

slowly, which shows the obvious elastic-plastic characteristics. After the peak stress, the curve continues to decline slowly and finally shows a horizontal trend. The confining effect of the steel tube on the core RAC can be simply considered as the actively confining effect.

(3) The axial strain-lateral strain curve

The axial strain is taken as the abscissa, and the lateral strain in the middle-height area of the steel tube (the average value of the strain values measured by the circumferential strain gauges) is taken as the ordinate, the relationship between the two strains is drawn in Fig. 6.29 to reflect the lateral expansion of the core RAC.

For the SRCC specimens, the lateral strain increases slowly and then rapidly with the increase in the axial strain. It shows that the lateral deformation of core

RAC is small in the initial loading stage. With the increase in the load, the lateral expansion of the core RAC is large, so the steel tube provides an obvious lateral confining effect. In the late loading stage, the lateral deformation increases significantly and the local buckling occurs in the middle-height area of the steel tube. Except for SRCC127-1, the axial strain-lateral strain curves of other specimens show basically the same trend before the point (0.005, −0.0005). Except for SRCC127-1 and SRCC140-1, the axial strain-lateral strain curves of other specimens are approximately in a straight line after the point (0.005, −0.0005). And the slopes of these lines are approximately equal, which can be expressed as $\Delta\varepsilon_l/\Delta\varepsilon = -0.8970$.

6.4.2 GFRP confined RAC

(1) The stress-axial strain curve

For the GRCC specimens, the axial load provided by the GFRP tube can be ignored. Therefore, the stress−strain curves can be directly computed from the load-axial strain curves. Then the stress−strain curves are shown in Fig. 6.30.

It can be found that if GFRP is thin (e.g., GRCC-1 and GRCC-2), the GFRP tube will have a less confining effect on the core RAC. And the stress−strain curve shows a decreasing trend after the peak stress. The RAC in these forms is under the weak passive confining effect. If GFRP is thick (e.g., GRCC-5 and GRCC-8), the confining effect on the core RAC will be excellent, which improves the axial compression bearing capacity and ductility of the core RAC. The curves show a bilinear increasing trend. The RAC in these forms is under the strong passive confining effect.

(2) The axial strain-lateral strain curve

The axial strain is taken as the abscissa, and the lateral strain in the middle-height area of the GFRP tube (the average value of the strain values measured by

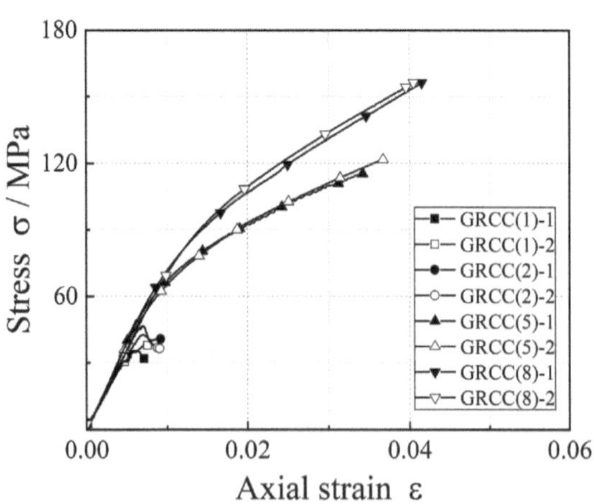

Figure 6.30 Stress−strain of GFRP confined RAC. Stress−strain relationship.

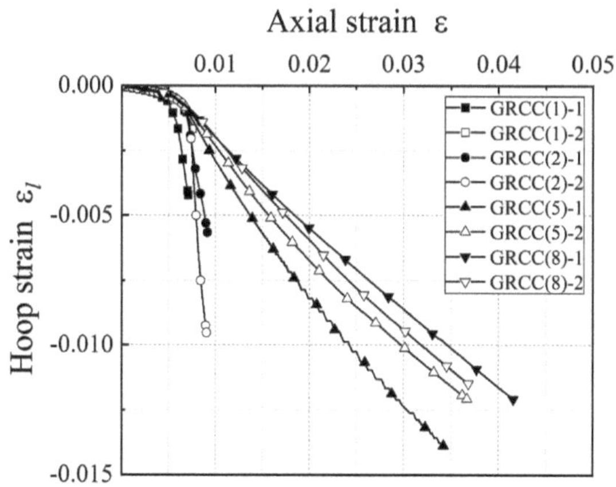

Figure 6.31 Axial strain-lateral strain curve of GRCC Axial strain-lateral strain relationship.

the circumferential strain gauges) is taken as the ordinate. The relationship between the two strains is drawn in Fig. 6.31 to reflect the lateral expansion of the core RAC.

Except for GRCC-1-1, the axial strain-lateral strain curves of other specimens basically coincide before points (0.0065, −0.0007). For GRCC-1-1,2 and GRCC-2-1,2, the lateral strain first increases slowly with the increase of axial strain. When the axial strain reaches 0.0065, the hoop strain begins to increase rapidly until the GFRP cloth rupture. And the slopes of these curves are approximately equal, which can be expressed as $\Delta\varepsilon_l/\Delta\varepsilon = -2.359$.

For GRCC-5-1,2 and GRCC-8-1,2, with the increase in the axial strain, the circumferential strain first increases slowly and then increases rapidly. In the later stage, the curves show an obvious linear relationship. Under the same axial strain, the lateral strain of the GRCC-8-1,2 is slightly smaller than that of the GRCC-5-1,2. Meanwhile, the circumferential strain gauge of GRCC-5-1 may slip in the later stage. Except for test piece GRCC-5-1, the slopes of axial strain circumferential strain curves of other test pieces are roughly the same, which can be expressed as $\Delta\varepsilon_l/\Delta\varepsilon = -0.3164$.

6.4.3 Double confined concrete

(1) The load-axial strain curve of the specimen without the inner steel tube

We take the DRCC105-5-1 as an example to show the process of deducting the bearing capacity of the steel tube: (1) the strain of DRCC102-5-1 corresponds to the average load-axial strain curve of the empty steel tube (the diameter is 102), then we can gain the bearing capacity of the empty steel tube under the same strain as DRCC102-5-1; (2) the bearing capacity data of the empty steel tube can be subtracted from the bearing capacity of the specimen; (3) the bearing capacity after deducting the steel tube can be obtained and drawn into the load-axial strain curve

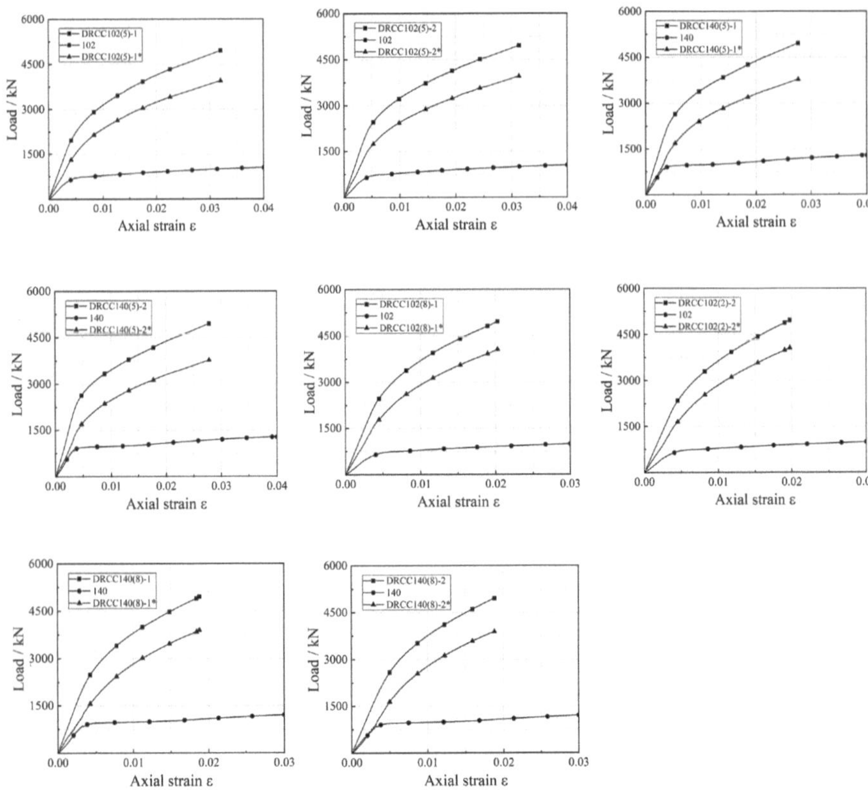

Figure 6.32 Load-axial strain curve of DRCC load-axial strain relationship.

as shown in Fig. 6.32. The black curve is the load-axial strain curve of the DRCC102-5-1, the red curve is the load-axial strain curve of the empty steel tube, and the blue curve is the load-axial strain curve after deducting the steel pipe, represented by "*". The results of the other specimens are shown in Fig. 6.32.

The comparisons between the load-axial strain curves of the DRCC specimens are shown in Fig. 6.33. It can be found that the curves of DRCC102-5-1,2* and DRCC140-5-1,2* are similar and the curves of DRCC102-8-1,2* and DRCC140-8-1,2* are similar. The curve comparisons show that the double confining effect in DRCC can further improve the mechanical behavior of the core RAC because the specimens with different inner steel tubes show similar bearing capacities. Meanwhile, the DRCC without the inner steel tube has a larger bearing capacity than GRCC with the same GFRP tube, which also shows the double confining effect in DRCC.

(2) The stress–strain curve

For the DRCC specimens, if we want to gain the stress–strain curve of the core RAC in the steel tube, both the bearing capacity of the steel tube and the bearing capacity of the sandwich RAC should be deduced. In the deducting process, the

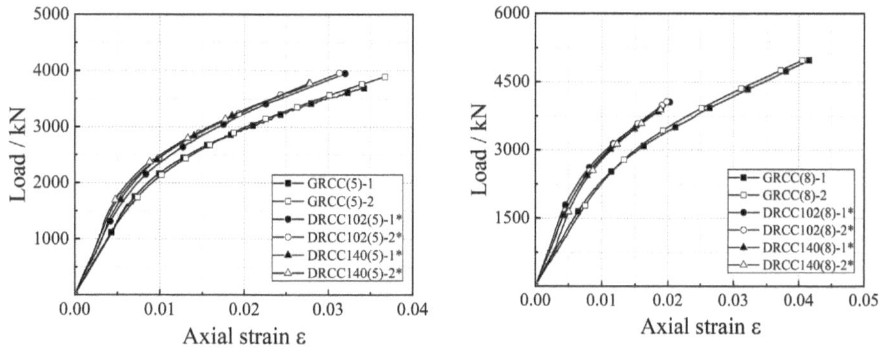

Figure 6.33 Comparisons between the load-axial strain curve of DRCC without steel tube comparisons between the load-axial strain curve.

load-strain curves of GRCC specimens should be adopted. In Fig. 6.30, it can be easily found that the curves of GRCC-5-1 and GRCC-5-2 are coincident and the curves of GRCC-8-1 and GRCC-8-2 are coincident, so the curves of GRCC-5-1 and GRCC-8-1 are used to reduce the bearing capacity of the sandwich RAC. We take the DRCC102-5-1 as an example to show the deducting process of the bearing capacity of the sandwich RAC: (1) the load-axial strain curve of sandwich RAC is gained by multiplying the ratio of the area of sandwich RAC to the area of core RAC in specimen GRCC-5-1; (2) the strain of DRCC102-5-1 corresponds to the average load-axial strain curve of the sandwich RAC is identified; (3) the corresponding bearing capacity is deduced from the bearing capacity of the sandwich RAC from the bearing capacity of DRCC102-5-1*, then the load-axial strain curve of core RAC in steel tube is gained. The stress−strain curves of core RAC in the steel tube of all DRCC specimens are shown in Fig. 6.34. The black curve is the load-axial strain curve of DRCC102-5-1* without an inner steel tube, the red curve is the load-axial strain curve of the sandwich RAC in the specimen, and the blue curve is the load-axial strain curve of the core RAC in the steel tube, which is represented by "**".

The stress−strain curves of the core RAC in the inner steel tube of the DRCC specimen are shown in Fig. 6.35. Based on comparisons of the stress−strain curves of the core RAC, it can be found that the curved shows a bilinear increasing trend, so the RAC under a double confining effect is a type of strong passively confined concrete. As the specimen has the same steel tube, the greater the thickness of GFRP has become and can lead to better mechanical properties of the core RAC.

(3) The axial strain-lateral strain curves

The axial strain is taken as the abscissa, and the lateral strain in the middle-height area of the GFRP tube (the average value of the strain values measured by the circumferential strain gauges) is taken as the ordinate. The relationship between the two strains is drawn in Fig. 6.36 to reflect the lateral expansion of the sandwich RAC. For the DRCC specimen, the variation law of the strain curves of the GFRP tube in DRCC specimens is similar to that of GRCC-5-1,2 and GRCC-8-1,2. The

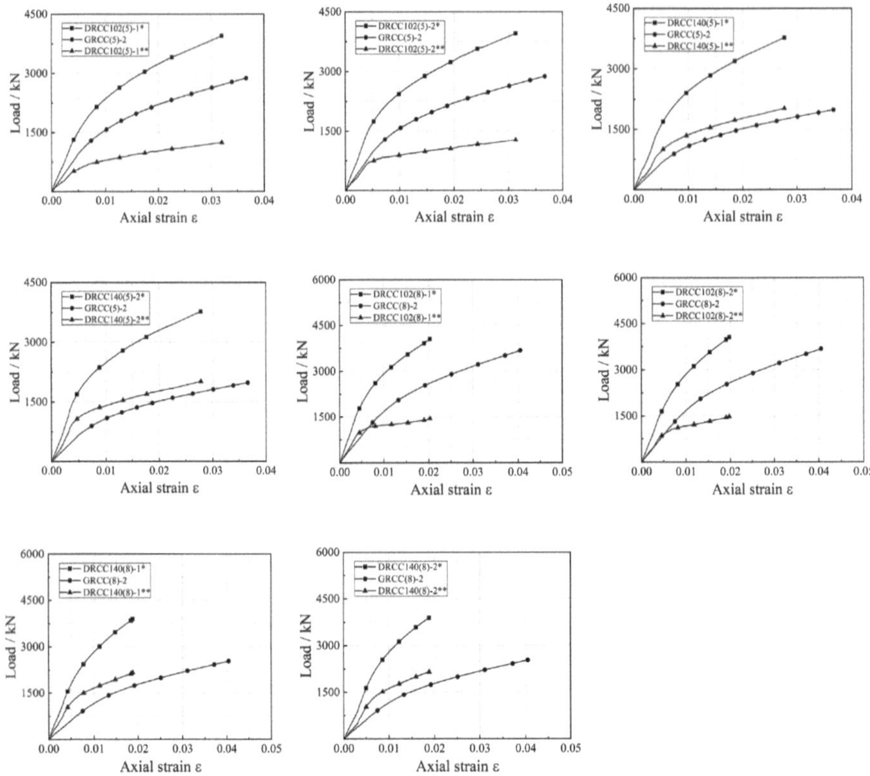

Figure 6.34 Load-strain curve of core RAC in steel tube load-strain relationship.

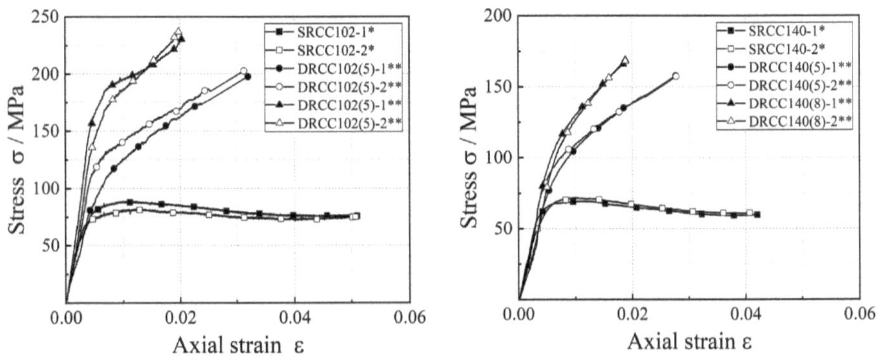

Figure 6.35 Stress−strain curves of core RAC stress−strain relationship.

axial strain-lateral strain curves basically coincide before the point (0.0045, −0.0006). The curves also show an obvious linear relationship in the later stage. Therefore, the axial strain during failure of the DRCC specimens can be inferred by the hoop breakage strain of the GFRP tube in the DRCC specimens. For the DRCC

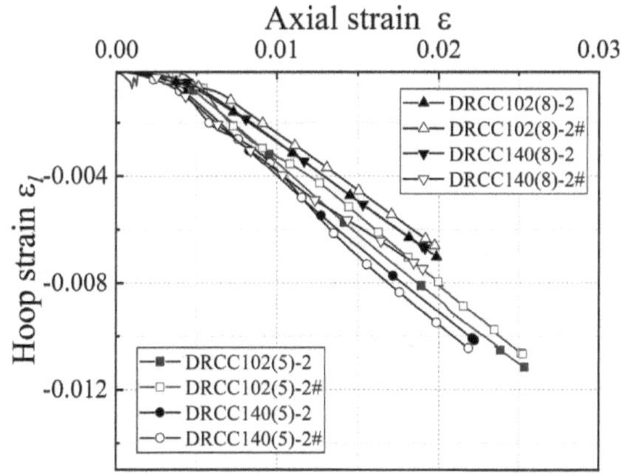

Figure 6.36 Axial strain-lateral strain curve of GFRP tube in DRCC specimen Axial strain-lateral strain relationship.

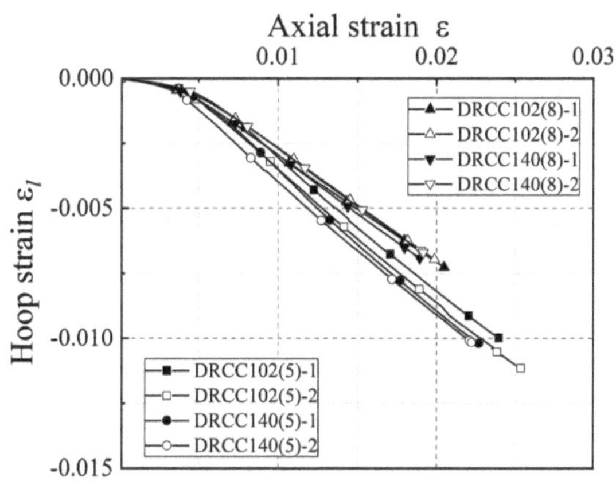

Figure 6.37 Axial strain-lateral strain curve of steel tube in DRCC specimen. Axial strain-lateral strain relationship.

specimens, the ratio can be expressed as: for DRCC102-5 and DRCC140-5 $\Delta\varepsilon_l/\Delta\varepsilon = -0.3164$; for DRCC102-8 and DRCC140-8 $\Delta\varepsilon_l/\Delta\varepsilon = -0.4400$.

In addition, the axial strain-lateral strain curves of the steel tube in DRCC specimens are shown in Fig. 6.37. Based on the comparisons between Figs. 6.36 and 6.37, the two curve show similar changing law.

6.5 Design-oriented constitutive model of the confined RAC

Based on the compressive test results in Chapter 3, it can be concluded that the steel tube confined RAC similarly belongs to the actively confined concrete and the

GFRP confined RAC belongs to the passively confined concrete. The mechanical properties of the confined concrete are mainly related to the form and the size of the lateral confining pressure. The lateral confining pressure of the actively confined concrete is unchanged, and the lateral confining pressure of the passively confined concrete changes with the lateral strain. For the establishment of the constitutive model, the following characteristics should be determined: the stress calculation equation, the strain calculation equation, and the full stress–strain curve calculation equation. The stress characteristics of actively confined concrete and passively confined concrete are relatively similar, but the strain characteristics are different. The two types of confining effects will be analyzed in this section, then the double confining effect is analyzed, and the corresponding constitutive model is established finally.

6.5.1 Constitutive model of the steel tube confined concrete

6.5.1.1 Stress and strain calculation equations

Based on the analysis in Chapter 2, the peak stress of the confined concrete can be expressed as follows:

$$f_{cc} = f_{co} + kf_l \tag{6.3}$$

where f_{cc} is the peak stress of the confined concrete, f_{co} is the peak stress of the unconfined concrete, f_l is the lateral confining pressure, and k is the lateral confining coefficient.

And the lateral confining pressure can be expressed as follows:

$$f_l = (2f_y t_s)/D_s \tag{6.4}$$

where f_y is the yield stress of confined concrete, t is the thickness of the steel tube, and D is the diameter of the steel tube.

Based on different confining materials and different section forms, the lateral confining coefficient (k) can be modified, and the stress calculation equation which is suitable for different confined concrete is proposed. The lateral confining coefficient (k) is related to the confining ratio (ω), $\omega = f_l/f_{co}$. Therefore, based on the analysis of the test data, the lateral confining coefficient (k) is fitted by the test results, which is given as follows:

$$k_s = 1.065 + 1/(2.7\omega^{7.2})\ 0.9 \le \omega \le 1.7 \tag{6.5}$$

By comparing Eq. (6.3), Eq. (6.4), and Eq. (6.5), we can compute the peak stress of the steel tube confined RAC. And the comparisons between computation results and test results are shown in Table 6.19. The comparisons show that the errors between computation results and test results are less than 5%, which shows the degree of coincidence between them is high.

Table 6.19 Verification of the peak stress.

Item	Test value	Computation value	Value ratio
SRCC102-1	88.088	86.128	0.978
SRCC102-2	84.169	86.128	1.023
SRCC127-1	79.285	78.404	0.989
SRCC127-2	77.523	78.404	1.011
SRCC140-1	71.471	72.626	1.016
SRCC140-2	73.781	72.626	0.984

Source: From Rong, C., Qv, Y. S., Shi, Q. X., Wang, P., & Li, C. F. (2022). Axial behaviors of recycled aggregate concrete cylinders under various confinement types: The experimental and theoretical analysis. *Construction and Building Materials, 340*, 127821.

The study results show that the relationship between $\varepsilon_{cc}/\varepsilon_{co}$ and f_{cc}/f_{co} is approximately linear. Based on the analysis of test results, the peak strain calculation equation of the steel tube confined RAC is fitted as follows:

$$\varepsilon_{cc}/\varepsilon_{co} = 5.5 - 1.35f_{cc}/f_{co} \tag{6.6}$$

6.5.1.2 Establishment of constitutive model

Xiao (2007) proposed the constitutive model of the unconfined RAC, which is given as follows:

$$y = \begin{Bmatrix} a_1x + (3-2a_1)x^2 + (a_1-2)x^3 0 \leq x \leq 1 \\ x/(b_1(x-1)^2 + x)x \geq 1 \end{Bmatrix} \tag{6.7}$$

where $y = \sigma_c/f_{co}$ and $x = \varepsilon_c/\varepsilon_{co}$, σ_c is the axial stress and ε_c is the corresponding strain, a_1 and b_1 are the coefficients related to the replacement rate (r) of the recycled coarse aggregate, which can be calculated in following equations or got by the linear interpolation in Table 6.20.

$$a_1 = 2.2(0.748r^2 - 1.231r + 0.975) \tag{6.8}$$

$$b_1 = 0.8(7.6483r + 1.142) \tag{6.9}$$

In this section, the replacement rate of recycled coarse aggregate is $r = 100\%$. When the stress drops to 85% of the peak stress, we take the corresponding strain as the ultimate strain. The full stress–strain curve is calculated as shown in Fig. 6.38. In addition, the test result is the average curve of the stress–strain curves of specimens RCC-1 and RCC-1. Fig. 6.38 shows that the constitutive model proposed by Xiao (2007) can provide an accurate prediction for the unconfined RAC in this study.

It has been known that the stress–strain curve form of steel tube confined RAC is similar to that of the unconfined RAC. However, the peak stress and the peak

Table 6.20 Coefficient values of a_1 and b_1.

r (%)	**0**	**30**	**50**	**70**	**100**
a_1	2.2	1.32	1.26	1.15	1.04
b_1	0.8	3.3	3.96	4.31	7.5

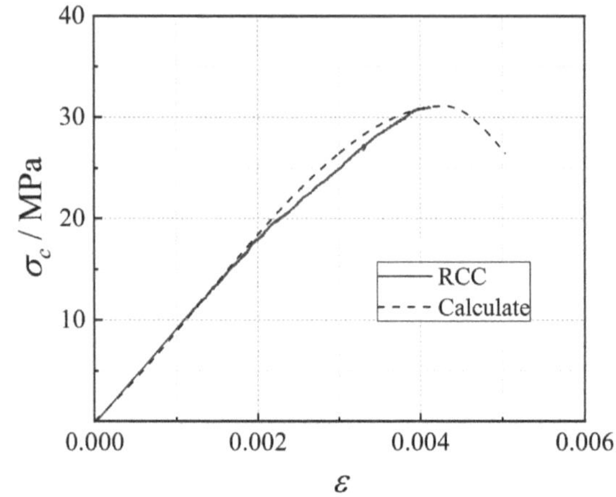

Figure 6.38 Verification of unconfined RAC. Stress–strain curve.

strain are improved by the confining effect. After the peak point, the stress–strain curve of the steel tube confined RAC decreases slowly. Therefore, in combination with Eq. (6.7), we introduce the increasing section coefficient (a_2), decreasing section coefficient (b_2), and decreasing section slope coefficient (n), then the full stress–strain curve equation can be expressed as follows:

$$y=\left\{\begin{array}{l} ax+(3-2a)x^2+(a-2)x^3, 0\leq x\leq 1 \\ x/(b(x-1)^n+x), x\geq 1 \end{array}\right\} \quad (6.10)$$

where $y = \sigma_c/f_{cc}$ and $x = \varepsilon_c/\varepsilon_{cc}$, f_{cc} is the peak stress of the confined RAC and ε_{cc} is the peak strain, a_1 is the increasing section coefficient which is the product of a_1 and a_2 ($a = a_1 \times a_2$), and b_1 is the decreasing section coefficient which is the product of b_1 and b_2 ($b = b_1 \times b_2$), and n is the decreasing section slope coefficient ($n = \omega$).

Based on the analysis of test results, the coefficients (i.e., a_1 and b_1) are as follows:

$$a_2 = 2.4 - 0.0125f_{cc} \quad (6.11)$$

$$b_2 = -0.126(\omega - 1.26)^2 + 0.03 \quad (6.12)$$

As mentioned above, the constitutive model can be gained by the combination of Eqs. (6.3), (6.6), (6.8), (6.11), and (6.12). Based on the curves verification in Fig. 6.39, the proposed model can provide accurate prediction results.

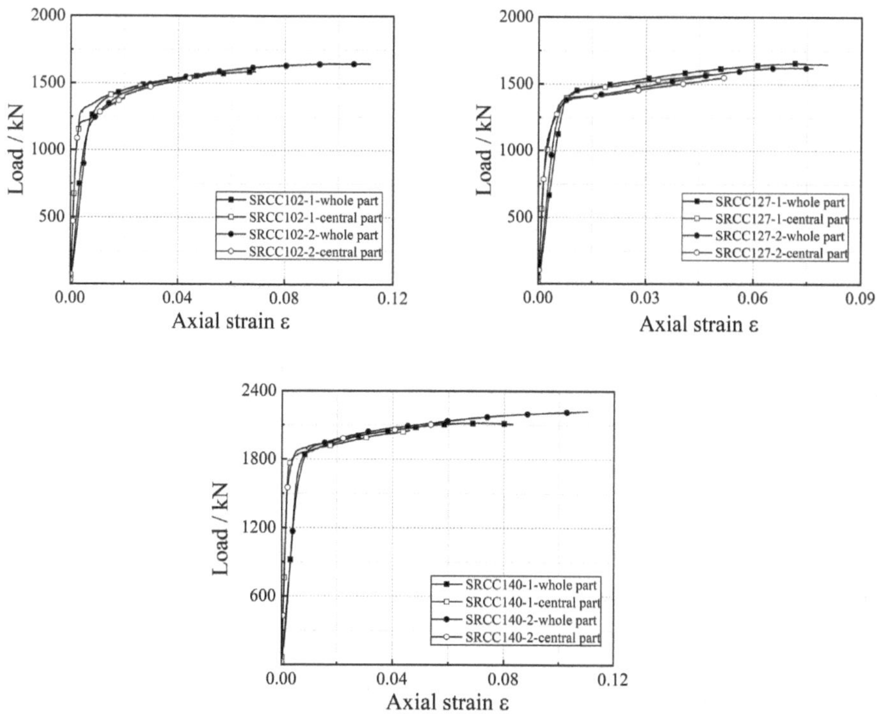

Figure 6.39 Verification of steel tube confined RAC. Stress−strain curves.

6.5.2 Constitutive model of the GFRP confined RAC

6.5.2.1 Stress and strain calculation equations

For the GFRP confined RAC, the lateral confining pressure:

$$f_l = (2f_f t_f)/D \tag{6.13}$$

where t_f is the thickness of the FRP tube, D is the diameter of the FRP tube, and f_f is the lateral tensile force.

Because the GFRP is an elastic linear material, the lateral tensile force is related to the lateral strain, there is $f_f = E_f \times \varepsilon_l$ (E_f is the elastic mode of the GFRP tube). When the GFRP tube breaks, the lateral confining pressure reaches the maximum value, then the following equation can be obtained:

$$f_{l,a} = (2E_f \varepsilon_{l,riup} t_f)/D \tag{6.14}$$

where $f_{l,a}$ is the actual lateral confining pressure, and $\varepsilon_{l,rup}$ is the circumferential rupture strain.

Based on the analysis in Chapter 2, the typical stress−strain curve of FRP confined concrete is shown in Fig. 6.40. When the confining effect is strong, there is

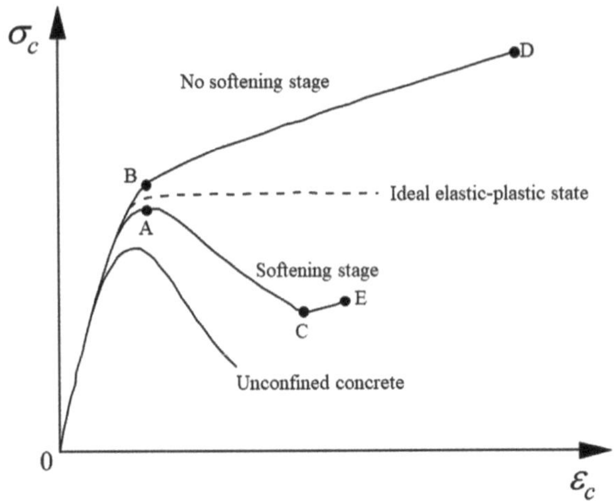

Figure 6.40 Typical stress–strain curve of FRP confined concrete. Typical curve.

Table 6.21 Confining ratio of each specimen.

Item	GRCC-1-1	GRCC-1-2	GRCC-2-1	GRCC-2-2	GRCC-5-1	GRCC-5-2	GRCC-8-1	GRCC-8-2
$f_{l,a} / f_{co}$	0.016	0.006	0.051	0.086	0.951	0.823	1.649	1.567
Average value	0.011		0.069		0.887		1.608	

no softening stage in the curve, and the stress reaches the maximum at the limit point (D), that is $f_{cc} = f_{cu}$. Meanwhile, the determination of the dividing point (B) between the bilinear increasing stage is very important. When the confining effect is weak, there is a softening stage in the curve and the stress reaches the maximum at the peak point (A).

The study by Wu et al. (2006) proposed a method to judge whether there is a softening section in the stress–strain curve of the FRP confined concrete by using the limit value (λ) of confining ratio: when $f_{l,a}/f_{co} > \lambda$, there is no softening stage in the curve; when $f_{l,a}/f_{co} < \lambda$, there is a softening stage in the curve. Meanwhile, based on the analysis of test results, it is suggested that for the FRP confined concrete with the ordinary elastic modulus (elastic modulus less than 250 GPa), the confining value ratio is $\lambda = 0.13$. The confining ratios in this experiment are shown in Table 6.21, and the test curve trends are consistent with the results judged by the method.

As given in the above discussion, for the stress–strain curve without the softening stage, the peak stress is mainly related to the actual maximum lateral confining stress. However, for the stress–strain curve with the softening stage, when the stress reaches the peak point, the GFRP has not broken, and the peak stress is mainly related to the lateral confining stiffness. Therefore, based on the model proposed by Lam and Teng

(2003), the confining ratio can be divided into the lateral confining stiffness ratio (ρ_k) and the lateral strain ratio (ρ_ε), which can be calculated as follows:

$$f_l/f_{co} = (2E_f\varepsilon_l t_f)/(f_{co}D) = \rho_K\rho_\varepsilon \tag{6.15}$$

$$\rho_K = 2E_f t_f/(Df_{co}/\varepsilon_{co}) \tag{6.16}$$

$$\rho_\varepsilon = \varepsilon_l/\varepsilon_{co} \tag{6.17}$$

Based on the analysis of the stress–strain curves (see Fig. 6.30) and axial strain-lateral strain curves (see Fig. 6.31) of the GFRP confined concrete, the curve with a softening stage reaches the peak point when the lateral strain is 0.002, and the curve without a softening stage reaches the dividing point (B) when the lateral strain is 0.002, and two kinds of curves reach the ultimate points (E or D) when GFRP breaks. In this experiment, the lateral rupture strain of the GFRP tube can be taken as 60% of the rupture strain in the GFRP cloth tensile test or 90% of the rupture strain in the GFRP tube circumferential tensile test. Based on the test results analysis, we can fit the stress calculation equation and strain calculation equation of GFRP confined RAC, which are given as follows:

$$f_c/f_{co} = 1 + 3.19\rho_K^{0.5}\rho_\varepsilon^{0.5} \tag{6.18a}$$

$$f_c/f_{co} = 1 + 3.19\rho_K^{0.75}\rho_\varepsilon^{0.75} \tag{6.18b}$$

$$\varepsilon_c/\varepsilon_{co} = 1 + 3.3\rho_K^{0.27}\rho_\varepsilon^{1.03} \tag{6.19}$$

For the curve with a softening stage, the peak stress is computed by Eq. (6.18a), and the peak strain is computed by Eq. (6.19), that is, $f_c = f_{cc}$, $\varepsilon_c = \varepsilon_{cc}$, and $\varepsilon_l = 0.002$. The ultimate stress is computed by Eq. (6.18b), and the ultimate strain is computed by Eq. (6.19), that is, $f_c = f_{cu}$, $\varepsilon_c = \varepsilon_{cu}$, and $\varepsilon_l = \varepsilon_{l,rup}$.

For the curve without softening section, the stress at the dividing point is computed by Eq. (6.18a), and the strain at the dividing point is computed by Eq. (6.19), that is $f_c = f_{ce}$, $\varepsilon_c = \varepsilon_{ce}$, and $\varepsilon_l = 0.002$. The peak stress is the ultimate stress which can be computed by Eq. (6.18b) and the peak strain is the ultimate strain which can be computed by Eq. (6.19), that is $f_c = f_{cu}$, $\varepsilon_c = \varepsilon_{cu}$, and $\varepsilon_l = \varepsilon_{l,rup}$.

The comparison between the computation results and the test results is shown in Table 6.22. The errors between the two values are within 10%, which reflects that the proposed equation can provide accurate prediction results.

6.5.2.2 Establishment of constitutive model

Based on the experiment results, the stress–strain curves of GRCC-1-1,2 and GRCC-2-1,2 are different. On one hand, the wrapped GFRP cloth is thin, which

Table 6.22 Verification of stress calculation equation.

Item	Test value	Computation value	Value ratio
GRCC-1-1	37.559	39.7	1.058
GRCC-1-2	37.071	39.7	1.072
GRCC-2-1	44.823	44.4	0.991
GRCC-2-2	48.735	44.4	0.911
GRCC-5-1	117.369	122.7	1.061
GRCC-5-2	123.666	122.7	1.007
GRCC-8-1	158.289	155.9	0.985
GRCC-8-2	158.234	155.9	0.985

Source: From Rong, C., Qv, Y. S., Shi, Q. X., Wang, P., & Li, C. F. (2022). Axial behaviors of recycled aggregate concrete cylinders under various confinement types: The experimental and theoretical analysis. *Construction and Building Materials, 340*, 127821.

leads to the weak confining effect, and it is influenced by the core RAC. On the other hand, it may be caused by errors in specimen fabrication. It can be concluded that if the confining effect is a weak passively confining effect and the elastic modulus is a constant value, the stress axial strain curve will first decrease and then level after the peak stress (e.g., GRCC-2-1,2). However, the subsequent curve is not measured after the peak point because the wrapped GFRP cloth was too thin to rupture early. It should be noted that the strength of the recycled concrete decreases continuously in softening stage, which can be regarded as new concrete. The corresponding stress of the changed RAC can be expressed as f'_{co}, which can be calculated by the descending stage of Eq. (6.1), mentioned as follows:

$$f'_{co}/f_{co} = x/(b_1(x-1)^2 + x) \tag{6.20}$$

where $x = \varepsilon_x/\varepsilon_{co}$, ε_x is the strain after peak strain.

The corresponding lateral confining pressure can be obtained by adding the lateral confining pressure corresponding to the peak strain and the lateral confining pressure increased from the peak strain to the point strain, that is:

$$f'_{l,cc} = f_{l,cc} + \Delta f_l \tag{6.21}$$

$$f_{l,cc} = 2E_f \varepsilon_{l,cc} t_f / D \tag{6.22}$$

$$\Delta f = 2E_f \Delta\varepsilon_l t_f / D \tag{6.23}$$

It can be concluded from the above analysis and Eq. (6.2), in which $\Delta\varepsilon_{l,cc} = 0.002$, $\Delta\varepsilon_l = 0.002$, $\Delta\varepsilon = 2.359(\Delta\varepsilon_x - \Delta\varepsilon_{cc})$, the judgment method ($f_{l,cc}/f'_{co} = \lambda = 0.13$) is used to gain the ε_x that the curve will not soften at the axial strain. It means that the axial strain can be used to judge when the curve drops to the lowest point. After substituting relevant data, the axial strains at the lowest point of GRCC-1-1,2 and specimen GRCC-2-1,2 are 0.00869 and 0.00785, respectively.

In this paper, we assume that the stress–strain curve of GFRP confined RAC with a softening stage is a trilinear equation, which is composed of an increasing stage, a decreasing stage, and a horizontal stage. Therefore, the full stress–strain curve equation can be expressed as follows:

$$\sigma_c = \begin{cases} f_{cc}\varepsilon_c/\varepsilon_{cc} 0 < \varepsilon_c < \varepsilon_{cc} \\ f_{cc} - (f_{cc} - f_{cu})/(\varepsilon'_{cc} - \varepsilon_{cc})\varepsilon_{cc} < \varepsilon_c < \varepsilon'_{cc} \\ f_{cu}\varepsilon'_{cc} < \varepsilon_c < \varepsilon_{cu} \end{cases} \tag{6.23}$$

In this paper, we assume that the stress–strain curve of the GFRP confined RAC without a softening stage is a bilinear equation, which is composed of two increasing stages. Therefore, the full stress–strain curve equation can be expressed as follows:

$$\sigma_c = \left\{ \begin{array}{l} f_{ce}/\varepsilon_{ce}\varepsilon_c, 0 < \varepsilon_c < \varepsilon_{ce} \\ f_{ce} + (f_{cc} - f_{ce})/(\varepsilon_{cc} - \varepsilon_{ce}), \varepsilon_{ce} < \varepsilon_c < \varepsilon_{cc} \end{array} \right\} \tag{6.24}$$

In summary, equations from Eqs. (6.15) to (6.23) are combined to make up the constitutive model of the GFRP confined RAC. The comparisons between the stress–strain curve calculated by this model and the test curve are shown in Fig. 6.41. The comparisons show that the proposed model can provide accurate prediction results.

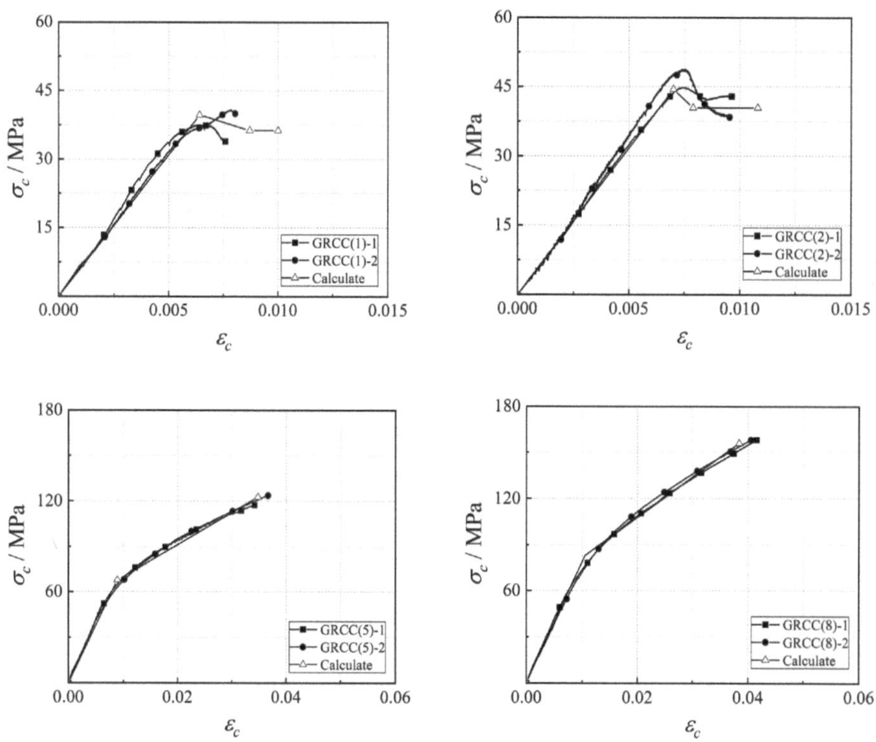

Figure 6.41 Verification of stress–strain curves of GFRP confined RAC. Model verification.

6.5.3 Constitutive model of the GFRP confined RAC

In the double confined RAC specimen, the steel tube is used as the inner tube and the GFRP tube is used as the outer tube. The double confining pressure provided by the two tubes shows a strong passively confining effect. And the stress–strain curve of the core RAC shows a bilinear increasing trend. Therefore, the constitutive model can be established in the same form as that of the GFRP confined RAC without the softening stage.

Based on Figs. 6.31 and 6.36, it can be found that when the lateral strain is the same, the axial strain of the DRCC specimen is less than that of the GRCC specimen. In this experiment, the DRCC specimen is not loaded to failure, which means that the stress and strain data at failure are not obtained. However, based on the analysis of Fig. 6.36, the axial strain during the failure of the DRCC specimen can be calculated by the existing test data and the axial strain-lateral strain relationship.

If the axial stress of the core RAC in the steel tube of the DRCC specimen is calculated by a simple superposition method, we can superimpose the lateral double confining pressures by the inner steel tube and the outer GFRP tube. Then the stress–strain curves of the DRCC specimens can be calculated by Eq. (6.25) and plotted in Fig. 6.42. Based on the comparisons between the combination of the computation results and the test results, the stress calculated by the simple superposition is larger than the test results. There is a certain reduction in the lateral confining pressure provided by GFRP in the transmission process, due to the local buckling of the steel tube.

$$f_c/f_{co} = 1 + [1.065 + 1/(2.7\omega^{7.2})]f_{l1}/f_{co} + 1 + 3.19\rho_K^{0.5}\rho_\varepsilon^{0.5} \tag{6.25}$$

where $\rho_k\rho_\varepsilon = f_{l2}/f_{co}$.

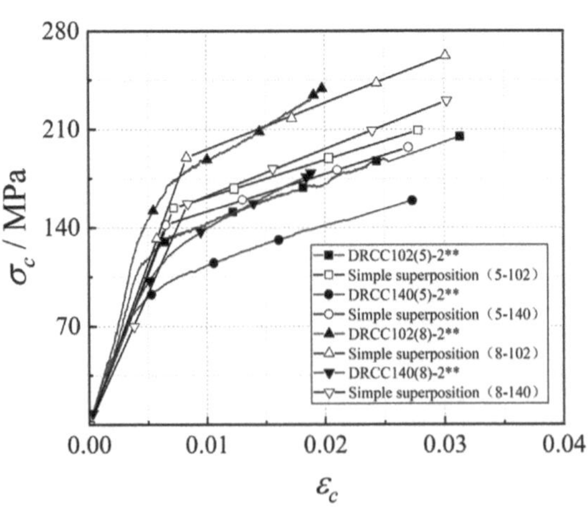

Figure 6.42 Curve comparisons between calculation results and test results. Curve analysis.

Therefore, the reduction coefficient (i) should be introduced, then we can get the stress calculation equation of the core RAC under the double confining effect, which is given as follows:

$$f_c/f_{co} = 1 + [1.065 + 1/(2.7\omega^{7.2})]f_{l1}/f_{co} + i\left(1 + 3.19\rho_K^{0.5}\rho_\varepsilon^{0.5}\right) \tag{6.26}$$

In the DRCC specimen, the inner diameter of the GFRP tube is 200 mm and the thickness of the steel tube is 6.5 mm. It can be seen from the curve trend in Fig. 6.42 that if the thickness of GFRP tube is greater or the diameter of steel tube is smaller, the difference between the simple superposition results and test results (e.g., DRCC103-8**) will be small. Therefore, it can be inferred that the reduction coefficient (i) is related to the following parameters: the thickness (t_f) and the diameter (D_f) of GFRP tube and the thickness (t_s) and the diameter (D_s) of steel tube. Meanwhile, the parameter (j) is determined that $j = t_f/D_f + t_s/D_s$. Based on the analysis of the test data, the relationship between i and j can be obtained, and then substituted into Eq. (6.26), $i = 9j$. In addition, we can get the strain calculation equation as follows:

$$\varepsilon_c/\varepsilon_{co} = 1 + 2.4\rho_K^{0.27}\rho_\varepsilon^{1.03} \tag{6.27}$$

For the core RAC under the double confining effect, we can calculate the dividing stress by Eq. (6.26), and calculate the dividing strain by Eq. (6.27), where $\varepsilon_l = 0.002$, $i = 9j$. The peak stress is the ultimate stress, which is calculated by Eq. (6.26), and the peak strain is the ultimate strain, which is calculated by Eq. (6.27), where $\varepsilon_l = \varepsilon_{l,rup}$, $i = 9j$. Then the stress–strain curve can be calculated by Eq. (6.24). In summary, the constitutive model of the double confined RAC is combined by Eqs. (6.24), (6.26), and (6.27) together. The comparisons between the stress–strain curves calculated by this model and the test curves are shown in Fig. 6.43. The comparisons show that the proposed model can provide accurate prediction results for double confined RAC.

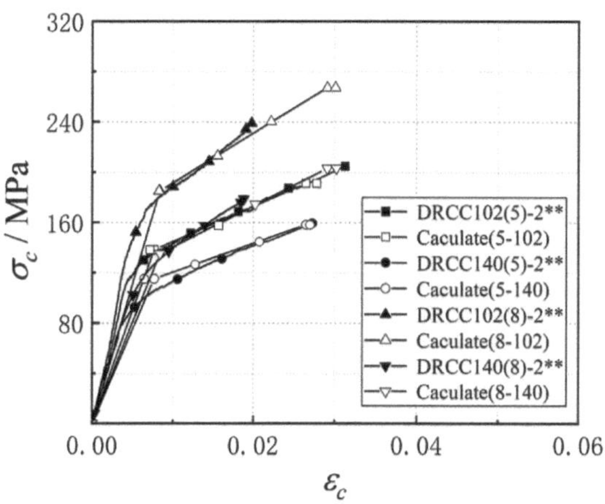

Figure 6.43 Curve comparisons between stress–strain curves of double confined concrete. Model verification.

6.6 Conclusions

In the front part of this chapter, we analyzed the influence factors for the compressive strength of RAC by the orthogonal test method. Further, we clarified the relationship between the compressive strength and the main influencing factors by the single factor test. Then we proposed a simple mix design method of RAC according to the required compressive strength. In the latter part of this chapter, we presented the experimental study on axial compressive behavior of the confined RAC, including the RAC filled steel tube specimens, the GFRP confined RAC specimens, and the double confined RAC specimens. Based on the experiment, we discussed the confining mechanism of different confining forms and identified the axial compressive performance characteristics of different confined RAC. Finally, we established and verified the corresponding constitutive models. The main conclusions of this chapter can be drawn as follows:

(1) The orthogonal experiment with three-factor and three-level was carried out. In this experiment, the 7-day and 28-day compressive strengths of RAC were taken as the property indexes, and the range analysis and the variance analysis were set for test results. The results show that the order of the influence factors on the compressive strength of RAC is the water–binder ratio > the fly ash content > the binder–aggregate ratio.

(2) For the single-factor experiment, the 28-day compressive strength of RAC is taken as the performance index, the binder–aggregate ratio is 0.30 and the fly ash content is 10%. And the water–binder ratio is taken as the variable to determine the mix proportion. The test results show that the 28-day compressive strength decreases linearly with the increase in the water–binder ratio. Based on the foundation, we can propose the relationship between the 28-day compressive strength and the water–binder ratio.

(3) We presented the axial compressive tests on the RAC columns, including one group of the unconfined RAC columns, 3 groups of the RAC filled steel tube columns, 4 groups of the GFRP confined RAC columns, and 4 groups of the double confined RAC columns. The test results show that the failure process, the failure mode, and the axial behavior of the confined RAC are similar to the corresponding ordinary confined concrete.

(4) The stress–strain curve form of the steel tube confined RAC is similar to that of the unconfined RAC. With the confining effect provided by the steel tube, the peak stress and the peak strain of RAC are improved, and the stress–strain curve decreases relatively slowly after the peak stress. Based on the Double Shear Strength Theory and the constitutive model of the unconfined RAC provided by Xiao (2007), we established the constitutive model of the steel tube confined RAC.

(5) If GFRP is thin, the confined effect will be weak, and there will be a softening stage in the stress–strain curve. If GFRP is thick, the confined effect will be strong, and there will be no softening stage in the stress–strain curve. The limit value of the confined ratio proposed by Wu et al. (2006) can be used to judge whether there is a softening stage in the stress–strain curve of the GFRP confined RAC. In this paper, we assume that the curve with a softening stage adopts a trilinear equation, and the curve without a softening stage adopts a bilinear equation. Based on the confining stiffness ration and the confining strain ratio, we established the constitutive models of the GFRP confined RAC.

(6) The stress–strain curve of the double confined RAC shows a bilinear increasing trend, and the double confined pressure shows a strong passively confining effect. Because of the inner steel tube, the confining pressure provided by GFRP will be reduced in the

lateral stress transmission process, and we introduced the reduction coefficient in the constitutive models. Based on the constitutive model of GFRP confined RAC without a softening stage, we established the constitutive model of double confined RAC.

Based on the experimental and theoretical studies, we analyzed the axial compressive properties of the confined RAC and established the constitutive models of different confined RAC. The research results have a certain significance, but there are still deficiencies that can be further studied in the future:

(a) The confining effect of each form on different RAC should be studied because the mechanical properties of RAC are influenced by many factors. However, the test data in this chapter are limited, and only the confining materials are considered.
(b) It is also necessary to study the square section of the recycled concrete columns to facilitate practical engineering applications. This is because this chapter only studies the confined RAC cylinders. However, there are multiple section forms in double confined RAC such as outer circle inner square, outer square inner circle, outer square inner square, and so on.
(c) The confining mechanism of double confined RAC is more complex and needs to be further studied. Especially for the lateral stress transmission method is not only important for the mechanical analysis of the double confined RAC but also can be the foundation to study the mechanical behavior of the multiple confined RAC in future study.

Acknowledgments

This work was supported by the National Natural Science Foundation of China [grant numbers 52108171; 51878540]. The authors gratefully acknowledge two Master's students (Mr. Qv Yunsong and Mr. Tian Wenkai) of Xi'an University of Architecture and Technology for the drawing works.

References

Behera, M, Bhattacharyya, SK, Minocha, AK, Deoliya, R, & Maiti, S (2014). Recycled aggregate from C&D waste its use in concrete-A breakthrough towards sustainability in construction sector A review. *Construction and Building Materials*, *68*, 501–516.

Dimitriou, G, Savva, P, & Petrou, MF (2018). Enhancing mechanical and durability properties of recycled aggregate concrete. *Construction and Building Materials*, *158*, 228–235.

Etxeberria, M, Vázquez, E, Marí, A, & Barra, M (2007). Influence of amount of recycled coarse aggregates and production process on properties of recycled aggregate concrete. *Cement and Concrete Research*, *37*, 735–742.

Evangelista, L., & De Brito, J. (206 C.E.). Mechanical behaviour of concrete made with fine recycled concrete aggregates. *Cement and Concrete Composites*, *29*, 397-401.

Fam, A, & Rizkalla, S (2001). Behavior of axially loaded concrete-filled circular fiber-reinforced polymer tubes. *ACI Structure Journal*, *98*, 280–289.

Gao, DY, Lv, MY, Yang, L, Tang, JY, Chen, G, & Meng, Y (2020). Experimental study of utilizing recycled fine aggregate for the preparation of high ductility cementitious composites. *Materials, 13*, 679.

GB/T 228.1-2010. (2010). *Metallic materials-tensile testing-Part 1: Method of test at room temperature*. Standards Press of China.

GBT 14685-2011. (1468). *Pebble and crushed stone for construction*. Standards Press of China.

GBT 17431.2-2010. (2010). *Lightweight aggregates and its test methods-Part 2 Test methods for lightweight aggregates*. Standards Press of China.

GBT 25177-2010. (2517). *Recycled coarse aggregate for concrete*. Standards Press of China.

He, Z, & Zhang, J (2014). Strength characteristics and failure criterion of plain recycled aggregate concrete under triaxial stress states. *Construction and Building Materials, 54*, 354−362.

JGJ 55-2011. (2011). *Specification for mix proportion design of ordinary concrete*. China Architecture Building Press.

Kaeseberg, S, Messerer, D, & Holschemacher, K (2020). Experimental study on concrete under combined FRP-steel confinement. *Materials, 13*, 4467.

Lam, L, & Teng, JG (2003). Design-oriented stress-strain model for FRP-confined concrete in rectangular columns. *Journal of Reinforced Plastics Composites, 22*(13), 1149−1186.

Li, W, Luo, Z, Tao, Z, Duan, W, & Shah, SP (2017). Mechanical behavior of recycled aggregate concrete-filled steel tube stub columns after exposure to elevated temperatures. *Construction and Building Materials, 146*, 571−581.

López-Uceda, A, Ló Ayuso, J, pez, M, Jimenez, J, Agrela, F, & Sierra, M (2016). Properties of non-structural concrete made with mixed recycled aggregates and low cement content. *Materials, 9*(2), 74.

Meng, D, Wu, XM, Quan, HZ, & Zhu, CJ (2021). A strength-based mix design method for recycled aggregate concrete and consequent durability performance. *Construction and Building Materials, 281*, 122616.

Meyer, C (2008). The greening of the concrete industry. *Cement and Concrete Composites, 31*, 601−605.

Nam, J, Kim, G, Yoo, J, Choe, G, Kim, H, Choi, H, Kim, Y, & Sarker, P (2016). Effectiveness of fiber reinforcement on the mechanical properties and shrinkage cracking of recycled fine aggregate concrete. *Materials, 9*, 131.

Pacheco-Torgal, F, Ding, Y, Miraldo, Z, Abdollahnejad, Z, & Labrincha, JA (2012). Are geopolymers more suitable than Portland cement to produce high volume recycled aggregates HPC. *Construction and Building Materials, 36*, 1048−1052.

Rong, C, & Shi, QX (2021). Analysis constitutive models for actively and passively confined concrete. *Composite Structures, 256*, 113009.

Rong, C, Shi, QX, & Zhao, HC (2020). Behaviour of FRP composite columns: Review and analysis of the section forms. *Advance in Concrete Construction, 9*, 125−137.

Torgal, FP, & Jalali, S (2011). *Eco-efficient construction and building materials*. Springer.

Wu, G, & Lv, ZT (2003). Study on the stress-strain relationship of FRP-confined concrete circular column without a strain-softening response. *Journal of Building Structures, 5*, 1−9.

Wu, G, Wu, ZS, & Lv, ZT (2006). Study of the stress-strain relationship of FRP-confined circular concrete column with a strain-softening response. *China Civil Engineering Journal, 11*, 7−14.

Xiao, JZ (2007). Experimental investigation on complete stress-strain curve of recycled concrete under uniaxial loading. *Journal of Tongji University (Natural Science)*, *11*, 1445–1449.

Xiao, JZ, Li, JB, & Zhang, CH (2004). Mechanical properties of recycled aggregate concrete under uniaxial loading. *Cement and Concrete Research*, *35*, 1187–1194.

Xu, J, Chen, Z, Xiao, Y, Demartino, C, & Wang, J (2017). Recycled Aggregate Concrete in FRP-confined columns: A review of experimental results. *Composite Structures*, *174*, 277–291.

Yu, Y, Zhao, X, Xu, J, Chen, C, Deresa, ST, & Zhang, J (2020). Machine learning-based evaluation of shear capacity of recycled aggregate concrete beams. *Materials*, *13*, 4552.

Hybrid double-skin tubular rectangular columns

7

7.1 Introduction

The double-skin tubular column (DSTC) is a special form of composite column proposed by Teng et al. (2007), which comprises the inner steel tube, the outer FRP tube, and the sandwiched concrete filled between two tubes. Since the existing studies put forward DSTC at the beginning of this century, several axial compressive tests have been presented on the excellent mechanical behaviors of DSTC (Khus et al., 2021; Wang et al., 2021; Zhang et al., 2015). The study proved that the outer FRP tube can provide both confining effect and anticorrosion protection effect for the core concrete, while the inner steel tube can act as longitudinal reinforcement and provide lateral support to distribute the lateral confining pressure on the core concrete (Wang et al., 2019; Yu et al., 2017). Therefore, even if there is a hollow area in the middle of the specimen, FRP can still give full play to the characteristics of strong tensile behavior, which improves the strength and deformation capacity of the core concrete. That means the DSTC is a stress-bearing member with superior performance. Therefore, the DSTCs have a series of advantages: the lightweight, the high axial strength, the high durability, the large section moment of inertia, the convenient construction, and so on. We can widely apply these columns in practical engineering, such as frame columns in buildings and pier columns in bridges.

In the study presented by Yu et al. (2010), researchers modified the constitutive model of the FRP confined concrete to establish the constitutive model of the core concrete in DSTC, and they introduced the influence of void ratio into the ultimate axial strain equation. Further, the study presented by Yu et al. (2012) presented the cyclic axial compressive experiment on the DSTC specimens and discussed the test data in other studies (Lam & Teng, 2003; Yu, 2007). Yu et al. (2012) found that the model proposed by Yu (2007) was too conservative in predicting the test curve, while the model proposed by Lam and Teng (2003) accurately predicted the test curve. And the conclusions can also be drawn: the DSTCs have good ductility under cyclic load; the cyclic axial stress−strain curve is almost the same as the monotonic axial stress−strain curve, but the repeated cycles have a cumulative effect on the concrete damage in DSTC; if it completely unloads the specimen from the axial strain exceeding the yield strain of the steel tube, the interface slip between steel tube and concrete will lead to the significant difference in the axial strain between them. In addition, many existing studies presented monotonic or cyclic axial compression experiments on DSTC specimens under different parameters to explore the effects of different parameters on the axial compression mechanical properties. Yu et al. (2017) and Ozbakkaloglu and Fanggi

Concrete Composite Columns. DOI: https://doi.org/10.1016/B978-0-323-85171-8.00007-0

(2013) conducted experimental studies on the DSTC confined by GFRP tube, PET FRP tube, and AFRP tube, respectively. These studies show that DSTC specimens confined by three types of FRP tubes have good strength and ductility. Among them, the curves of the DSTC specimens with PET FRP tube show the developing trend of first increasing, then decreasing, and then increasing. The curves of DSTC specimens with GFRP tube and AFRP tube show an almost monotonically increasing trend. Meanwhile, it is found that the ultimate strength and the corresponding strain of concrete in DSTC increase with the increase of the thickness of the FRP tube and steel tube, and the ultimate axial strain of concrete increases with the increase in the diameter of the steel tube.

For the DSTCs with the square outer FRP tubes. The study presented by Yu and Teng (2013) shows that the square DSTC has good ductility and the axial stress—-strain curve of the core concrete is similar to that in the square FRP confined concrete. The study presented by Fanggi and Ozbakkaloglu (2015) shows that the DSTCs with circular inner steel tubes have greater ultimate axial stress and ultimate axial strain than DSTCs with square inner steel tubes. In addition, if we fill the circular inner steel tube with concrete, the ultimate axial stress of concrete in DSTCs will increase and the ultimate axial strain will decrease. However, if we fill the square inner steel tube with concrete, the ultimate axial stress and strain of concrete will both increase significantly.

Ozbakkaloglu and Fanggi (2014, 2015) presented the axial compression experiments on the DSTC specimens by using ordinary strength concrete and high-strength concrete. The study presented by Ozbakkaloglu and Fanggi (2014) shows that if the confining rate (f_l/f_{co}) is a constant value, the higher strength of the core concrete in DSTC will lead to a lower strength enhancement rate (f_{cc}/f_{co}) and the lower strain enhancement rate ($\varepsilon_{cu}/\varepsilon_{co}$). The study presented by Ozbakkaloglu and Fanggi (2015) shows that the DSTC specimens with ordinary strength concrete show a similar strength enhancement rate (f_{cc}/f_{co}) and a similar strain enhancement rate ($\varepsilon_{cu}/\varepsilon_{co}$) under monotonic axial compression and cyclic axial compression, and the DSTC specimens with high-strength concrete show a larger strength enhancement rate (f_{cc}/f_{co}) and a larger strain enhancement rate ($\varepsilon_{cu}/\varepsilon_{co}$) under cyclic axial compression than that under monotonic axial compression.

For the double-skin tubular column with an inner ribbed steel tube (R-DSTC), the studies presented by Peng et al. (2018) and Huang et al. (2020) discussed the square and circle R-DSTC specimen confined by PEN FRP, compared the mechanical behavior of R-DSTC specimen with traditional DSTC specimen and analyzed the best combination of stiffeners (including size, shape, etc.). The study results show that the R-DSTC specimen has higher bearing capacity and higher ductility than the DSTC specimen. This is because the vertically ribbed stiffener can effectively delay or prevent the inward local buckling of the steel tube. In the experimental study (Zeng et al., 2021), Zeng studied the axial compressive behavior of the R-DSTC specimens with the Q690 high-strength steel tube and the ultra-high strength concrete, and they drew the following conclusions: the ribbed stiffener on the steel tube improve the interaction effect between the steel tube and the core concrete, which leads to the better bearing capacity than ordinary DSTC specimens; the axial stress–strain curve of the core concrete in R-DSTC specimen shows the developing trend of first increasing, then decreasing, and then increasing.

It is well known that researchers have studied a series of mechanical behavior of DSTCs in the existing studies. However, few studies are presented on rectangular columns, which are also widely used in practical engineering. Against this background, we propose a new form of a composite column in this paper, namely hybrid FRP-concrete-steel double-skin tubular rectangular columns (hybrid DSTRCs), which comprises a rectangular outer FRP tube with round corners, two inner steel tubes, and the sandwiched concrete filled in the between two tubes. We have completed an experimental study on the behavior of hybrid DSTRCs with various section forms under concentric compression or eccentric compression. The detailed experimental program and the experimental analysis will be presented and discussed in the following sections.

7.2 Experimental program

In order to study the mechanical behavior of DSTRCs in detail, the axial and eccentric compressive tests are presented in the section. The axial compressive test focuses on the size of the inner steel tube, the thickness of the FRP tube, and the strengthening measures of the inner steel tube. The eccentric compressive test focuses on the strengthening measures of the inner steel tube, including the concrete filled in the steel tube and ribbed stiffener welded on the steel tube.

7.2.1 Test specimens

A total of 26 specimens were prepared and tested in the experiment, including seven pairs of specimens prepared for the axial compressive test and six pairs of specimens prepared for the eccentric compressive test. We describe the design details of the hybrid DSTRC specimens in the design scheme, as shown in Table 7.1. And we showed three types of cross sections of hybrid DSTRC specimens, including the basic typical specimen, the specimen with full concrete, and the specimen with ribbed stiffener. In these specimens, six pairs of specimens are the typical hybrid DSTRC specimens, as shown in Fig. 7.1. Three pairs of specimens are the hybrid FRP and steel tubular concrete column which is the concrete filled in DSTRCs. Three pairs of specimens are the hybrid DSTRC specimens with ribbed stiffener, which have four ribbed stiffeners (i.e., the thickness is 3mm and the width is 25 mm) that point to four corners of the FRP tube and one ribbed stiffener (i.e., the thickness is 3.2 mm and the width is 25 mm) which connects two inner steel tubes, as shown in Fig. 7.1. Besides, we prepared and tested one pair of rectangular FRP confined concrete columns to reflect the confining effect provided by the outer FRP tube in the hybrid DSTRC specimens. And each pair included two identical specimens, which reflect the accuracy of the test results.

Table 7.1 Design details of hybrid DSTRC specimens.

Item	Specimen ID	FRP plies	Size of steel tube D/t (mm)	Load eccentricity (mm)	Remark
1	AENN6-1, 2	6	—	0	—
2	AEMN6-1, 2	6	101.6/3.2	0	—
3	EEMN6-1, 2	6	101.6/3.2	50	—
4	EEMN6-1, 2	6	101.6/3.2	75	—
5	AEMN9-1, 2	9	101.6/3.2	0	—
6	AEBN6-1, 2	6	101.6/5	0	—
7	AESN6-1, 2	6	88.9/3.2	0	—
8	AFMN6-1, 2	6	101.6/3.2	0	Full of concrete
9	EFMN6(50)-1, 2	6	101.6/3.2	50	Full of concrete
10	EFMN6(75)-1, 2	6	101.6/3.2	75	Full of concrete
11	AEMY6-1, 2	6	101.6/3.2	0	With ribbed stiffeners
12	EEMY6(50)-1, 2	6	101.6/3.2	50	With ribbed stiffeners
13	EEMY6(75)-1, 2	6	101.6/3.2	75	With ribbed stiffeners

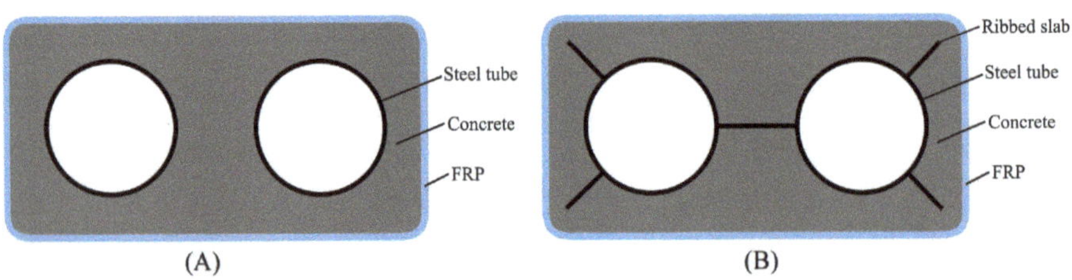

Figure 7.1 Cross sections of hybrid DSTRC specimens (A) typical hybrid DSTRC specimens; (B) the hybrid DSTRC specimens with ribbed stiffener.

All specimens are rectangular columns with the outer rectangular GFRP tube. And all specimens have the same column height of 600 mm and the same inside section sizes of FRP tube (i.e., the inner length is 300 mm, the inner width is 150 mm, and the radius of each rounded corner is 30 mm). Except for the same size, the main study parameters include the size of the inner steel tube, the thickness of the outer FRP tube, the strengthening measures of the inner steel tube, and the load eccentricity. Four pairs of the hybrid DSTRC specimens covered two thicknesses of FRP tubes (i.e., comprising six and nine plies of FRP cloth, respectively), two thicknesses of steel tubes (i.e., 3.2 and 5 mm), and two diameters of steel tubes (i.e., 101.6 and 88.9 mm). Four pairs of specimens covered no inner steel tube in column, empty steel tube in column, concrete filled steel tube in column, and ribbed stiffener welded on steel tube. Six pairs of specimens covered three load eccentricities (i.e., 0, 50, and 75 mm) of the three kinds of section forms (i.e., empty steel tube, concrete filled in steel tube, and ribbed stiffener welded on steel tube). In order to identify the characteristic parameter, we gave each specimen a name, which is according to the following rules:

(1) the first letter is "A" or "E" which is used to represent the type of compressive test, where "A" represents the specimen is under axial compressive test, and "E" represents the specimen is under eccentric compressive test;
(2) the second letter is "E" or "F" which is used to represent the filling condition of the inner steel tube, where "E" represents the inner steel tubes are empty (i.e., the DSTRC specimen and the R-DSTRC specimen), and "F" represents the inner steel tubes are filled by concrete;
(3) the third letter is "S", "M" or "B" which is used to represent three sizes of inner steel tube, where "S" represents the small size of steel tube (the diameter is 88.9 mm and the thickness is 3.2 mm), "M" represents the middle size of steel tube (the diameter is 101.6 and the thickness is 3.2 mm), and "B" represents the big size of steel tube (the diameter is 101.6 and the thickness is 5 mm);
(4) the fourth letter is "N" or "Y" which is used to represent the ribbed stiffeners welded on steel tube, where "N" represents no ribbed stiffener on steel tube, while "N" represents ribbed stiffeners welded on steel tube;
(5) the first number is "6" or "9" which is used to represent the thickness of the FRP tube, where "6" represents 6-plies of FRP cloth and "9" represents 9-plies of FRP cloth;
(7) the number in the bracket is used to represent load eccentricity, where "50" represents the load eccentricity is 50 mm and "75" represents the load eccentricity is 75 mm;
(6) the number after dash is used to differentiate the two identical specimens of each pair (i.e., 1 or 2).

Based on the design scheme, the AEMN6-1 and the AEMN6-2 are set as basic comparison specimens.

7.2.2 Preparation of specimens

For all specimens, we used the rectangular FRP tubes as the outer formwork in the preparation. These FRP tubes were prefabricated by wrapping the GFRP cloth around a plastic form column which had the same sizes as the core concrete column. In order to lead out the connecting line of the inner strain gauges, we opened a small hole at about 10 mm from the top of the GFRP tube.

In order to fix the inner steel tubes, we welded several short steel bars on steel tubes to form the steel tube frame. For R-DSTRC specimens, the ribbed stiffeners should be welded on a pair of steel tube frame by the gas welding process. After that, the strain gauge should be pasted on the outer surface of the steel tube. In addition, a layer of sealant is evenly applied on the surface of the strain gauge, then the gauze with soaked epoxy resin is evenly applied to prevent the damage of strain gauges during pouring concrete.

At last, we should finish the centering and fixing of the prefabricated GFRP tube and the steel tube in the flowing steps: (1) the GFRP tube should be placed on the bottom wood plate; (2) the level laser should be used to align the top, the location of GFRP tubes should be fixed by the wood bars, and the tube bottom should be fixed by the epoxy resin; (3) the steel tube frame should be in GFRP tube and the steel bars should be used to ensure the location of steel tubes; (4) the plastic foam should be filled into a steel tube in DSTRC specimens and R-DSTRC specimens. The formwork before concrete pouring is shown in Fig. 7.2.

We used the commercial concrete to pour all specimens three times and fully vibrated once per one-third height. After pouring, we leveled the concrete surface of all specimens, then watered them for curing at the normal temperature for about 28 days. At last, we removed the bottom wood plate after 7 days.

Figure 7.2 Formwork of hybrid DSTRC specimens a picture.

7.2.3 Material properties

(1) Tensile properties of the GFRP tube

As mentioned above, we made the GFRP tubes with the GFRP cloth, which is wrapped in the prefabricated plastic form in a wet-layup process. And the overlapping zone was on one short side, the overlapping length was 80 mm. It should be noted that the GFRP cloth is unidirectional glass fiber cloth, so the glass fibers are wrapped in the hoop direction of the GFRP tubes. The design scheme in the above sections shows that two types of GFRP tubes are used in specimens (i.e., 6-ply GFRP tube and 9-ply GFRP tube). Therefore, the GFRP coupons are made for each type of GFRP cloth, each with a width of 25 mm and a length of 250 mm. Based on the multiple measures, the nominal thickness is 0.165 mm per ply. We presented the tensile tests on these FRP coupons, based on the standard (GB/T 228.1-2010, 2010). In the material test, two strain gauges were used on the two sides of each coupon to measure the longitudinal strains of the central zone, respectively. Based on the average values of the test results, the rupture strain of the GFRP cloth is 0.0149 and the tensile strength of the GFRP cloth is 82.84 GPa.

(2) Compressive properties of the concrete

As we knew, the commercial concrete was poured at one time in this test. In order to ensure the quality of concrete pouring, we used self-compacting concrete with a maximum aggregate size of 10 mm to pour the specimens in preparation. Because the minimum distance between the steel tube and the GFRP tube is less than 25 mm. We prepared 9 standard concrete cylinders (150 mm × 300 mm) for the compressive properties test, and these cylinders were under water curing for about 28 days. We presented the compressive test on these cylinders to get the mechanical properties of concrete. In addition, three cylinders were tested after 28 days, three cylinders were tested before the specimen test, and three cylinders were tested after the specimen test. Based on the material test, the compressive strengths of the latter 6 cylinders are similar. Therefore, the mechanical properties of concrete during the specimen test can be obtained from these test results. The average compressive peak stress is 34.5 MPa and the corresponding peak strain is 0.0029.

(3) Tensile properties of the ribbed stiffeners.

The two types of ribbed stiffeners (i.e., the width is 25 mm or 50 mm) were cut from the same steel plate, so they have the same nominal mechanical properties. Therefore, we prepared and tested three standard tensile coupons for the mechanical properties of the ribbed stiffeners, based on the standard (GB/T 228.1-2010, 2010). Based on the coupon test results, the average elastic modulus is 229 GPa, the average yield stress is 346 MPa, and the average tensile strength is 522 MPa.

(4) Tensile properties of the steel tube

The design scheme in the above section shows that three kinds of steel tubes are used in specimens, i.e., 88.9/3.2 (namely S-tube), 101.6/3.2 (namely M-tube) and 101.6/5 (namely B-tube). So we prepared and tested three standard tensile coupons for each type of steel tube, based on the standard (GB/T 228.1-2010, 2010).

It can be found from the test results that the differences in the mechanical properties of different steel tubes were small. Therefore, the average values are used for all steel tubes, the average elastic modulus is 205 GPa, the average yield stress is 335 MPa, and the average tensile strength is 455 MPa.

(5) Compressive properties of the steel tube

It is easily found that the steel tube has the obvious axial load capacity, which means the compressive properties of steel tubes can play an important effect. It is necessary to test the axial compressive behavior of the empty steel tubes in this study. Therefore, we tested the concentric compressive behaviors of all steel tubes which were filled in specimens, including three types of steel tubes and one pair of steel tubes with ribbed stiffeners. In the compressive test, the steel tubes showed a large plastic deformation. The failure mode is a combination of the plane bending buckling and the local buckling, as shown in Fig. 7.3. The details of the four steel tubes and their ultimate load got from the compression tests are summarized in Table 7.2.

Load–strain curves of the steel tubes are summarized in Fig. 7.4, where the axial strain is the ratio of total axial displacement to the column height. Based on Fig. 7.4, we can conclude that the curves of the steel tube can be divided into three loading stages: the elastic stage, the force characteristics of the steel tube is in the elastic state; the hardening stage, the force characteristics of the steel tube are in yield and hardening state; the softening stage, the force characteristics of steel tube are in the buckling state. Then the developing trend of curves in Fig. 7.4 shows the following conclusions: the bearing capacity depends on the section area of the steel tube; the local buckling of the steel tube depends on the diameter–thickness ratio; the ribbed stiffeners partly restrict the local buckling. And we will discuss further their axial compressive behaviors in the following sections.

7.2.4 Test set-up and instrumentation

All specimens were loaded by the displacement controlling method on the 500-tonne Denison Compression Test Machine in the mechanical library at the University of Wollongong. And we show the loading diagram of axial and eccentric compressive tests in Fig. 7.5. The loading was controlled by axial displacement, the displacement control rate of 0.6 mm/min was adopted at the beginning of each test, and then the displacement control rate was gradually increased to 1 mm/min after the yield state of the steel inner tube. All the test data are recorded simultaneously by the data logger, including strain, load, and displacement.

(1) Axial compressive test

For axial compressive specimens, it should measure the strain of a different zone in specimen, due to the uneven expansion. Therefore, two pairs of strain gauges (including one axial strain gauge and one circumferential strain gauge) with a gauge length of 20 mm were installed on the side center of the middle-height surface. And two circumferential strain gauges with a gauge length of 20 mm were installed on the diagonal corners of the middle-height surface. Because the steel tube will bear both the axial load and the lateral confining pressure under

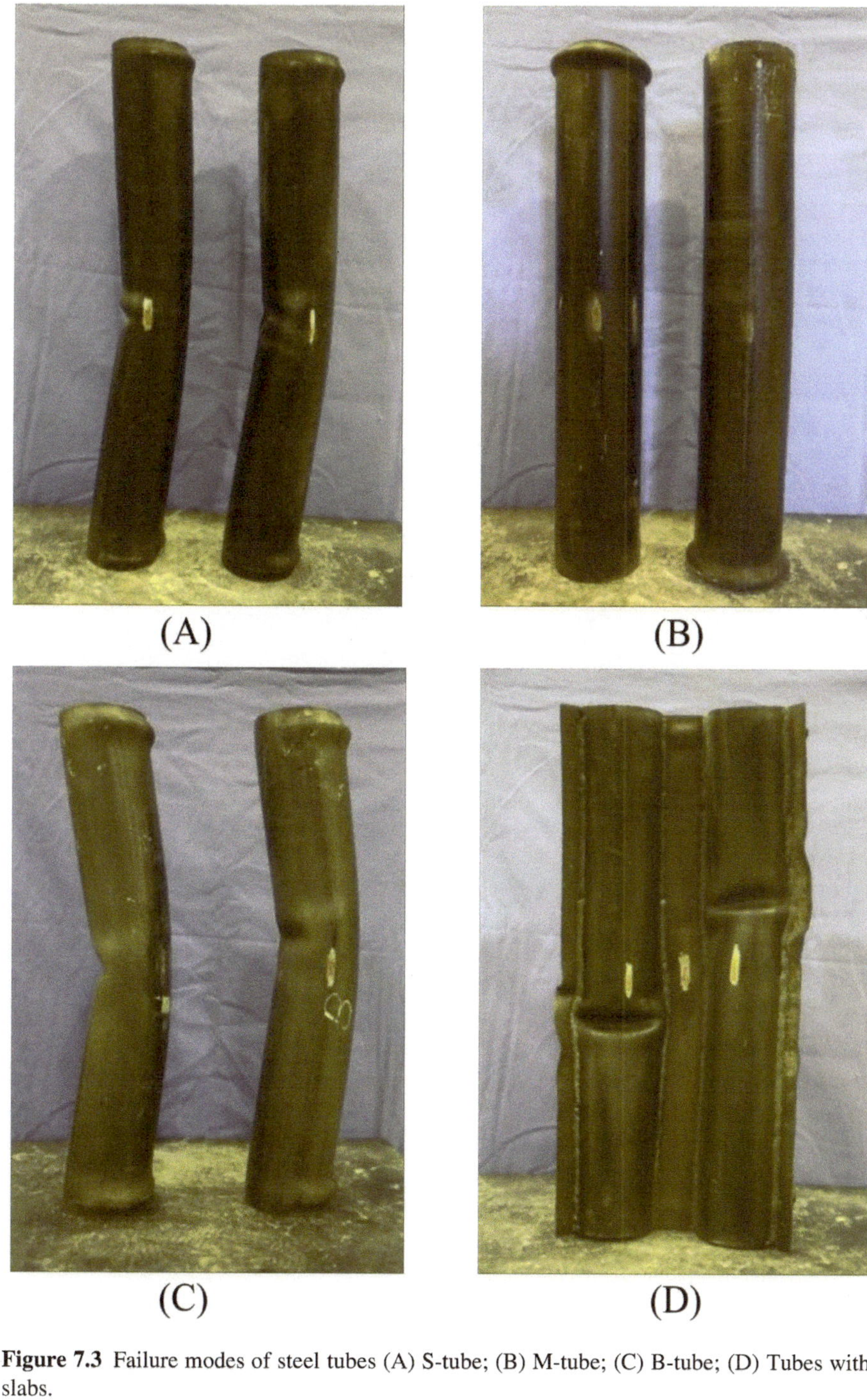

(A) (B) (C) (D)

Figure 7.3 Failure modes of steel tubes (A) S-tube; (B) M-tube; (C) B-tube; (D) Tubes with slabs.

Table 7.2 Details and test results of steel tubes.

Type of steel tube	Diameter/ thickness (mm)	Width of stiffeners (mm)	Thickness of stiffeners (mm)	Number of stiffeners	Ultimate load (kN)
S-tube	88.9/3.2	—	—	—	328
M-tube	101.6/3.2	—	—	—	367
B-tube	101.6/5	—	—	—	538
Tubes with slabs	101.6/3.2	One 50 Four 25	3	5	974

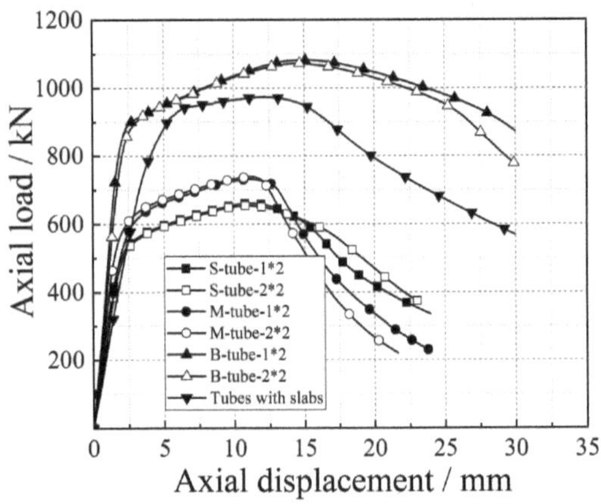

Figure 7.4 Load–strain curves of steel tubes. Load–strain relationship.

axial loading. The local buckling of inner steel tubes will occur on the specified positions, where they should be measured. Several axial strain gauges, each with a gauge length of 10 mm, were attached at the mid-height of the steel tube and the ribbed stiffeners. The distribution of all strain gauges is shown in Fig. 7.6, where the long red strips represent hoop strain gauges, and the small red squares represent axial strain gauges.

In the test process, two steel pinned supports are used to set the loading process on two ends of the specimen to ensure the fixed condition of the two-column ends. Therefore, we should measure the total displacement and the local displacement of the specimen in the axial compressive test. Two pairs of liner variable displacement transducers (LVDTs) were used for each specimen under the axial compressive test: one pair of LVDTs was used to measure the total axial displacement of the specimen; another pair of LVDTs was installed on LVDT holders and used to measure the axial displacement of the mid-height region (about 200 mm height) of the specimen. The distribution of all LVDTs is shown in Fig. 7.5.

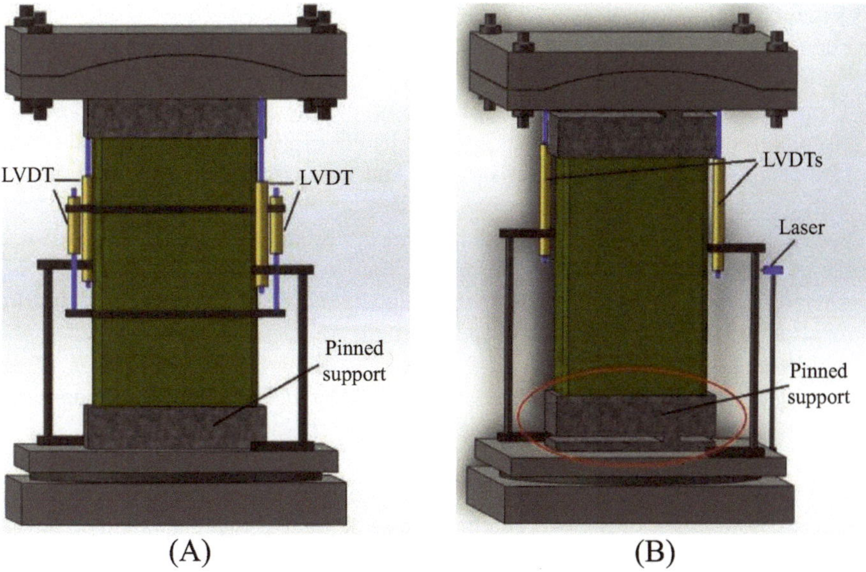

Figure 7.5 Loading diagram of axial and eccentric compressive tests (A) axial test; (B) eccentric test.

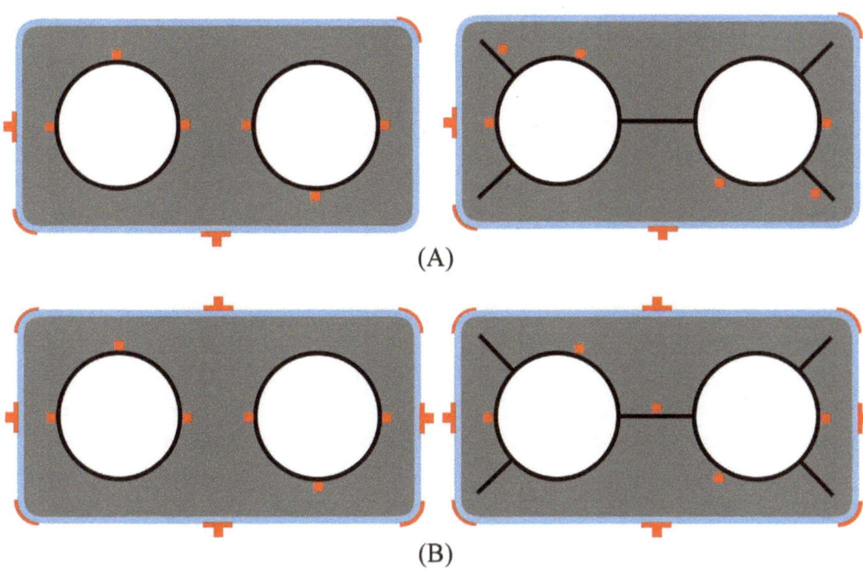

Figure 7.6 Distribution of all strain gauges in specimens (A) axial compression; (B) eccentric compression.

(2) Eccentric compressive test

For eccentric compressive specimens, we should measure the strain of different zones in the specimen, due to the section deformation and the uneven expansion. Therefore, four pairs of strain gauges (including one axial strain gauge and one circumferential strain gauge) with a gauge length of 20 mm were installed on the side center of the middle-height surface. And four circumferential strain gauges with a gauge length of 20 mm were installed on the four corners of the middle-height surface. Under eccentric compression, the lateral confining pressure and the section deformation will influence the axial strain of the steel tube. Therefore, the axial strain of the specified positions should be measured. Several axial strain gauges, each with a gauge length of 10 mm, were attached at the mid-height of the steel tube and the ribbed stiffeners. The distribution of all strain gauges is shown in Fig. 7.6, where the long red strips represent hoop strain gauges, and the small red squares represent axial strain gauges.

In the test process, two steel pinned supports were used to set the loading process on two ends of the specimen to maintain the load eccentricity and ensure the fixed condition of the two-column ends. For each specimen, axial and lateral displacements should be measured. One pair of LVDTs was used to measure the total axial displacement of the test machine. And one laser was used to measure the lateral displacement of the specimen at the mid-height section, which was still parallel to the loading surface of the test machine. It showed the test instrumentations in Fig. 7.5.

7.3 Axial compressive test results

7.3.1 The failure mode

For the specimens with concrete-filled steel tubes (i.e., AFNN6 and AFMN6), the specimens failed by the hoop rupture of the GFRP tube at or near one of the middle-height corners, which are similar to the FRP confined concrete square column. And the failure is accompanied by loud noise, as shown in Fig. 7.7. For the AFMN6 specimens, there was no obvious local buckling on the inner steel tubes, so the combination between the outer FRP tube and the core concrete on both sides can restrict the local buckling completely.

For the specimens with empty steel tubes (i.e., AESN, AEMN, AEMY, and AEBN), the concrete can expand inward, because of the local buckling of the inner empty steel tube. Therefore, these specimens have less outward expansion at the corners, larger outward expansion at the middle of the long side, and larger axial displacement than that of AFNN and AFMN specimens. And several specimens stopped testing because the long side touched the LVDT holder, as shown in Fig. 7.7. In order to get the failure mode under the large axial displacement, we tested one of the same specimens without an LVDT holder. These specimens failed by the hoop rupture of the GFRP tube corner, as shown in Fig. 7.7.

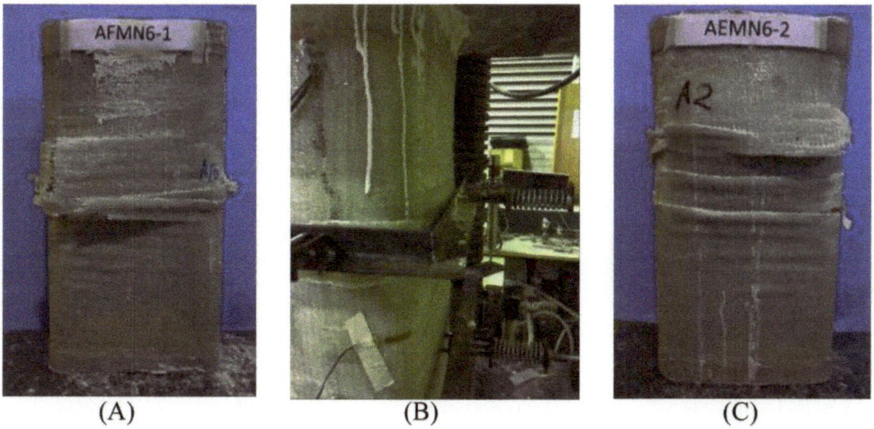

Figure 7.7 Failure modes of specimens (A) AFMN6; (B) AEMN6-1; (C) AEMN6-2.

As shown in the section about material properties, the empty steel tube decreases show a considerable load reduction, due to the local buckling and the plane bending buckling. However, the inner empty steel tube in the specimens only experienced significant local buckling, so the load reduction in test results of DSTRC specimens was significantly less magnitude and started at a much larger axial displacement than that of the empty steel tube. The typical local buckling on the steel tube of the DSTRC specimen (i.e., AESN, AEMN, and AEBN) after the test is shown in Fig. 7.7. In order to restrict the local buckling, the ribbed stiffeners were welded on the steel tube (i.e., AEMY). And the AEMY specimen has a larger bearing capacity than the AEMN specimen. The typical local buckling on the steel tube of the AEMY specimen is also shown in Fig. 7.7.

7.3.2 Typical axial compressive behavior

In order to study the axial compressive behavior of the specimens with the same thickness as the GFRP tube, we summarize the key test results in Table 7.3. In this table, we selected the axial peak load in the first increasing stage of the specimen, which is the rapid growth branch, namely the P point. And the P_s point is the summarized axial peak load of two steel tubes, corresponding to the endpoint in the first increasing stage. Then we used the two load points to gain the axial peak load of concrete at the endpoint in the first increasing stage, namely, the axial load $(P-P_s)$ which was subtracted by the load of the inner steel tubes. After that, we computed the compressive peak stress of core concrete in the specimen by the ratio of the load to the section area, namely, $(P-P_s)/A_c$. And the Ac is the area of the core concrete section. Based on the test results in Table 7.3, we can draw the following conclusions: the concrete stress changes of the specimens with variable steel tubes are similar, which means the concrete stress does not depend on the steel tube; the concrete stress changes of AFNN6 and AFMN6 are similar and larger than that of

Table 7.3 Test results of axial compressive specimens.

Specimen ID	Load at the first branch end P (kN)	Load of two steel tubes P_s (kN)	Area of concrete section A_c (mm^2)	$(P-P_s)/A_c$ (MPa)
AFNN6-1	1439	0	44,807	32.1
AFNN6-2	1418	0	44,807	31.6
AFMN6-2	2079	600	42,945	34.4
AFMN6-2	1909	600	42,945	30.5
AESN6-1	1450	560	32,393	27.5
AESN6-2	1540	560	32,393	30.3
AEMN6-1	1474	600	28,528	30.6
AEMN6-2	1460	600	28,528	30.1
AEBN6-1	1747	880	28,528	30.4
AEBN6-2	1637	880	28,528	26.5
AEMY6-1	1360	580	28,078	27.8
AEMY6-2	1447	580	28,078	30.8

specimens with the empty steel tubes, so that the concrete stress depends on the FRP tube. In summary, the inner steel tubes can play a supporting role in the lateral confining pressure distribution, so they only influence the bearing capacity of DSRCTs. However, the empty steel tubes reduce the concrete area in the effectively confining zone, so they will affect the average stress of the core concrete.

(1) AEMN6

We take AEMN6-1,2 specimens as examples in this section to analyze the typical axial compressive behavior of DSTRCs and the summary of two steel tubes, as shown in Fig. 7.8. For the curves in Fig. 7.8, we take the axial strains as the ratio of the average overall axial displacement measured by two overall LVDTs to the entire height of the specimen. And we used this computation for axial strains instead of the local strains measured by the axial strain gauges attached to the specimens. As discussed in the above sections, the steel tube showed the local buckling at the two ends, as shown in Fig. 7.4. And the inner steel tube in DSTRC showed the local buckling near the middle-height zone, as shown in Fig. 7.4, because the two-column ends are confined by the loading machine. Therefore, the local strain may show an uneven distribution along the height of the column. The average strains measured by strain gauges at the middle-height zone may differ slightly from the actual axial strains of the specimens. In order to compare the axial compressive behavior of the specimens, the axial strains are all computed from the total displacements measured by the two LVDTs, unless otherwise specified. Meanwhile, we defined the compressive strains are the positive values and the tensile strains are the negative values, unless otherwise specified.

Fig. 7.7 shows the load–strain curves of three types of specimens (i.e., AEMN6, AFNN6, and steel tubes) to reflect the compressive behavior of DSTRCs. Based on the curves of the AEMN6-1,2 specimens, the following conclusion can be

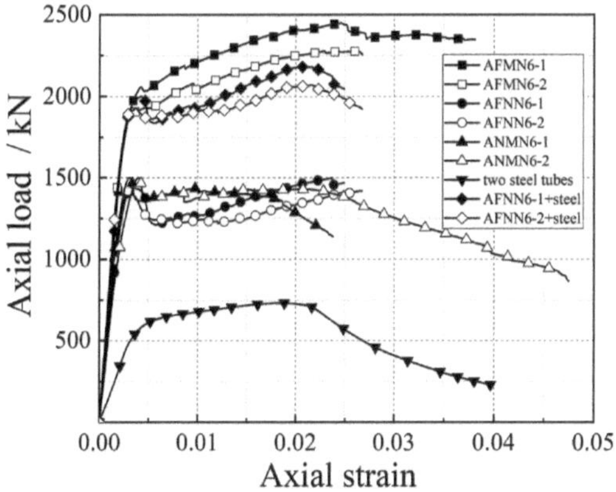

Figure 7.8 Typical axial compressive behaviors of specimens. Stress–strain relationship.

drawn: (1) the load–strain curve has four stages; (2) the first stage is an increasing curve; (3) the second stage is a descending curve; (4) the third stage is an increasing curve, which increases slower than the first stage; (5) the last stage is a gradually descending curve, which is caused by the inward buckling of the inner steel tube and sustained before the specimen failed; (6) the later loading stages have some little suddenly descending branches which are caused by the dislocation along the height direction of the FRP tube; (7) the ultimate strain of AEMN6-1 is much smaller than that of AEMN6-1, because the AEMN6-1 stops by the long side of specimen touched the LVDTS holder, as shown in Fig. 7.6B, while the AEMN6-2 stops by the FRP tube rupture.

In addition, Fig. 7.7 shows that the load–strain curves of AFNN6 have three stages without the last decreasing stage of AEMN6, and the load–strain curves of steel tube have three stages without the second decreasing stage of AEMN6. Therefore, the characteristics of the load–strain curves of AEMN6 include that of AFNN6 and steel tube. It means that the compressive behavior of DSTRCs is influenced by the GFRP confined concrete square column and the inner steel tubes. There are several differences between these curves, which are as follows: (1) compared to the AFNN6, the AEMN6 has a shorter second decreasing stage and a smaller increasing load in the third increasing stage, due to the smaller confined concrete section; (2) compared to the steel tube, the AEMN6 has a smaller strain and a smaller decreasing rate when the load decreases in the last stage, so the confining pressure make the local buckling occur early and the sandwiched concrete restrict the local buckling of the steel tube.

(2) AFMN6

In practical engineering, the solid rectangular column also has a wide application. The FRP confined double concrete-filled steel tube column (i.e., AFNN6-1,2 specimens) has a higher bearing capacity and higher durability, so it has a wide application future. Therefore, we study the solid rectangular columns (i.e., AFNN6-1,2 specimens) in this section. And we compare the axial load–strain curves of two pairs of the solid specimens (i.e., AFNN6-1,2 and AFMN6-1,2 specimens), as shown in Fig. 7.8. In order to consider the influence of the inner steel tubes, we also draw the axial load–strain curves of summarized compressive behavior between the inner steel tube and AFNN6-1,2, as shown in Fig. 7.8. We get the load–strain curve of the inner steel tube from the material compression test. Meanwhile, the axial load of the FRP tube is negligible compared to those of the concrete and the steel tube, so the contribution of the FRP tube is only a lateral confining effect.

Fig. 7.8 shows the load–strain curves of the three types of specimens (i.e., AFMN6, AFNN6, steel tube, and computational model) to reflect the compressive behavior of the proposed column. Based on the curves of AEMN6-1,2 specimens, the following conclusion can be drawn: (1) the load–strain curve has two approximate bilinear increasing stages; (2) the second increasing stages have some little suddenly descending branches which are caused by the dislocation along the height direction of the FRP tube; (3) the specimen stops by FRP tube rupture, as shown in Fig. 7.6; (4) the ultimate strain of AFMN6-2 is much smaller than that of AFMN6-1, mainly because of the fabrication error of the thick FRP tube.

In addition, Fig. 7.8 shows that the load–strain curves of AFNN6 have three approximate stages with a decreasing stage between two the increasing stages; the load–strain curves of steel tube have three approximate stages with a decreasing stage after the two increasing stages; the load–strain curves of the computational model have three stages which are similar to that of AFNN6. It means that the inner steel tube can provide confining effect on the core concrete to improve the compressive behavior of the proposed rectangular column. And there are several differences between these curves, which are as follows: (1) the two increasing stages in load–displacement curves of AFMN6-1,2 and the computational model show the approximate trends, while AFMN6 has higher bearing capacity in the later loading stage, due to the second decreasing stage in load–strain curves of the computational model; (2) there is no descending stage in AFMN6. The above conclusions show that the inner steel tubes can provide an additional confining effect on the core concrete, which contributes to the early loading stage and makes no decreasing stage after the first increasing stage. And the sandwiched concrete on both sides of the steel tube can resist the local buckling of steel tubes, which makes no decreasing stage after the second increasing stage. It should be noted from Fig. 7.8 that the first increasing stages of AFMN6-1,2 and the computational model are similar, which reflects that the confining effect provided by the steel tubes can play an important role after the peak strain of the core concrete. In addition, the above conclusions also prove that empty steel tubes can influence the compressive behavior at the later loading stage.

In summary, the local buckling on the inner empty steel tube significantly affected the compressive behavior of DSTRCs. Fig. 7.9 shows the deformation shapes of the steel tubes in AEMN6 specimens after the compressive test. Based on the failure model of the empty steel tubes in Fig. 7.9, we can conjecture the following conclusions: the concrete can restrict the outer local buckling, which makes the local buckling occur near the middle-height zone; the inner local buckling shows

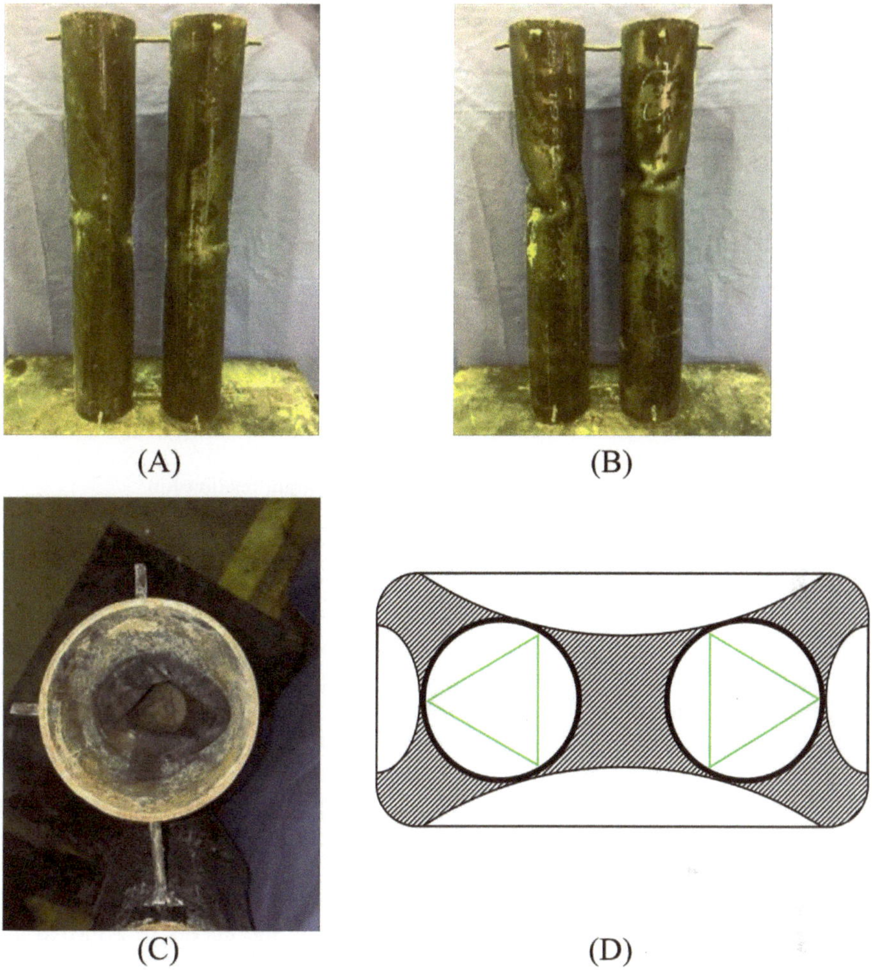

Figure 7.9 Local buckling on inner empty steel tube (A) AEMN6-1; (B) AEMN6-2; (C) inside of steel tube; (D) effective confining area.

an approximate triangle, which is caused by the well-confining effect at the corner, and the effective confining area is shown in Fig. 7.9. Therefore, we will study the influence factors on the local buckling of the inner steel tubes in the following sections, including the diameter–thickness ratio and the ribbed stiffeners.

7.3.3 Axial strain-hoop strain curves

In order to reflect the lateral expansion of the specimens, we draw the axial strain-hoop strain curves of specimens in Fig. 7.10. For the strain curves in Fig. 7.10, the lateral strains were averaged from the strain circumferential gauges, and the axial

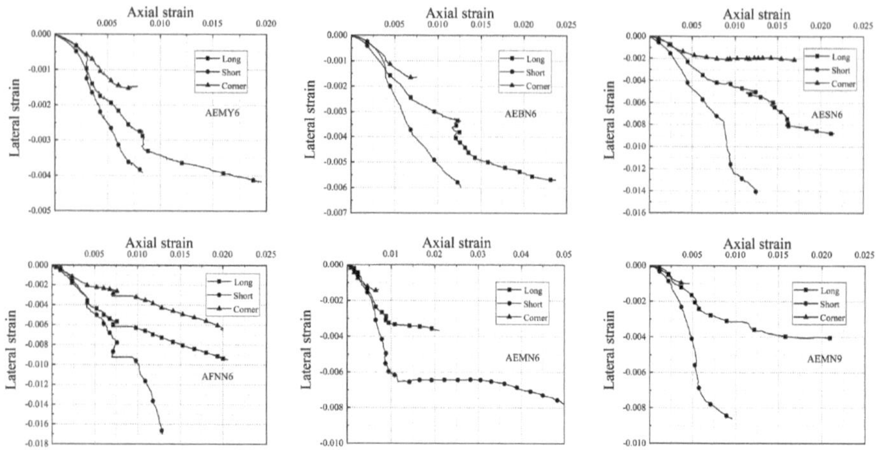

Figure 7.10 Axial strain-hoop strain curves of specimens. Strain relationship.

strains were the ratio of the displacement measured by two LVDTs at the middle-height zone to the spacing of the LVDT holders. As discussed in the test process, we tested three specimens (i.e., AEMY6-2, AEMN6-2, and AEMN9-2) without the middle-height LVDT holders, so as to make the specimens reach the ultimate load with the FRP tube rupture. Meanwhile, there are three types of lateral strains on the middle-height surface of the FRP tube, including the corner strain, the middle strain of the long side, and the middle strain of the short side. Therefore, there are three axial strain-lateral strain curves in one figure: (1) the labeled "Corner" curves have the lateral strains averaged from the two strain gauges at the diagonal corners; (2) the labeled "Long" curves have the lateral strains from the strain gauge at the middle point of the long side in the middle-height section; (3) the labeled "Short" curves have the lateral strains from the strain gauge at the middle point of the short side in the middle-height section. It is also noted that the curves of AFMN6 are absent because of the damaged data. In addition, we just select one of the two same specimens to draw the axial strain-lateral strain curve, so as to simplify the curve comparison.

As shown in the failure mode of specimens in Fig. 7.7, the rupture of the FRP tube occurred at or close to one corner of the middle-height zone. But the curves in Fig. 7.9 show that the three types of lateral strains corresponding to the same axial strain are in a large to small order: the "Short" strain, the "Long" strain, and the "Corner" strain. It is clear that the concrete expansion in the middle of the short sides and the long sides are much larger than that at the corners. And these curves show two approximate linear trends, so the growth rate of concrete expansion is fast at first and then becomes slow. Although the existing studies on the FRP confined concrete square column reported the lateral strains at the corners are smaller than that at the flat sides, because of the well-confining pressure provided by the FRP tube which resists the concrete expansion. However, the difference between the lateral strains at the short sides and the long sides, which is shown in Fig. 7.10, appears to be much larger. And the "Short" strain is larger than the "Long" strain, where the confining effect on the short side is larger than that on the long side, so the above explanation is not suitable for these test results.

The section analysis shows that not only the confining effect provided by the FRP tube can restrain the concrete expansion, but also the bending effect can influence the local lateral strain under the same concrete expansion at the flat sides. And we believed that the thick FRP tube would make the bending effect become more pronounced. Especially for the specimens, the thickness of the GFRP tube is relatively large: the thickness of the 6-ply GFRP tube is around 10 mm and the thickness of the 6-ply GFRP tube is around 13 mm. The bending effect of the GFRP tube makes the outer surface have smaller lateral strains than the inner surface, which leads to a strain gradient along the thickness direction. Meanwhile, the strain gradient at the corners is larger than that at the flat sides in a square tube with rounded corners. Because the circular strain gauges attach to the surface of the outer GFRP tube to measure the lateral strains of the specimens. The lateral strains at the corners are much smaller than that at the middle point of the flat sides under the combined effect of the lateral confining pressure and the strain gradient. In addition, the lever effect would increase the bending effect, so the measured point far from the corner has a large strain gradient. Therefore, the increasing strain gradient leads to lateral strains at the middle point of the short sides being much larger than that of the long sides.

Except for the measured lateral strain, the strain gradient would affect the properties of the FRP tube at the corners when the FRP rupture occurred. The inner surface reaches the rupture strain first, then leads to failure of the specimen. Therefore, the rupture strains measured at the outer surface can be significantly lower than those got from flat coupon tests for a relatively thicker FRP tube. And several existing studies introduced a rupture strain reduction coefficient to reflect that the actual rupture strain of FRP tube is smaller than that of the material properties test.

Based on the curve comparisons of the specimens with the different steel tubes, the larger diameter–thickness ratio leads to higher lateral strain at the corners (namely the larger outer expansion of the core concrete), so the steel tube with a large diameter–thickness ratio has a small local buckling. It is found in Fig. 7.10 that the axial strain-lateral strain curves of AEMY6 and AEBN6 are similar, although the inner steel tubes of AEMY6 differ from that of AEBN6. It means that the welding ribbed stiffeners play an obvious role in both resisting the lateral confining pressure provided by the GFRP tube and restricting the local bucking of the steel tubes, so the welding ribbed stiffeners can increase the outer lateral expansion of the core concrete. We can suggest that increasing the diameter–thickness ratio of the inner steel tubes and welding ribbed stiffeners on the inner steel tubes are two effective methods to restrict the local buckling of the steel tube and improve the compressive behavior of the specimens with the inner empty steel tubes.

7.3.4 Parameter analysis

As discussed in the above section, the different size of each component in the section influences the compressive behavior of DSTCSs. Therefore, we will discuss the influence factors in the following section, including the GFRP thickness, the diameter and the thickness of steel tube, and the ribbed stiffeners.

(1) Thickness of the outer GFRP tube

In order to study the effect of the thickness of the outer FRP tube on the compressive behavior of DSTRCs, we summarize the load–displacement curves of AEMN6 and AEMN9 specimens in Fig. 7.11. Based on the curve comparisons, the following compressive characters of the AEMN9-1,2 can be drawn: (1) the curves have three stages without the load-drop branch between the two increasing stages in the curves of AEMN6; (2) there might be a little level branch between the two increasing stages; (3) the initial stiffness is the same as that of AEMN6; (4) the increasing ratio of the second increasing stage is larger than that of the third increasing stage of AEMN6; (5) the last decreasing ratio is smaller than that of AEMN6; (6) the AEMN9 specimens have the large axial bearing capacity and large ductility. We easily found that the curves of the two same specimens, for both AEMN6 and AEMN9, show a different trend in the later loading stage, mainly due to the confining effect provided by the LVDT holders. Based on the axial strain-lateral strain curves, the AEMN9-1,2 had a higher bearing capacity and a smaller lateral expansion than AEMN6-1,2. We believe that the more thickness FRP can provide larger confining effect on the sandwiched concrete, which can increase the axial compressive behavior and resist the concrete expansion. Meanwhile, the displacement of the last decreasing stage, caused by the local buckling of the empty steel tubes, is approximately similar. It means that the thicker GFRP tube has little effect on the local buckling of the steel tubes in specimens. In other words, the local buckling is a type of self-property of the steel tube.

In summary, the thicker FRP tube can provide well-confining effect, which contributes to the increase in the axial compressive behavior and eliminates the descending branch between two increasing stages. However, the inward local buckling of

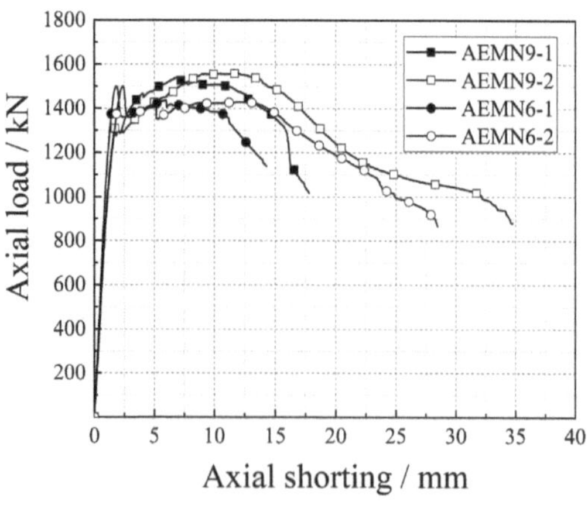

Figure 7.11 Load–displacement curves of AEMN6 and AEMN9. Load–displacement relationship.

the inner steel tube buckling is determined by the properties of the steel tube and the section form and is not significantly influenced by the FRP tube.

(2) Diameter and thickness of the inner steel tube

In order to study the effect of diameter and thickness of the inner steel tubes on the compressive behavior of DSTRCs, we summarize the load–displacement curves of specimens with variable steel tubes (i.e., AESN6, AEMN6, and AEBN6) in Fig. 7.11. In addition, AESN6 has a steel ratio of 0.039 and a diameter–thickness ratio of 27.8, AEMN6 has a steel ratio of 0.045 and a diameter–thickness ratio of 31.8, AEBN6 has a steel ratio of 0.069 and a diameter–thickness ratio of 20.3. As is shown in Fig. 7.12, the load–displacement curve of AEMN6-2 is not discussed, because the test is without the LVDT holders. All specimens listed in Fig. 7.12 stopped tests by the long section side touching the LVDT holders. Meanwhile, all curves of these specimens show similar trends, including four stages: the first increasing stage, the second decreasing stage, the third linear increasing stage, and the fourth linear decreasing stage. However, there are many differences in these load–displacement curves, and the curve comparisons showed that: (1) the curves of specimens show similar trends; (2) the load of AESN6 is larger than AEMN6, mainly because the AESN6 has a larger concrete section than AEMN6; (3) based on the ultimate point, the bigger diameter–thickness ratio has a larger axial displacement; (4) AESN6 has a larger increasing rate of the third increasing stage than AEMN6 and AEBN6, which means the confining effect provided by the FRP tube is obvious in AESN6; (5) the specimen with a smaller diameter–thickness ratio has the gentler fourth decreasing stage. In summary, the steel tubes have a big effect on the latter mechanical behavior, but they are not the only influential factor.

(3) Ribbed stiffeners

In order to study the effect of ribbed stiffeners on the compressive behavior of DSTRCs, we summarize the load–displacement curves of specimens with

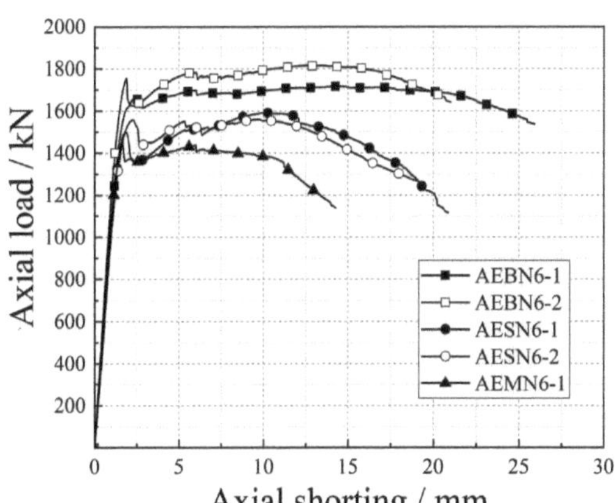

Figure 7.12 Load–displacement curves of specimens with variable steel tubes. Load–displacement relationship.

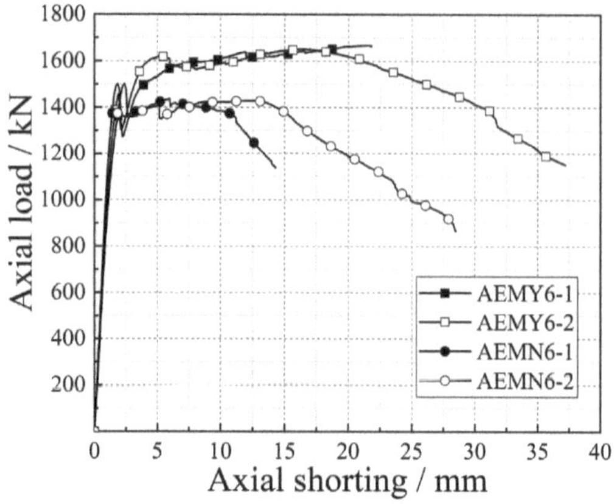

Figure 7.13 Load–displacement curves of specimens with reinforced steel tubes. Load–displacement relationship.

reinforced steel tubes (i.e., AEMN6 and AEMY6) in Fig. 7.13. Based on the curve comparisons, the following compressive characters of the AEMY6-1,2 can be drawn: (1) the curves have five stages that can describe the curves of AEMY6 with an addition rapid increasing stage than AEMN6; (2) the inner steel tubes do not occur the local buckling when the AEMY6-1 stopped test by the specimen touched LVDT holders; (3) based on the measured strain results of AEMY6-2, the stiffeners occur the local buckling earlier than the inner steel tubes; (4) the ribbed stiffeners can improve the mechanical properties of steel tube.

Based on load–displacement curves of steel tubes in Fig. 7.4, the ribbed stiffeners can restrict the local buckling of steel tubes, which leads to good ductility of specimens. And the restricting effect provided by the ribbed stiffeners makes the local buckling occur near the middle-height section of the steel tube. However, the inner buckling shapes are the same as the steel tubes for both the steel tube test results and the specimen test results, as shown in Fig. 7.3.

7.4 Eccentric compressive test results

The existing studies showed the section form is an important factor to influence the failure modes and the load–displacement curve of the eccentric compressive specimens. Therefore, the compressive test in the chapter selected three types of section forms: AEMN6, AEMY6, and AFMN6 specimens.

7.4.1 Failure modes

For the DSTRCs under the eccentric compression (i.e., EEMN6(50) and EEMN6 (70) specimens), the tests were stopped by the rupture of the FRP tube near the middle-height zone on the compressive short side of the specimens. The columns fail with large bending deformation in the middle-height regions, which leads to the rupture of the FRP tube in this zone. In the loading process, the load decreased in the later loading stage and dropped after the failure of specimens, as shown in Fig. 7.14.

After the test, we removed the FRP tube and the sandwiching concrete to see the typical failure of the inner steel tube, as shown in Fig. 7.15. We can find that the local buckling would occur close to or at the rupture of the FRP tube. In addition, the test results don't include the EEMN6(50)-1 specimen. Because the pinned supports are too weak to transmit the load, the specimen failed by the destruction on the column end and the bending deformation of the pinned supports.

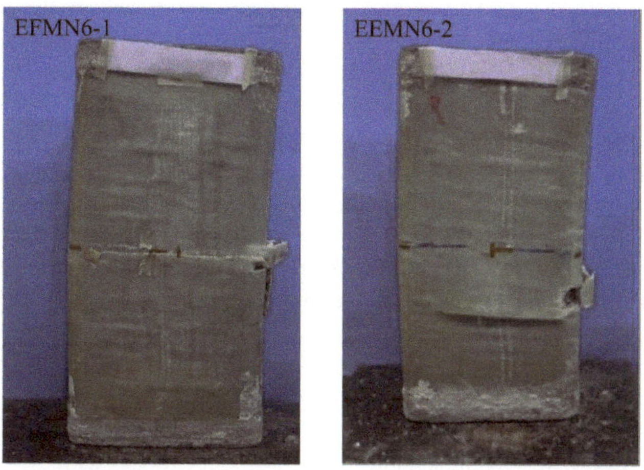

Figure 7.14 Failure modes of eccentric compressive specimens. Failure modes.

Figure 7.15 Failure modes of the inner steel tubes in eccentric compressive specimens. Failure modes.

For the R-DSTRCs under eccentric compression (i.e., EEMY6(50) and EEMY6 (75) specimens), the outer GFRP tube of EEMY6(75) don't show the obvious lateral rupture, which means the GFRP tube can provide a sustainable confining effect on the sandwiched concrete with the gentle decreasing load. And the specimen stopped by the load was less than 60% of the peak load. The tests of EEMY6(50) were stopped by the rupture of the FRP tube, which is similar to EEMN6(50). After the test, we removed the FRP tube and the sandwiching concrete to see the typical failure of the inner steel tube, as shown in Fig. 7.15. There is also local buckling on the ribbed stiffeners and the steel tube on the compression side. The local buckling would occur close to or at the rupture of the FRP tube. And the strain gauges also show the stiffeners show the local buckling earlier than the inner steel tubes. Unlike the inner steel tubes, the ribbed stiffeners can provide lateral pressure distribution, even in the buckling state. Therefore, the load decreased slowly in the last stage.

For the specimens with the solid steel tubes under eccentric compression (i.e., EFMN6(50) and EFMN6(75) specimens), the tests were stopped by the rupture of the FRP tube near the middle-height section on the compressive short side of the specimens. The columns failed with large bending deformation at the middle-height zone, which led to the rupture of the FRP tube in this region. In the loading process, the load remained the same value in the later loading stage and dropped after the failure of specimens. After the test, we removed the FRP tube and the sandwiching concrete to see the typical failure of the inner steel tube, as shown in Fig. 7.15. There is no local buckling on steel tubes of EFMN6, due to the filling concrete on both sides of the steel tube.

7.4.2 Axial load–displacement curve

Fig. 7.16 shows load–displacement curves of specimens with various section forms. However, there is only one curve of EEMN6(50) specimens, due to the failure of the pinned supports.

Fig. 7.16 reflects the effect of the load eccentricity on the load–displacement curves in various section forms. Based on the curve comparisons, the peak load decreases with the increase in load eccentricity. Compared to the peak load of the axial compressive specimens, the reduction ratios of the peak load under 50 mm-load eccentricity are about 23.3% (EEMN6), 17.6% (EEMY6), and 26.5% (EFMN6); the reduction ratios of the peak load under 75 mm-load eccentricity are about 34.8% (EEMN6), 28.2% (EEMY6), and 38.1% (EFMN6). The reduction ratios of EFMN6 are larger than that of EEMN6, which reflects the confining effect provided by GFRP tube obviously decreases with the increase of the load eccentricity. The reduction ratios of EEMY6 are smaller than that of EEMN6 and EFMN6, which reflects the ribbed stiffeners can restrict the local buckling of the steel tubes and improve the confining effect on the sandwiching concrete.

Meanwhile, the following conclusion can be drawn from load–displacement curves in Fig. 7.16: (1) the curves of EEMN6 show approximate linear decreasing trends after the peak point; (2) the curves of EEMY6 show decreasing trends after the peak point, the decrease rates are smaller than that of EEMN6; (3) the curves of

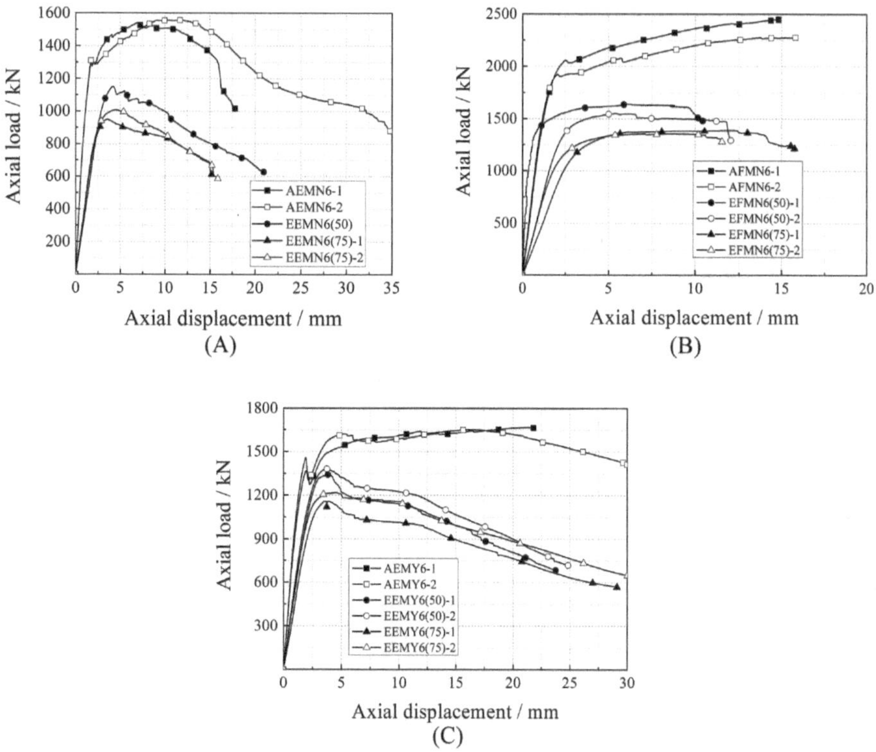

Figure 7.16 Load–displacement curves of specimens with various section forms (A) EEMN6; (B) EFMN6; (C) ENMY6.

EEMY6 show a complex developing trend, due to the ribbed stiffeners; (4) the decreasing curves of EEMY6 can be divided into three stages, including the rapid decreasing stage, the approximate level stage and the gentle decreasing stage; (5) the curves of EFMN6 show level trends after the peak point; (6) the effect of load eccentricity is obvious in EFMN6, due to the strong confining effect provided by the inner steel tubes and the outer GFRP tube; (7) the ribbed stiffeners can decline the effect of load eccentricity.

Fig. 7.17 reflects the effect of the section form on the axial load-axial displacement curves under the same load eccentricity. It is easily found that the curves of EFMN6s have the largest stiffness, the highest load, and the lowest ductility. Because both the inner steel tubes and the outer FRP tubes provided the largest confining effect on the concrete, which leads to higher compressive strength and higher lateral expansion of concrete. Compared with the EEMN6s, EEMY6s have a larger stiffness, a higher load, and higher ductility. Therefore, the steel tubes with the ribbed stiffeners can improve the mechanical properties of specimens, because the ribbed stiffeners can not only improve the steel ratio of the column section to

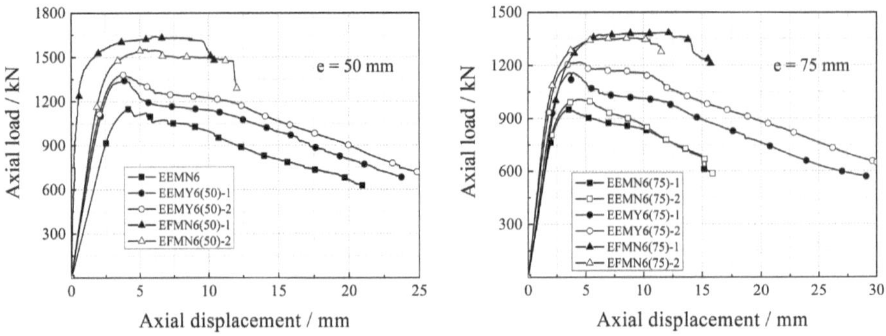

Figure 7.17 Load-axial displacement curves under the same load eccentricity. Load-axial displacement relationship.

increase the load but also restrict the local buckling of the inner steel tubes to increase the ductility.

7.4.3 Axial load-lateral deflection curve

Fig. 7.18 shows the axial load-lateral deflection curves of the eccentric compressive specimens. In addition, the lateral deflection was measured by the Laser at the middle-height section of the specimens. The middle-height lateral deflection can reflect the total rotational deformation of the specimens. However, the curves of some specimens are not shown in Fig. 7.18, due to the abnormal test data. The test results show the specimens have well rotational deformation capacity, especially for the specimens with the ribbed stiffeners. Based on the curve comparisons in Fig. 7.18, we also find the following conclusions: (1) EEMN6 and EEMY6 have an approximate linear decreasing curve after the peak load; (2) EFMN6 has the largest eccentric compressive load and an approximate leveling curve after the peak load; (3) the lateral deflections of EFMN6 are similar to that of EEMN6; (4) EEMY6 has higher eccentric compressive load and higher lateral deflection than EEMN6; (5) the decreasing rates of EEMY6 and EEMN6 are similar; (6) the specimens under the larger eccentricity show the smaller initial stiffness, the larger middle-height lateral deflection, and the gentler and longer descending branch. However, the curves are not smooth, probably because of the following reasons: the bending effect of the GFRP tube, the lateral expansion of the sandwiching concrete, and the possible rotational deformation of the middle-height section.

As per the above analysis, the confining effect and the steel ratio would lead to different rotational deformation capacities. The former makes the specimens have the higher bearing capacity and higher load holding capacity, and the latter makes the specimens have the higher bearing capacity and higher displacement ductility. In addition, the confining effect is more obvious under the smaller load eccentricity. The application of the ribbed stiffeners can obviously contribute to

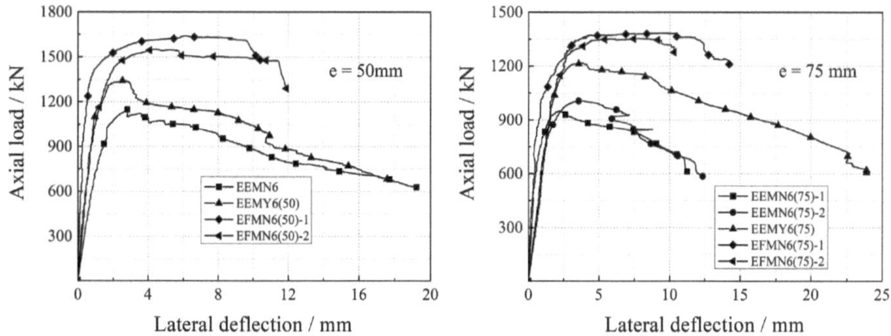

Figure 7.18 Load-lateral deflection curves. Load-lateral deflection relationship.

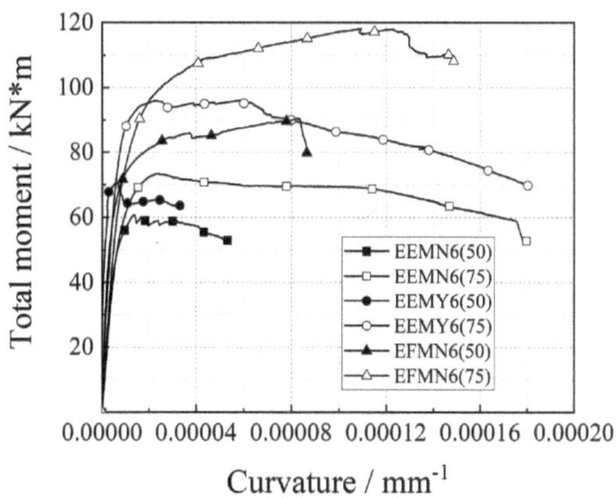

Figure 7.19 Section moment–curvature curves. Section analysis.

both restricting the local buckling of the inner steel tubes and balancing the stress distribution in the sandwiching concrete.

7.4.4 Moment–curvature curves

Fig. 7.19 showed the section moment–curvature curves of the specimens under eccentric compression specimens. For the curves, the middle-height section of specimens is the foundation for calculating the section moments and the section curvatures. The section moment can be divided into two parts: the first part is the moment caused by eccentric load, which is the product of the axial load and the original load eccentricity; the second part is the moment caused by the lateral displacement of the middle-height section, which is the product of the axial load and the lateral displacement. The section curvature can be computed by the equation: $(\varepsilon_c - \varepsilon_t)/L$, where ε_c is the compressive strain, which is measured by the axial strain gauge at the middle on the surface of the compression side; ε_t is the tensile strain, which is measured by the axial strain gauge at the middle on the surface of the tensile side; L is the length of the long side.

Based on the curve comparisons in Fig. 7.19, we can draw the following conclusions: (1) under the larger load eccentricity, the specimens have larger sections and larger curvatures; (2) with the ribbed stiffeners on steel tube or the concrete filled in the steel tube, the specimens have larger stiffness; (3) the curves of EFMN6(50) and EFMN6(75) show the approximate bilinear increasing trends; (4) the curves of EEMN6(75) and EEMY6(75) show the similar developing trends, while EEMY6 (75) have a larger section moment; (5) the curve of EEMY6(75) has an obvious moment drop and a small curvature, possibly due to the test errors.

7.4.5 Strain of the steel section

In order to measure the section deformation, some strain gauges were attached at the mid-height of the inner steel tube in the eccentric compression specimens to monitor the axial strain at various positions, as shown in Fig. 7.6. And we draw the relationships between the axial strain and the section position in Fig. 7.20. In addition, one curve represents one loading displacement, and six loading displacements (from 1.5 to 9.0 mm) are selected for each specimen.

Based on the distribution of axial strains of the inner steel tubes, we can draw the following conclusions: (1) the curves are intersected into one point near to zero strain, so the section is rotated around an axis; (2) with the larger load eccentricity, the rotation axis is near to the central axis; (3) the axial strains of all specimens are approximately linearly distributed along the section rotation direction, which means the "plane section assumption" is valid for the eccentric compression specimens; (4) the strain distribution of steel tube was not suitable for the "plane section

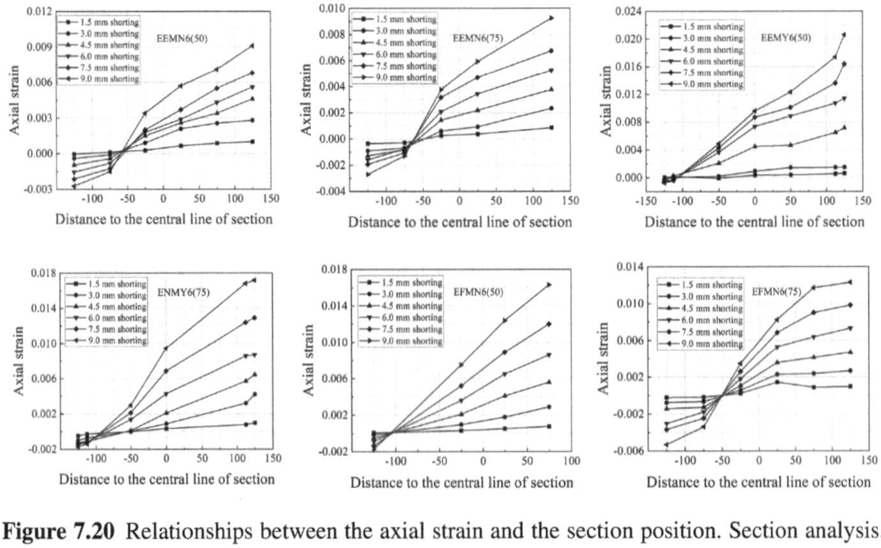

Figure 7.20 Relationships between the axial strain and the section position. Section analysis.

assumption" in the later loading stage, due to the local buckling of steel tube; (5) the strain of steel flanges in the compressive side of all specimens reached the yield strain before the peak loading displacement; (6) the strain of steel flanges in the tensile side of all specimens under 50 mm-load eccentricity reached the yield strain after the peak loading displacement; (7) the strain of steel flanges in the tensile side of EEMN6(75) and EFMN6(75) reaches the yield strain before the peak loading displacement.

The above conclusions illustrate that the EEMN6(50), EFMN6(50), EEMY6(50), and EEMY6(75) belong to the small eccentric compression state. The ribbed stiffeners can restrict the local buckling on the steel tubes, which increases the load eccentricity of the specimen between small and large eccentric compression states. Therefore, the section form has an obvious effect on the eccentric compression condition. These conclusions are important for the development of the theoretical analysis for the composite columns.

7.4.6 Strain of the FRP tube

In order to illustrate the concrete expansion and the confining effect provided by the GFRP tube under the eccentric compression, we draw the lateral strain-lateral deflection curves, as shown in Fig. 7.21, to study the effect of the section form on the behavior of the FRP tubes. In these curves, the lateral strains were measured by the strain gauges on the outer FRP tube in the middle-height section, and the lateral deflections were measured by the Laser. In Fig. 7.21, each specimen has five types of curves and their names are as follows: the "Compression corner" represents the average lateral strain at the corners of the compression side; the "Tensile corner" represents the average lateral strain at the corners of the tensile side; the "Mid compression short-side" represents the average lateral strain at the middle of the

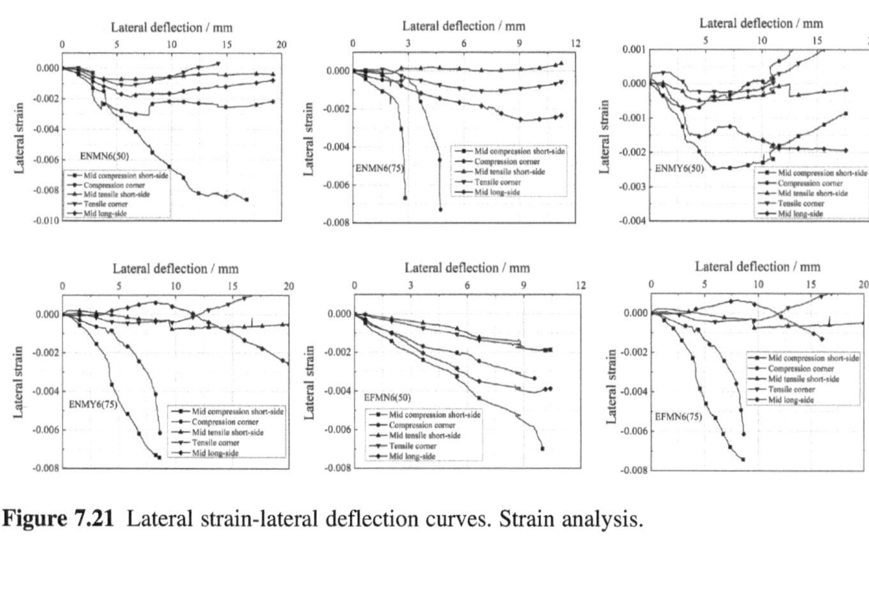

Figure 7.21 Lateral strain-lateral deflection curves. Strain analysis.

compression short-side; the "Mid tensile short-side" represents the average lateral strain at the middle of the tensile short-side; the "Mid long-side" the average lateral strain at the middle of the long side perpendicular to the line load.

Based on the curve comparisons in Fig. 7.21, we can find the following observations. First, the strains at the compression side show an obvious developing trend, because of the concrete expansion under compressive pressure. The "Mid compression short-side" showed the largest value under the entire loading process, which shows the middle concrete at the compression short side expands significantly, and mostly. The "Compression corner" is smaller than the "Mid compression short-side", which shows the strain at the compression corner is smaller than that at compression short side because the well-confining effect provided by the GFRP tube corners can restrain the concrete expansion. Second, the lateral strain at the tensile side (including the corners and the short side) has little tensile change and even has compressive change. This is because the concrete on the tensile side bears the tensile stress. And the strain changes are mainly caused by the deformation of the outer FRR tube. Third, the "Mid long-side" middle strain of the long side has obvious changes which are larger than that of the compression corners in some specimens (i.e., EEMY6(50) and EFMN6(50)). Meanwhile, the middle strains of the long side of the specimens under 50 mm-load eccentricity are more obvious than that of the specimens with 75 mm-load eccentricity. This is probably because of the different rotation axis location. In the specimens under 50 mm-load eccentricity, the rotation axis location is far away from the central axial, which is attached by the circular strain gauge, so the central axial is in the compressive zone and has the obvious concrete expansion.

7.5 Conclusions

This section presents the concept of a hybrid FRP-concrete-steel double-skin tubular rectangular column, namely hybrid DSTRCs. Then this section presents the axial and eccentric compressive experiments to study the compressive behaviors of the hybrid DSTRCs. We present the experimental study on a total of 26 specimens (i.e., 13 pairs), including 7 pairs of specimens for the axial compressive tests and 6 pairs for the eccentric compressive tests. For the axial compressive test, the test variables include the sizes of the steel tubes, the thickness of the FRP tube, the welding rib stiffeners, and the concrete filled in steel tubes. For the eccentric compressive, the test variables include the load eccentricity, the welding rib stiffeners, and the concrete filled in steel tubes. Based on the discussions presented in the paper, we can draw the following conclusions:

(1) The inner empty steel tubes effectively enhance the section area for the hybrid DSTRCs with a similar axial load capacity, which leads to high ductility and high bending load capacity.
(2) The sandwiched concrete can restrict the plane bending buckling and the outward local buckling of the steel tubes, which can improve the mechanical behavior of the inner steel tubes in the local buckling state.

(3) The empty steel tubes occur the inward local buckling in the DSTRC specimens under both the axial compression and the eccentric compression. Meanwhile, the load–displacement curves of specimens showed an obvious decreasing trend when the steel tubes occurred the local buckling, showing that the local buckling of the inner steel tubes is a critical parameter for the later mechanical behavior.
(4) The increase in the diameter-to-thickness ratio can restrict the development of steel tubes, which is an effective way to improve the ductility of the columns.
(5) The ribbed stiffeners welded on the steel tube can effectively delay the local buckling of the steel tube and continuously support the distribution of lateral confining pressure provided by the outer GFRP tube. Therefore, the ribbed stiffeners can improve the mechanical behavior of the hybrid DSTRCs.
(6) The large thickness of the FRP tube provides a large confining effect on the sandwiched concrete, which increases the mechanical behavior of the hybrid DSTRCs.
(7) For the solid specimens (i.e., concrete filled in steel tubes), the concrete on both sides of the steel tube prevents the local buckling of the steel tubes, and the steel tubes provide the confining effect on the core concrete to further improve the mechanical behavior of the column.
(8) With the large confining effect, the solid specimens have good mechanical behaviors under both concentric and eccentric load, including the high bearing capacity and no softening stage in the load–displacement curve. But the solid specimens have low ductility, due to the concrete expansion.
(9) The deformation of the hybrid DSTRCs under the eccentric compression satisfies the "plane section assumption". And the rotation axis is changed with the load eccentricity.
(10) The section form can obviously influence types of eccentric compression, only EEMN6 (75) and EFMN6(75) show a large eccentric compression failure.

Acknowledgments

This work was supported by the National Natural Science Foundation of China [grant numbers 52108171; 51878540]. The authors gratefully acknowledge two Master's students (Mr. Qv Yunsong and Mr. Tian Wenkai) of Xi'an University of Architecture and Technology for the drawing works.

References

Fanggi, BL, & Ozbakkaloglu, T (2015). Square FRP-HSC-steel composite columns: Behavior under axial compression. *Engineering Structures*, *92*, 156–171.

GB/T 228.1-2010. (2010). *Metallic materials-Tensile testing-Part 1: Method of test at room temperature*. Standards Press of China.

Huang, L, Zhang, SS, Yu, T, & Peng, KD (2020). Circular hybrid double-skin tubular columns with a stiffener-reinforced steel inner tube and a large-rupture-strain FRP outer tube: Compressive behavior. *Thin-Walled Structures*, *155*, 106946.

Khus, RS, Fawzia, S, & Thambiratnam, DP (2021). Experimental testing of novel hybrid rubberised concrete double skin tubular columns with filament wound CFRP tube under axial compressive loading. *Composite Structures*, *4*, 114568.

Lam, L, & Teng, JG (2003). Design-oriented stress-strain model for FRP-confined concrete. *Construction and Building Materials*, *17*, 471–489.

Ozbakkaloglu, T, & Fanggi, BL (2013). Compressive behavior of aramid FRP-HSC-steel double-skin tubular columns. *Construction and Building Materials*, *48*, 554–565.

Ozbakkaloglu, T, & Fanggi, BL (2014). Axial compressive behavior of FRP-concrete-steel double-skin tubular columns made of normal- and high-strength concrete. *Journal of Composites for Construction*, *18*(1), 46–47.

Ozbakkaloglu, T, & Fanggi, BL (2015). FRP-HSC-steel composite columns: Behavior under monotonic and cyclic axial compression. *Materials and Structures*, *48*, 1075–1093.

Peng, KD, Yu, T, Muhammad, NSH, & Huang, L (2018). Compressive behavior of hybrid double-skin tubular columns with a rib-stiffended steel inner tube. *Composite Structures*, *204*, 634–644.

Teng, JG, Yu, T, & Wong, YL (2007). Hybrid FRP-concrete-steel tubular columns: Concept and behavior. *Construction Building Materials*, *21*(4), 846–854.

Wang, WQ, Wu, CQ, Yu, Y, & Zeng, JJ (2021). Dynamic responses of hybrid FRP-concrete-steel double-skin tubular column (DSTC) under lateral impact. *Structures*, *32*, 1115–1144.

Wang, W, Wu, C, & Liu, Z (2019). Compressive behavior of hybrid double-skin tubular columns with ultrahigh performance fiber-reinforced concrete. *Engineering Structures*, *180*, 419–441.

Yu, T (2007). *Behavior of hybrid FRP-concrete-steel double-skin tubular columns*. Hong Kong Polytechnic University.

Yu, T, & Teng, JG (2013). Behavior of hybrid FRP-concrete-steel double-skin tubular columns with a square outer tube and a circular inner tube subjected to axial compression. *Journal of Composites for Construction*, *17*(2), 271–279.

Yu, T, Teng, JG, & Wong, YL (2010). Stress-strain behavior of concrete in hybrid FRP-concrete-steel double-skin tubular columns. *Journal of Structural Engineering*, *136*(4), 379–389.

Yu, T, Zhang, B, Cao, YB, & Teng, JG (2012). Behavior of hybrid FRP-concrete-steel double-skin tubular columns subjected to cyclic axial compression. *Thin-Walled Structures*, *61*, 196–203.

Yu, T, Zhang, SS, Huang, L, & Chan, C (2017). Compressive behavior of hybrid double-skin tubular columns with a large rupture strain FRP tube. *Composite Structures*, *171*, 10–18.

Zeng, JJ, Zheng, YZ, & Long, YL (2021). Axial compressive behavior of FRP-concrete-steel double skin tubular columns with a rib-stiffened Q690 steel tube and ultra-high strength concrete. *Composite Structures*, *268*, 113912.

Zhang, SS, Teng, JG, & Yu, T (2015). Experimental behavior of hybrid FRP-concrete-steel double-skin tubular columns under combined axial compression and cyclic lateral loading. *Engineering Structures*, *99*, 214–231.

Seismic behavior of steel frame confined concrete column

8

8.1 Introduction

In recent decades, steel reinforced concrete (SRC) columns have been wildly applied in practical engineering and further studied in existing researches. The SRC comprises the concrete column and the encased section steel. Many existing studies (Chen & Wu, 2017; Denavit et al., 2018; Hong et al., 2019; Zenzai et al., 2018) showed the SRC structures have the following characteristics: (1) high bearing capacity; (2) great seismic performance; (3) the steel volume ratio is much lower than that of the steel structure; (4) the steel ratio volume is slightly higher than that of the reinforced concrete (RC) structure. However, the need for reinforcing bars in the SRC structure will lead to the difficult construction capacity. In addition, non-compacting sandwiched concrete between steel bars and section steel would deduce the mechanical behavior and the seismic performance. Therefore, the reasonable section form would improve the material utilization and the construction convenience of the SRC structure, which can lead to great mechanical behavior and seismic performance.

The steel frame confined concrete column (SFC) is improved from the steel-caged RC column. The SFCs comprise the outer concrete cover, the core confined concrete and the embedded steel frame. The steel frame is welded by four angle steels and several steel battens, as is shown in Fig. 8.1. In the loading process of

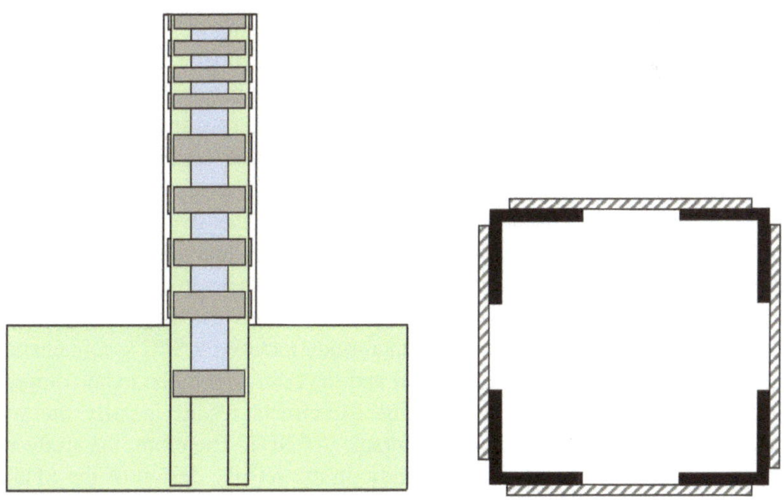

Figure 8.1 Sketch of STC specimen.

Concrete Composite Columns. DOI: https://doi.org/10.1016/B978-0-323-85171-8.00008-2

SFC, the angle steel section can provide vertical force to bear the axial load and lateral confining pressure to confine the core concrete, the steel battens can transfer shear force and confine the core concrete. The distribution of angle steel ensures steel frame can play the full material properties. In many existing studies (Julio, Joaquín, et al., 2011; Julio, Jose, et al., 2011; Montuori & Piluso, 2009; Nagaprasad et al., 2009), the seismic behaviors of steel-caged RC columns have been analyzed, and the following conclusions can be drawn: the wrapping steel frame can improve the seismic performance of the reinforced construction structures; steel batten is the most effective connection method between four angle steels; the anchoring method is also an important influence factor for seismic behavior. The study results of steel-caged RC columns show that the steel frame can instead steel bars and section steel to be a component in new construction, because of the following reasons: (1) the connection between the angle steels is the steel batten, which can effectively transfer the shear stress in SFC, confine core concrete, and restrict the local buckling on angle steel; (2) the embedded method makes the steel frame and the concrete column become a unit member; (3) the steel frame makes SFC have no steel bars, which provides great construction convenient and seismic performance. Some studies have been presented on the seismic behavior of the composite columns with the angle steel. The study presented by Zhou and Shi (1993) showed the strength and ductility of the composite columns are higher than that of the RC columns, and the horizontal batten rather than the oblique batten is the effective connection between the angle steels. The study presented by Zhao (2007) also shows that the oblique batten would provide lower confining effect and transmit a small shear force. The influence factors of seismic behavior were studied by Feng (2014), including the axial compression ratio, the volume steel ratio and the distribution of the angle steel, but the specimen number is insufficient for the quantitative study. In the experiment studies presented by Shi and Bai (2000) and Tang et al. (2004), the hysteretic curves would show obvious pinching phenomenon, due to the low volume steel ratio of the angle steel. A serious of seismic behavior studies on the SFCs presented by Ji (2008) and Zheng and Ji (2008a, 2008b) analyzed the following influence factors: the angle steel ratio and the steel batten ratio, the axial compression ratio and the shear span ratio. However, there are some defects in existing studies: (1) the specimens in most studies had the lower angle steel ratio, which means these specimens have less representative; (2) the specimens have bigger spacing between the steel battens, which would make the steel frame cannot play its own effect; (3) few studies focused on the theoretical analysis.

Against the above background, this chapter presents the cyclic lateral load experimental study on the seismic behaviors of SFCs, which have a large angle steel ratio and a large steel batten ratio. The experimental study includes 8 SFC specimens to analyze size and arrangement of the angle steel and the steel batten, the axial compression ratio and the addition confining effect. The experiment results identify the seismic mechanism and evaluate the seismic performance of SFC. Therefore, the study results can provide a foundation for the further study on the seismic design of the SFC composite structure and improve the application of SFCs in practical engineering.

8.2 Experimental design

8.2.1 Specimen design

We designed and prepared a total of 8 STC specimens, and we assigned the specimen numbers from SFC-1 to SFC-8. For all specimens, the form was the inverted T-shaped cantilever columns, the column section was square and section size was 300 mm × 300 mm, the column height was 1200 mm, the shear span ratio was 3.67. In order to prevent the large bond slip between the steel frame and the concrete column, the following measures were adopted in all specimens: the steel frame should be inserted into the bottom beam; the zone above 350 mm height of the steel frame should adopt the layout method of the small batten and the small spacing; the bottom of the steel frame should be welded with the reinforcement; the bottom layer of batten should be embedded in the bottom beam. The measures are shown in Fig. 8.1. The strength grade of the steel frame was Q235, the strength grade of the concrete was C40, and the thickness of the concrete cover is 20 mm. In order to resist the local buckling of the angle steel, the length of batten plate was increased to make the steel battens fully contact with the angle steels, as shown in Fig. 8.2. The longitudinal reinforcement in the bottom beam was HRB400 grade, and the diameter was 16 mm. The strength grade of the hoops was HPB300 grade, and the diameter was 8 mm. The variation parameters include the size of the angle steel, the arrangement of the steel battens, the width of the steel battens, the addition of spiral hoops, and the axial compression ratio. The detailed design parameters are shown in Table 8.1.

8.2.2 Specimen preparation

In the specimen preparation, the four longitudinal steel bars and hoops were first bound up to form a reinforced frame in the bottom beam. Then the angle steel frame was inserted into the reinforced frame, and followed with the colligation of the whole reinforcement cage. After that, the bottom of the angle steel was welded with the longitudinal reinforcement after adjusting the position of the steel

Figure 8.2 Steel form in specimens.

Table 8.1 Detailed design parameters of specimens.

Item	Angle steel bars (mm)		Steel batten (mm)			Axial compression ratio	Remarks
	Size	Content (%)	Size	Spacing	Content (%)		
SFC-1	L60 × 8	3.98	240 × 100 × 6	100	3.20	0.3	
SFC-2	L60 × 8	3.98	240 × 100 × 6	100	3.20	0.42	
SFC-3	L60 × 8	3.98	240 × 50 × 6	50	3.20	0.3	Different arrangements of steel battens
SFC-4	L60 × 8	3.98	240 × 50 × 6	50	3.20	0.42	Different arrangements of steel battens
SFC-5	L60 × 8	3.98	240 × 100 × 6	100	3.70	0.3	Adding spiral hooping
SFC-6	L60 × 8	3.98	240 × 100 × 6	100	3.70	0.42	Adding spiral hooping
SFC-7	L80 × 8	5.40	240 × 100 × 6	100	3.20	0.3	Wide angle steel bars
SFC-8	L60 × 8	3.98	240 × 135 × 6	65	4.30	0.3	Wide steel battens

Source: From Rong, C., Wang, Q. X., & Wang, B. (2021). Seismic performance of angle steel frame confined concrete columns: Experiments and FEA model. *Engineering Structures*, *235*, 111983.

frame. The specimen was placed vertically, and the wood formwork was used. After the formwork was closed and the position was adjusted, the concrete was poured from both sides of the column and the top of the column. The steel skeleton formed by the steel frame and the reinforcement, and the pouring of the specimens are shown in Fig. 8.3.

8.2.3 Material properties

In this test, the strength grades of the angle steels and the steel battens were Q235 grade, but there were two thickness types (i.e., 6 and 8 mm). The strength grade of the spiral hoops equipped in the column was HRB400 grade and the diameter was 8 mm. The strength grade of the longitudinal bar in the bottom beam was HRB400 grade, and the diameter is 16 mm. The strength grade of the steel hoops, which were equipped in the bottom beam, was HPB300 and the diameter was 8 mm. The detailed mechanical properties indexes of the angle steels, the steel battens, and the reinforcement are shown in Table 8.2.

Figure 8.3 Production process of specimens.

Table 8.2 Detailed mechanical properties.

Shape	Thickness (mm)	Elastic modulus (MPa)	Yield stress (MPa)	Tensile strength (MPa)
Steel batten	4	2.1×10^5	340	502
Steel batten	6	2.1×10^5	319	434
Angle steel bar	5	2.1×10^5	343	505
Angle steel bar or steel batten	8	2.1×10^5	254	376
Spiral hooping	8	2.1×10^5	417	620

Source: From Rong, C., Wang, Q. X., & Wang, B. (2021). Seismic performance of angle steel frame confined concrete columns: Experiments and FEA model. *Engineering Structures, 235*, 111983.

We poured a batch of the commercial concrete at one time. After pouring, the specimens were cured in the outdoor natural conditions. In the pouring process, we reserved 9 standard cubes (150 mm × 150 mm × 150 mm) and cured under the cubes in the same conditions as the specimen. Then we used them to test and determine the mechanical properties (GJ 101-96, 1997). Based on the test results, the average value of the cube compressive strength was 40.5 MPa. And the corresponding axial prism compressive strength was 27.0 MPa.

8.2.4 Test set-up

We tested all specimens by the Lateral Load System of MTS hydraulic servo-machine in the Key Laboratory of Structure Engineering and Seismic Resistance Education at the Xi'an University of Architecture and Technology. In the experiment, a cantilever loading device was used, a jack with 20,000 kN was used to apply the vertical load and an actuator with 500 kN was used to apply a reciprocating load in the horizontal direction, as shown in Fig. 8.4. The testing process includes the following steps: (1) the specimen is installed in the MTS hydraulic servo-machine; (2) the top section is leveled by sand in axial preload; (3) the lifting jack is installed on the top section and applied the axial compression; (4) the lateral load is applied.

The loading system is shown in Fig. 8.5. The lateral loading method follows several steps: (1) the load cycle is controlled by the force before the yield state, each stage has one cycle and the force increment is 40 kN; (2) the load cycle is controlled by the displacement after the yield state, each stage has three cycles and the displacement increment is 5 mm; (3) the test is stopped once the specimen cannot bear axial constant loading, or the force is less than 70% of the peak load. It should

Figure 8.4 Test of specimens.

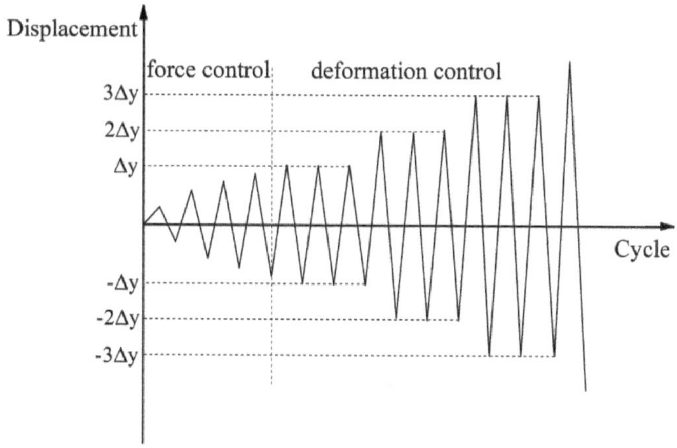

Figure 8.5 Loading system of specimens.

be noted that the loading and unloading process should keep the same loading speed.

8.2.5 Test instrumentation

In the experiment, the test content includes that: (1) the failure process and the failure modes of the specimens; (2) the lateral force at top of the specimens; (3) the lateral displacement at top of the specimens; (4) the lateral displacement at middle-height of the specimens; (5) the lateral displacement of the bottom beam; (6) the axial strain and the lateral strain of the angle steel; (7) the lateral strain of the steel batten.

The horizontal force was measured by the MTS system of the hydraulic servo-machine. The top lateral displacement was measured by two linear variable displacement transducers (LVDTs) at the same height. The middle lateral displacement was measured by one LVDT. The bottom displacement was measured by one LVDT, which is placed at the cross-section of the bottom beam. And the bottom displacement is used to measure the whole movement of the specimen. In fact, the bottom beam should be fixed on the ground. All placements of LVDTs were shown in Fig. 8.6.

In order to investigate the stress condition of the steel section, 20 strain gauges with a gauge length of 5 mm were installed on two side surfaces that were perpendicular to the loading direction. On each side surface, six axial strain gauges were installed on the angle steel along the height direction, and four strain gauges were installed on the steel battens in the horizontal direction. The layout positions of the strain gauges with a gauge length of 5 mm were shown in Fig. 8.6. All data were collected automatically by the data acquisition instrument in the loading process.

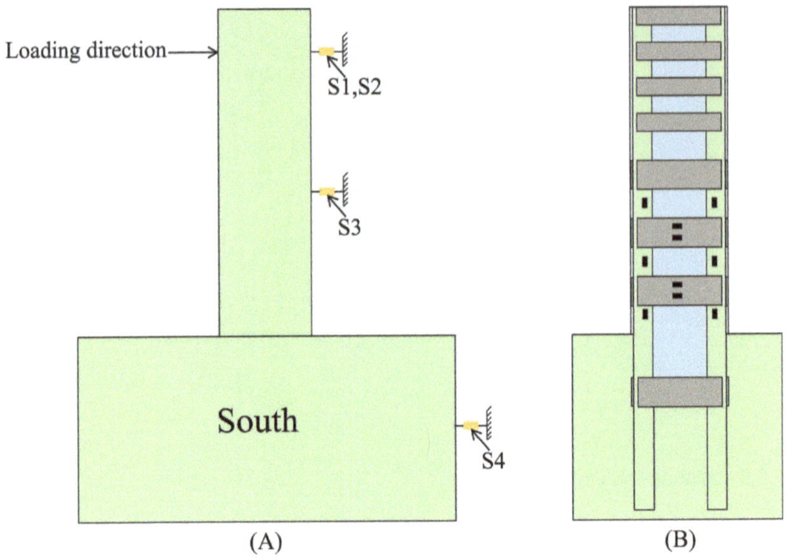

Figure 8.6 Test instrumentation on specimens. (A) LVDTs; (B) strain gauges.

8.3 Failure process and failure mode

8.3.1 Failure process

During the loading process, the actuator pushed the specimen from west to east and pulled the specimen from east to west. The axial cracks first appeared at the corner, the horizontal cracks first appeared on the east and west sides, and then horizontal cracks appeared on the four sides. In the final failure stage, the bottom concrete cover completely peeled off. The failure processes of specimens are described in the following paragraphs. Because the specimen is loaded symmetrically. For simple description, it is specified that the horizontal load and displacement angle are the average values in forward and reverse loading. The displacement angle is the ratio of the lateral top displacement to the column height, $\theta = \Delta/H$.

(1) SFC-1

The axial compression ratio of test piece SFC-1 was 0.30, and the size of angle steel was L60 × 8 mm. Both the width and the spacing of steel battens were 100 mm. The cracking load was 160 kN, the displacement angle was $\theta = 1/121$, the microcracks appeared at 13 cm height above the bottom on the left side of the south surface and 14 cm from the column bottom on the left side of the north surface. These cracks were about 10 cm long and developed horizontally. The yield load was 200 kN, the corresponding displacement angle was $\theta = 1/140$, and many inverted octagonal oblique cracks appeared on the south surface. In addition, there were lateral through cracks on the four-column surfaces. The north surface appeared the oblique cracks which ranged from the right boundary at the 33 cm height to the middle section at the 15 cm height. After that, the loading was controlled by the

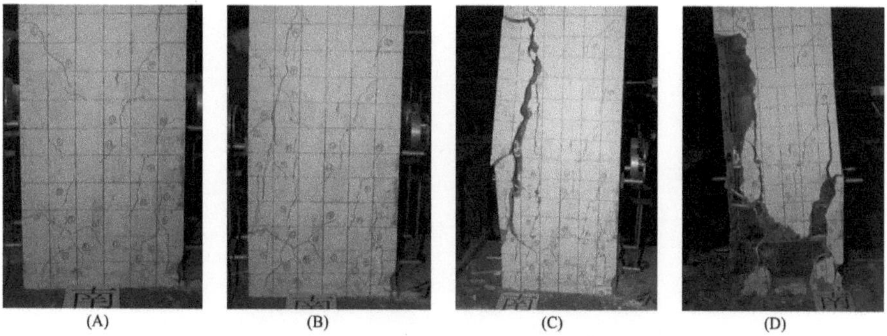

Figure 8.7 Failure process of SFC-1. (A) 10 mm; (B) peak load; (C) 25 mm; (D) ultimate load.

displacement, and the displacement increment was $\Delta = 5$ mm, each loading stage had 3 cycles. When the lateral displacement was 10 mm ($\theta = 1/110$), the lateral load was 216 kN, the cracks on the specimen surfaces developed quickly, the second lateral through crack appeared at 35 cm height on the east and west surfaces, and the map cracking appeared in the area below 50 cm height on the south and north surfaces, as shown in Fig. 8.7. When the lateral displacement was 20 mm ($\theta = 1/110$), the SFC-1 reaches the peak load (245 kN), the cracks on specimen surfaces developed quickly, partial concrete covers crushed below the 15 cm height area on the east and west surfaces, and longitudinal long cracks appeared in the southwest corner and northwest corner, as shown in Fig. 8.7. When the lateral displacement was 25 mm ($\theta = 1/44$), the lateral load was 240 kN, a large part of concrete covers occurred peeling in the area from 15 cm height to 70 cm height of the southwest corner, as shown in Fig. 8.7. When the lateral displacement was 30 mm ($\theta = 1/37$), the lateral load was 231 kN, and the concrete covers occurred peeling in the area below 15 cm height of the northwest corner. When the lateral displacement was 40 mm ($\theta = 1/28$), the lateral load was 215 kN, and the concrete cover occurred the overall peeling in the areas below 20 cm height of the four surfaces. When the lateral displacement was 50 mm ($\theta = 1/22$), the angle of steel occurred with the great outward expansion, the core concrete crushed, and the bearing capacity decreased to less than 70% of the peak load. Therefore, the loading was stopped. And the failure mode of SFC-1 is shown in Fig. 8.7.

The failure mode of the bottom part of SFC-1 is shown in Fig. 8.8. Based on the observation, it can be found that the bottom part of the angle steel show serious compression and local buckling, the bottom batten shows obvious outward expansion, the core concrete at the bottom part shows serious crushing, and the junction part of the bottom beam shows serious concrete crushing. Therefore, the deformations of the steel batten and the angle steel mainly occur at the bottom part of the column. Meanwhile, the concrete in this part is most severely crushed, and the inner steel frame also has a significant influence on the concrete cover of the bottom beam.

Figure 8.8 Failure mode of the bottom part of SFC-1.

(2) SFC-2

The axial compression ratio of SFC-2 is 0.42, and the size of angle steel and the distribution form of steel batten are the same as that of SFC-1. The cracking load is 160 kN, the displacement angle $\theta = 1/240$, the microcracks appeared at 13 cm height above the bottom on the right side of the west surface and the left side of the south surface. These cracks were about 5 cm long and develop horizontally. And there was one horizontal crack and one oblique crack (both of them have a length of 10 cm) at 15 cm height above the bottom on the right side of the north surface. The yield load is 200 kN and the corresponding displacement angle $\theta = 1/149$, one horizontal crack appeared at 14 cm height above the bottom on the west surface and 15 cm height above the bottom on the east surface. As the loading obviously enters the yield stage, so the loading was changed into the displacement control. The displacement increment was $\Delta = 5$ mm and each loading stage had 3 cycles. When the lateral displacement is 10 mm ($\theta = 1/110$), the lateral load is 235 kN, the cracks on specimen surfaces developed quickly, the second lateral through crack appeared at 35 cm height on the west surface, and several oblique cracks appeared below 15 cm height on the south surface and the map cracking appeared in the area below 55 cm height on the north surfaces, as shown in Fig. 8.9. When the lateral displacement was 15 mm ($\theta = 1/73$), the lateral load was 256 kN, the through crack on the east surface widened, and the long oblique crack appeared on the west surface from the

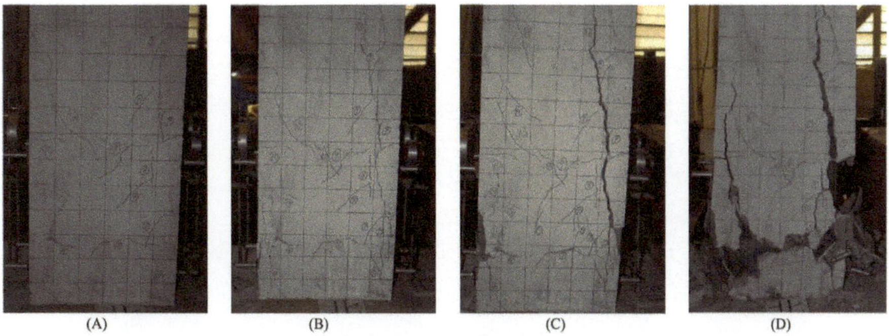

Figure 8.9 Failure process of SFC-2. (A) 10 mm; (B) 15 mm; (C) peak load; (D) ultimate load.

35 cm height on the left to the right bottom. There were many vertical cracks below the 35 cm height on the south surface, and several vertical long cracks on the right of the north surface, as shown in Fig. 8.9. When the lateral displacement was 20 mm ($\theta = 1/55$), the SFC-2 reached the peak load (264 kN), the map cracking appeared in the area below 30 mm height on the south surface, and part of the concrete covers occurred to peeling in the south corners. A vertical wide crack appeared at 10 cm from the left on the west surface and the concrete cover on the north surface peeled off, as shown in Fig. 8.9. When the lateral displacement was 25 mm ($\theta = 1/44$), the lateral load was 261 kN, the vertical crack on the right side of the north surface continued to expand, and the concrete cover on the right side peeled off. The concrete cover below the 15 cm height on the east surface peeled off. When the lateral displacement was 30 mm ($\theta = 1/37$), the lateral load was 256 kN, the concrete cover below the 30 cm height on the east surface peeled off, and the concrete in the area below 40 cm in the west peeled off. When the lateral displacement was 50 mm ($\theta = 1/22$), the angle of steel occurred with the great outward expansion, the core concrete crushed, and the bearing capacity decreased to less than 70% of the peak load. Therefore, the loading was stopped. And the failure mode of SFC-2 is shown in Fig. 8.9.

The failure mode of the bottom part of SFC-2 is shown in Fig. 8.10. Based on the observation, it can be found that the area between two bottom steel battens shows serious compression local buckling on the angle steel and serious crushing of the core concrete, the second steel batten at the bottom shows obvious outward expansion, and the junction part of the bottom beam has serious concrete crushing. Therefore, the deformation of the steel batten and the angle steel mainly occurs below the second bottom steel batten. And the concrete in this part shows most severe crush. The inner steel frame has a significant effect on the concrete cover of the bottom beam.

(3) SFC-3

The axial compression ratio of SFC-3 was 0.30, and the size of angle steel was the same as that of SFC-1, but the small batten (50 mm) and small spacing (50

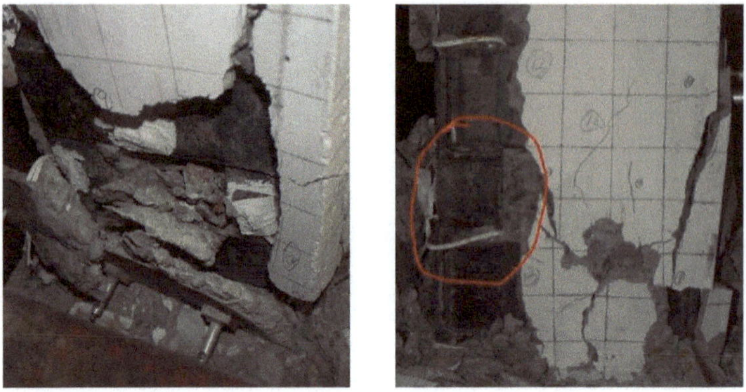

(A) (B)

Figure 8.10 Failure mode of the bottom part of SFC-2. (A) West side; (B) east side.

mm) were adopted in SFC-3. The cracking load was 160 kN, the displacement angle was $\theta=1/203$, the crossing microcracks of about 8 cm appeared on the lower left bottom corner of the south surface, and two horizontal cracks of about 4 cm appeared separately at 13 cm height and 28 cm height on the left side of the north surface. The yield load was 200 kN, the displacement angle was $\theta=1/145$, the similar cracks of about 5 cm appeared separately at 15 cm height on the right side, 5 cm height, and 25 cm height on the left side of the south surface. Two through horizontal cracks appeared at 8 cm height on the right side and 15 cm height on the left side of the west surface. Map cracking appeared below the 40 cm height on the north surface. As the loading obviously enters the yield stage, so the loading was changed into the displacement control. The displacement increment was $\Delta=5$ mm and each loading stage has 3 cycles. When the lateral displacement was 10 mm ($\theta=1/110$), the lateral load was 217 kN, the crack width at the southeast root enlarged and showed the peeling trend. When the lateral displacement was 15 mm ($\theta=1/73$), the lateral load was 237 kN, the map cracking appeared below 40 cm height on the south surface, and the concrete cover at the left root peeled off, as shown in Fig. 8.11. The through horizontal cracks appeared at 15 cm height and 30 cm height on the west surface, and the concrete cover below 10 cm on the east surface peeled off. When the lateral displacement was 20 mm ($\theta=1/55$), the SFC-3 reached the peak load (243 kN), the concrete covers peeled off at the left and right roots on the south surface, and the concrete cover at the 10 cm height also slightly peeled off. The long vertical crack about 60 cm in length appeared on right side of the north surface, as shown in Fig. 8.11. When the lateral displacement was 25 mm ($\theta=1/44$), the lateral load was 234 kN, the concrete covers on the left side of the north surface peeled off, and the oblique cracks generated on the west surface from the 40 cm height to the right bottom. When the lateral displacement is 35 mm ($\theta=1/32$), the lateral load is 213 kN, the concrete cover below the 40 cm height on the west surface is seriously cracked and accompanied by the concrete peeling, the concrete on the right root of the north surface peels off, and a through horizontal

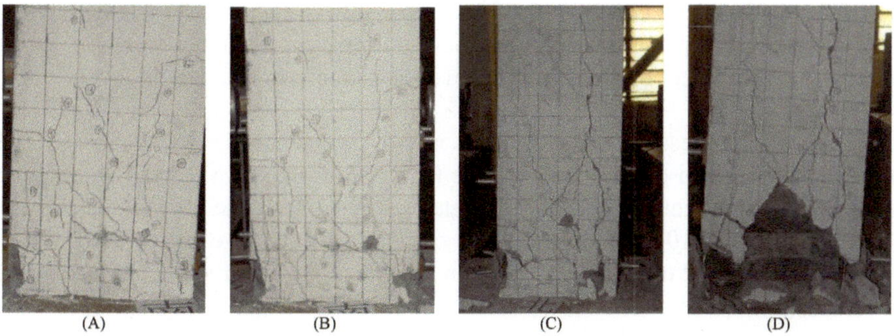

Figure 8.11 Failure process of SFC-3. (A) 15 mm; (B) peak load; (C) 35 mm; (D) ultimate load.

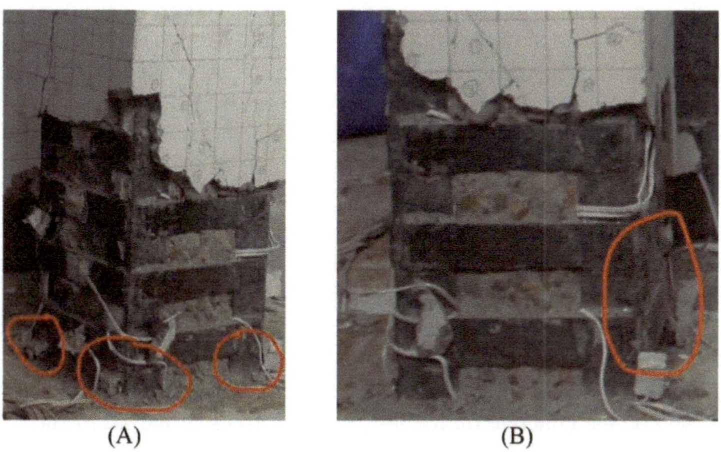

Figure 8.12 Failure mode of the bottom part of SFC-3. (A) West side; (B) east side.

crack formed at the 10 cm height, as shown in Fig. 8.11. With the continuous loading, the peeling of the concrete cover appeared on each column side. There was no obvious change after the lower concrete cover completely peeled off. When the lateral displacement was 50 mm ($\theta = 1/22$), the angle of steel occurs a great outward expansion, the core concrete was crushed, and the bearing capacity decreases to less than 70% of the peak load, so the loading stopped. And the failure mode of SFC-3 is shown in Fig. 8.11.

The failure mode of the bottom part of SFC-3 is shown in Fig. 8.12. Based on the observation, it can be found that the bottom part of the angle steel show serious outward expansion, the bottom batten shows obvious outward expansion, the core concrete at the bottom part shows serious crushing, and the junction part of the bottom beam has a serious concrete crush. It should be noted that there is no obvious local buckling on the angle steel. Therefore, the deformation of the steel batten and the angle steel mainly occurs at the bottom part of the column. At the same time, the concrete in this part shows the most severe crush, and the inner steel frame also has a significant effect on the concrete cover of the bottom beam.

(4) SFC-4

The axial compression ratio of SFC-2 was 0.42, and the size of angle steel and the distribution form of steel batten were the same as that of SFC-3. The cracking load was 200 kN, the displacement angle was $\theta=1/164$, the through horizontal crack appeared at 10 cm height on the east surface, and two microcracks about 4 cm appeared at 12 cm height and 35 cm height on the north surface. Meanwhile, the loading obviously enters the yield stage, so the loading was changed into displacement control. The displacement increment was $\Delta=5$ mm and each loading stage has 3 cycles. When the lateral displacement was 10 mm ($\theta=1/110$), the lateral load was 236 kN, and the through horizontal crack appeared at 10 cm height on the west surface. The vertical crack of about 11 cm extends from the column bottom, which is at 2 cm to the left side of the east surface. The vertical cracks appeared simultaneously on the left and right sides of the south surface. In addition, the horizontal crack of about 10 cm appeared at the 10 cm height on the left side and the horizontal crack of about 4 cm appeared at the 35 cm height. Many cracks appeared on the left and right sides of the north surfaced, as shown in Fig. 8.13. When the lateral displacement was 15 mm ($\theta=1/73$), the lateral load was 251 kN, the through horizontal cracks on the east and west surface grew, and the concrete covers at column bottom showed the peeling trend. The map cracking appeared below the 40 cm height on the north surface, and the through horizontal crack was generated at 10 cm height. When the lateral displacement was 20 mm ($\theta=1/55$), the SFC-4 reached the peak load (251 kN), many oblique cracks appeared below the 45 cm height on the south surface, and the concrete cover peeling occurred at the roots, as shown in Fig. 8.13. When the lateral displacement was 30 mm ($\theta=1/37$), the lateral load was 246 kN, the concrete on the left and right roots of the north surface peeled off, and the concrete cover below 10 cm height on the east and west surface peeled off, the bottom concrete cover on the south surface peeled off, and the width of inclined cracks increased, as shown in Fig. 8.13. With the continuous loading, the

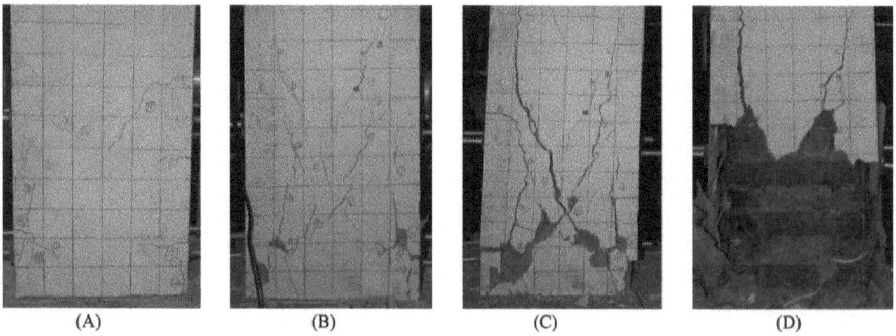

(A) (B) (C) (D)

Figure 8.13 Failure process of SFC-4. (A) 10 mm; (B) peak load; (C) 30 mm; (D) ultimate load.

cracks on each surface continued to expand, and the peeling of the concrete cover became serious. When the lateral displacement was 30 mm ($\theta = 1/37$), the bottom concrete cover peeled off completely. When the lateral displacement was 50 mm ($\theta = 1/22$), the lateral load was 193 kN, and the bearing capacity decreased to less than 70% of the peak load, so the loading stopped, and the bottom core concrete was crushed. The failure mode of SFC-3 is shown in Fig. 8.13.

The failure mode of the bottom part of SFC-4 is shown in Fig. 8.14. Based on the observation, it can be found that the peeling area of the concrete cover is large, the bottom part of the angle steel show serious outward expansion, the four bottom battens show obvious outward expansion, the core concrete at the bottom part shows serious crush, and the junction part of the bottom beam has a serious concrete crush. It should be noted that there is no obvious local buckling on the angle steel. Therefore, the deformation of the steel battens mainly occurs in the low-height area, and the deformation of the angle steel mainly occurs at the column bottom. That means the confining effect area is expanding. And the bottom concrete shows the most severe crush, and the inner steel frame also has a significant effect on the concrete cover of the bottom beam.

(5) SFC-5

The axial compression ratio of SFC-5 was 0.30, the size of angle steel and the distribution form of steel batten were the same as that of SFC-1, and the spiral hoops were added to the steel frame. The cracking load was 120 kN, the displacement angle was $\theta = 1/266$, the horizontal crack of about 5 cm appeared at 5 cm height on the left of the west surface, and a vertical crack about 10 cm upward from the root appeared at 8 cm from the left edge. When the load was 160 kN, the displacement angle was $\theta = 1/177$, and three horizontal cracks of about 10 cm appeared at 10 cm height, 15 cm height, and 34 cm height on the right of the north surface, as shown in Fig. 8.15. An oblique crack of about 7 cm appeared at the 15 cm height in the middle of the west surface. Two oblique cracks and one vertical

(A)

(B)

Figure 8.14 Failure mode of the bottom part of SFC-4. (A) West side; (B) east side.

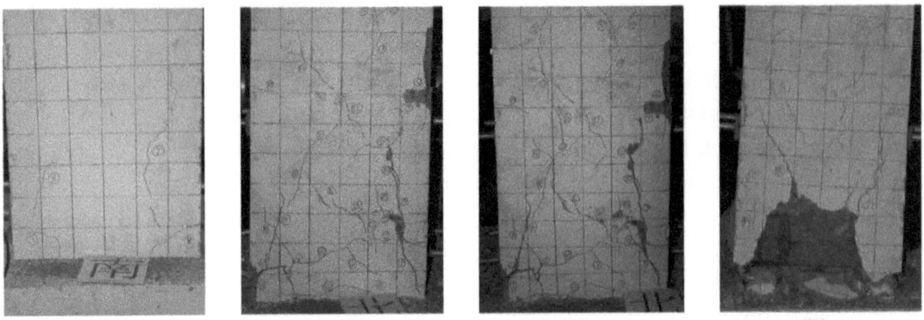

(A) (B) (C) (D)

Figure 8.15 Failure process of SFC-5. (A) 16 t; (B) peak load; (C) 30 mm; (D) ultimate load.

crack appeared at the bottom of the south surface. The yield load was 200 kN, the displacement angle $\theta=1/111$, the through horizontal crack appeared at 35 cm height on the west surface, the vertical crack about 34 cm upward from the root appeared at 2 cm from the right edge, and the map cracking appeared below 30 cm height in the north surface. Meanwhile, the loading obviously entered the yield stage, so the loading was changed into displacement control. The displacement increment was $\Delta=5$ mm and each loading stage had 3 cycles. When the lateral displacement was 15 mm ($\theta=1/73$), the lateral load was 223 kN, the through horizontal crack appeared at 30 cm height on the east surface, the concrete in the right corner spilled out, and another through horizontal crack appeared at 20 cm height on the west surface. When the lateral displacement was 20 mm ($\theta=1/55$), the lateral load was 235 kN, the concrete cover at the left and right bottom of the south surface showed a peeling trend, and the concrete cover at the right bottom of the west surface peeled off. the concrete at the 40 cm height on the right side of the north surface cracked and then peeled out. When the lateral displacement was 25 mm ($\theta=1/44$), the SFC-5 reached the peak load (238 kN), and the concrete cover at 20 cm height on the east surface peeled off. The map cracking appeared below the 25 cm height on the north surface, and the concrete cover at the left bottom peeled off, as shown in Fig. 8.15. When the lateral displacement was 30 mm ($\theta=1/37$), the lateral load was 234 kN, the concrete cover at the root on the south surface peeled off, the crack widened at the 20 cm height on the east and west surfaces, and the concrete cover below the crack obviously separated from the steel batten, the wide vertical crack appeared on the right side of the north surface, as shown in Fig. 8.15. When the lateral displacement was 35 mm ($\theta=1/31$), the lateral load was 225 kN, the cracks below 30 cm height on the south surface widened, and concrete cover below 20 cm height on the east surface peel off. When the lateral displacement was 45 mm ($\theta=1/24$), the concrete cover in the low-height area peeled off completely. When the lateral displacement was 60 mm ($\theta=1/18$), the lateral load was 161 kN, and the bearing capacity decreased to less than 70% of the peak load, and thus the loading stopped. And the bottom core concrete shows serious crush. The failure mode of SFC-5 is shown in Fig. 8.15.

(A) (B)

Figure 8.16 Failure mode of the bottom part of SFC-5. (A) West side; (B) east side.

The failure mode of the bottom part of SFC-5 is shown in Fig. 8.16. Based on the observation, it can be found that the peeling area of the concrete cover is below 25 cm height, the bottom part of the angle steel show serious local buckling, the bottom steel batten shows obvious outward expansion, and the core concrete at the bottom part shows serious crush, and the junction part of the bottom beam has a serious concrete crush. Therefore, the deformation of the steel batten and the angle steel mainly occurs at the column bottom. The bottom concrete shows the most severe crush, and the inner steel frame also has a significant effect on the concrete cover of the bottom beam.

(6) SFC-6

The axial compression ratio of SFC-6 was 0.42, the size of angle steel and the distribution form of steel batten were the same as that of SFC-5, and the spiral hoops were also added. The cracking load was 160 kN, the displacement angle was $\theta = 1/199$, the horizontal crack of about 5 cm appeared at the root on the left side of the south surface, and a vertical crack extended from 15 cm to 35 cm at 8 cm from the right edge. The vertical crack about 20 cm upward from the column root at 8 cm from the right edge on the north surface, and an oblique crack about 7 cm appeared at 45 cm height on the right side. Meanwhile, the loading obviously enters the yield stage, so the loading was changed into displacement control. The displacement increment was $\Delta = 5$ mm and each loading stage had 3 cycles. When the lateral displacement was 10 mm ($\theta = 1/110$), the lateral load was 211 kN, the horizontal crack of about 17 cm appeared at 13 cm height on the east surface, two through horizontal cracks appeared at 13 cm height and 27 cm height separately, the vertical crack about 60 cm extended upward from the root on south surface, and many vertical cracks appeared at the right bottom of the north surface, as shown in Fig. 8.17. When the lateral displacement was 15 mm ($\theta = 1/73$), the lateral load was 237 kN, further cracks appeared on the north surface, many vertical cracks appeared on the west surface, and the vertical long cracks on the south surface extended. When the lateral displacement was 20 mm ($\theta = 1/55$), the SFC-6 reached the peak load (253 kN), map cracking appeared below 55 cm height on the north surface, and the concrete cover peeled off. The concrete cover below 35 cm height on the

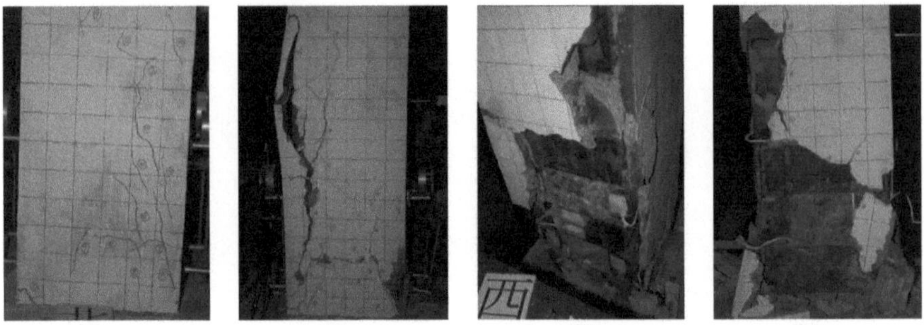

(A) (B) (C) (D)

Figure 8.17 Failure process of SFC-6. (A) 10 mm; (B) peak load; (C) 30 mm; (D) ultimate load.

east surface partially peeled off. The concrete cover separated from the steel batten in the area between 35 cm height and 65 cm height on the west surface. The vertical long cracks on the south surface obviously extended, and parts of the concrete cover peel off, as shown in Fig. 8.17. When the lateral displacement was 30 mm ($\theta = 1/37$), the lateral load was 240 kN, the concrete cover on the east bottom peeled off, and the concrete cover below 55 cm height on the west surface peeled off completely, as shown in Fig. 8.17. When the lateral displacement was 45 mm ($\theta = 1/24$), the lateral load was 214 kN, the concrete cover at the 15 cm height on the east surface peeled off, and the concrete cover below the 35 cm height on the north and south surfaces peeled off. When the lateral displacement was 55 mm ($\theta = 1/20$), the lateral load was 189 kN, and the bearing capacity decreased to less than 70% of the peak load, so the loading stopped. And the bottom core concrete showed an obvious crush. The failure mode of SFC-6 is shown in Fig. 8.17.

The failure mode of the bottom part of SFC-6 is shown in Fig. 8.18. Based on the observation, it can be found that the peeling area of the concrete cover is in the low-height area, the local buckling appears on the angle steel between two bottom steel battens, and the junction part of the bottom beam has a serious concrete crush. In addition, the bottom steel batten shows slight outward expansion, and the core concrete at the bottom part shows a slight crush. That means the inner spiral hoops can provide an obvious confining effect to restrict the concrete expansion. Therefore, the deformation of the angle steel mainly occurs at the column bottom, the deformation of the steel batten is not obvious, and the bottom concrete still has a certain integrity.

(7) SFC-7

The axial compression ratio of test piece SFC-7 was 0.30, and the size of angle steel was L 80 × 8 mm. Both the width and the spacing of steel battens were 100 mm. The cracking load was 200 kN, the displacement angle was $\theta = 1/168$, and a vertical crack about 25 cm extended upward from the column root at 5 cm from the right edge on the south surface. A horizontal crack of about 15 cm appeared at 15 cm height on the right side of the north surface. The yield load was 235 kN, the

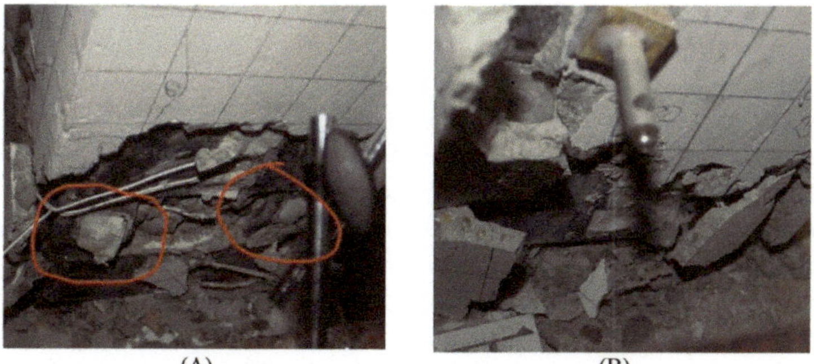

(A) (B)

Figure 8.18 Failure mode of the bottom part of SFC-6. (A) West side; (B) east side.

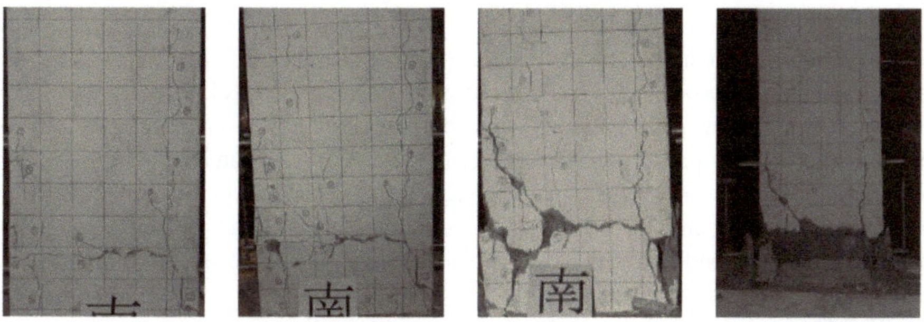

(A) (B) (C) (D)

Figure 8.19 Failure process of SFC-7. (A) 15 mm; (B) peak load; (C) 30 mm; (D) ultimate load.

displacement angle was $\theta=1/102$, the through horizontal crack appeared at 15 cm height on the south surface. The through horizontal crack appeared at 15 cm height on the north surface, and the oblique crack appeared from 35 cm height to 60 cm height. The oblique crack of about 10 cm appeared at the left root on the east surface, and the vertical crack of about 10 cm appeared at 25 cm height. At this moment, the loading obviously entered the yield stage, so the loading was changed into displacement control. The displacement increment was $\Delta=5$ mm and each loading stage had 3 cycles. When the lateral displacement was 15 mm ($\theta=1/73$), the lateral load was 241 kN, the through horizontal crack appeared at 15 cm height on the east surface, and the oblique crack about 33 cm extended upward from the root of the left edge. The through horizontal crack appeared at 15 cm height on the west surface. At this time, the four through horizontal cracks connected. The map cracking appeared below the 30 cm height on the north surface. Several vertical cracks appeared on the south surface, as shown in Fig. 8.19. When the lateral displacement was 20 mm ($\theta=1/73$), the SFC-7 reaches the peak load (245 kN), and parts of the concrete cover peeled off at the 15 cm height on the left side of the west and south surfaces, as shown in Fig. 8.19. When the lateral displacement was

30 mm (θ=1/37), the lateral load was 237 kN, the concrete in the four corners peeled off, and the concrete cover below 15 cm on the west surface separated from the steel batten. The through horizontal crack on the south increased and the concrete cover below the crack separated, as shown in Fig. 8.19. When the lateral displacement was 35 mm (θ=1/31), the lateral load was 232 kN, the concrete cover on the west surface peeled off, and the concrete cover below 15 cm on the east surface separated from the steel batten. With the continuous loading, the cracks on each side of the column continued to expand, and the peeling part of the concrete covered on each side intensifies. When the lateral displacement was 50 mm (θ=1/22), the lateral load was 216 kN, the concrete cover below 15 cm height on the south surface and the east surface peeled off completely. When the lateral displacement was 65 mm (θ=1/17), the lateral load was 184 kN, the bearing capacity decreased to less than 70% of the peak load, so the loading stopped. And the core concrete showed an obvious crush. The failure mode of SFC-7 is shown in Fig. 8.19.

The failure mode of the bottom part of SFC-7 is shown in Fig. 8.20. Based on the observation, it can be found that the peeling area of concrete cover is below 15 cm height of four surfaces, the bottom part of the angle steel show serious local buckling, the bottom steel batten shows obvious outward expansion, the core concrete at the bottom part shows serious crush, and the junction part of the bottom beam has a serious concrete crush. Therefore, the deformation of the steel batten and the angle steel mainly occurs at the column bottom. The bottom concrete is most severely crushed, and the inner steel frame also has a significant effect on the concrete cover of the bottom beam.

(8) SFC-8

The axial compression ratio of test piece SFC-8 was 0.30, and the size of angle steel was L 60 × 8 mm. The width of steel battens was 135 mm, and the spacing between steel battens was 65 mm. The cracking load was 180 kN, the displacement angle was θ=1/184, the horizontal crack of about 8 cm appeared at 13 cm height on the right side of the south surface, the vertical oblique about 20 cm upward from the root at 5 cm from the left edge. The oblique crack of about 15 cm appeared at

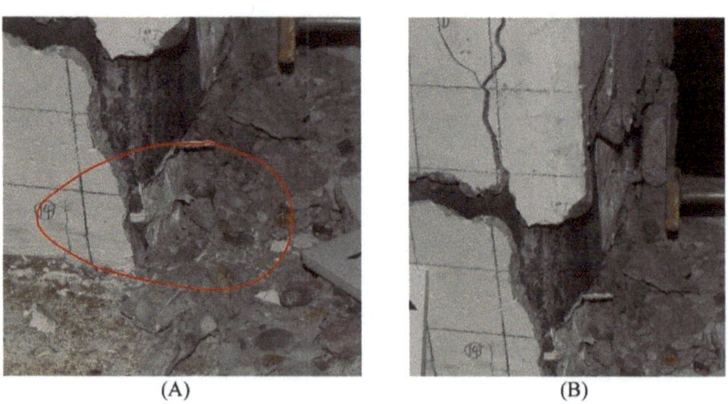

(A) (B)

Figure 8.20 Failure mode of the bottom part of SFC-7. (A) West side; (B) east side.

40 cm height on the left side of the north surface. Meanwhile, the loading obviously entered the yield stage, so the loading was changed into displacement control. The displacement increment was $\Delta=5$ mm and each loading stage had 3 cycles. When the lateral displacement was 10 mm ($\theta=1/110$), the lateral load was 218 kN, the vertical crack of about 6 cm appeared from the root on the left side of the east surface, the map cracking appeared below 25 cm height on the north surface, and the map cracking appeared below 45 cm height on the left side of the south surface. When the lateral displacement was 15 mm ($\theta=1/73$), the lateral load was 236 kN, the through horizontal crack appeared at the 8 cm height on the east surface, the through horizontal crack also appeared at the 13 cm height on the west surface, and the through cracks appeared at the junction of beams and columns on the east and west surfaces. The map cracking appeared below 60 mm on the north surface. The cracks on the left side of the south surface widened, and the concrete cover on the right side peeled off, as shown in Fig. 8.21. When the lateral displacement was 20 mm ($\theta=1/55$), the SFC-8 reached the peak load (242 kN), the cracks on the north surface widened, the concrete at the right root on the east surface peeled off, several vertical cracks appeared in the height range of 20–45 cm on the west surface, and the concrete cover on the right side of the south surface cracked, as shown in Fig. 8.21. When the lateral displacement was 25 mm ($\theta=1/44$), the lateral load was 233 kN, the concrete at the 18 cm height of the west surface peeled off, the cracks on the south surface widened, the vertical crack about 60 cm upward from the root widened on the right side of the north surface, and the vertical crack about 60 cm upward from the root appeared on the left side of the east surface. When the lateral displacement was 30 mm ($\theta=1/37$), the lateral load was 228 kN, the concrete at the southeast corner peeled off, and the crack about 30 cm long upward from the root appeared on the left side of the north surface. The concrete below 20 cm height on the west surface shows serious crack, and the concrete cover below the crack is separated from the steel batten. When the lateral displacement was 40 mm ($\theta=1/28$), the lateral load was 214 kN, the concrete cover below 10 cm height on the east surface completely peeled off, the concrete cover on the north

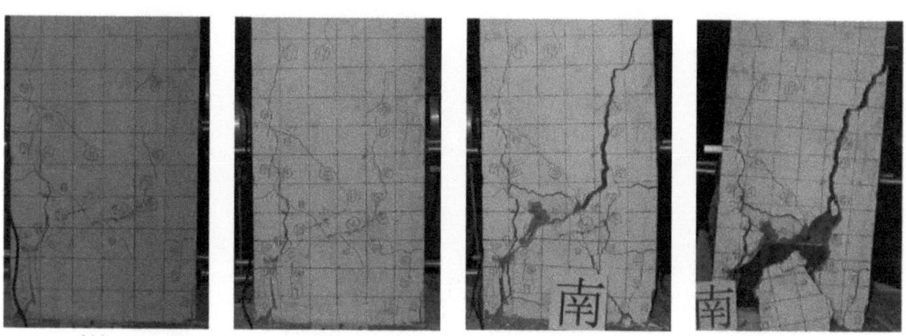

(A) (B) (C) (D)

Figure 8.21 Failure process of SFC-8. (A) 15 mm; (B) peak load; (C) 30 mm; (D) ultimate load.

and west surfaces continuously peeled off, and the oblique crack appeared from the left root to the right edge on the south surface, as shown in Fig. 8.21. When the lateral displacement was 45 mm ($\theta = 1/24$), the lateral load was 204 kN, the concrete cover below 10 cm height on the west surface completely peeled off, and the concrete cover below 15 cm height on the north surface completely peeled off. When the lateral displacement was 55 mm ($\theta = 1/22$), the lateral load was 195 kN, and the concrete cover below 10 cm height on the south surface completely peeled off. When the lateral displacement was 50 mm ($\theta = 1/20$), the lateral load was 176 kN, the bearing capacity decreased to less than 70% of the peak load, so the loading stopped. And the bottom core concrete showed an obvious crush. The failure mode of SFC-8 is shown in Fig. 8.21.

The failure mode of the bottom part of SFC-8 is shown in Fig. 8.22. Based on the observation, it can be found that the peeling area of the concrete cover is below 20 cm height of the four surfaces, the bottom part of the angle steel shows serious local buckling, the bottom steel batten shows obvious outward expansion, the core concrete at the bottom part shows serious crush, and the junction part of the bottom beam has serious concrete crushing. Therefore, the deformation of the steel batten and the angle steel mainly occurs at the column bottom. The bottom concrete shows the most severe crush, and the inner steel frame also has a significant effect on the concrete cover of the bottom beam.

8.3.2 Analysis of failure modes

Based on the failure modes of all SFC specimens, it can be concluded that all specimens show the large eccentric failure mode, and the failure of the specimens mainly occurs in the lower part of the columns, including the complete peeling off of the lower concrete cover, the complete crushing of the outer side of the core concrete

Figure 8.22 Failure mode of the bottom part of SFC-8.

at the bottom, the local buckling of the low-height angle steel and the tensile expansion of the low-height steel batten. Based on the observation of tested specimens, it can be concluded that the decrease in bearing capacity may be mainly, because of the local buckling of angle steel and the crushing of confined concrete. The above conclusions prove that the core concrete is effectively confined by the steel frame.

The axial compression ratio is the main influence factor for the bearing capacity of the SFC specimens. With the increase in the axial compression ratio, the bearing capacity of the specimen increases obviously, the ultimate displacement (the corresponding load drops below 70%) decreases obviously, and the peeling off of the concrete cover is more serious. Mainly because the larger vertical load can make the inner steel frame provide a larger confining effect on the core concrete, so the bearing capacity of the specimen shows an upward trend. Meanwhile, with the increase in the axial compression ratio, the local buckling position of the angle steel moves up from the lowest area to the area above the lowest steel batten, and the most obvious tensile expansion of the steel batten changes from the lowest batten to the low second batten. Therefore, the larger axial compression ratio leads to a larger plastic hinge zone, and the corresponding deformation form is shown in Fig. 8.23. The main reason is that the lowest steel batten is 50 mm from the bottom beam and the spacing between two steel battens is 100 mm. If the axial compression ratio is small, the steel frame will be mainly influenced by the lateral force, and the local buckling will mainly occur at the bottom. If the applied vertical load is larger, the steel frame can provide an obvious confining effect on core concrete, which can improve the properties of the lowest concrete. At this time, the weak part of the

(A) (B)

Figure 8.23 Failure modes of specimens under different axial compression ratios. (A) n = 0.30; (B) n = 0.42.

column is between the two bottom battens. Therefore, the local buckling position of the angle steel moves upward.

The steel ratio of the angle steel is an important factor in the seismic performance of the SFC specimen. Similar to the steel reinforced concrete column, the seismic performance of the specimen significantly increases with the increase in the steel ratio. For example, the seismic behaviors of the SFC specimen can be significantly improved by using large-size angle steel.

The good arrangement of small steel batten and small spacing can effectively restrict the local buckling of the angle steel, and the outward expansion of the steel frame is weakened. Therefore, the force on the concrete cover is small, and the peeling off of the concrete cover in the specimen is not as serious as that of other specimens.

8.4 The seismic experiment results

In this section, we analyze the seismic experiment results of 8 SFC cantilever specimens, including the hysteretic performance, the lateral bearing capacity, the lateral ductility, and the energy consumption capacity. We analyze the influence factors of the lateral bearing capacity and ductility, including the effects of confining form, the steel ratio of steel batten, the steel ratio of angle steel, and the axial compression ratio. Furthermore, we study the strain distribution of the angle steel and the steel batten under different transfer displacements, which can provide a scientific foundation for the seismic performance analysis and the further application of the SFC specimens.

8.4.1 Hysteretic curve

The hysteretic curve is the relationship between force and displacement of the SFC specimen under the lateral repeated load. It is not only the comprehensive embodiment of structural seismic behaviors but also the main foundation for analyzing the structural elastic-plastic dynamic responses. The hysteresis curves of 8 SFC specimens are shown in Fig. 8.24, which is the relationship between the seismic bearing capacity (P) and the column end displacement (Δ).

Based on Fig. 8.24, the hysteretic curves have the following characteristics:

(1) In the initial loading stage, there are few cracks on the SFC column surfaces, and the load–displacement curve is linear, so the covered area of the hysteretic curve of each specimen is very small. Under lateral cyclic loading, the column is at the elastic load stage, so the stiffness of the specimen has no obvious change. The residual deformation is also tiny.

(2) With the increase in the loading, several bending cracks appear at the column end, the covered area of the hysteretic curve increases, and the stiffness of the column gradually decreases. When the lateral load is unloaded to zero, the displacement of the column end will no longer return to zero, so the specimen has large residual deformation.

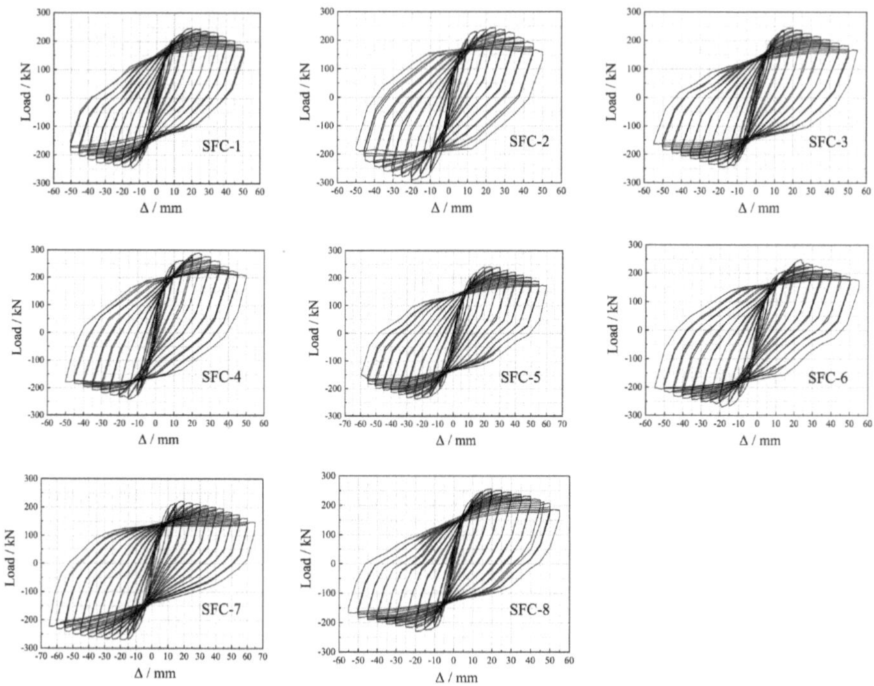

Figure 8.24 Hysteresis curves of SFC specimens.

(3) After the yield load, the bearing capacity continues to improve with the increase in the lateral displacement, and the hysteresis curve is much fuller.
(4) After the peak load, the cracks on the concrete surface develop rapidly. However, the specimen can still have a stable bearing capacity, which shows that the steel frame has a good confining effect on the core concrete, and the magnificent combination between the angle steel and the core concrete can lead to a good load-bearing capacity and energy dissipation capacity.
(5) Because of the influence of the damage accumulation, the degradation of strength and stiffness occur in the same displacement cycles. Meanwhile, a large part of the concrete cover gradually peels off with continuous loading. And the stress redistribution occurs in the specimen: part of the core concrete gradually loses working capacity due to the crushing, then the internal force is changed from the crushing concrete to the angle steel.
(6) With the increase in the loading cycle number and the displacement amplitude, the hysteretic curves of all specimens are always arched, and the load decreases slowly. When the ultimate load drops below 70% of the peak load, the lateral displacement angle is greater than 1/22, which shows that the SFC specimens have good ductility.
(7) Compared with the hysteretic curve of the ordinary steel reinforced concrete column, the seismic behavior of the SFC specimens is significantly improved, especially for the load-bearing capacity and the energy dissipation capacity.
(8) The higher axial compression ratio leads to a higher load-bearing capacity, a fuller hysteresis curve, and stronger energy dissipation capacity, but the lower ductility. The test

results show that the steel frame can provide a great confining effect on the core concrete. The larger the axial compression ratio leads to the greater confining effect provided by the steel frame.

(9) The thicker steel batten, the intensive configuration of the steel batten, or the addition of the internal spiral hoops can lead to the fuller hysteretic curves and higher energy consumption capacity.

(10) The big size of angle steel (i.e., SFC-7) leads to fuller hysteretic curves of SFC specimens. The cover area of each cyclic hysteretic curve is larger, and the ductility is also higher, which means that the angle of steel plays an important role in the energy consumption and ductility of SFC specimens.

8.4.2 Skeleton curve

The Skeleton curve is to be expressed as the envelope curve got by connecting the extreme load points of each load–displacement circle curve in the same direction (tension or compression) on the hysteretic curve. The Skeleton curve is the track of the maximum value of each cyclic loading, which reflects the different conditions and characteristics of the component stress and deformation (e.g., strength, stiffness, ductility, energy consumption, and collapse resistance). It is also an important foundation to determine the characteristic points in the hysteresis model.

The influence of axial compression ratio on the skeleton curve of the SFC specimens with different structural forms (i.e., small batten, small spacing, and spiral hoops) is shown in Fig. 8.25. The influence of axial compression ratio on the skeleton curve of unique specimens is basically the same. Under the large axial compression ratio, the skeleton curve of each specimen in the elastic stage is large, which shows that the increase in the axial compression ratio can make the column have greater initial stiffness and higher bearing capacity. For the decreasing stage, the increase of the axial compression ratio will speed up the decreasing speed of the skeleton curve and reduce the ductility of the column. Based on the curve comparison, it can be found that the decline of the decreasing stage in the skeleton curve of SFC-4 and SFC-6 is gentler than that of SFC-2, which shows that the mechanical behavior of SFC specimens can be effectively improved by using the arrangement of small battens and small spacing or adding the spiral hoops.

The influence of different batten arrangements on the skeleton curves of SFC specimens is shown in Fig. 8.26. It can be concluded that the arrangement of steel battens has little effect on the skeleton curve of the SFC specimen in the increasing stage. In addition, the arrangement of small batten and small spacing can make the decreasing stage of the skeleton curve gentler, so the seismic performance becomes better. Furthermore, the above phenomenon is more obvious under a high axial compression ratio.

The influence of the steel ratio of angle steel on the skeleton curve of the SFC specimen is shown in Fig. 8.27. Based on the curve comparisons, it can be concluded that the skeleton curve of SFC-7 in the elastic stage is basically similar to that of SFC-1, but the SFC-7 has the higher bearing capacity, the smoother decreasing curve, and the better seismic performance. Therefore, the seismic behaviors of

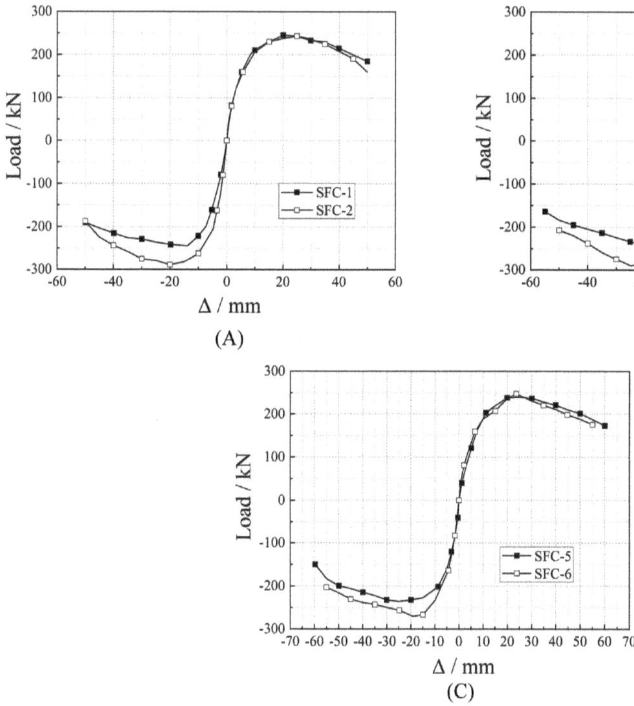

Figure 8.25 Skeleton curve of SFC specimens with different structural forms. (A) Basic form; (B) small batten and small spacing; (C) adding spiral hoops.

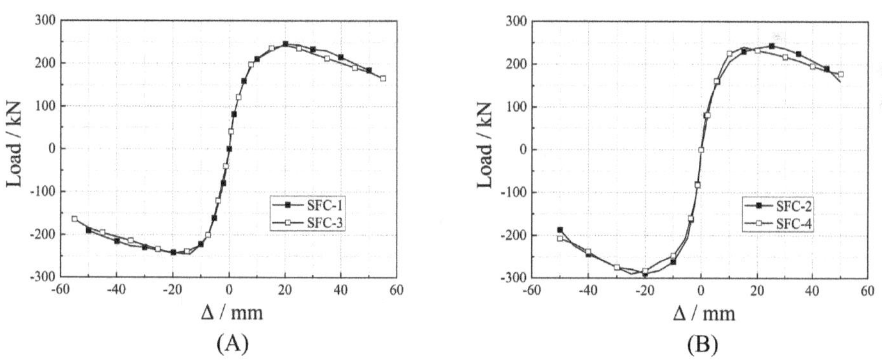

Figure 8.26 Influence of different batten arrangements. (A) $n = 0.30$; (B) $n = 0.42$.

the SFC specimens increase with the increase in the steel ratio of angle steel, including the bearing capacity and the ductility. The influence of the steel ratio of steel batten on the skeleton curve of the SFC specimen is shown in Fig. 8.27. It can be concluded that changing the steel form of the steel batten would influence the

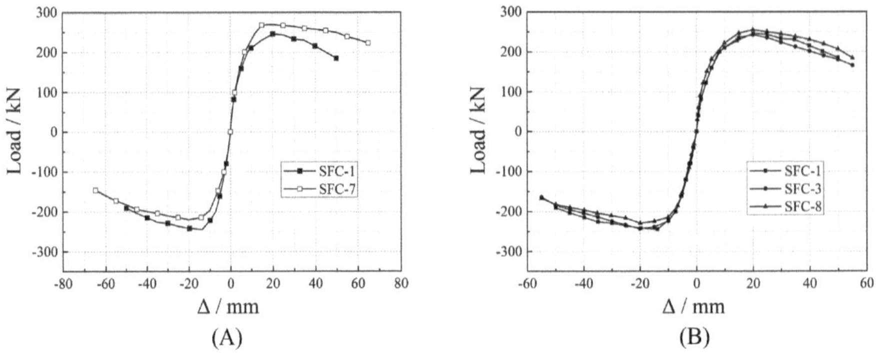

Figure 8.27 Influence of the steel ratio. (A) Angle steel; (B) steel batten.

confining effect on core concrete. The larger confining effect leads to the larger ultimate displacement, well ductility, and the slower degradation of the bearing capacity of the specimen. In addition, the configuration method of wide batten and small spacing has a better effect than that of small batten and small spacing.

In summary, based on the analysis of the skeleton curve of the SFC specimens, it can be concluded that the main influence factor on the bearing capacity is the axial compression ratio, and the main influence factors on the ductility are the steel ratio of angle steel and steel batten. Increasing the confining effect on the core concrete can slow down the degradation of the bearing capacity and increase the ultimate displacement and ductility of the specimen. Therefore, under the same axial compression ratio, the influence factors of specimen ductility can be in the following order: increasing the steel ratio of angle steel, arranging spiral hoops, and changing the size or spacing of steel battens.

In order to analyze the seismic behaviors of the SFC specimens in detail, the measured data of each specimen in different load stages are shown in Table 8.3, including the bearing capacity, displacement, and displacement angle. In the Table 8.3, P_c is the cracking load, P_y is the yield load (based on the measured load–displacement skeleton curve of the column, it is determined by the equal energy method), P_m is the peak load, P_u is the ultimate load (it is the corresponding value once the load below 70% of the peak load), and the Δ_c, Δ_y, Δ_m, Δ_u are the displacement values corresponding to the P_c, P_y, P_m, P_u, separately, and the θ_c, θ_y, θ_m, θ_u are the corresponding displacement angle values, separately.

Based on the result analysis of Table 8.3, it can be seen that the SFC-7 has a large peak load and a large peak displacement, while the peak loads in pull or push direction of other specimens are relatively closed. Therefore, the steel ratio of the angle steel is the factor that has a significant effect on the bearing capacity. Meanwhile, the bearing capacity of specimens shows obvious asymmetry under a large axial compression ratio. However, the SFC-7 and the SFC-8 also show large load asymmetry. Based on comparisons of the peak displacements, the confining effect (e.g., changing the arrangement of steel batten or adding spiral hoops) can improve the asymmetry. In summary, it can be concluded that the influence factors

Table 8.3 Parameter point analysis of test results.

Item	Loading direction	Cracking point			Yield point			Peak point			Failure point			Ductility coefficient	
		P_c (kN)	Δ_c (mm)	θ_c	P_y (kN)	Δ_y (mm)	θ_y	P_m (kN)	Δ_m (mm)	θ_m	P_u (kN)	Δ_u (mm)	θ_u	μ	$\bar{u}$
SFC-1	Positive	160	5.28	1/208	208	9.5	1/154	246	20	1/55	185	50	1/22	4.85	5.19
	Negative	160	5.26	1/209	211	8.62	1/170	245	15	1/73	191	50	1/22	5.57	
SFC-2	Positive	160	5.74	1/192	211	10.41	1/106	243	25	1/44	160	50	1/22	4.21	4.79
	Negative	160	3.43	1/321	254	7.75	1/142	289	20	1/55	188	50	1/22	5.56	
SFC-3	Positive	160	5.34	1/206	205	8.98	1/122	243	20	1/55	166	55	1/20	4.82	5.01
	Negative	160	5.51	1/200	212	8.73	1/126	243	20	1/55	164	55	1/20	5.2	
SFC-4	Positive	200	7.72	1/142	211	8.72	1/126	240	15	1/73	178	50	1/22	4.78	4.69
	Negative	200	5.69	1/193	239	9	1/122	290	25	1/44	207	50	1/22	4.61	
SFC-5	Positive	120	5.18	1/212	208	12.3	1/89	240	25	1/44	173	60	1/18	4.33	5.08
	Negative	120	3.1	1/355	202	8.67	1/127	236	25	1/44	150	60	1/18	6.15	
SFC-6	Positive	160	6.68	1/165	200	13.01	1/85	248	25	1/47	175	55	1/20	3.43	4.11
	Negative	160	4.38	1/251	234	9.96	1/110	270	20	1/58	203	55	1/20	5	
SFC-7	Positive	200	6.96	1/158	225	9.47	1/116	269	20	1/55	223	65	1/17	7.3	6.28
	Negative	200	10.09	1/109	196	10	1/110	220	20	1/55	146	65	1/17	5.31	
SFC-8	Positive	180	5.31	1/207	210	8.56	1/129	255	20	1/55	184	55	1/20	5.9	6.02
	Negative	180	6.64	1/166	197	8.02	1/137	229	20	1/55	167	55	1/20	6.15	

Source: From Rong, C., Wang, Q. X., & Wang, B. (2021). Seismic performance of angle steel frame confined concrete columns: Experiments and FEA model. *Engineering Structures*, *235*, 111983.

of the bearing capacity are mainly the axial compression ratio and the steel ratio of angle steel, and the influence factors of the displacement are mainly the arrangement mode of steel batten and the internal confining effect in the specimens.

8.4.3 Deformation capacity

Deformation capacity is a very important index to reflect the seismic performance of specimens. The ductility coefficient is usually used to quantify the deformation capacity of the specimens. The ductility coefficient is the ratio of the effective ultimate displacement (Δ_u) to the yield displacement (Δ_y). In this chapter, we take the effective ultimate displacement as the lateral displacement corresponding to the lateral bearing capacity decreases to 85%, which can be gained by the interpolation calculation in the skeleton curve. The yield displacement is the top lateral displacement corresponding to the initial yield condition of the outermost longitudinal reinforcement. Generally, the yield displacement is gained by the following methods: the energy equivalent area method, the yield moment method, and the simulated yield process method (Xiong & Wang, 2015). Moreover, we use the energy equivalent area method in this chapter. Based on the computation results, the ductility coefficient of each specimen is shown in Table 8.1. The left column of the ductility coefficient represents the values in positive or negative directions (in which, the positive direction is the pull direction), and the right column represents the average value between positive and negative directions. The following formula can calculate it:

$$u = (|\Delta_u^+| + |\Delta_u^-|)/\left(|\Delta_y^+| + |\Delta_y^-|\right) \tag{8.1}$$

Based on Table 8.3, it can be concluded that all SFC specimens have high ductility, which reflects the superior seismic behaviors. Based on three pairs of comparisons (SFC-1 and SFC-2, SFC-3 and SFC-4, SFC-5 and SFC-6), it can be concluded that: the ductility decreases with the increase of axial compression ratio; the addition of spiral hoops has little effect on the ductility coefficient, but the lateral displacement of yield point, peak point and failure point are increased simultaneously; the increase in the steel ratio of angle steel or steel batten can improve obviously the ductility of the specimen.

8.4.4 Strength reduction

Strength reduction is a very important index to reflect the influence of the damage accumulation of specimens. In the low cycle repeated lateral loading process, the bearing capacity of the specimen will decrease with the increase in loading cycle times at the same displacement amplitude, which is the strength reduction of the specimen. Generally, the strength reduction ratio (P_n/P_1) is used to characterize the reduction of the bearing capacity, which is the ratio of the peak load of the last two cycles to the peak load of the first cycle under the same displacement amplitude.

The relationship between strength reduction and displacement amplitude of each specimen is shown in Fig. 8.28.

As can be seen from Fig. 8.28, the strength reduction ratio of each specimen is small, all strength reduction ratios are basically less than 10%. The strength reduction ratio of each specimen decreases with the increase in the displacement amplitude, which shows that strength reduction increases with the increase in the displacement amplitude. Based on the strength reduction relationship of specimens under the same axial compression ratio, it is found that the strength reduction of specimens with a better confining effect is smaller and more stable. Among all the specimens, the strength reduction in SFC-7 is the most stable. Therefore, the strength reduction mainly depends on two factors: the confining effect on the core concrete and the local buckling of the angle steel. The strength reduction of specimens with a strong confining effect or a large steel ratio of angle steel is small.

8.4.5 Stiffness reduction

With the increase in the cyclic displacement and the number of cycles, the stiffness of the specimen will also decrease. In this section, we study the variation law of the secant stiffness of each SFC specimen by using the stiffness in each loading stage. The secant stiffness is the ratio of the maximum bearing capacity to the corresponding displacement under the same displacement amplitude, as shown in Eq. (8.2),

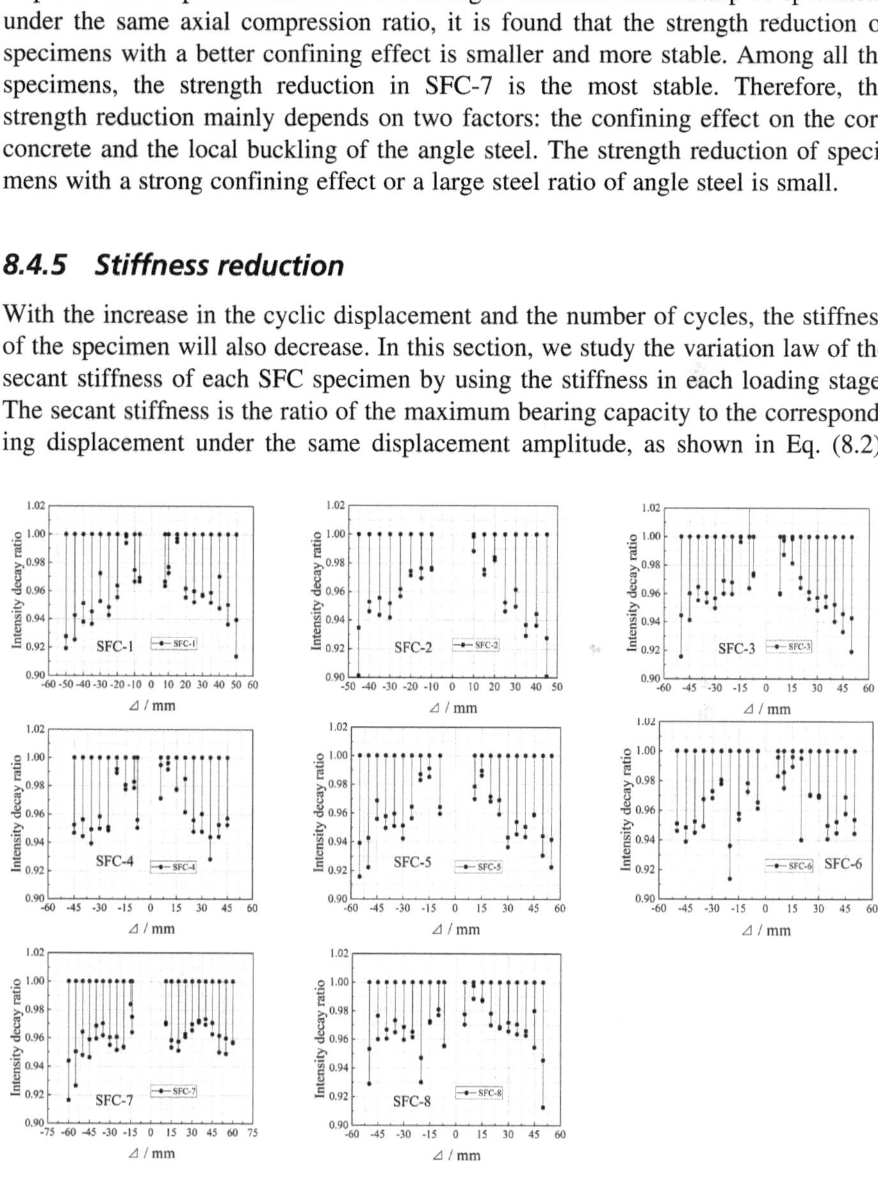

Figure 8.28 Strength reduction of specimens.

which is used to reflect the stiffness reduction (ACI 374.1, 2005). The smaller reduction rate of secant stiffness leads to the better stability of the hysteretic curve and the better energy dissipation capacity of specimens.

$$K_i = P_i / \Delta_i \tag{8.2}$$

where K_i is the secant stiffness of the i-th cycle; P_i is the lateral maximum load in the i-th cycle; Δ_i is the corresponding displacement of the i-th cycle.

The relationship between secant stiffness and displacement amplitude of all SFC specimens in the test is shown in Fig. 8.29. It can be concluded from the curves in Fig. 8.29 that the specimens under two axial compression ratios show a relatively similar stiffness reduction trend: with the cracking of the concrete, the stiffness of the SFC specimens decreases rapidly; then the developing law is from fast to slow in the loading process. In addition, the axial compression ratio has a significant effect on the stiffness of the SFC specimens, and the larger axial compression ratio leads to the larger stiffness of each loading stage.

8.4.6 Energy consumption capacity

Energy consumption capacity is the energy absorption capacity of members under repeated load. It can reflect the nonlinear mechanical performance of the members or structures. Therefore, the energy consumption capacity can be an important index to measure the seismic performance of the members or structures. Therefore, under a certain strength guarantee, if the component has good energy consumption capacity, it can absorb a large part of seismic energy in the earthquake, so as to achieve the better seismic effect. Under repeated lateral loading, the change in energy consumption of each specimen during loading can be expressed by the cumulative energy consumption (E) at each characteristic point. The energy consumption of each specimen under a low cycle repeated load is listed in Table 8.4. And the relationship between the cumulative energy consumption and the displacement

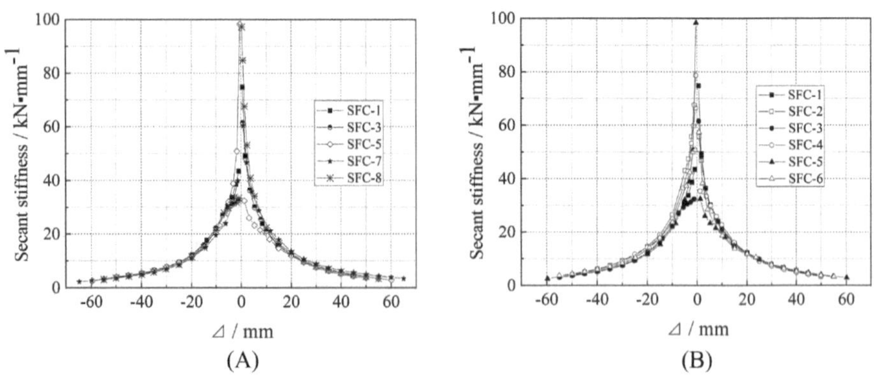

Figure 8.29 Secant stiffness of specimens.

Table 8.4 Energy consumption of specimens.

Cycle number	SFC-1		SFC-2		SFC-3		SFC-4		SFC-5		SFC-6		SFC-7		SFC-8	
	Displacement (mm)	Energy consumption (kN\cdotmm)	Displacement (mm)	Energy consumption (kN\cdotmm)	Displacement (mm)	Energy consumption (kN\cdotmm)	Displacement (mm)	Energy consumption (kN\cdotmm)	Displacement (mm)	Energy consumption (kN\cdotmm)	Displacement (mm)	Energy consumption (kN\cdotmm)	Displacement (mm)	Energy consumption (kN\cdotmm)	Displacement (mm)	Energy consumption (kN\cdotmm)
1	0.7	24.2	0.7	28.9	1	37.3	0.8	30	0.8	34.1	0.7	22.4	1	38.1	0.6	19.1
2	1.9	96.5	1.5	71.6	2.1	102	1.8	81.9	2.3	112.2	1.9	95.5	2.7	181.6	1.3	56.9
3	3.5	224.7	2.8	242.8	3.6	219	3	189.3	4.1	275.6	3.5	263.8	5.2	566.8	2	97.5
4	5.3	454.4	4.6	507	5.4	428.4	4.6	396.8	6.2	554.1	5.5	643.5	8.5	1267	3	187.6
5	7.9	1007.3	4.6	389.5	7.7	830.6	6.9	968.7	10	1343.1	5.6	509	13	2667.7	4.4	378
6	7.9	749	4.7	370.4	7.6	577.3	6.9	669.6	10	996.3	5.6	466.6	13.1	2052.8	6.1	660.9
7	7.9	702.6	10	2012.4	7.6	519	6.9	618.2	10	899.9	10	1764.3	13.1	1881.5	5.9	504
8	10	1452.1	10	1441.9	10	1347.4	10	1892.8	15	2918.4	10	1279.1	15	2937.2	5.9	482.4
9	10	1290.4	10	1308.1	10	1431.6	10	1494.4	15	2523.9	10	1176.2	15	2679.1	10	1918.3
10	10	1222.6	15	3784.1	10	1314.2	10	1432.3	15	2370.2	15	3438.5	15	2555.2	10	1391.8
11	15	3662.5	15	3116.6	15	3808.4	15	3975.5	20	4938.2	15	2825.5	20	5532.1	10	1327.1
12	15	3127.4	15	2955.4	15	3430.9	15	3288.7	20	4609.3	15	2672	20	4865.4	15	3890.3
13	15	2964.9	20	5732.7	15	3315.2	15	3156.1	20	4475	20	5636.8	20	4646.8	15	3423.9
14	20	5949.2	20	5032.3	20	6478.5	20	6230.8	25	7446.8	20	4778	25	7923.1	15	3290
15	20	5739.2	20	4850.6	20	6097	20	5519.2	25	7367.5	20	4799.9	25	7279.5	20	6292.3
16	20	5671.2	25	7919.6	20	5956.8	20	5322.5	25	7168.2	25	8025.5	25	7069.7	20	5938.9
17	25	8772.1	25	8081.3	25	9288.8	25	8755.3	30	10203.1	25	7607.7	30	10396.8	20	5880
18	25	8644.6	25	8005.3	25	8971.5	25	8674.2	30	10019.5	25	7430.1	30	9768.2	25	9066
19	25	8504	30	11,528.1	25	8777.9	25	8527.4	30	9957.6	30	10666.5	30	9567.6	25	8685.1
20	30	11,700.6	30	11,479.3	30	12,074.5	30	12,009.9	35	12957.1	30	10696.4	35	13061.3	25	8542.8
21	30	11,512.9	30	11,387.6	30	11,451.4	30	11,952.9	35	12924.6	30	10636.7	35	12405.1	30	11873.2
22	30	11,509.1	35	14,762	30	11,228.4	30	11,858.8	35	12780.9	35	13788.8	35	12192.6	30	11522
23	35	15,298.3	35	14,522.2	35	14,704.1	35	15,389.4	40	15848.5	35	13695.3	40	15808.4	30	11337.9
24	35	14,832.1	35	14,438.4	35	14,091.5	35	15,306.5	40	15708.2	35	13970.6	40	15305.6	35	14617.5
25	35	14,541.1	40	18,123.1	35	13,858.1	35	15,233.2	40	15479.8	40	17281.5	40	15096.3	35	14248.6
26	40	17,899.5	40	17,912.3	40	17,081	40	18,775.1	45	18580.3	40	17000.5	45	18770.7	35	13966.7
27	40	17,398.6	40	18,072.9	40	16,624.3	40	19,154.8	45	18522.4	40	16838.7	45	18095.9	40	17188.4

(*Continued*)

Table 8.4 (Continued)

Cycle number	SFC-1		SFC-2		SFC-3		SFC-4		SFC-5		SFC-6		SFC-7		SFC-8	
	Displacement (mm)	Energy consumption (kN\cdotmm)	Displacement (mm)	Energy consumption (kN\cdotmm)	Displacement (mm)	Energy consumption (kN\cdotmm)	Displacement (mm)	Energy consumption (kN\cdotmm)	Displacement (mm)	Energy consumption (kN\cdotmm)	Displacement (mm)	Energy consumption (kN\cdotmm)	Displacement (mm)	Energy consumption (kN\cdotmm)	Displacement (mm)	Energy consumption (kN\cdotmm)
28	40	17,124.5	45	22,412.4	40	16,316.2	40	19,151.1	45	18438.6	45	20822.5	45	17887.8	40	16738.8
29	45	20,075.5	45	21,963.3	45	19,409.1	45	22,924	50	21375.4	45	20648.1	50	21761.3	40	16474.4
30	45	19,612.8	45	22,803.9	45	18,883	45	23,228	50	21077.4	45	20250.7	50	21206.5	45	19821.3
31	45	19,193.6	50	25,868.5	45	18,514.6	45	23,114.2	50	20593.1	50	23709.4	50	20816.7	45	19691.5
32	50	22,221.2			50	22,262.7	50	26,231.7	55	23237.6	50	23336.5	55	24205.6	45	19754.3
33	50	21,379.9			50	21,614.9			55	22836.8	50	23356.4	55	23492.5	50	22060.6
34	50	17,844.5			50	20,993.3			55	22334	55	27641.9	55	23159.3	50	21365.4
35					55	23,440.7			60	25605.4			60	26555.8	50	20699.2
36													60	25724.1	55	23283
37													60	25916.3		
38													65	29136.8		
Total	312,403.1		281,124.5		335,479.6		295,553.3		376513.1		337774.9		464474.8		336675.7	

Source: From Rong, C., Wang, Q. X., & Wang, B. (2021). Seismic performance of angle steel frame confined concrete columns: Experiments and FEA model. *Engineering Structures*, *235*, 111983.

amplitude is shown in Fig. 8.30. When each specimen reaches the characteristic point (e.g., the yield load, the peak load, and the ultimate load), the corresponding cumulative energy consumption is listed in Table 8.5, in which the ultimate energy consumption of the specimen can effectively reflect the maximum energy consumption capacity of the SFC specimens.

The cumulative energy consumption of the specimen at each displacement amplitude is shown in Fig. 8.31. Then the following conclusions can be drawn: when the axial compression ratio is small (i.e., 0.3), the cumulative energy consumption of each displacement amplitude is basically closed, and the cumulative energy consumption can be slightly increased by using small steel batten or increasing the width of steel batten; increasing the steel ratio of angle steel or adding stirrups will reduce the cumulative energy consumption at each displacement amplitude, because of smaller lateral force under the same lateral displacement; for

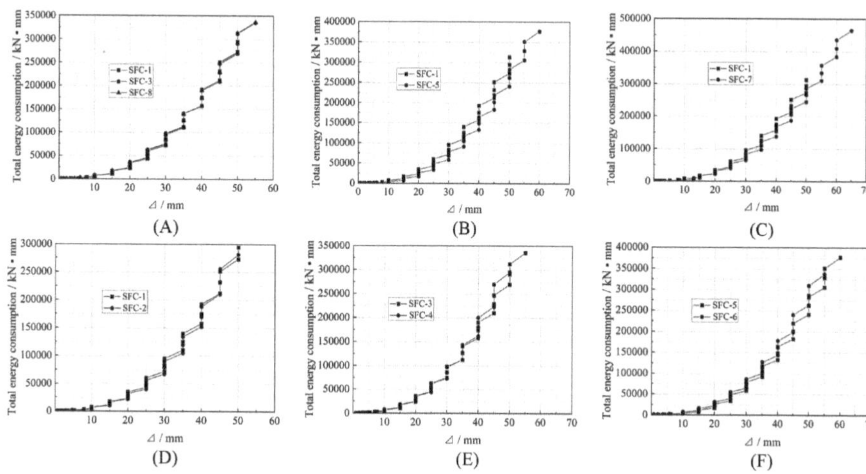

Figure 8.30 Cumulative energy consumption of specimens.

Table 8.5 Cumulative energy consumption of each parameter point.

Item	Energy dissipation E (kN · mm)		
	Yield point	**Peak point**	**Failure point**
SFC-1	4710.8	22,927.8	288,647.5
SFC-2	3622.6	21,961.4	281,124.5
SFC-3	4061	23,839.8	335,479.6
SFC-4	4847.3	24,425.1	295,553.3
SFC-5	2319.1	33,497.1	376,513.1
SFC-6	3765.1	38,396.6	337,774.9
SFC-7	4721.2	22,359.1	464,474.8
SFC-8	4304.7	23,920.1	336,675.7

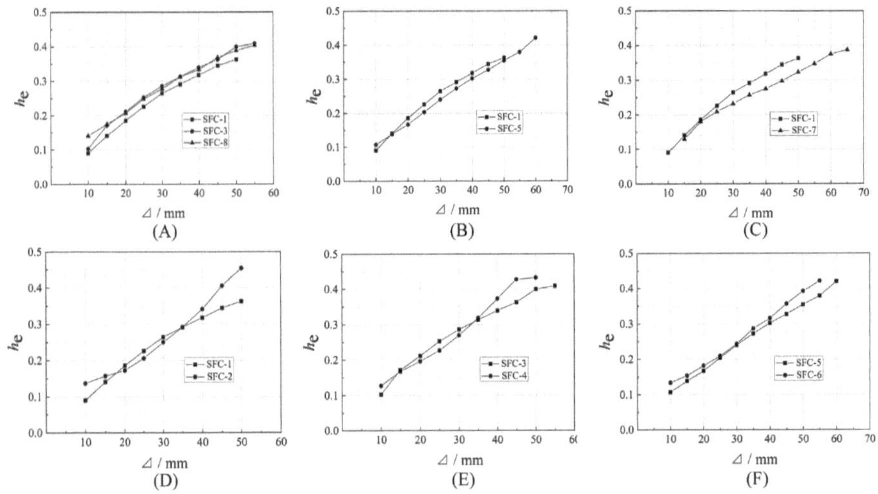

Figure 8.31 Cumulative energy consumption at each displacement amplitude.

the SFC specimen with the basic parameters, the axial compression ratio has little effect on the cumulative energy consumption, which shows that the confining effect of the steel frame is not obvious; for the SFC specimens with larger confining effect, the cumulative energy consumption at each displacement amplitude significantly increases with the increase of the axial compression ratio.

The cumulative energy consumption corresponding to each characteristic point of the specimen is summarized in Table 8.5. It can be concluded that: the larger confining effect (i.e., addition of the internal spiral hoops or the assembly mode of small batten and small spacing) can lead to the higher cumulative energy consumption at the yield point and the peak point, but the cumulative energy consumption at the ultimate point is small, due to the low ductility of the specimen; For the SFC specimen without sufficient confining effect, the increase of the axial compression ratio would reduce the cumulative energy consumption at the three characteristic points; the section form or the steel ratio of the angle steel has little effect on the cumulative energy consumption at the yield point and the peak point, but the cumulative energy consumption at the ultimate point increases with the increase in the steel ratio or the reasonable section form, which shows that the several measures can slow down the decline of lateral bearing capacity and improve the seismic performance of composite columns; the cumulative energy consumption of the three characteristic points is increased by adding internal spiral hoops, which shows that the addition of internal confining effect can improve the seismic performance of composite columns; however, increasing the steel ratio of the angle steel has the greatest cumulative energy consumption at the ultimate point, so as to have the best seismic performance.

Under the low cyclic loading, the energy dissipation capacity of members can also be expressed by the equivalent viscous damping coefficient, $h_e = E/$

($2\pi \times P \times \Delta$), where E is the covered area by a hysteresis curve, and P is the load at the peak point of the hysteresis curve, Δ is the corresponding displacement, the larger value of h_e leads to the fuller hysteretic curve and the stronger the energy dissipation capacity.

The relationship between the equivalent viscous damping coefficient (h_e) and the deformation (Δ) in the last cycle under each displacement amplitude is shown in Fig. 8.31. Based on the analysis, the following conclusions can be drawn: the value of h_e increases with the increase of displacement, the specimen can still maintain good energy dissipation capacity after yielding condition, due to the bending failure; the value of h_e can be improved by changing the steel distribution form or adding the spiral hoops, so as to have the well seismic performance; the increase of axial compression ratio can increase the value of h_e, which leads to the higher seismic performance of the SFC specimens, so the steel frame can play a good confining role and make the specimens have good ductility.

8.4.7 Stress state of angle steel and steel batten

(1) Angle steel

The relationship between the longitudinal strain of the angle steel and the lateral load is shown in Fig. 8.32. And the strain belongs to the angle steel in the plastic region. It can be concluded that the longitudinal strain of the angle steel increased with the increase in the lateral load of the SFC specimen, and the increasing ratio

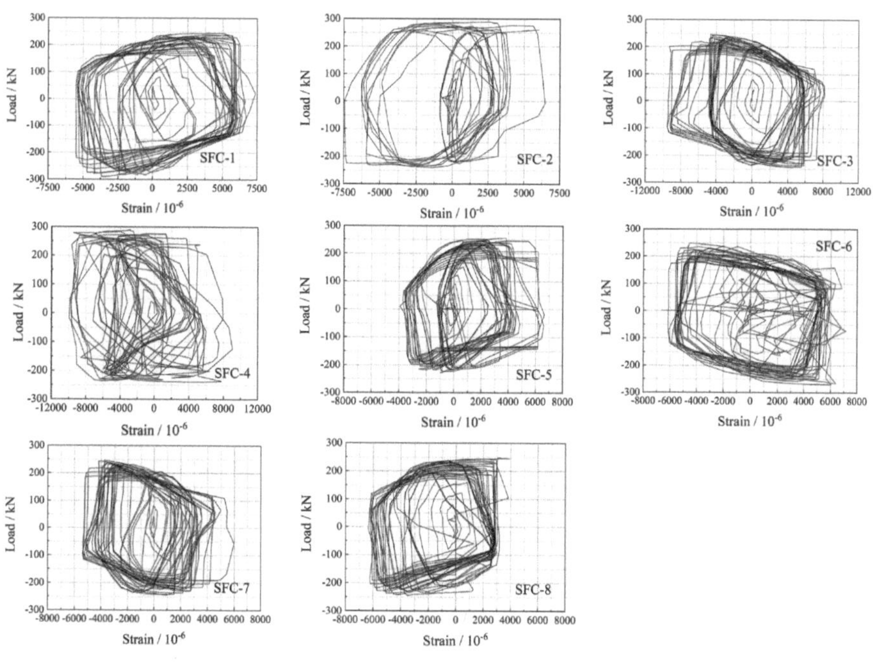

Figure 8.32 Longitudinal strain of angle steel.

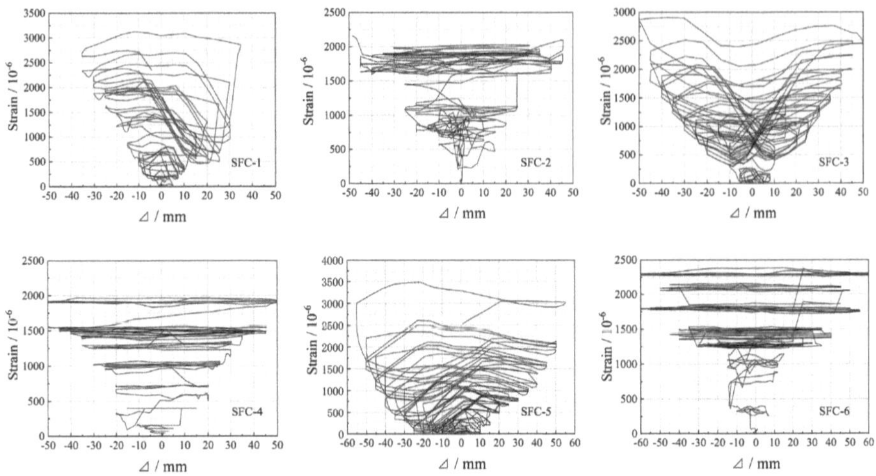

Figure 8.33 Lateral strain of steel batten.

of strain shows a trend of first fast and then slow; when the specimen reaches the peak load, parts of angle steel in the plastic hinge zone have reached the yield state; when the specimen reaches the ultimate load, all angle steels in the plastic hinge zone have reached yield state and also have a large strain; based on the strain analysis, when the specimen reaches the ultimate load, most of the angle steel above the plastic hinge zone does not reach yield state.

(2) Steel batten

The stress condition of steel batten can reflect the confining effect on the core concrete, so it is an important index for the seismic performance of the SFC specimen. The relationship between the lateral strain of the steel batten and the lateral displacement is shown in Fig. 8.33. Based on the variation curve, it can be concluded that the lateral strain of steel batten increases with the increase of lateral displacement, and the increasing ratio of strain shows a trend of first slow and then fast. Before the specimen reaches the peak load, the lateral strain of steel batten is always in a small condition, and the increasing rate of the batten strain is basically similar. After the peak load, the increasing rate of the batten strain is large. When the specimen reaches the ultimate load, all the steel battens in the plastic hinge zone exceed the yield strain. From the peak condition to the ultimate condition, the steel batten continues to confine the core concrete, and the specimen enters the limit state gently.

8.5 Conclusions

In this chapter, we design and test 8 SFC specimens, including the design, test scheme, the failure process, and failure characteristics of the specimens. After the test, the test results are analyzed, and the influence factors include the assembly mode of steel battens, the internal spiral hoops, the steel ratio of angle steel and the axial compression ratio, the analysis contents are the lateral bearing capacity, and the ductility. In summary, the following main conclusions can be drawn:

(1) All the SFC specimens show the bending failure, which occurs at the column bottom. And the concrete cover peels off in varying degrees at the failure condition, but the peeling off of the concrete cover does not affect the mechanical properties of the SFC specimens. All the SFC specimens show the large eccentric compression failure and have a large ductility. With the increase in the axial compression ratio, the bearing capacity of SFC specimens gradually increases, but the ductility decreases.

(2) The failure modes of SFC specimens include the peeling off of the concrete cover at the low part of the column, the crushing of the bottom core concrete, the expansion of the bottom steel batten, and the local buckling of the bottom angle steel. These failure modes fully show that the steel frame has an effective confining effect on the core concrete. Although the failure modes of 8 SFC specimens are similar, the local buckling of angle steel, the expansion of steel batten, and the peeling off of the concrete cover are influenced by the following factors, including the combination mode of steel batten, the size of steel batten, the size of angle steel, the axial compression ratio.

(3) The hysteretic curves of the SFC specimens are very full, and the degradation of bearing capacity after peak load is small, which shows the large ductility coefficient, the great plastic deformation, and the great energy dissipation capacity. It shows that the steel frame can provide an obvious confining effect on the core concrete, so the SFC specimen with a high steel ratio shows many advantages.

(4) Improving the arrangement of steel battens or adding the spiral hoops can slow down the decline of bearing capacity, make the hysteretic loop fuller and enhance the energy dissipation capacity. The higher axial compression ratio leads to the fuller hysteresis curves and the stronger energy dissipation capacity of the SFC specimens because the steel frame can provide an obvious confining effect on the core concrete. In addition, the larger the axial compression ratio can increase the influence of the confining effect provided by the steel frame.

(5) The most important factor for the mechanical properties of the SFC specimen is the steel ratio of angle steel. The larger steel ratio of the angle steel leads to the better ductility and the better energy dissipation capacity of the SFC specimen.

(6) The strength reduction of all SFC specimens is small, basically less than 10%. The strength reduction increases with the increase in the displacement, due to the internal concrete damage. And the good confining effect can restrict strength degradation.

(7) The stiffness reduction of all SFC specimens shows the law from fast to slow. With the cracking of the concrete, the stiffness of specimens decreases rapidly. And the axial compression ratio has a significant effect on the stiffness. The larger axial compression ratio leads to the larger the stiffness in each loading stage.

(8) When the specimen reaches the peak load, parts of the angle steels in the plastic hinge zone have basically reached the yield state. When the specimen reaches the ultimate load, all of the angle steel in the plastic hinge zone have reached yield condition. Only a few of the angle steel above the plastic hinge zone reach the yield state before the peak load.

(9) The strain of the steel batten in the plastic hinge zone is small before the peak load. The strain of the steel batten in the plastic hinge zone increases rapidly after the peak load. When the specimen reaches the ultimate load, the steel battens in the plastic hinge zone reach the yield state, while the steel battens above the plastic hinge zone do not reach the yield state.

Acknowledgments

This work was supported by the National Natural Science Foundation of China [grant numbers 52108171; 51878540]. The authors gratefully acknowledge two Master's students (Mr. Qv Yunsong and Mr. Tian Wenkai) of Xi'an University of Architecture and Technology for the drawing works.

References

ACI 374.1. (2005). *Acceptance criteria for moment frames based on structural testing and commentary*. Farmington Hills: American Concrete Institute.

Chen, S, & Wu, P (2017). Analytical model for predicting axial compressive behavior of steel reinforced concrete column.. *Journal of Constructional Steel Research*, *128*, 649–660.

Denavit, MD, Hajjar, JF, Perea, T, & Leon, RT (2018). Elastic flexural rigidity of steel-concrete composite columns. *Engineering Structures*, *160*, 293–303.

Feng, G (2014). *Study on seismic performance of steel reinforced concrete columns with inner space steel frame*. Suzhou University of Science and Technology.

GJ 101-96. (1997). Specification of test methods for earthquake resistant building. China Architecture& Building Press.

Hong, HP, Yuan, H, Deng, L, & Bai, Y (2019). Axial capacity of steel tube-reinforced concrete stub columns. *Engineering Structures*, *183*, 523–532.

Ji, J (2008). *Research on seismic performance and design method of jacketing steel reinforced concrete frame prestressed with bonded tendons*. Harbin Institute of Technology.

Julio, GR, Joaquín, RP, Adam, JM, & Pedro, AC (2011). An experimental study on steel-caged RC columns subjected to axial force and bending moment. *Engineering Structures*, *33*(2), 580–590.

Julio, GR, Jose, MA, & Pedro, AC (2011). Behaviour of RC columns strengthened by steel caging under combined bending and axial loads. *Construction and Building Materials*, *25*, 2402–2412.

Montuori, R, & Piluso, V (2009). Reinforced concrete columns strengthened with angles and battens subjected to eccentric load. *Engineering Structures*, *31*(2), 539–550.

Nagaprasad, P, Sahoo, DR, & Rai, DC (2009). Seismic strengthening of RC columns using external steel cage. *Earthquake Engineering and Structural Dynamics*, *38*, 1536–1586.

Shi, J, & Bai, G (2000). An experiment study on restoring force characteristics of lattice type steel reinforced concrete frame columns. *Journal of Xi'an Highway University*, *20*(4), 94–97.

Tang, X, Liu, W, & Qian, Y (2004). Experimental research on seismic behavior of concrete column with spatial steel frame subjected to low cyclic reversed load. *Journal of Building Structures*, *25*(5), 22–28.

Xiong, Z, & Wang, S (2015). *Civil engineering structural test*. China Architecture& Building Press.

Zenzai, S, Shimizu, S, & Chikahiro, Y (2018). Behavior of a concrete filled steel box column with considering detachment under seismic load. *Thin-Walled Structures*, *124*, 98–106.

Zhao, S (2007). Investigation on the seismic performance of open-web trussed steel reinforced concrete column. *Industrial Construction*, *37*(2), 93–95.

Zheng, WZ, & Ji, J (2008a). Dynamic performance of angle-steel concrete columns under low cyclic loading-I: experimental study. *Earthquake Engineering and Engineering Vibration*, *1*, 67–75.

Zheng, WZ, & Ji, J (2008b). Dynamic performance of angle-steel concrete columns under low cyclic loading-II: Parametric study. *Earthquake Engineering and Engineering Vibration*, *2*, 137–146.

Zhou, X, & Shi, Q (1993). On the aseismic behavior of concrete encased space angle under horizontal repeated loads. *Journal of Harbin University of Civil Engineering and Architecture*, *26*(8), 190–197.

Design method of steel frame confined concrete column

9

9.1 Introduction

Based on the analysis in Chapters 5 and 8, the ordinary steel-reinforced concrete column has the following disadvantages: (1) the combination of steel bars and inner section steel leads to the hard construction capacity; (2) the shapes of section steel just provide a weak confining effect on the core concrete; (3) the material utilization ratio of section steel is insufficiency, especially for the seismic behavior. Therefore, the reasonable section form has an important effect on various behaviors, including the mechanical behavior, the seismic behavior, the material utilization, and the construction convenience. Against this background, we designed and proposed the steel frame confined concrete column (SFC). Based on the systematic experimental studies on the mechanical behavior and the seismic behavior, the SFC specimens have great mechanical behavior because of the confining effect provided by the steel frame, and have great seismic behavior because of the reasonable distribution of the angle steel.

However, the experimental study on seismic behaviors of the SFC specimens in Chapter 8 is limited, due to the fewer specimens. It is necessary to establish the finite element models by a software and verify the proposed model by the experimental results. Then, the proposed model can further analyze the seismic behaviors of SFCs. Therefore, the influence factors for seismic behavior of SFCs can be further studied by the simulation analysis, including the size and arrangement of angle steel and steel battens, the axial compression ratio, and the composite confining effect.

Based on the experimental and theoretical studies, it can be found that SFC has the advantages of high bearing capacity, good ductility, and good seismic behavior. Meanwhile, the SFC has good convenient construction, so it can be widely applied in high-rise buildings, industrial buildings, and bridge structures. Based on existing studies and experiment studies in this book, it can be found that: the failure mode of SFC specimen belongs to the compression bending type; the steel ratio of angle steel and the axial compression ratio have a significant effect on the seismic behavior of SFCs; the confining effect will influence the relationship between the axial compression ratio and the seismic behavior. However, the above studies belong to the qualitative analysis, which only identifies the influence factors of the seismic behavior and the effects of various factors. In order to analyze the seismic behavior of SFCs quantitatively, a section analysis of the SFC specimens should be set to establish the axial force-moment-curvature computation model. A section numerical analysis of the SFC specimens can refer to the corresponding studies of the concrete

Concrete Composite Columns. DOI: https://doi.org/10.1016/B978-0-323-85171-8.00009-4

filled steel tubular and the steel-reinforced concrete column (Chen et al., 2006, 2014; Hosseini & Abbas, 2013; Julio et al., 2012; Mostafaei & Kabeyasawa, 2007; Parant et al., 2007; Wu et al., 2005). However, large parts of these studies are limited by oversimplified models or incomplete proposed parameters, so these models are inconsistent with the test results. Therefore, it is necessary to present the scientific section analysis of the SFC specimen, so as to establish the moment-curvature hysteresis model for the engineering application of SFCs.

Based on the section moment-curvature analysis and the seismic experiment results, the damage condition of the SFC specimen under the lateral cyclic loading, so as to gain the damage index model. We can not only use the damage index model to describe quantitatively the damage degree of the structures or members after earthquake action, but also quantify the behavior objectives which make up the important theoretical foundation for the safety evaluation of the damaged structures or members.

Many existing studies have proposed a series of damage index models, based on different parameters and different analysis methods. The damage index model proposed by Kratzig et al. (1989) introduced the energy method to analyze the loading process in the positive and negative directions. The study presented by Diao et al. (2008) proposed the cumulative damage model by the combination between the principle of energy dissipation and the characteristics of a hysteretic curve of specimens. In the damage index model proposed by Mehanny and Deierlein (2001), the influence of the loading process is considered by the cumulative deformation. The damage index model proposed by Park et al. (1985) is the double parameter model which simultaneously introduced the energy and the deformation. The study of Zhang et al. (2017) proposed the damage index model by the frequency parameter, so the damage degree can change with the lateral displacement rate.

Based on the above review, most of the existing damage index models are only proposed for the traditional reinforced concrete structures or members. And the cumulative damage caused by the loading process is an important part of these structures. However, there are many fundamental differences between the confined concrete member and the traditional reinforced concrete member. The studies presented by both Ji et al. (2013) and Qian et al. (2014) showed that the number of cycles and the cumulative damage have little effect on the bearing capacity and the damage condition of the concrete filled steel tube column. By reference to the seismic behavior of the concrete filled steel tube column, we presented the seismic test results of the SFC specimens in Chapter 8. It can be concluded that the energy method and the cumulative damage method are not suitable for the establishment of the damage index model of the SFC specimens, because the great confining effect on core concrete leads to the small reduction of bearing capacity and stiffness of SFC under the lateral cyclic loading. However, most existing damage index models are established by the displacement parameter and have poor theory and accuracy. Based on the experimental analysis, it can be concluded that the confining effect can improve the mechanical properties of the core concrete, so as to improve the seismic behavior of the compressive bending members, especially for the high axial compression ratio. It means that the control factors on the yield state and the ultimate state are different under variable axial compression ratio. The indefinite parameters make the damage index model based on the displacement cannot reasonably evaluate the confined concrete composite columns. It is necessary to fully study the mechanical properties of each component material in the loading process of the SFC specimens, then introduce a reasonable parameter to establish the accuracy damage index model.

Against this background, we establish the finite element model in OpenSEES and verify it by the seismic test results first. Then, we present the simulation analysis by the proposed model to further study the seismic behavior of the SFC specimens, including the main influence factor, the deformation characteristics, and the section deformation law. After that, we propose the section analysis method, based on the analysis results. And the section analysis method is used to define the stress state of each component material in each loading stage, identify the influencing factors for the damage degree of the SFC specimens, and accurately distinguish the deformation state of the compression bending members in the loading process. Based on the analysis results, we propose the calculation methods of section curvature and ductility. Finally, we establish the damage index model by the combination between the section ultimate curvature and the test results of seismic behavior of the SFC specimens in Chapter 8.

9.2 Finite element model establishment

Based on the test results, the various behaviors of the SFC specimens have been preliminarily analyzed, including the ductility, energy dissipation capacity, stiffness reduction, strength reduction and other seismic behavior indexes in Chapter 8, the mechanical behavior, and the damage developing process in Chapter 5. However, it is unrealistic to analyze comprehensively all influence factors by the experiment method, because of the limitations of the workload, the test condition, the research funds, and other conditions. With the rapid development of computer technology, the application of the finite element numerical analysis method in structural engineering is extensive and reasonable. The combination of the numerical simulation and the experimental study is one of the effective methods for the influence analysis on the seismic behavior of the composite columns.

At present, large general commercial finite element software (e.g., ABAQUS, ANSYS, and OpenSEES) has been widely used by scholars and achieved fruitful study results. The finite element model established by the ABAQUS and ANSYS has high calculation accuracy and can reflect the indexes such as stress, strain, and damage distribution of each specimen, especially for the strong postprocessing visualization ability. However, the two software has a long calculation time, low calculation efficiency, a heavy workload, and poor convergence. Especially for that, the simulation effect of seismic behavior of the reinforced concrete is not ideal (Wang, 2017). Based on the OpenSEES finite element analysis software (i.e., Open System for Earthquake Engineering Simulation), the finite element model can gain a powerful nonlinear solution ability, rich material models, and element types, so as to have high calculation efficiency. The OpenSEES has become one of the most commonly used finite element analysis software for structural engineers.

Similar to other finite element analysis software, for the establishment process in OpenSEES, it is necessary to define the constitutive model of each material, the element type, the section properties, the boundary conditions, the constraint forms, the node and element location coordinates, the loading method, the analysis iteration method, and so on. Because there is no mature preprocessing and postprocessing software, the model establishment and analysis need to be defined in the specific form of code writing. We will introduce these steps in the following sections.

9.2.1 Constitutive models of materials

(1) Constitutive model of concrete

The OpenSEES provides seven kinds of constitutive models to reflect the concrete properties (Chen, 2012). Based on the studies (Ji, 2008; Montuori & Piluso, 2009), it can be concluded that the confining effect of the steel frame on the core concrete is similar to that of stirrup hoops confined concrete, and the corresponding parameter calculation method can be modified based on the Mander model. Therefore, the "uniaxial Material Concrete 04" is used to reflect the SF confined concrete in the material library of the OpenSEES. Furthermore, the constitutive model is improved from the "Concrete 04" model based on the confining mechanism analysis of SFCs in Chapter 5. The equation of the axial compression skeleton curve is given as:

$$y = \left\{ \begin{array}{l} 2x + x^2 x \leq 1 \\ x/(\psi \alpha_d (x-1)^n + x) x \geq 1 \end{array} \right\} \tag{9.1}$$

where $y = f_c / f_{cc}$, f_c is the axial stress, f_{cc} is the peak stress of SF confined concrete; $x = \varepsilon_c / \varepsilon_{cc}$, ε_c is the axial strain, ε_{cc} is the peak strain of SF confined concrete; n and α_d are the decreasing parameters, respectively.

The peak stress (f_{cc}), the peak strain (ε_{cc}), and the corresponding decreasing parameters of SF confined concrete can be obtained in Chapter 5, based on two equivalent lateral confining stresses (f_{lx} and f_{ly}), which can be computed as follows:

$$f_{lx} = f_{ly} = k_e \rho_{sv} f_{yh} / 2 \tag{9.2}$$

where ρ_{sv} is the volume ratio of steel battens; f_{yh} is the yield strength of steel batten; k_e is the effective confining coefficient. It can be computed by confining mechanism analysis of SF confined concrete in Chapter 5, the equation is given as follows:

$$k_e = (1 - \sum_{i=1}^{4} w_i^2 / (6 b_c d_c))(1 - s/(2b_c))(1 - s/(2d_c))/(1 - \rho_{cc}) \tag{9.3}$$

where b_c is the distance between two angle steel centroids in x direction, d_c is the distance between two angle steel centroids in y direction, w_i is the clear distance between two-angle steel, s is the clear distance between two steel battens, and ρ_{cc} is the ratio of the section area of angle steel to the area of core concrete surrounded by the steel frame.

For the tension properties, the stress increasing stage is simplified as a linear curve, the elastic module at this stage is equal to the origin tangent modulus of the compressive constitutive model, and the peak tensile stress (f_t) is f_t=0.6228f_{co}. For the stress decreasing stage, the mathematical equation can be expressed as follows:

$$f = f_t((\beta f_t)/f_t)^{(\varepsilon-\varepsilon_t)/(\varepsilon_{tu}-\varepsilon_t)} \tag{9.4}$$

where f_t is the peak tensile strain, ε_t is the peak tensile strain, ε_{tu} is the ultimate tensile strain, and β is the ratio of ultimate tensile stress (corresponding to the ultimate strain) to peak tensile stress, which is usually 0.1.

(2) Constitutive model of steel

In the OpenSEES material library, two constitutive models of steel are provided. In this section, the "steel 02" model is chosen, and the command code in OpenSEES is "uniaxial material steel 02". The parameters of this model are simple and in good agreement with the basic properties of steel. In addition, the model can reflect the isotropic strain hardening and the Bauschinger effect of steel. The stress–strain relationship of the transition curve depends on the unloading point (σ_r, ε_r) under the stress state and the parameter R (Eom et al., 2014), and the mathematical expression is given as follows:

$$\sigma^* = b\varepsilon^* + ((1-b)\varepsilon^*)/(1-\varepsilon^{*R})^{1/R} \tag{9.5}$$

where b is the hardening coefficient of steel, R is the curvature coefficient of the transition curve, which can reflect the Bauschinger effect of steel, σ^* and ε^* are the normalized stress and the normalized strain. And the parameters (i.e., σ^*, ε^* and R) are calculated as follows:

$$\left\{\begin{array}{l} \sigma^* = (\sigma-\sigma_r)/(\sigma_0-\sigma_r) \\ \varepsilon^* = (\varepsilon-\varepsilon_r)/(\varepsilon_0-\varepsilon_r) \\ R = R_0 - a_1\xi/(a_2+\xi) \end{array}\right\} \tag{9.6}$$

where, R_0, a_1, and a_2 are the material constant, and ξ is the absolute value of cyclic plastic strain of upward displacement.

At the same time, the model can also describe the isotropic hardening of steel. Based on the study (Chen, 2014), the stress translation of linear yield asymptote is used to describe the isotropic hardening of steel. The translation value depends on the maximum plastic strain of the material. The relevant calculation equation can be found in the study (Zhu, 2013).

In "steel 02" model, the input parameters from elasticity to plasticity mainly include the steel yield strength, the elastic modulus and the strain hardening rate (i.e., $R_0, $cR_1 and $cR_2) of steel, and the isotropic hardening parameters (i.e., $a_1, $a_2, $a_3, $a_4, etc.). For these parameters, the yield strength and the elastic modulus are taken by the material test results, the ($R_0, $cR_1, $cR_2, $a_1, $a_2, $a_3, $a_4) are taken as the (19, 0.925, 0.15, 0, 1, 0, 1).

9.2.2 Element selection and section division

OpenSEES provides three kinds of nonlinear beam-column elements, including the beam-column element based on the stiffness method, the beam-column element based on the flexibility method, and plastic hinge element based on the flexibility method (Chen et al., 2011). The beam-column element based on the stiffness method has high calculation efficiency, so the model has a strong convergence rate after matching with the constitutive relations of various materials. Based on the experiment results, the failure mode of the SFC specimen is the bending failure. Therefore, the beam-column fiber element based on the stiffness method ("dispbeamcolumn") can be used in the numerical simulation (Gu & Huang, 2016), and the influence of shear deformation can be ignored. In order to gain the load–displacement relationship of the member, the following calculation steps are adopted: (1) the member is divided into several integral segments by the beam-column fiber element; (2) the displacement of the section at each integral point is obtained by the cubic Hermit polynomial interpolation; (3) the corresponding section deformation can be obtained by driving the shape function; (4) the section resistance force vector and the stiffness matrix are obtained by the section deformation and the section force-displacement relationship; (5) the element stiffness matrix is obtained by the integration along the length of the section stiffness matrix; (6) the element resistance force vector is obtained by the integration along the length of the section resistance force vector; (7) finally, the relationship between the force and the deformation of the member is obtained.

It should be noted that the subdivision element method is adopted in this chapter to reduce the error of the simulation model. Because the section displacement at each integral point of the element is obtained by cubic Hermit polynomial interpolation in the beam-column fiber element based on the stiffness method, it cannot well describe the curvature distribution of the element after the column end is in the yield state (Eom et al., 2014). Based on the experiment results, it can be found that the bottom of the composite column is the concentrated area of the elastic-plastic deformation, which should be subdivided by elements. And the size of element is an important factor for the simulation accuracy and calculation efficiency. In comprehensive consideration of the element convergence and the calculation efficiency, the SFC model is divided into seven elements along the height, the seven elements are the 0–100 mm height, the 100–200 mm height, the 200–300 mm height, the 300–450 mm height, the 450–600 mm height, the 600–800 mm height, and the 800–1050 mm height. Each height element has five Gauss integral points.

Except for the element division in the height direction, the column section needs to be divided into several fiber elements. Then the uniaxial stress–strain relationship is defined for each fiber element. The influence of shear deformation can be ignored, as each element fiber should follow the defined uniaxial stress–strain relationship. Based on the material properties in each section part of SFC, the concrete section can be divided into the core concrete and the concrete cover. The former should adopt the constitutive model of the SF confined concrete, and the latter should adopt the constitutive model of the unconfined concrete. It should be noted that the divided steel section just includes the angle steel, and there is no steel batten in the divided steel section because the steel battens cannot bear the axial load directly. And the confining effect provided by the steel battens is reflected in the constitutive model of the SF confined concrete. In OpenSEES, the command "section fiber" is used to define the position and size of the element fiber (Gu & Huang, 2016). The divided sections are shown in Fig. 9.1, where the fiber elements of the core concrete is named "CoreC", the fiber elements of the concrete cover is named "CoverC" and the fiber elements of the angle steel is named "SA".

9.2.3 Boundary conditions and loading steps

After the definition of material parameters and the division of section element fiber, the calculation process of the whole numerical simulation also includes the definitions of the boundary conditions, the loading steps, and the output variables.

In order to simulate the actual loading, this paper adopts two-dimensional fiber element section modeling: the command code "fix" is used to constrain the freedom degrees in all directions of the bottom node, which is kept as a fixed end; the three freedom degrees (i.e., the axial direction, the lateral direction, and the in-plane rotation) of the top node are not constrained.

In the finite element model, the analysis type of "gravity" + "pushover (cycle)" is used to simulate the composite columns under the low cyclic reciprocating load.

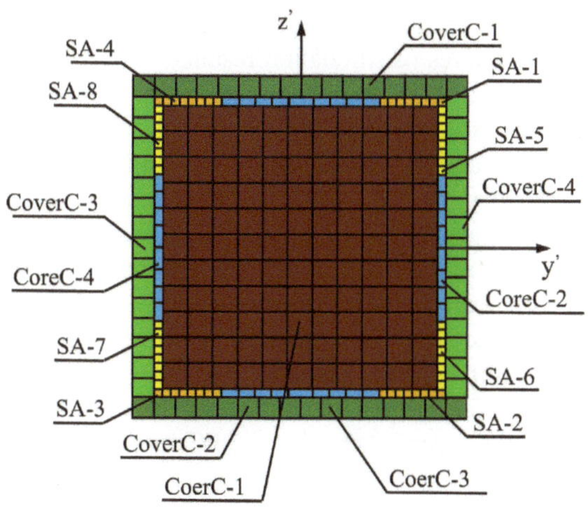

Figure 9.1 Divided sections of specimen diagrammatic sketch.

The following steps are adopted: (1) the constant axial load is applied at the top of the column to simulate the axial compression; (2) the axial load is controlled by the "loadconst" command code to keep constant in the subsequent analysis; (3) the lateral displacement is applied according to the loading history in the seismic experiments of SFC specimens, based on the study (Li, 2010); (4) it is necessary to use the "source" command to call two relevant subroutine codes, namely "LibAnalysisStaticParameters.tcl" and "LibGeneratePeaks.tcl", which are respectively used to define and calculate the corresponding parameters, such as the iterative algorithm, the loading mode, and the convergence criterion, so as to get the characteristic points on each hysteresis curve; (5) the "cycletype" of the calculation analysis type is defined as "full" to reflect a complete hysteretic analysis; (6) the command code "geomTransf PDelta 1" is used to reflect the influence of the second-order effect (P-Delta) of the axial force in the coordinate transformation.

The output variables of the simulation model are the lateral displacement of the top element and the internal force of the bottom element. Then the hysteretic curve of the composite column can be obtained by the two simulation calculation results.

9.2.4 Verification of the proposed finite element model

Based on the specimen data (i.e., specimen size, section forms, and material properties) in Chapter 8, the corresponding finite element models can be established in OpenSEES. After the simulation calculation, the comparisons between the simulation results and the test results will verify the proposed finite element model, including the hysteretic curve and the skeleton curve. It should be noted that six specimens, excluding the SFC-5 and SFC-6, are used to verify the proposed model, because of the simple and regular section forms of these specimens.

(1) Hysteretic curve

All SFC specimens are simulated in OpenSEES, and the hysteretic curves are used to verify first. The simulation results and test results are compared in Fig. 9.2. Based on the comparison of the hysteretic curve, we can conclude that: (1) the shapes of the hysteretic curves of simulation results and test results show a similar trend; (2) the characteristic behavior of the finite element models is little different from the corresponding experimental specimens, including the structure deformations, the lateral stiffness, and the bearing capacity; (3) the computational hysteretic curves have well symmetry between push and pull directions, but the experimental specimens under large axial compression ratio (i.e., SFC-2 and SFC-4) show obvious asymmetry between push and pull directions. The reason for the above simulation errors is that the experimental specimen cannot reach the ideal state in software, including the model size, the material properties, the boundary conditions, and the contact relationship between two different materials. Therefore, the simulation results can basically fit with test results, and the finite element models can be used to further analyze the seismic behavior of SFC specimens.

(2) Skeleton curve

The comparisons of the load–displacement skeleton curve between simulation results and test results are shown in Fig. 9.3. It can be seen that the developing

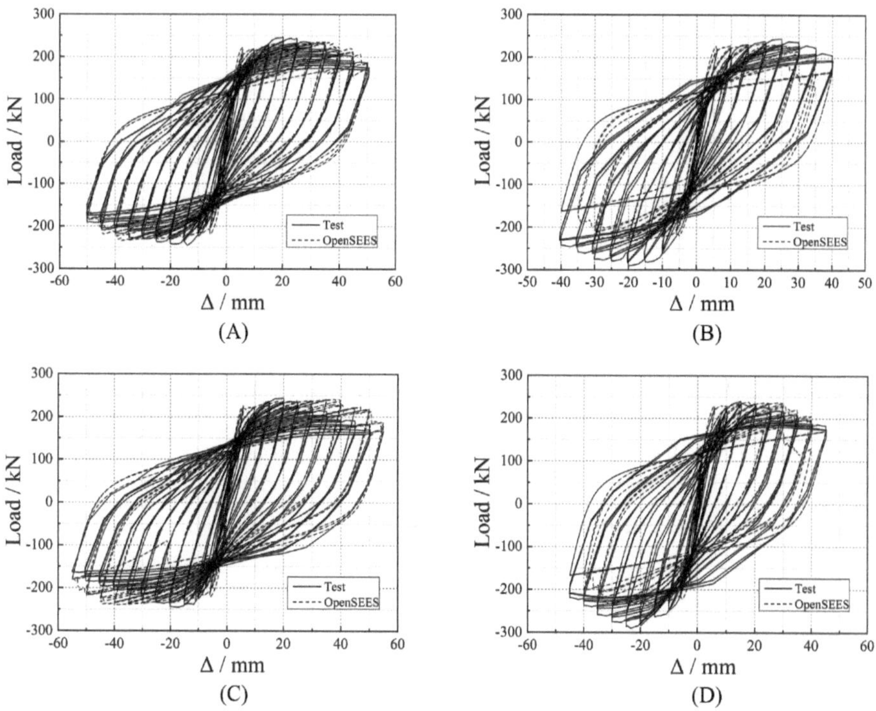

Figure 9.2 Model verification based on hysteretic curves. (A) SFC-1; (B) SFC-2; (C) SFC-3; (D) SFC-4.

trend of simulation curves is similar to that of test curves, and the bearing capacity of the finite element models is basically consistent with that of the specimens (the error is less than 15%). However, there are still some differences between the two results: the initial stiffness of the simulation curve is slightly larger than the test curve; the experimental specimens with a large axial compression ratio show asymmetry behavior between the push and pull directions. The main reason is that the finite element model selects the idealized parameters which differ from the actual state. And the complete fixation of the bottom beam cannot be achieved during the test. In summary, the skeleton curves obtained by simulation results are in good agreement with that of the test results. Therefore, the accuracy and effectiveness of the proposed model are verified, and further parameter analysis can be carried out.

9.3 Finite element model analysis

Based on the finite model, further analysis of the SFCs can be carried out to identify the mechanical mechanism and the influence factors. Therefore, the stress state of the angle steel and the strain state of the core concrete are analyzed first to

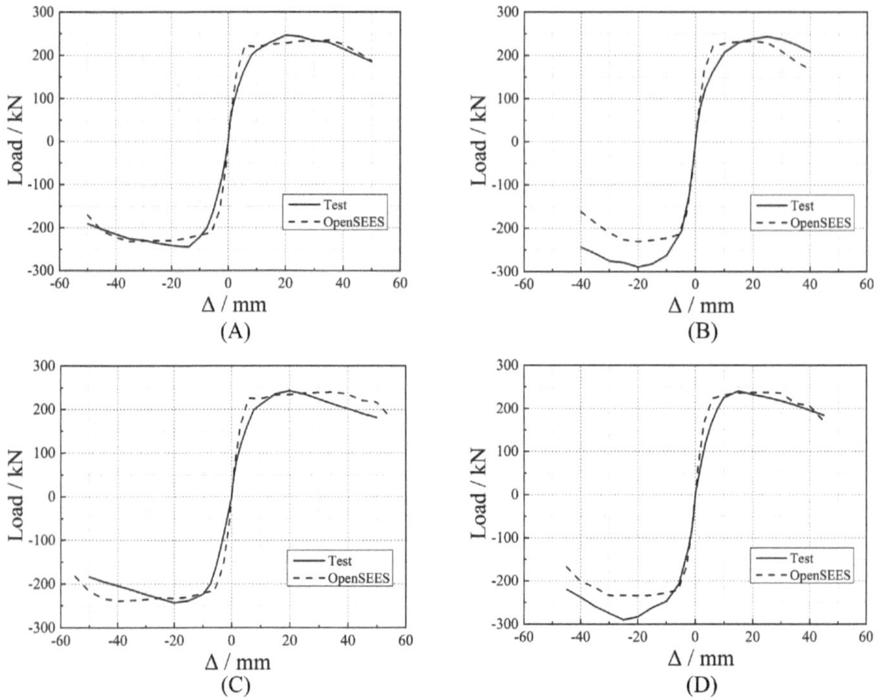

Figure 9.3 Model verification based on skeleton curves. (A) SFC-1; (B) SFC-2; (C) SFC-3; (D) SFC-4.

identify the mechanical mechanism of the SFC member under seismic loading. Then the section moment-curvature relationship is studied, which can be the foundation for analyzing the equivalent plastic hinge length of the SFCs. Finally, the parameter analysis is carried out for the seismic behavior of the SFCs. Furthermore, these study results can be the foundation for the section analysis and the damage evaluation in the following sections.

9.3.1 *Strain state of the angle steel*

In order to analyze the stress−strain state of each component at each loading stage. We take the SFC-1 specimen as an example to analyze the strain state of angle steel, so the relationships between the fiber strain of angle steel and the lateral top displacement of the column are shown in Fig. 9.4, in which the fiber strain is measured at the centroid of the angle steel section at 150 mm height or 300 mm height. In Fig. 9.4, the triangle point represents that the column reaches the lateral displacement corresponding to the yield load, the square point represents that the column reaches the lateral displacement corresponding to the peak load, the circle point represents that the column reaches the lateral displacement corresponding to the ultimate load, and the dotted line represents the yield strain of the angle steel.

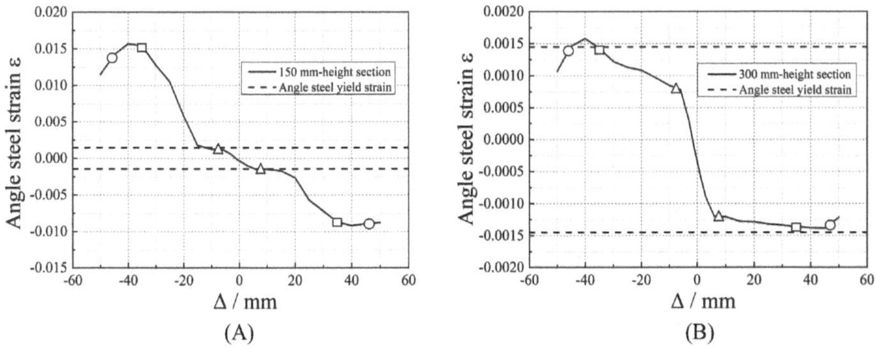

Figure 9.4 Fiber strain of angle steel. (A) 150 mm height; (B) 300 mm height.

Based on the comparisons in Fig. 9.4, it can be concluded that the strain of the angle steel at 150 mm height is much larger than that at 300 mm height. When the column reaches the yield displacement, the strain at 150 mm height has reached or is close to reaching the yield state, but the strain at 300 mm height is much less than the yield strain of the angle steel. When the column reaches the peak displacement, all strains of the angle steel below 300 mm height reach the yield state. Therefore, the strain analysis results show that the angle steel can play the full mechanical properties in the section form, which means SFCs have high material utilization. It should be noted that the tensile strain of the angle steel in each SFC model decreases slightly before the ultimate displacement and the compressive strain at 300 mm height also decreases slightly after the ultimate displacement. The main reason is that the stress–strain state develops in the following steps: (1) the edge of the bottom compressive section of the core concrete reaches the ultimate compressive strain in the late loading stage; (2) the compressive stress provided by the core concrete decreases rapidly and failure to bear the axial load, based on the constitutive model of "Concrete 04" in OpenSESS; (3) then the local buckling of the angle steel occurs at the bottom compressive area, and the curvature of the bottom section increases rapidly which leads to the large variation rate of displacement at top section; (4) the decreasing compressive stress leads to the decreasing tensile stress provided by the angle steel, due to the plan section assumption, so the tensile strain of angle steel in each SFC model decreases; (5) in the late loading stage, the compressive strain of the angle steel above the plastic hinge zone will decrease, because of the decrease of bearing capacity of SFC specimen.

9.3.2 Stress state of core concrete

We also take the SFC-1 specimen as an example to analyze the stress state of the core concrete. The relationships between the fiber stress of the core concrete and the lateral top displacement of the column are shown in Fig. 9.5, in which the fiber stress is measured at the edge fiber of the core concrete section at 50 mm height or

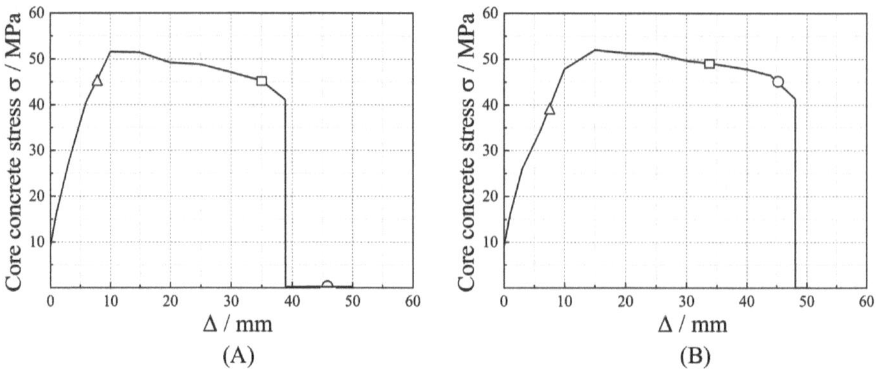

Figure 9.5 Fiber stress of core concrete. (A) 150 mm height; (B) 300 mm height.

100 mm height. In Fig. 9.5, the triangle point, square point, and circle point represent the same meaning as the points in Fig. 9.4.

Based on the comparisons in Fig. 9.5, it can be concluded that the stress of the core concrete still increases before the yield state of the column, and the stress basically maintains the stress range from 45 to 52 MPa before the peak state of the column. Therefore, the properties of the core concrete have been improved by the confining effect of the steel frame, which contributes to the seismic behavior of SFC specimens. After the peak load, the edge of core concrete at 50 mm height reaches the ultimate compressive strain, followed by the failures before the lateral ultimate displacement. But the core concrete at 100 mm height doesn't reach the failure strain during the whole loading process. Meanwhile, the simulation results can satisfy the failure process and the failure mode of the SFC specimen in Chapter 8.

9.3.3 Section moment-curvature relationship

The curvature of the column bottom at each loading stage cannot be effectively measured, because the concrete cover peels off rapidly. Therefore, the finite element model is used to gain the moment-curvature hysteretic curve and skeleton curve at the bottom section. We take the SFC-1 and SFC-2 specimens as two examples to analyze and summarize the corresponding curves in Fig. 9.6. In the existing studies (Li et al., 2016; Su et al., 2014; Yang et al., 2013; Zhang et al., 2001), the curvature and the equivalent plastic hinge length are taken the average value of the bottom section. Therefore, we take the average curvature from the two-element heights at the column bottom.

It can be concluded from Fig. 9.6 that the moment-curvature hysteretic curves are similar to the load–displacement hysteretic curves. The hysteretic curves are relatively full, due to the inner confining effect. Therefore, the plastic hinge of the SFC column has good energy dissipation capacity and rotation capacity. And the rotational capacity becomes worse with the increase in axial compression ratio.

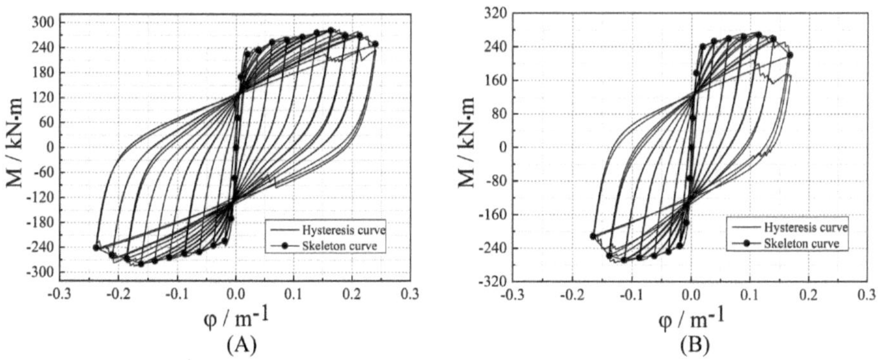

Figure 9.6 Section moment-curvature curves. (A) SFC-1; (B) SFC-2.

9.3.4 Equivalent plastic hinge length

We calculate the equivalent plastic hinge length by the combination between the lateral displacement and the curvature integration method along the height direction (Yang et al., 2013). For the cantilever column, the lateral top displacement can be integrated, and the equation is given as follows:

$$\Delta = \int_0^l \varphi(x)xdx \tag{9.7}$$

When the column is in yield state, the curvature shows the nearly linear distribution along column height. Therefore, the yield top displacement can be computed as follows:

$$\Delta_y = \phi_y l^2/3 \tag{9.8}$$

When the column is in the ultimate state, the curvature shows the nonlinear distribution along column height. Based on the study presented by Park and Paulay (1995), the equivalent plastic hinge length (l_p) is introduced. It is assumed that the distribution of curvature can be divided into two parts: the yield curvature (φ_y) which is linearly distributed along column height; the plastic curvature (φ_p) which is uniformly distributed at the equivalent plastic hinge zone. Therefore, the ultimate top displacement can be computed as follows:

$$\Delta_u = \Delta_y + \Delta_p = 1/3\phi_y l^2 + (\phi_u - \phi_y)l_p(l - l_p/2) \tag{9.9}$$

Then the plastic top displacement can be calculated as:

$$\Delta_p = \Delta_u - \Delta_y = (\phi_u - \phi_y)l_p(l - l_p/2) \tag{9.10}$$

In the calculation equation of the equivalent plastic hinge length, the displacement and curvature in the yield state and the ultimate state can be obtained in Chapter 8. Based on the computational method, we computed the equivalent plastic hinge length of all SFC specimens, all values are about 230 mm. In addition, the existing study (Jiang & Su, 2008) concluded that the composite column has a linear relationship between the equivalent plastic hinge length and the equivalent section height. Therefore, the equivalent plastic hinge length (l_p) of SFC is 0.88 times of the equivalent section height, $l_p = 0.88h_0$.

9.3.5 Parameter analysis

In order to expand the study results of seismic behavior, we presented the parameter analysis using the finite element model. The influence factors include the axial compression ratio, the angle steel size, the arrangement of steel battens and the bottom steel batten size. Each model is given a name (e.g., SFC-S1) which starts with letters ("SFC" for the column type), a letter after the dash ("S" for simulation), and then followed by a number to represent each specimen.

9.3.5.1 Axial compression ratio

(1) Steel distribution form of ASTC-1

We establish four models which have the same steel distribution form as the SFC-1 specimen in Chapter 8, and the axial compression ratios are 0.30, 0.42, 0.52, and 0.6, namely from SFC-S1 to SFC-S4. The skeleton curves of the four models are shown in Fig. 9.7.

For these established models, the SFC-S1 and SFC-S2 have higher symmetry between pull and push directions, and the SFC-S3 and SFC-S4 have lower

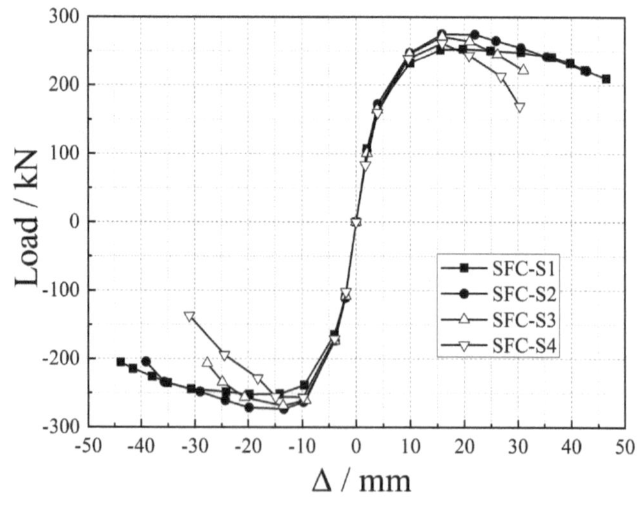

Figure 9.7 Effect of axial compression ratio on models for SFC-1. Parameter analysis.

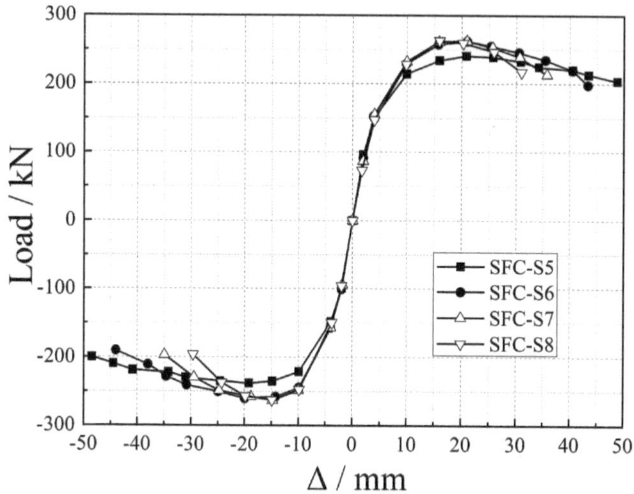

Figure 9.8 Effect of axial compression ratio on models for SFC-3. Parameter analysis.

symmetry. Based on the simulation results, the following conclusions can be drawn: (1) the higher axial compression ratio leads to lower ductility; (2) with the increase in the compression ratio, the peak load increases before the ratio is 0.42; (3) the corresponding peak loads of ratios are 0.42 and 0.52 are similar; (4) with the increase in the compression ratio, the peak load decreases after the ratio is 0.52. The conclusions show that the dividing axial compression ratio of the maximum peak load is between 0.42 and 0.52.

(2) Steel distribution form of SFC-3

In order to analyze the specimen under high confining effect, we establish four models which have the same steel distribution form as the SFC-3 specimen, and the axial compression ratios are 0.30, 0.42, 0.52, and 0.6, namely from SFC-S5 to SFC-S8. The skeleton curves are shown in Fig. 9.8.

Fig. 8 shows all models have high symmetry between pull and push directions, which shows that the well-confining effect can improve the stability of the column. Based on the simulation results, the following conclusions can be drawn: (1) the higher axial compression ratio leads to lower ductility; (2) the peak load still increases with the increase in axial compression ratio; (3) the ratio range from 0.52 to 0.6, the increase in peak load is small. The conclusions show that the dividing axial compression ratio is about 0.6. The comparisons of the above models show that the well-confining effect can lead to the good symmetry of the skeleton and the higher dividing axial compression ratio.

9.3.5.2 *Angle steel size*

We establish four models to study the effect of the angle steel size, namely from SFC-S9 to SFC-S12. The design details of these models are shown in Table 9.1 and the axial compression ratio is 0.3.

Table 9.1 Design details of SFC-S9 ~ SFC-S12.

Item	Angle steel (mm)			Steel batten (mm)		
	Length	Thickness	Ratio (%)	Width	Spacing	Thickness
SFC-S9	80	8	5.41	100	100	6
SFC-S10	100	8	6.83	100	100	6
SFC-S11	60	10	4.89	100	100	6
SFC-S12	60	12	5.76	100	100	6

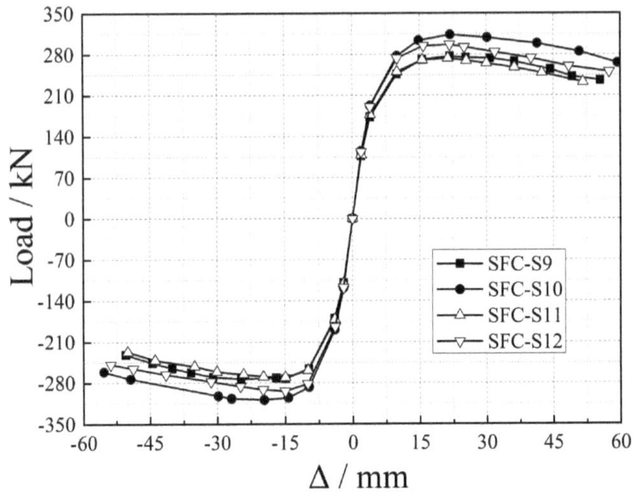

Figure 9.9 Effect of angle steel size. Parameter analysis.

The skeleton curves are shown in Fig. 9.9. The simulation results show higher steel ratio of angle steel has more serious asymmetry and higher symmetry between pull and push directions. Meanwhile, it can be concluded that: (1) with the same steel ratio of the angle steel, the bearing capacity and ductility of SFC-S10 are higher than that of SFC-S12; (2) although the steel ratio of the angle steel in SFC-S11 is lower than that in SFC-S9, the two skeleton curves have basically coincided. Therefore, the increase in the angle steel thickness is an effective measure to improve the seismic behavior of SFCs.

9.3.5.3 Arrangement of the steel battens

In order to study the seismic behavior of SFCs with different arrangements of steel battens, we establish another two models (namely, SFC-S13 and ASFC-S14), then compared them with two established models. The design details of the four models are shown in Table 9.2 and the axial compression ratio is 3.0. It should be noted that the arrangement of the steel battens will influence the confining effect provided

Table 9.2 Design details of SFC-S13 ~ SFC-S14.

Item	Angle steel (mm)		Steel batten (mm)		
	Length	Thickness	Width	Spacing	Thickness
SFC-S13	60	8	135	65	6
SFC-S14	60	8	150	50	6

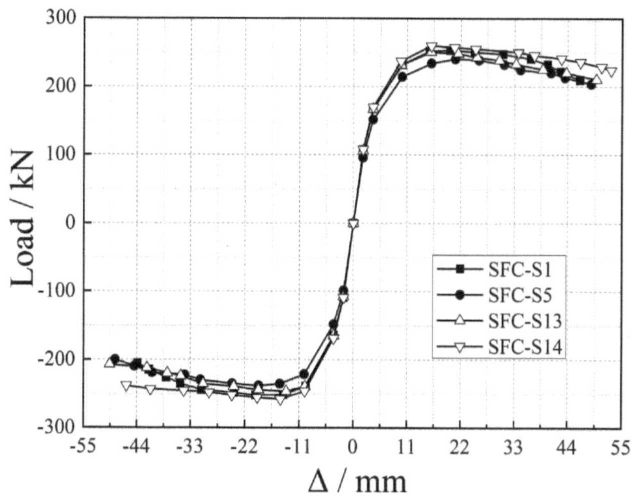

Figure 9.10 Effect of steel batten. Parameter analysis.

by the steel form, and they can be in increasing order: SFC-S1, SFC-S5, SFC-S13, and ASFC-S14.

The skeleton curves are shown in Fig. 9.10. The simulation results show higher confining effect leads to good symmetry between pull and push directions. Meanwhile, it can be concluded that: the confining effect has little effect on the bearing capacity under the same axial compression ratio; the model in a higher confining effect has the higher ductility. In addition, the higher confining effect can lead to the higher dividing axial compression ratio of SFCs, which means that the confining effect can influence the bearing capacity indirectly.

9.4 Section analysis model

In this section, the section analysis model is used to analyze the seismic behavior of SFCs quantitatively. We can establish the section analysis model by the layered finite element method. And the layered finite element in the SFC section is shown in Fig. 9.11. And the establishment process includes the following steps: (1) the

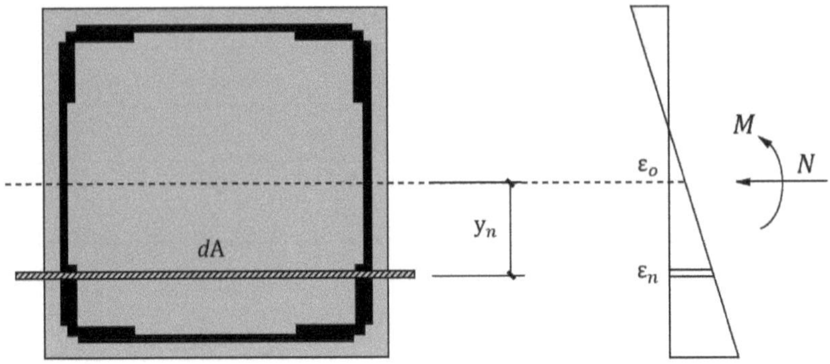

Figure 9.11 Layered finite element in SFC section. Diagrammatic sketch.

section is divided into several element layers; (2) the strain of each element layer is computed; (3) the axial force and the bending moment of each element layer are computed: (4) the axial force and the bending moment of the section are summarized and then drawn into the target curves, including the moment-curvature curves and the axial force-moment curves.

Furthermore, several basic assumptions should be followed to simplify the analysis method: (1) the plane section assumption, which means the axial deformation is linearly distributed along the loading direction which is perpendicular to the central line; (2) the shear slip between the steel frame and concrete is ignored, which means the section is in the whole; (3) the shear deformation is ignored in the compression-bending analysis, because of the typical bending failure of the SFC specimens.

9.4.1 Constitutive models of materials

In the section analysis models of SFCs, three types of materials should be defined: the angle steel adopts the constitutive model of steel, the concrete cover adopts the constitutive model of the unconfined concrete, and the core concrete adopts the constitutive model of the SF confined concrete.

(1) Steel constitutive model

The constitutive model of the steel is the bilinear increasing model, which has an elastic stage with the initial elasticity modulus (E_s) and a hardening stage with the hardening modulus which is equal to 0.01 E_s. And the elasticity modulus (E_s) can be attained in the material test.

(2) Constitutive model of unconfined concrete

The constitutive model of the unconfined concrete is proposed by Guo and Shi (2003), the compressive stress–strain curve can be calculated as follows:

$$y = \left\{ \begin{array}{l} \alpha_a x + (3 - 2\alpha_a)x^2 + (\alpha_a - 2)x^3 x \leq 1 \\ x/(\alpha_d(x-1)^2 + x)x \geq 1 \end{array} \right\} \tag{9.11}$$

where $x=\varepsilon/\varepsilon_{co}$, $y=\sigma/f_{co}$, f_{co} is the axial compressive strength of concrete, ε_{co} is the corresponding strain, α_a is the increasing parameter, α_d is the decreasing parameter, and the calculation method is showed in study (Guo & Shi, 2003).

In addition, the tensile stress–strain curve can be calculated as follows:

$$y=\begin{Bmatrix} 1.2x+0.2x^6 x\leq 1 \\ x/(\alpha_t(x-1)^{1.7}+x)x\geq 1 \end{Bmatrix} \tag{9.12}$$

where $x=\varepsilon/\varepsilon_t$, $y=f/f_t$, f_t is the axial tensile strength of concrete, ε_t is the corresponding strain, α_t is the parameter of decreasing section, and the calculation method is shown in study (Guo & Shi, 2003).

(3) Constitutive model of SF confined concrete

It has been known that the confining effect provided by the steel frame can improve the mechanical behavior of the core concrete. The compressive constitutive model has been improved and proposed by the author in Chapter 5. The compressive stress–strain curves can be calculated as follows:

$$y=\begin{Bmatrix} \alpha_a x+(3-2\alpha_a)x^2+(\alpha_a-2)x^3 x\leq 1 \\ x/(\psi\alpha_d(x-1)^n+x)x\geq 1 \end{Bmatrix} \tag{9.13}$$

where $x=\varepsilon/\varepsilon_{cc}$, ε is strain, $y=\sigma/f_{cc}$, σ is stress, α_a is the increasing parameter ($\alpha_a=2.4-0.0125\sigma_{cc}$), n is the decreasing slope coefficient ($n=1+\exp(-3\omega)$), ψ is the correction coefficient for the nonuniform confining effect ($\psi=2.3/\omega$), ω is the confining coefficient ($\omega=p/f_{co}$), and α_d is the decreasing parameter, which can be calculated as follows:

$$\begin{Bmatrix} \alpha_d=3d/(17(d-1)^n) \\ d=0.0033[1+2.18\times 10^5(2\omega f_c\varepsilon_c c/f_y)^{1.2}]/\varepsilon_{cc} \end{Bmatrix} \tag{9.14}$$

where f_{cc} is the compressive strength of the confined concrete, ε_{cc} is the corresponding strain, and they can be calculated as follows:

$$\begin{Bmatrix} f_c c=\gamma_u f_c+kf_l \\ \gamma_u=1.67L^{-0.112} \\ k=1+1/(3.5\omega^{0.7}) \end{Bmatrix} \tag{9.15}$$

$$\begin{Bmatrix} \varepsilon_{cc}=\varepsilon_c+\varepsilon_p \\ \varepsilon_c=[700+172\sqrt{f_{co}}+(k-1)p]\times 10^{-6} \\ \varepsilon_p=2400\times(2\omega)^{0.2}\times 10^{-6} \end{Bmatrix} \tag{9.16}$$

In the tensile state, the steel frame has no confining effect on the core concrete. Therefore, the tensile constitutive model of the core concrete is the same as the unconfined concrete.

9.4.2 Section analysis model of axial force-moment-curvature (N-M-φ)

9.4.2.1 Model establishment

The calculation sketch of the section analysis model is shown in Fig. 9.11. Each transient state can be calculated by the following process. First, a known axial strain of the central line (ε_o) is assumed. Second, an assumption curvature (φ) is given. Third, the axial strain of the n-th section element layer (ε_n) is calculated, based on the plane section assumption and the known parameters, including the known axial strain (ε_o), the rotation angle (θ), and the corresponding curvature radius (ρ). There is a relationship between these parameters, which is given as follows:

$$d\theta = ds/\rho = (\varepsilon_n - \varepsilon_o)ds/y_n \tag{9.17}$$

Meanwhile, the curvature equation is $\varphi = d\theta/ds$. Therefore, the strain of each section element layer can be computed as follows:

$$\varepsilon_n = \varepsilon_o + y_n\phi \tag{9.18}$$

It should be noted that the area of each section element layer (dA) can be computed as follows:

$$dA = dA_c + dA_s \tag{9.19}$$

where dA_c is the concrete area in one section element layer, and dA_s is the angle steel area in one section element layer.

Fourth, the stress of each material in one section element layer can be calculated by the axial strain (ε_n) and the corresponding constitutive models. Fifth, the axial force (N) can be computed as follows:

$$N = \sum\nolimits_1^n (f_c dA_c + f_s dA_s) \tag{9.20}$$

If the computation axial force is not equal to the application axial compression, the calculation process will turn back to the second step, and the new assumption curvature (φ) will be set and tried again. If the computation axial force is equal to the application axial compression, the calculation process will go to the next step. Sixth, the moment (M) can be computed as follows:

$$M = \sum\nolimits_1^n (f_c y_c dA_c + f_s y_s dA_s) \tag{9.21}$$

Based on the above calculation process, the assumption curvature (φ), the axial force (N), and the moment (M) can be gained and then drawn into the target curves, including the moment-curvature curves and the axial force-moment curves.

9.4.2.2 Model verification

Based on the proposed model, the section moment-curvature ($M-\varphi$) curves can be calculated for six SFC specimens in the seismic experiment. And the detail of each SFC specimen is shown in Chapter 8. After that, the comparisons between computation curves and test curves are shown in Fig. 9.12.

Based on the comparisons of the section moment-curvature curves, the proposed model can provide accurate calculation predictions for most SFC specimens. However, a few models do not fit well with the specimens. For the SFC-4 and the SFC-7, the proposed model curves cannot fit with test curves in the pull direction, the most error of SFC-4 is 13%, and the most error of SFC-7 is 14%. This is because there are many errors in the specimen production and the loading process. Therefore, the proposed model is basic and accurate and can be used for parameter analysis. And the parameter analysis of the moment-curvature relationship and the axial force-moment relationship will be presented in the following sections.

9.4.2.3 Parameter analysis of moment-curvature relationship

(1) Basic behavior

We study the moment-curvature curves of the SFC model under different axial compression ratios in this section. The material properties and section form of the established models are the same as that of SFC-1. Based on simulation results, the moment-curvature curves are shown in Fig. 9.13. Then, the following conclusions can be drawn: (1) all moment-curvature curves have two stages, the first stage is an increasing curve, and the second stage shows different trends under various axial compression ratios; (2) under the low axial compression ratio, the second stage shows a gently increasing trend; (3) under the high axial compression ratio, the

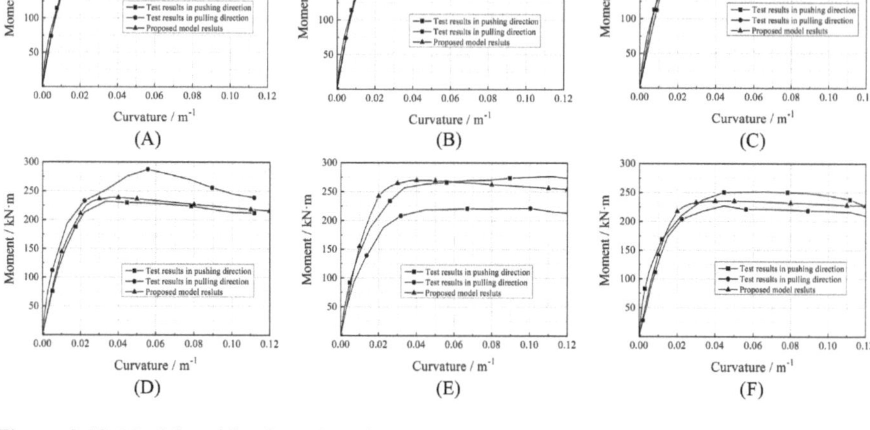

Figure 9.12 Model verification of section moment-curvature curves. Model verification.

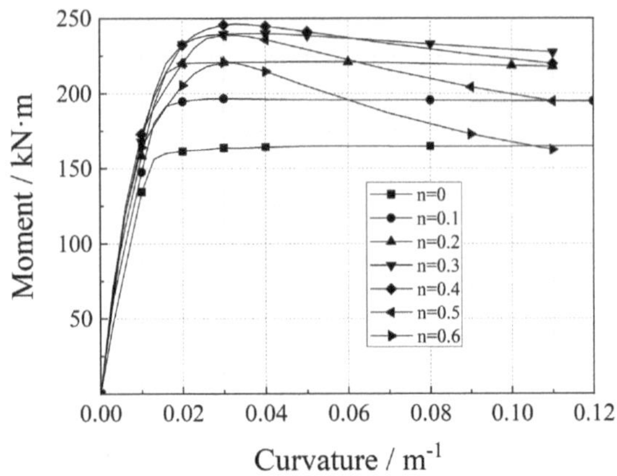

Figure 9.13 Moment-curvature curves of SFC-1 model under different axial compression ratios. Parameter analysis.

second stage shows a gently decreasing trend, and the slope increases with the increase in the axial compression ratio; (4) the peak moment increases with the increase in the axial compression ratio at first, and the peak moment reaches maximum peak moment at the ratio of 0.4, namely dividing ratio, the peak moment is 1.49 times than the peak moment at 0-ratio; (5) with the increase in the axial compression ratio, the curvature corresponding to the peak moment increases gradually and tends to a stable value.

(2) Steel ratio of angle steel

Based on the seismic test results in Chapter 8, the steel ratio of the angle steel is the most important factor for the seismic behavior of SFCs. Therefore, the SFC models with more steel ratio should be studied. And the thickness of the angle steel in the established model is changed from 8 to 12 mm. The moment-curvature curves of the established models under different axial compression ratios are shown in Fig. 9.14. Based on the curve comparisons, the following conclusions can be drawn: (1) the SFC model with a higher steel ratio has a higher section comment; (2) the high steel ratio of the angle steel can improve the whole seismic behavior, but has little effect on the trend of the curves; (3) the dividing axial compression ratio is 0.4, the peak moment is 1.34 times than 0-ratio, which means the increase in steel ratio of the steel angle can reduce the effect of axial compression ratio on section moment.

(3) Confining effect

Based on experimental studies in Chapters 5 and 8, the steel frame can provide well-confining effect on the core concrete to improve the axial mechanical behavior of the core concrete, then the core concrete can further influence the seismic behavior of the SFC specimens. Therefore, we study the confining effect in this section, the confining coefficient (ω) of the established model is changed from 0.52 to 0.99. Based on the simulation results, the moment-curvature curves of established models under different axial compression ratios are shown in Fig. 9.15. Based on the curve comparisons, the following conclusions can be drawn: (1) the increase in the confining effect has little

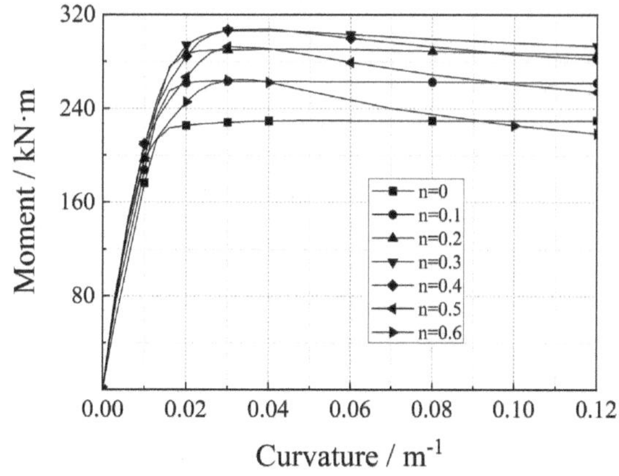

Figure 9.14 Moment-curvature curves of the model with large angle steel under different axial compression ratios. Parameter analysis.

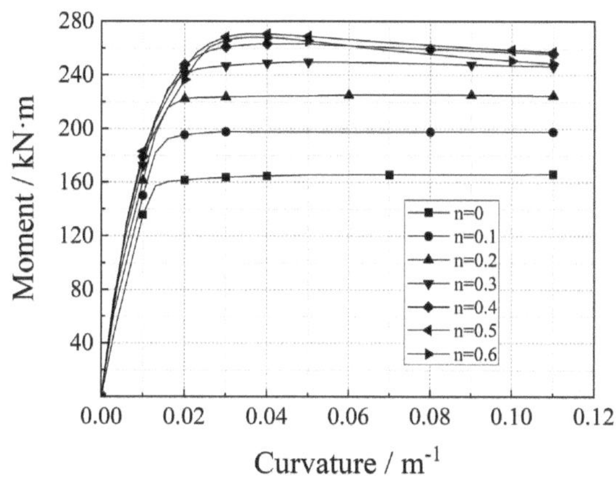

Figure 9.15 Moment-curvature curves of model with large confining effect under different axial compression ratios. Parameter results.

effect on the peak section moment, but it can obviously improve the development trend of the section moment and increase the axial compression ratio of the curves without decreasing stage; (2) the larger confining effect leads to the higher dividing axial compression ratio, the dividing ratio is 0.5 in the established models; (3) the increase in the confining effect can slow down the slope of the decreasing stage in moment-curvature curves and reduce the affecting of axial compressive ratio.

9.4.2.4 Parameter analysis of axial force-moment relationship

We can also get the axial force-moment relationship using the proposed section analysis model. It should be noted that the section moment corresponding to an axial force is determined by a judgment criterion. The section moment should select the minimum value in the following values: (1) the peak moment of the member section; (2) the moment corresponding to the peak curvature of the pure bending model. Based on the simulation results by OpenSEES, we can conclude that: under the low axial compression ratio, the model belongs to the large eccentric compression state, and the section moment will increase with the increase in the axial compression ratio; under the high axial compression ratio, the model belongs to the small eccentric compression state, and the section moment will decrease with the increase in the axial compression ratio. We will study the above developing trend in the following sections, and the influence factors are the steel ratio of angle steel and the confining effect.

(1) Steel ratio of angle steel

Under the same steel ratio, the increase in the thickness of angle steel can lead to higher seismic behavior of SFCs than that in the side length of angle steel, based on the simulation results by OpenSEES. Therefore, we study the thicknesses of angle steel in this section, and the thicknesses of angle steel in the established models are 8, 10, and 12, respectively. The axial force-moment curves of the three models are shown in Fig. 9.16. Based on the curve comparisons, the following conclusions can be drawn: (1) under low axial compression ratio, the axial force-moment curve shows the quadratic function trend; (2) under high axial compression ratio, the axial force-moment curve shows the linear trend; (3) the three kinds of axial force-moment curves show similar trend; (4) higher steel ratio of angle steel leads to the larger covered area, which means the SFC with higher steel ratio has the higher peak moment at the same axial compression ratio; (5) the dividing axial compression ratios between large and small eccentric compression types of the three SFC models are similar, ($n = 0.4$).

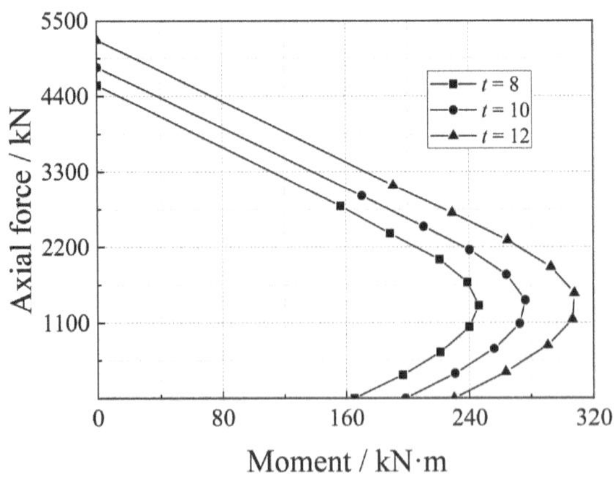

Figure 9.16 Axial force-moment curves of models with different angle steels. Parameter analysis.

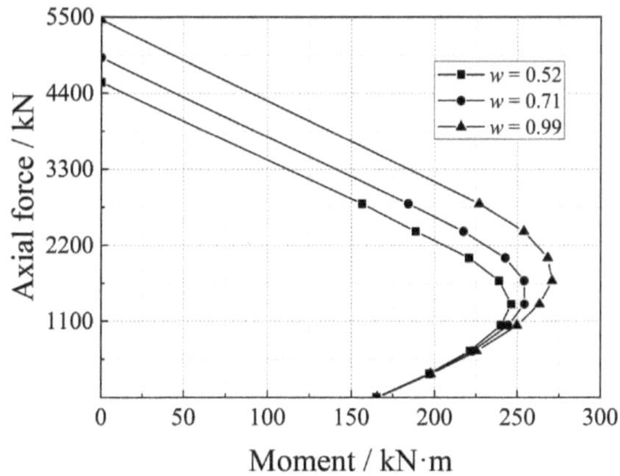

Figure 9.17 Axial force-moment curves of models with different confining effects. Parameter analysis.

(2) Confining effect

It has been known that the confining effect has an obvious effect on the SFC under a higher axial compression ratio. To study the confining effect, we establish three models in this section, the various confining coefficients are 0.52, 0.71, and 0.99, separately. Based on the simulation results, the axial force-moment curves of the three models are shown in Fig. 9.17. Based on the curve comparisons, the following conclusions can be drawn: (1) the confining effect has an obvious effect on the axial force-moment curve, especially for the high axial compression ratio, because of the second-order effect of axial load (i.e., $P-\Delta$ effect) and the improved mechanical behavior of the core concrete; (2) the confining effect can improve the axial bearing capacity but has no effect on the pure bending moment; (3) under the high axial compression ratio, the confining effect can obviously improve the section moment; (4) the dividing compression ratio increases with the increase in the confining effect, which means SFCs with well-confining effect are likely to stay in the large eccentric compression state in the seismic loading.

9.5 Calculation method of the axial force-moment-curvature ($N-M-\varphi$)

In the elastic-plastic seismic analysis of composite columns, it is necessary to calculate the characteristic values of member M-hinge and PMM-hinge, based on the axial force-moment-curvature relationship in the column section and the corresponding section hysteresis model (Guo & Shi, 2003; Lin, 2013; Qi et al., 2013). Therefore, it is necessary to propose the calculation method of the axial force-moment-curvature relationship of the SFC section, then establish the section moment-curvature hysteresis model, which can be the foundation for the nonlinear finite element analysis of the SFC structure.

9.5.1 Calculation method of the axial force-moment curve

Based on all studies on seismic behavior of SFCs in this book, the axial compression ratio has a great effect on the peak section moment of SFC under the compression-bending load: when the axial compression ratio is small (the section is in a small eccentric compression state), the peak section moment increases with the increase of axial compression ratio; when the axial compression ratio is large (the section is in large eccentric compression state), the peak section moment decreases with the increase of axial compression ratio; with the increase of confining effect, SFCs are much likely in the large eccentric compression state. Based on the analysis of the moment-curvature relationship and the axial force-curvature relationship, the influences of the steel ratio and the confining effect are contemplated, and then the calculation method of axial force-moment of SFCs is proposed. According to the shape of the axial force-moment curves, the limit point between the large and small eccentric compression state is defined as the dividing axial compression ratio, which can be the foundation to establish the piecewise equations for the axial force-moment curve. In order to simplify the calculation method, we introduce the axial compressive force (N_u) and the peak moment (M_u) to be the foundation points to get the normalized curves. The two parameters have a relationship with section form and material properties, and we will study them in this section.

(1) Computation method of the axial compressive force

Similar to the section analysis model, the SFC section should be divided into three parts: angle steel part, confined concrete part and unconfined concrete part. Angle steel has been in a hardening state before the peak strength of confined concrete, due to the large strain of confined concrete. Therefore, the axial compressive force of SFCs can be computed as follows:

$$N_u = f_{cc}A_{cc} + k_{co}f_{co}A_{co} + f_yA_s \tag{9.22}$$

where A_{cc} is the section area of confined concrete, A_{co} is the section area of unconfined area, A_s is the section area of angle steel, f_{cc} is the compressive strength of confined concrete, f_{co} is the compressive strength of unconfined concrete, f_y is the yield stress of angle steel, and k_{co} is the reduction factor of peak strength to reflect the difference between the peak strains of confined and unconfined concrete. k_{co} can be calculated as follows:

$$k_{co} = \begin{Bmatrix} \varepsilon_{co}/(4\varepsilon_{cc})\varepsilon_{co}/\varepsilon_{cc} \geq 0.4 \\ 0.1\varepsilon_{co}/\varepsilon_{cc} \leq 0.4 \end{Bmatrix} \tag{9.23}$$

(2) Computation method of the pure bending moment

Based on the axial force-moment-curvature ($N-M-\varphi$) relationships, there is no decreasing stage in the moment-curvature curves of the SFC under low axial compression ratio. Therefore, it can be assumed that angle steel has reached peak stress before the section reaches the peak moment in a purely bending state. Because of the complex mechanical properties and stress discussion of the core concrete, the peak moment of the steel frame in a pure bending state can be a foundation to compute the peak moment, which can be computed as follows:

$$M_u = \gamma_s M_s \tag{9.24}$$

where M_u is the peak moment of the STC section, and M_s is the peak moment of the steel frame, which can be computed as follows:

$$M_s = f_y(A_{s1}l_1 + A_{s2}l_2) \tag{9.25}$$

where f_y is the yield stress of angle steel, A_{s1} is the section area of angle steel parallel to the central line, l_1 is the corresponding length, A_{s2} is the section area of angle steel perpendicular to the central line, and l_2 is the corresponding length. The four parameters can be computed as follows:

$$\left\{\begin{array}{l} A_{s1} = 4(b-t)tA_{s2} = 4bt \\ l_1 = (L-t)/2 - s \\ l_2 = (L-b)/2 - s \end{array}\right\} \tag{9.26}$$

where L is the side length, s is the thickness of concrete cover, b is the side length of angle steel, and t is the thickness of angle steel.

In addition, the γ_s in Eq. (9.25) is the influence coefficient of section moment, which is influenced by the steel ratio of angle steel and the concrete strength. The equation can be proposed by the fitting analysis in an extensive database:

$$\gamma_s = 0.004\rho_s^2 - 0.06\rho_s + 1.3 + 0.0002f_{cc} \tag{9.27}$$

where ρ_s is the steel ratio of angle steel, and f_{cc} is the compressive strength of confined concrete.

(3) The axial force-moment model

Based on the normalization process for the axial force-moment curves by the axial compressive force and the peak moment, the curve analysis shows the dividing axial compression (N_b) between the large and small eccentric compression is about 0.3 times the axial compressive force (N_u), N_b=0.3N_u. When SFC is in the large eccentric compression state ($N < N_b$), the curve can be seen as the quadratic function. When STC is in a small eccentric compression state ($N > N_b$), the curve can be seen as the linear function. The axial force-moment curve can be expressed as follows:

$$M/M_u = \left\{\begin{array}{l} -0.45(N/N_u)^2 + a(N/N_u) + 1 N \le N_b \\ a = 4.24exp(-0.38\omega) - 6.63\rho_s + 5.37 \\ M_b(N-N_b)/(N_b - N_u) + M_b N > N_b \end{array}\right\} \tag{9.28}$$

where ρ_s is the steel ratio of angle steel, ω is the confining coefficients, and M_b is the dividing moment between large and small eccentric compression, which can be computed by Eq. (9.28).

9.5.2 Skeleton model of the section moment-curvature curves

Based on the parameter analysis, the typical moment-curvature curve of SFC is shown in Fig. 9.18. In order to simplify the calculation method, the moment-curvature curve can be expressed as the trilinear equation, including the following stages: (1) the first stage is the elastic stage; (2) the second stage is the hardening stage after the yield state; (3) the third stage is the stable stage which is a horizontal line under low axial compression ratio or a decreasing line under high axial compression ratio.

Based on simulation results and experiment results, the influence factors of the section moment are the axial compression ratio, the steel ratio of the angle steel, and the confining effect. The computation equation can be expressed as follows:

$$M = \left\{ \begin{array}{l} K_e\phi 0 \leq \phi \leq \phi_y \\ K_e\phi_y + K_p(\phi - \phi_y)\phi_y \leq \phi \leq \phi_p \\ M_u - K_d(\phi - \phi_p)\phi_p \leq \phi \end{array} \right\} \tag{9.29}$$

where $\varphi_y = M_y/K_e$, M_y is the yield section moment ($M_y = 0.85M_u$), K_e is the stiffness of the elastic stage, K_p is the stiffness of the hardening stage, and K_d is the stiffness of the stable stage. The three kinds of stiffnesses can be referred to existing studies on the concrete filled steel tube (Ding et al., 2009; Zhou et al., 2014, 2015).

The stiffness of the elastic stage can be calculated as:

$$\left\{ \begin{array}{l} K_e = (0.578M_u)/((1.1 + n)\phi_e) \\ \phi_e = f_y/(E_sD) \end{array} \right\} \tag{9.30}$$

The stiffness of the hardening stage can be calculated as:

$$K_p = [(1.7 + \omega)n^2 + (0.6 - 3\omega)n + \omega - 0.3]K_e \tag{9.31}$$

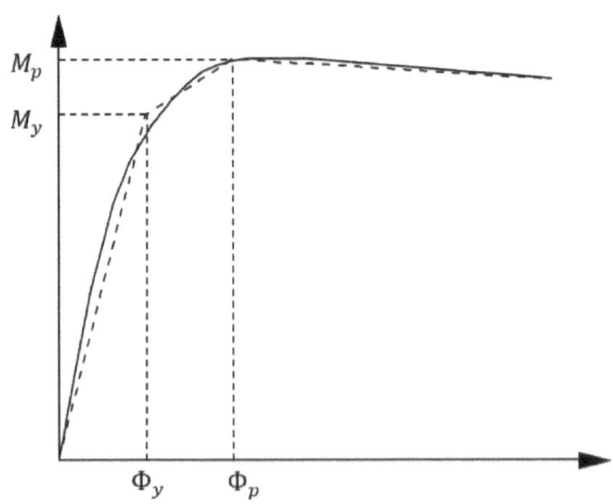

Figure 9.18 Typical moment-curvature curve of SFC. Typical moment-curvature curve.

The stiffness of the stable stage can be calculated as:

$$K_d = [(0.75\omega - 1.16)n - 0.11\omega - 0.224]K_p \leq 0 \quad (9.32)$$

9.5.3 Section moment-curvature hysteresis model

In the establishment of the moment-curvature hysteresis model, the following steps should be adopted: (1) the peak moment is computed by the known axial compression ratio in the axial force-curvature equation; (2) the moment-curvature skeleton curve is established; (3) the moment-curvature hysteresis curve is drawn, based on the hysteresis rule including the unloading stiffness and the reloading path. Based on the analysis results, the unloading stiffness is similar to the loading stiffness before the yield moment, then it reduces with the increase in the curvature after the yield moment. In addition, the reloading path is the peak point to reverse another peak point in the next loop. Therefore, the moment-curvature hysteresis model of SFCs can be simplified as the degenerate trilinear model, and the path is shown in Fig. 9.19.

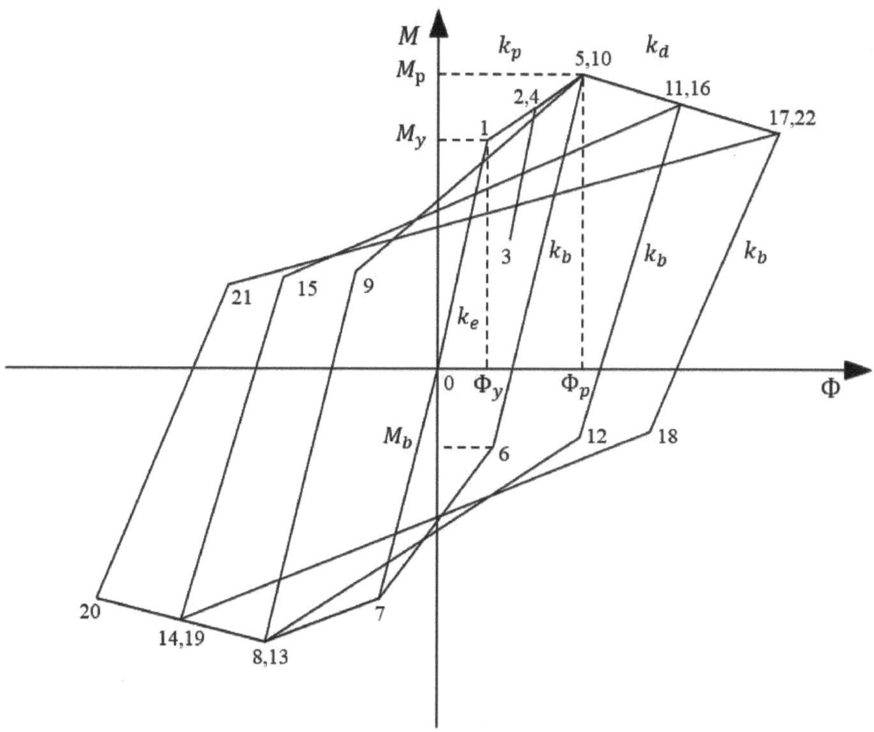

Figure 9.19 Moment-curvature hysteresis model of SFC. Typical curve.

For the unloading path in Fig. 9.19, the unloading stiffness (K_b) is equal to the elastic stiffness (K_e) before the yield moment, and the unloading stiffness is smaller than the elastic stiffness after the yield moment. Meanwhile, the reverse moment (M_b) of the unloading path can choose the value which is 0.25 times the peak moment before unloading.

As shown in the above section, the unloading stiffness should be computed by the elastic stiffness and the yield curvature. Based on the analysis results, the relationship between the unloading reduction coefficient and the curvature is shown in Fig. 9.20.

Then the unloading stiffness (K_b) can be computed as follows:

$$K_b = \beta K_e \beta = \left\{ \begin{array}{l} 1 \phi/\phi_y < 1.54 \\ -0.3 ln(\phi/\phi_y) + 1.13 \phi/\phi_y > 1.54 \end{array} \right\} \tag{9.33}$$

Based on the above analysis, the moment-curvature hysteresis model can be established in the following steps: (1) the peak moment is computed by the axial force-moment model; (2) the moment-curvature skeleton curve is proposed; (3) the first curvature is defined to compute the corresponding moment, then the first point is connected with the starting point; (4) the reduction stiffness and the reduction moment are computed by the previous curvature and comment; (5) the reduction curvature is computed to gain the reduction point, then the reduction point is used to form the first unloading path; (6) the symmetry point is defined to the first loading point about the origin; (7) the symmetry point is linked to the first unloading point to form the second loading path; (8) the symmetry point is used as the starting point to begin the next cycle. The detailed calculation method is shown in Fig. 9.21.

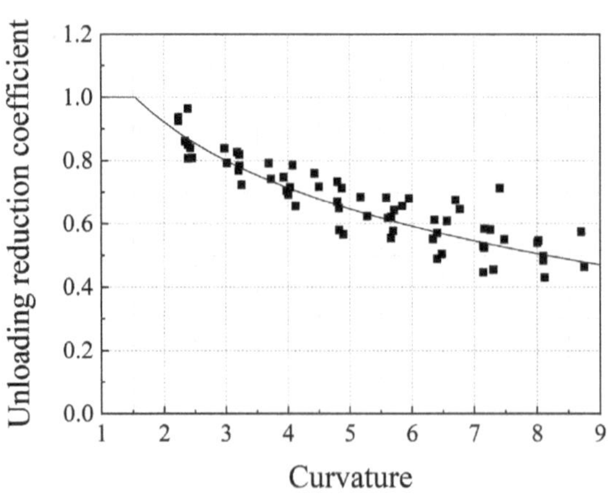

Figure 9.20 Relationship between unloading reduction coefficient and curvature. Data analysis.

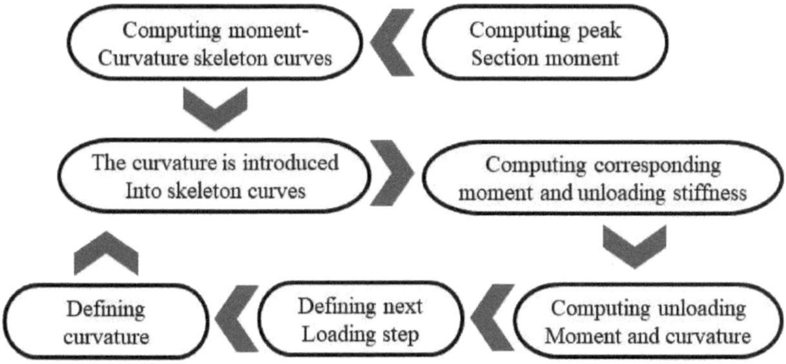

Figure 9.21 Detailed calculation method of moment-curvature hysteresis model. Flow chart.

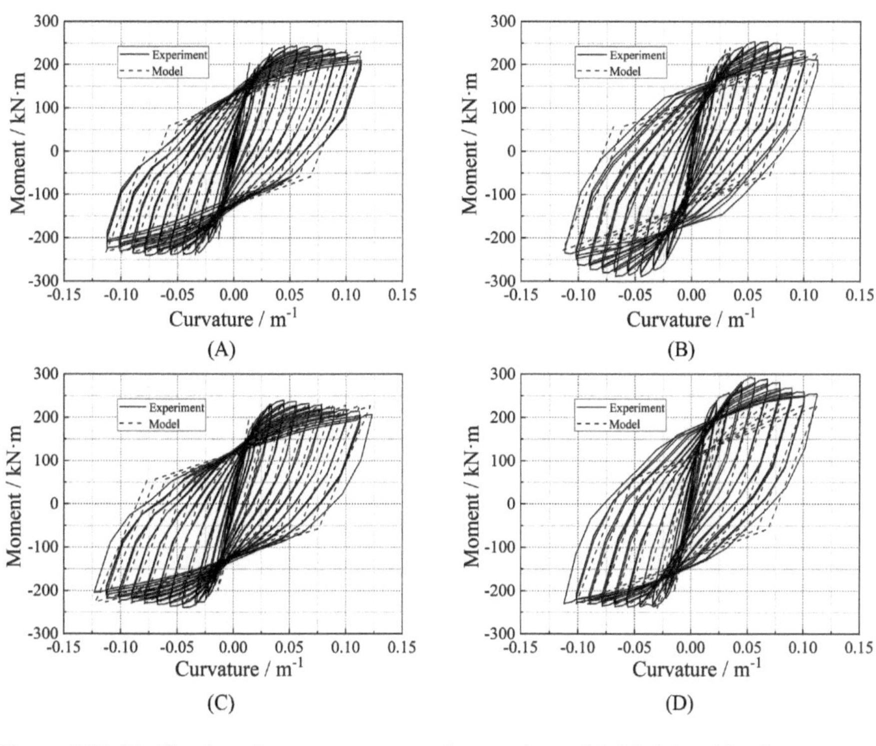

Figure 9.22 Verification of moment-curvature hysteresis model. Model verification.

In order to verify the proposed model, the moment-curvature hysteresis curves of four specimens (i.e., from SFC-1 to SFC-4) are calculated. And the comparisons between calculation curves and test curves are shown in Fig. 9.22. The curve comparisons show that the proposed model can provide accurate prediction results for the SFC specimens. Therefore, the accuracy and effectiveness of the proposed model are verified.

9.6 Damage evaluation method of SFCs

Based on the section analysis, the deformation condition of compression bending members in the stress process can be distinguished accurately, and the influence factors of the damage degree can be identified, so as to accurately establish the damage parameter model of SCFs. In order to propose the damage evaluation method, the yield curvature and ultimate curvature of the SFC section are analyzed first in this section. Then the calculation method of the corresponding section curvature and ductility of the compression bending members is established. Finally, based on the section ultimate curvature and the failure process in the seismic experiment in Chapter 8, the damage index model of SFCs is established.

9.6.1 Analysis of yield section curvature

Based on the existing design method by using bearing capacity, we can assume that the basic characteristic of the section belongs to the stiffness section, so the yield curvature of the section is proportional to the yield strength directly. However, many studies show that the section moment is also related to the axial compression ratio and the steel ratio of the angle steel, which means that although the stiffness is proportional to the strength, the change of yield curvature is relatively small. Therefore, the definition of section yielding by using bearing capacity is not accurate enough. The yield curvature should be accurately defined by the section characteristics. Then the yield curvature can be taken as the basic characteristics of the section. In order to ensure the accuracy of the structural design, it is necessary to analyze the yield curvature at first.

(1) Definition of the yield section curvature

The main influence factors for the mechanical properties and the seismic behaviors of the SFC section, including the section size, the steel ratio of angle steel, the confining effect, and the axial compression ratio. But for the yield state of the compression bending members, the behavior is mainly influenced by two factors: the angle steel and the core concrete. All the above factors can be reflected by the basic characteristics and the stress state of the two materials. And the effects of the two influence factors are controlled by the height of the compressive zone: if the angle steel is the controlling factor, the yield curvature will increase with the increase in the compression zone height; if the core concrete is the controlling factor, the yield curvature will decrease with the increase in the compression zone height (Li et al., 2016). Therefore, the yield criterion for the SFC section is that the angle steel in the tensile zone of the SFC section reaches a yield state or the stress of the core concrete in the compressive zone of the SFC section reaches the compressive strength.

(2) Parameter analysis

In order to simplify the analysis method, we introduce the normalization yield curvature coefficient to analyze the influence factors of the section yield curvature of SFC:

$$k_y = (\phi_y L_s E_s)/f_y \tag{9.34}$$

where L_s is the distance between two-angle steel.

Based on the finite element model analysis, the relationship between the yield curvature coefficient and the axial compression ratio can be obtained, as shown in Table 9.3. The table shows the variation of the normalization yield curvature coefficient under different axial compression ratios with different steel ratios of the angle steel and different confining effects.

Based on the relationship analysis, the following conclusions can be drawn: (1) when the axial compression ratio is small, the angle steel is the controlling factor for the section yield curvature; (2) when the axial compression ratio is large, the core concrete is the controlling factor for the section yield curvature; (3) the normalization yield curvature coefficient increases first and then decreases with the increase in the axial compression ratio, and the maximum curvature coefficient corresponds to the initial stage where the core concrete is the control factor; (4) the steel ratio has little effect on the relationship between the axial compression ratio and the normalization yield curvature coefficient; (5) the confining effect has a significant influence on the relationship between the axial compression ratio and the normalization yield curvature coefficient.

(3) Simplified calculation equation of the yield curvature

Based on the data analysis of each group, it can be found that the relationship between the axial compression ratio and the normalization yield curvature coefficient can be expressed as the quadratic curve function. In addition, the confining effect has an obvious effect on the relationship trend, and the steel ratio of angle steel linearly increases the normalization yield curvature coefficient. Then the multivariate function is used for the fitting analysis, and the calculation equation of the normalization yield curvature coefficient of SFCs can be proposed:

$$k_y = -(5.9\omega + 0.4)n^2 + (2.7\omega + 3.5)n + 0.28\rho_s + 1.2 \tag{9.35}$$

By substituting the above equation into Eq. (9.35), the calculation equation of the section yield curvature of SFC can be proposed:

$$\phi_y = [-(5.9\omega + 0.4)n^2 + (2.7\omega + 3.5)n + 0.28\rho_s + 1.2]f_y/(L_s E_s) \tag{9.36}$$

The correlation coefficient of the proposed equation is 0.91, which shows that the correlation of the result is high and it is reasonable to use the equation form for fitting analysis. The proposed equation in this section has involved the effect of the axial compression ratio, the confining effect, and the mechanical properties of steel

Table 9.3 Relationship between yield curvature coefficient and axial compression ratio.

Steel ratio of angle steel (%)	Confining coefficient	Axial compression ratio	Yield curvature coefficient	Controlling material
3.98	0.52	0	1.51	Angle steel
		0.1	1.78	Angle steel
		0.2	1.99	Angle steel
		0.3	2.40	Angle steel
		0.4	2.88	Angle steel
		0.5	3.15	Concrete
		0.6	2.88	Concrete
		0.7	2.33	Concrete
		0.8	2.26	Concrete
4.89	0.56	0	1.78	Angle steel
		0.1	1.92	Angle steel
		0.2	2.05	Angle steel
		0.3	2.47	Angle steel
		0.4	3.01	Angle steel
		0.5	3.29	Concrete
		0.6	2.74	Concrete
		0.7	2.74	Concrete
		0.8	2.47	Concrete
5.76	0.59	0	1.64	Angle steel
		0.1	1.85	Angle steel
		0.2	2.08	Angle steel
		0.3	2.47	Angle steel
		0.4	3.08	Angle steel
		0.5	3.15	Concrete
		0.6	2.88	Concrete
		0.7	2.60	Concrete
		0.8	2.40	Concrete
3.98	0.71	0	1.51	Angle steel
		0.1	1.78	Angle steel
		0.2	1.99	Angle steel
		0.3	2.33	Angle steel
		0.4	2.74	Angle steel
		0.5	3.29	Angle steel
		0.6	3.08	Concrete
		0.7	2.84	Concrete
		0.8	2.53	Concrete
3.98	0.99	0	1.51	Angle steel
		0.1	1.75	Angle steel
		0.2	1.95	Angle steel
		0.3	2.26	Angle steel
		0.4	2.60	Angle steel
		0.5	3.01	Angle steel
		0.6	3.42	Concrete
		0.7	3.08	Concrete
		0.8	2.81	Concrete

and confined concrete. Meanwhile, it is fitted by numerous calculation results to ensure the reliability and applicability of the proposed equation. Furthermore, the form of the proposed equation is relatively simple, which is convenient for engineering application.

9.6.2 Analysis of the ultimate section curvature

Based on the existing design method using the bearing capacity, we can assume that the specimen reaches the ultimate displacement once the bearing capacity decreases to 85%. However, this definition method does not involve the effect of each material. Therefore, the definition of the section's ultimate state using the bearing capacity is not accurate and theoretical. We can define the section's ultimate state using the section curvature analysis. Therefore, we study the ultimate section curvature of SFC by the moment-curvature analysis model. The analysis parameters include the effect of the axial compression ratio, the steel ratio of angle steel, and the confining effect. Finally, the calculation equation of the ultimate section curvature is proposed by fitting the analysis results.

(1) Definition of the ultimate section curvature

Under different section sizes and section distribution forms, the ultimate state of the SFC section is controlled by various factors. Therefore, the ultimate section curvature should be defined as the minimum curvature corresponding to the following three cases: the bending capacity of the SFC section decreases to 85% of the maximum bending moment of the SFC section; the core concrete reaches the ultimate compressive strain, which can be defined by the axial compression test of SFC and the constitutive model of SF confined concrete, and the ultimate compressive strain can take that $\varepsilon_{ccu}=0.02$; the angle steel in the tensile zone reaches the ultimate tensile strain, which can be defined by the experimental and theoretical studies (Eleni et al., 2013; Hou, 2007), and the ultimate tensile strain of angle steel can be taken as $\varepsilon_{su}=0.015$.

(2) Parameter analysis

In order to simplify the analysis method, we introduce the normalization ultimate curvature coefficient to analyze the influence factors of the section ultimate curvature of SFC:

$$k_u = 1000\phi_u L_s \tag{9.37}$$

where L_s is the distance between two-angle steels.

Based on the finite element model analysis, we can get the relationship between the ultimate curvature coefficient and the axial compression ratio, as shown in Table 9.4. The table shows the variation of the normalization ultimate curvature coefficient under different axial compression ratios with different steel ratios of the angle steel and different confining effects.

Based on the relationship analysis, the following conclusions can be drawn: (1) when the axial compression ratio is small, the angle steel is the controlling factor for the section yield curvature; (2) when the axial compression ratio is large, the

Table 9.4 Relationship between ultimate curvature coefficient and axial compression ratio.

Steel ratio of angle steel (%)	Confining coefficient	Axial compression ratio	Ultimate curvature coefficient	Controlling material
3.98	0.52	0	17.94	Angle steel
		0.1	19.76	Angle steel
		0.2	22.1	Angle steel
		0.3	28.6	Angle steel
		0.4	33.8	Concrete
		0.5	24.44	Bending
		0.6	17.94	Bending
		0.7	14.56	Bending
		0.8	10.66	Bending
4.89	0.56	0	18.46	Angle steel
		0.1	20.28	Angle steel
		0.2	22.88	Angle steel
		0.3	31.2	Angle steel
		0.4	33.8	Concrete
		0.5	28.6	Concrete
		0.6	24.96	Bending
		0.7	20.02	Bending
		0.8	14.3	Bending
5.76	0.59	0	18.72	Angle steel
		0.1	20.28	Angle steel
		0.2	23.4	Angle steel
		0.3	30.42	Angle steel
		0.4	32.5	Concrete
		0.5	27.3	Concrete
		0.6	24.96	Concrete
		0.7	20.54	Bending
		0.8	15.08	Bending
3.98	0.71	0	17.94	Angle steel
		0.1	19.76	Angle steel
		0.2	21.58	Angle steel
		0.3	25.74	Angle steel
		0.4	32.24	Angle steel
		0.5	32.5	Concrete
		0.6	27.82	Concrete
		0.7	22.88	Bending
		0.8	16.9	Bending
3.98	0.99	0	17.94	Angle steel
		0.1	19.5	Angle steel
		0.2	21.06	Angle steel
		0.3	23.92	Angle steel
		0.4	27.82	Angle steel
		0.5	34.32	Angle steel
		0.6	32.76	Concrete
		0.7	27.82	Concrete
		0.8	25.22	Concrete

core concrete is the controlling factor for the section yield curvature; (3) with the stronger confining effect, there is a larger range of the axial compression ratio corresponding to the control of the angle steel; (4) the normalization ultimate curvature coefficient increases first and then decreases with the increase in the axial compression ratio, and the maximum curvature coefficient corresponds to the initial stage where the concrete is the control factor; (5) similarly to the developing trend of the section yield curvature, the confining effect has a significant influence on the relationship between the axial compression ratio and the normalization ultimate curvature coefficient.

(3) Simplified calculation equation of the ultimate curvature

Based on the data analysis of each group, we can find that the quadratic curve function can reflect the relationship between the axial compression ratio and the normalization ultimate curvature coefficient. In addition, the confining effect has an obvious effect on the relationship trend, and the steel ratio of the angle steel linearly increases the normalization ultimate curvature coefficient. Then the multivariate function is used for the fitting analysis, and the calculation equation of the normalization ultimate curvature coefficient of SFCs can be proposed:

$$k_u = -(51.7\omega + 40.9)n^2 + (73.5 - 8.9\omega)n - 10\rho_s + 55.5 \tag{9.38}$$

By substituting the above equation into Eq. (9.28), the calculation equation of the section ultimate curvature of SFC can be proposed:

$$\phi_u = (-(51.7\omega + 40.9)n^2 + (73.5 - 8.9\omega)n + 10\rho_s - 24.1)/(1000L_s) \tag{9.39}$$

The correlation coefficient of the proposed equation is 0.90, which shows that the correlation of the result is high and it is reasonable to use the equation form for the fitting analysis. The proposed simple calculation model can provide the accurate prediction for the section ultimate curvature of SFC member, so as to make SFC can be applied in the actual engineering.

9.6.3 Calculation method of section curvature ductility coefficient

The calculation methods of the section yield curvature and the section ultimate curvature of SFCs are proposed in the above sections, respectively. Based on the calculation equations, the section curvature ductility coefficient (μ_φ) can be computed as follows:

$$\mu_\phi = \phi_u/\phi_y = \frac{E_s[-(51.7\omega + 40.9)n^2 + (73.5 - 8.9\omega)n + 10\rho_s - 24.1]}{1000f_y[-(5.9\omega + 0.4)n^2 + (2.7\omega + 3.5)n + 0.28\rho_s + 1.2]} \tag{9.40}$$

Based on the above computation method, the ductility coefficients of the SFC specimens in Chapter 8 can be computed, respectively. The calculation results are summarized in Table 9.5 and compared with the test results.

Based on the result comparisons in Table 9.5, we can conclude that the proposed model provides accurate values for the SFC specimens. For SFC-2, SFC-4, SFC-6, and SFC-8 specimens, the calculation results provided by the proposed model have a minor error from the test results. However, there are large errors in specimens SFC-1, SFC-3, SFC-5, and SFC-7 because the concrete in the specimens is under a strong confining effect. That means the above four specimens cannot take the bending moment reduction as the control factor to determine the section's ultimate curvature. However, the determination method of the bearing capacity reduction in the test method will lead to the calculated confining coefficient being too large, which is suitable for the practical situation. In general, the proposed calculation method in this paper can provide the accurate ductility coefficient of SFC. Therefore, the proposed calculation method can as the foundation for subsequent studies.

9.6.4 Damage index model

The SFC is a novel section of the composite column. We should evaluate accurately the damage degree in the seismic design of the SFC structure. At present, many studies have been presented on the damage evaluation method of the reinforced concrete columns and the concrete filled steel tube columns, and most studies focused on the quantification of damage indicators (Si & Wang, 2008; Su et al., 2018). A large number of damage index models have been established by introducing the influence factors, such as the displacement amplitude, the energy accumulation, and the loading times. However, for the confined concrete composite columns, it has been known that the energy accumulation and loading times have little influence on the seismic behavior, because of the strong confining effect on the core concrete. Therefore, the parameters in the loading process cannot be the foundation

Table 9.5 Verification of ductility coefficients of SFC specimens.

Item	Basic parameters		Test result	Computation result	Error (%)
	Confining coefficient	Steel ratio (%)			
SFC-1	0.52	3.98	5.19	4.66	10.23
SFC-2	0.52	3.98	4.79	4.64	3.06
SFC-3	0.57	3.98	5.01	4.58	8.50
SFC-4	0.57	3.98	4.69	4.55	3.06
SFC-5	0.8	3.98	5.08	4.39	13.54
SFC-6	0.8	3.98	4.11	4.20	−2.17
SFC-7	0.64	3.98	5.3	4.48	15.46
SFC-8	0.62	5.4	6.02	5.96	1.05

to establish the damage index model. Based on the section analysis, the internal angle steel and confined concrete play a common role in the loading process, which means the displacement amplitude is too subjective to be the damage index. Therefore, it is necessary to select a reasonable damage model by the accurate section analysis.

(1) Damage index model based on the curvature

Based on the review and the analysis of the existing damage index models and the seismic characteristics of SFC, the maximum historical lateral displacement and the corresponding curvature of SFC under the lateral repeated loading can reflect the inner damage. Therefore, the ratio of the maximum historical curvature to the section's ultimate curvature can be the foundation to reflect the deformation damage, the equation is as follows:

$$D = \phi/\phi_u \tag{9.41}$$

where φ is the maximum historical curvature, and φ_u is the section's ultimate curvature, which can be computed by Eq. (9.40).

(2) Damage index

In the judging method for the damage degree of SFCs based on the damage condition, the influence of subjective factors is inevitable, but the judgment usually affects the design decision. In order to reduce the error caused by the subjective factors, we divide four important characteristic damage conditions in this section, including the cracking condition (i.e., the cracks appear on the concrete cover), the crushing condition (i.e., the concrete cover crushes), the falling condition (i.e., the concrete cover peels off), and the buckling condition (i.e., the local buckling appears on the angle steel which is accompanied by the partial crushing of the core concrete). Based on the analysis of the frequency distribution of the four characteristic damage conditions, the discrimination limits of different damage degrees are determined, as shown in Table 9.6. It should be noted that the sum of the buckling frequencies of the angle steel in Table 9.6 is not equal to 1. It can be seen that the angle steel in the compression zone does not appear the local buckling in some

Table 9.6 Discrimination limits of different damage degrees.

Damage degree	**[0, 0.1]**	**[0.1, 0.2]**	**[0.2, 0.3]**	**[0.3, 0.4]**	**[0.4, 0.5]**
Cracking condition	0.12	0.88			
Crushing condition				0.5	0.5
Falling condition					
Buckling condition					
Damage degree	[0.5, 0.6]	[0.6, 0.7]	[0.7, 0.8]	[0.8, 0.9]	[0.9, 1]
Cracking condition					
Crushing condition					
Falling condition	0.12	0.76	0.12		
Buckling condition				0.25	0.5

Table 9.7 Qualitative description and damage index range.

Performance level	Qualitative description	Damage index range
Normal use	Concrete covers appear cracks. No concrete crushing	[0, 0.2]
Temporary use	Concrete covers appear many through cracks. Concrete covers appear crushing	[0.2, 0.5]
Use after repair	Concrete covers appear crushing. Concrete covers peel off	[0.5, 0.8]
Life safety	Concrete covers peel off largely. Angle steels appear local buckling partly. Core concrete appears crushing	[0.8, 1]
Prevent collapse	Angle steel reaches ultimate tension strain. Angle steels appear local buckling seriously. Core concrete appears serious crushing. The section bending capacity decreases by 15%. (Take the first occurrence as the failure condition)	>1

specimens. This is because the angle steel in the tension zone is the controlling factor of the section's ultimate curvature under the small axial compression ratio.

Based on the analysis of Table 9.6, it can be seen that the damage index of the above four characteristic damage conditions is basically between [0, 1], and only some cases would exceed the range. However, based on the section analysis, it can be seen that the ultimate curvature of the compression bending member is affected by many factors, so the absence of the failure characteristics of the composite column does not mean that it is still in a safe condition. Based on the section ductility coefficient in the above sections, the following conclusions can be drawn: when the damage index is about 0.2, the section reaches the yield state, there is almost no damage in the damage index range of [0, 0.2], and the concrete cover appears some cracks in this state; the concrete cover is crushed in the damage index range of [0.3, 0.5]; the concrete cover shows obvious peeling off in the damage index range of [0.5, 0.8]; in the damage index range of [0.9, 1], the local buckling of the angle steel and the crushing of core concrete will occur in most SFC specimens. Based on the values in Table 9.6 and the section analysis results, the qualitative description and the design suggestions of the damage index range by the proposed model are shown in Table 9.7.

9.7 Conclusions

Based on the seismic behavior test of the SFC specimens, we establish the finite element analysis models using the finite element analysis software (OpenSEES). Then the effects of the axial compression ratio, the angle steel size, and the steel distribution form on the seismic behavior of SFCs are further studied and analyzed.

Based on the above analysis, the section analysis model of SFCs under compression and bending load are established and verified. Based on the proposed model, the axial force-moment-curvature relationship of SFCs is analyzed, and then the moment-curvature hysteresis model is established. Based on the hysteresis model, the influence factors of the section yield curvature and the section ultimate curvature of SFCs are analyzed to propose the calculation equations of the two section curvatures. Then the section curvature ductility coefficient is obtained. Finally, the damage index model of SFCs is established. Based on the above study results, the following conclusions can be drawn:

(1) The axial compression ratio has a significant effect on the seismic behavior of SFCs. With the increase in the axial compression ratio, the peak bearing capacity increases first and then decreases, but the ductility will always decrease.

(2) When the arrangement of small batten and small spacing is adopted, the ductility decreases with the increase in the axial compression ratio, and the bearing capacity shows the trend of first increasing and then decreasing. But the dividing axial compression ratio increases with the increase in the axial compression ratio, which reflects that the confining effect can play an obvious role under a large axial compression ratio.

(3) Under the same axial compression ratio, the steel ratio of the angle steel has a significant effect on the seismic behavior. The larger the steel ratio leads to the fuller hysteretic curve, the greater the bearing capacity and the greater the ductility of the composite columns. Meanwhile, the increase in the thickness of the angle steel has the larger effect than the increase in the length.

(4) The unique arrangement of the steel battens can provide different confining effects on the core concrete in the SFCs. The greater confining effect can lead to the good symmetry of the hysteretic curve, the fuller hysteretic curve, and the good ductility. However, the bearing capacity does not change significantly with the increase in the confining effect.

(5) The moment-curvature curve of SFC mainly presents a two-stage developing trend, where the first stage is the increasing curve, the second stage is the gentle increasing curve under a small axial pressure ratio, and the second stage is the gentle decreasing curve under large axial pressure ratio. Meanwhile, the larger axial compression ratio leads to the larger slope of the decreasing stage.

(6) The axial force-moment curve of the SFC section has the following variation law. When the axial compression ratio is small (the composite column is under a large eccentric compression), the peak moment increases with the increase in the axial compression ratio. When the axial compression ratio is large (the composite column is under small eccentric compression), the peak moment decreases with the increase in the axial compression ratio. The moment corresponding to the same axial force increases with the increase of the steel ratio of angle steel, but the variation trend of axial force-moment curve is basically the same. The confining effect can effectively improve the axial compression bearing capacity but has little effect on the pure bending moment. With the increase in the confining effect, the dividing axial compression ratio will increase, which means the composite columns are under a large eccentric compression.

(7) Based on experimental results and theoretical analysis, the trilinear model is proposed for the moment-curvature skeleton curve of SFCs. Then, the directional hysteretic model of the maximum point for SFC is established. The calculation curve of the proposed model is in good agreement with the experiment curve.

(8) The main influence factors of the section yield curvature are the axial compression ratio, the steel ratio of angle steel, and the confining effect. When the axial compression ratio is small, the angle steel is the controlling factor, and the yield curvature increases with the increase in the axial compression ratio. When the axial compression ratio is large, the concrete is the controlling factor, and the yield curvature decreases with the increase in the axial compression ratio. When the curvature coefficient is at the peak point, the core concrete can be the control factor at the initial stage. The yield curvature always increases with the increase in the steel ratio of angle steel. The constraint effect has a significant effect on the developing trend of the axial compression ratio and the yield curvature coefficient.

(9) The main influence factors of the ultimate curvature are the axial compression ratio, the steel ratio of angle steel, and the confining effect. When the axial compression ratio is small, the angle steel is the controlling factor, and the ultimate curvature increases with the increase in the axial compression ratio. With the increase in the axial compression ratio, the controlling factor is gradually changed to concrete. Then the ultimate curvature decreases with the increase in the axial compression ratio. When the curvature coefficient is at the peak point, the core concrete can be the control factor at the initial stage. With the increase of confining effect, the range of axial compression ratio, which belongs to the angle steel controlling.

(10) Based on the calculation results, the calculation equations of the yield curvature and the ultimate curvature are proposed, and the curvature ductility calculation equation is further proposed and verified. These equations can be the foundation for the calculation of the section deformation capacity of SFCs.

(11) Based on the analysis of the section curvature, the damage index model of SFCs is established by the ultimate curvature. Then, the failure characteristics of the specimens in the seismic test are analyzed. The behavior level of the composite column is divided into five degrees. And the characteristic damage state and damage limit corresponding to each degree are determined, which can lay the foundation for the application in the actual engineering and theoretical design.

Acknowledgments

This work was supported by the National Natural Science Foundation of China [grant numbers 52108171; 51878540]. The authors gratefully acknowledge two Master's students (Mr. Qv Yunsong and Mr. Tian Wenkai) of Xi'an University of Architecture and Technology for the drawing works.

References

Chen, W (2012). *Application of concrete hysteretic constitutive model developed based on OPENSEES platform in structural analysis*. Chongqing University.

Chen, XW (2014). *Structural nonlinear analysis program OpenSEES theory and tutoral*. China Architecture & Building Press.

Chen, XW, Han, XL, & Sun, SW (2011). Research and development of three nonlinear beam-column elements. *Engineering Mechanics*, *28*(S1), 5–11.

Chen, X, Zhou, DH, Zhang, SP, Wang, P, & Li, LQ (2014). Analytical method to determine elato-plastic moment-curvature relationship of bending section. *Engineering Mechanics, 31*(11), 175–182.

Chen, ZY, Yi, WJ, Zhao, GF, Lin, LY, & Huang, XH (2006). Numerical analysis of load-moment interaction diagram for high strength concrete column reinforced with concrete filled steel tube. *Industrial Construction, 36*(4), 75–77.

Diao, B, Li, SH, & Ye, YH (2008). Analysis and experiment of cumulated damage of RC structures with special columns under cyclic loading. *Journal of Building Structures, 29* (1), 57–63.

Ding, FX, Zhang, P, Yu, ZW, & Ou, FP (2009). Practical calculation method of axial force-moment- curvature relationship for concrete- filled circular steel tubular beam-columns. *Journal of Harbin Institute of Technology, 12*, 142–146.

Eleni, S, Timothy, S, & Nigel, P (2013). Sectional response of T-shaped RC walls. *Bulletin of Earthquake Engineering, 11*(4), 999–1019.

Eom, TS, Hwang, HJ, Park, HG, Lee, CN, & Kim, HS (2014). Flexural test for steel-concrete composite members using prefabricated steel angles. *Journal of Structural Engineering, 140*(4), 1–12.

Gu, Q, & Huang, SR (2016). *OpenSEES practical tutorial*. Science Press.

Guo, ZH, & Shi, XD (2003). *Principle and analysis of reinforced concrete*. Tsinghua University Press.

Hosseini, M, & Abbas, H (2013). Strain hardening in M-P interaction for metallic beam of I-section. *Thin-Walled Structures, 62*, 243–256.

Hou, ZM (2007). *Study on mechanical behavior of steel reactive powder concrete simply supported composite beams*. Beijing Jiaotong University.

Ji, J (2008). *Research on seismic performance and design method of jacketing steel reinforced concrete frame prestressed with bonded tendons*. Harbin Industrial University.

Ji, XD, Zhang, ML, Liu, P, Kang, HZ, Qian, JR, Zhou, ZZ, & Hu, HS (2013). Experimental study of cumulative damage performance of steel tube-reinforced concrete composite columns. *Journal of Building Structures, 34*(12), 35–44.

Jiang, R, & Su, XZ (2008). Comparative study on emprical plastic hinge length formulas. *Industrial Construction, S1*, 425–430.

Julio, GR, Adam, JM, Calderón, PA, & Valente, IB (2012). Finite element modelling of steel-caged RC columns subjected to axial force and bending moment. *Engineering Structures, 40*(7), 168–186.

Kratzig, W. B., Meyer, I. F., & Meskouris, K. (1989). Damage evolution in reinforced concrete members under cyclic loading (pp. 795-802).

Li, GQ (2010). *Experimental study and numerical analysis on seismic performance of reinforced concrete pier*. Chongqing Jiaotong University.

Li, GQ, Tang, GW, & Zheng, G (2016). Equivalent plastic hinge length of circular reinforced concrete bridge columns. *China Civil Engineering Journal, 49*(2), 87–97.

Li, RZ, Tan, H, & Zhao, M (2016). Analysis and application of the correlation curve of axial force bending moment for the beam-columns of circular steel tube concrete. *ZheJiang Construction, 33*(5), 11–15.

Lin, HY (2013). Research on plastic hinge property of concrete-filled steel tubular members based on pushover theory. *Journal of Zhengzhou University (Engineering Science), 34* (5), 7–11.

Mehanny, SS, & Deierlein, F. GG (2001). Seismic damage and collapse assessment of composite moment frames. *Journal of Structural Engineering, 127*(9), 1045–1053.

Montuori, R, & Piluso, V (2009). Reinforced concrete columns strengthened with angles and battens subjected to eccentric load. *Engineering Structures*, *31*(2), 539–550.

Mostafaei, H, & Kabeyasawa, T (2007). Axial-shear-flexure interaction approach for reinforced concrete columns. *Aci Structural Journal*, *104*(2), 218–226.

Parant, E, Rossi, P, & Jacquelin, E (2007). Strain rate effect on bending behavior of new ultra-high-performance cement-based composite. *ACI Materials Journal*, *104*(5), 458–463.

Park, R, & Paulay, T (1995). *Reinforced concrete structures*. John Wiley & Sons.

Park, YJ, Alfredo, HS, & Ang, F (1985). Mechanistic seismic damage model for reinforced concrete. *Journal of Structural Engineering*, *111*, 822–839.

Qi, QQ, Jiang, LZ, & Zhou, WB (2013). Restoring force model of steel-concrete composite frame structure. *Journal of Central South University (Natural Science Edition)*, *9*, 3863–3872.

Qian, JR, Li, NB, Ji, XD, & Cao, WL (2014). Experimental study on cumulative damage performance of composite-sectioned high strength concrete-filled steel tubular columns. *China Civil Engineering Journal*, *2*, 30–40.

Si, WF, & Wang, C (2008). Damage analysis of circular-section concrete pile under pressure and bending load. *Journal of Chongqing Jiaotong University (Nature Science Edition)*, *27*(S1), 892–894.

Su, JS, Wang, JJ, Wang, WB, Dong, ZF, & Liu, B (2014). Comparative experimental research on seismic performance of rectangular concrete columns reinforced with high strength steel. *Journal of Building Structures*, *35*(11), 20–27.

Su, JZ, Liu, BQ, & Xing, GH (2018). Comparative analysis on seismic damage models for reinforced concrete columns. *World Earthquake Engineering*, *34*(2), 80–88.

Wang, B (2017). *Research on seismic behavior and calculation method of deformation capacity of reinforced concrete shear walls with flange*. Xi'an University of Architecture and Technology.

Wu, B, Wang, WJ, & Wang, F (2005). Analyses of the moment-curvature relationship of RC columns strengthened with carbon fiber sheets. *Journal of South China University of Technology (Natural Science Edition)*, *1*, 10–15.

Yang, K, Shi, QX, & Zhao, JH (2013). Plastic hinge length of high-strength concrete columns confined by high-strength stirrups. *Engineering Mechanics*, *30*(2), 254–259.

Zhang, GW, Gao, YC, Wang, S, Wu, JF, & Zhao, ZW (2017). Damage assessment of concrete-filled steel tube column based on deterioration of frequency. *Structural Engineers*, *6*, 116–122.

Zhang, K, Yue, RQ, Fu, CW, & Ye, LP (2001). Moment and curvature relation analysis of RC concrete column strengthened with CFS. *Industrial Construction*, *31*(6), 20–23.

Zhou, WB, Jiang, LZ, & Li, SJ (2015). Moment-curvature restoring force model research of steel-concrete composite box-beams. *Journal of Building Structures*, *36*(1), 78–84.

Zhou, WB, Jiang, LZ, Li, SJ, & Kong, F (2014). Finite element time history analysis on seismic response of steel-concrete composite frame structure. *Journal of Huazhong University of Science and Technology (Nature Science Edition)*, *42*(7), 61–65.

Zhu, HZ (2013). *Nonlinear finite element analysis of lattice steel reinforced concrete column based on OpenSees*. Chongqing University.

Engineering application of steel frame confined concrete column

10

10.1 Introduction of the static elastic-plastic analysis

10.1.1 Basic assumptions of static elastic-plastic analysis

The static elastic-plastic analysis is also known as the static pushover analysis or the pushover analysis. It is a simplified elastic-plastic analysis method that has less calculation time than the elastic-plastic time history analysis. The pushover analysis method can evaluate the seismic performance of structures conveniently and accurately. Therefore, FEMA273, ATC-40, National Institute of Standards (NIST), and other documents in the United States introduce this method. Japan's building benchmark law and China's seismic standard also advocate using this method in the nonlinear analysis of structures. Compared with the traditional static calculation method, the Pushover analysis method belongs to nonlinear analysis, so the process in the seismic performance analysis of the structure considers the elastic-plastic properties of materials.

We can get the theoretical basis for the pushover analysis from the equivalent single freedom system. In the equivalence process of the multi-freedom system, the pushover analysis needs to follow two assumptions: (1) the seismic response of the structure is related to the equivalent single freedom system, which means the seismic response of the structure can be controlled only by the first vibration mode, and the influence of other vibration modes on the seismic response of the structure can be ignored; (2) in the whole seismic response process, the displacement response mode along the height of the structure is stable, the influence of lateral deformation can be ignored.

The dynamic balance equation of the multi-freedom elastic structure system under earthquake is as follows:

$$[M]\{\ddot{x}\} + [C]\{\dot{x}\} + \{Q\} = -[M]\{1\}\ddot{x}_g \tag{10.1}$$

where $\{x\}$ is the lateral displacement of each floor of the multidegree freedom structure; $[M]$, $[C]$, $[Q]$ is the mass matrix, the damping matrix, and the interlayer restoring force, respectively; $\ddot{x}_g$ is the time history of seismic acceleration.

The above equations are decomposed by the mode decomposition method, then the displacement response of the system can be obtained as follows:

$$\{x\} = \sum_{j=1}^{n} \{\psi_j\} q_j \tag{10.2}$$

Concrete Composite Columns. DOI: https://doi.org/10.1016/B978-0-323-85171-8.00010-0

where $\{\psi_j\}$ is the j-th structural vibration mode, and q_j is the vibration mode coordinate contribution coefficient corresponding to $\{\psi_j\}$.

Based on the two assumptions in the pushover analysis, the displacement of the system can be expressed by the single shape vector $\{\phi\}$. Therefore, the lateral displacement vector $\{x\}$ of the multi-freedom system can be expressed as $\{x\}=\{\phi\}$ x_T, where x_T is the displacement of the fixed point in the structure, and Eq. (10.1) can be transformed into the following form:

$$\{\phi\}^T[M]\{\phi\}\{\ddot{x}\}+\{\phi\}^T[C]\{\phi\}\{\dot{x}\}+\{\phi\}^T\{Q\}=-\{\phi\}^T[M]\{1\}\ddot{x}_g \tag{10.3}$$

The reference displacement of the equivalent single freedom system is given as follows:

$$x_{equ}=\frac{\{\phi\}^T[M]\{\phi\}}{\{\phi\}^T[M]\{1\}}x^T \tag{10.4}$$

Further, Eq. (10.3) can be transformed into the dynamic equilibrium equation of the equivalent single freedom system, given as follows:

$$M_{equ}\ddot{x}_{equ}+C_{equ}\dot{x}_{equ}+Q_{equ}=-M_{equ}\ddot{x}_g \tag{10.5}$$

where M_{equ}, C_{equ} and Q_{equ} are the equivalent mass, the equivalent damping, and the equivalent restoring force, respectively. They can be computed as follows:

$$\begin{cases} M_{equ}=\{\phi\}^T[M]\{1\}\ Q_{equ}=\{\phi\}^T\{Q\} \\ C_{equ}=\{\phi\}^T[C]\{\phi\}\dfrac{\{\phi\}^T[M]\{1\}}{\{\phi\}^T[M]\{\phi\}} \end{cases} \tag{10.6}$$

We assume the vector ($\{\phi\}$) as a known value, then the restoring force–displacement relationship (i.e., $Q_{equ}-x_{equ}$ curve) of the corresponding equivalent single freedom system can be calculated by the above equations.

10.1.2 Specific steps of the pushover analysis

The Pushover analysis should adopt the following steps:

(1) The detailed information of the structural model should be determined in advance, such as the structural form, the member size, the material properties, and so on, so as to establish the calculation model of the structure.
(2) The load mode of the structure should be determined, the calculated vertical load is applied to the structure, and the vertical load is kept constant in the subsequent pushing process.
(3) The lateral load is applied to the structure step by step, and the performance parameters of structural members combined between the vertical load and the lateral load will be

calculated in each loading process, including the stiffness, internal force, elasticity, and elastic-plastic deformation. In the loading process, when the member is close to the yielding condition, the performance condition of the member will be corrected, and the vertical hinge method will consider this change of structural members.

(4) The performance condition of the member is corrected with the constantly loading, and repeat the operation until there are too many plastic hinges on the final member, the structure becomes a mechanism (i.e., variable system), or the lateral displacement of the structure is close to the expected failure level, then loading stops.

(5) The data of each loading step are summarized, and the relationship curve between the fixed-point displacement and the basal shear force is obtained to reflect the nonlinear characteristics of the structure.

(6) Several methods can calculate the displacement demand of the structure under the design earthquake, including the capacity spectrum method (converting pushover curve into capacity spectrum curve), the equivalent displacement coefficient method, or other methods.

(7) The elastic-plastic performance and seismic performance of the structure are evaluated.

10.1.3 Lateral loading mode of the pushover analysis method

Lateral loading mode is the distribution mode of lateral force applied step by step along the height of the structure. The choice of the lateral force distribution mode is very important for the analysis results. In order to ensure the rationality of pushover analysis results, it is necessary to select the lateral loading mode reasonably based on the characteristics of the structure. At present, the lateral force modes of the pushover analysis are mainly as follows:

(1) Uniform distribution mode of lateral force

The uniformly distributed distribution model of lateral force assumes that the lateral force of each layer of the structure is directly proportional to the mass of the layer. Meanwhile, this mode also assumes that the accelerations of all floors along are the same under the seismic loading, so the acceleration of each floor of the structure is the same under the earthquake. It is also called the acceleration loading mode. If the quality of each floor is the same, the mode becomes uniform distribution, and its expression is:

$$P_i = \frac{M_i}{\sum_{j=1}^{n} M_j} V_b = M_i a \tag{10.7}$$

where p_i is the lateral load of i-th floor, V_b is the bottom shear force of the structure, M_i is the mass of i-th floor, and a is the acceleration.

(2) Inverted triangle distribution model of lateral force

The inverted triangle distribution model of lateral force assumes that the lateral force is linear with the height of the structure, so the distribution of the lateral force along the height is proportional to the product of the floor mass and the height (i.e., the bottom shear method model). The expression of the lateral force of each floor is given as follows:

$$P_i = \frac{M_i H_i}{\sum_{j=1}^{n} M_j H_j} V_b \tag{10.8}$$

where is the height of the i-th floor from the ground

(3) Exponential distribution model of lateral force

The exponential distribution model of lateral force considers the difference in structural period, so as to assume that the lateral force is distributed in an exponential shape along the height of the structure. FEMA356 in the United States recommends this lateral force distribution mode, and the expression of the lateral force of each floor is given as follows:

$$P_i = \frac{M_i H_i^k}{\sum_{j=1}^{n} M_j H_i^k} V_b \tag{10.9}$$

where k represents the parameters related to the basic period of the structure, which can be calculated according to the following equation:

$$k = \left\{ \begin{array}{cc} 1 & T < 0.5\text{ s} \\ 1 + (T - 0.5)/2 & 0.5s \leq T \leq 2.5\text{ s} \\ 2 & 2.5\text{s} \leq T \end{array} \right\} \tag{10.10}$$

where T is the basic period of the structure.

(4) Vibration mode distribution model of lateral force

This lateral force distribution model considers the influence of high-order vibration modes, and the pushover analysis results are close to those of the dynamic time history analysis. First, the response spectrum of the structure is analyzed. Then the interlayer shear force of each layer of the structure is calculated by the square and the square root method in the mode-superposition response spectrum method, and then the lateral load of each layer is calculated inversely. The expressions are given as follows:

The lateral force and interlayer shear force of k-mode and i-th layer of the structure can be obtained by the following equation:

$$F_{ki} = \gamma_k x_{ki} m_i S_a(\xi_k, T_k) \tag{10.11}$$

$$Q_{ik} = \sum_{j=1}^{n} F_{kj} \tag{10.12}$$

where x_{ik} is the displacement value of i-particle in structural k-mode, γ_k is the influence coefficient of structural k-mode, and $S_a(\xi_j, T_k)$ is the acceleration of structural k-mode.

The interlayer shear force of each vibration mode of the structure is combined by the square and the square root method in the mode-superposition response spectrum method. The interlayer shear force and the lateral force of the i-layer in the structure can be calculated as follows:

$$Q_i = \sqrt{\sum_{j=1}^{n} Q_{ji}} \tag{10.13}$$

$$F_i = Q_i - Q_{i+1} \tag{10.14}$$

(5) Adaptive distribution model of the lateral force

In the pushover analysis, the dynamic characteristics of the structure will change, because of the change in component stiffness and other factors. In this condition, if the analysis adopts the initial unchanged lateral force distribution mode, the analysis results may contravene the results of the structure under seismic loading. Therefore, some researchers have proposed an adaptive distribution model, so the loading mode of the lateral force in the next step can be adjusted by the analysis results in the previous step of the pushover analysis.

10.1.4 Target displacement

The target displacement of the structure is the maximum displacement that the structure may reach under the input of seismic loading. Generally, it refers to the top displacement of the structure. The capacity spectrum method is one of the common calculation methods of target displacement, which is also the calculation method adopted by the American Applied Technology Committee (ATC-40) and the new building benchmark method (BSL2000) in Japan. This calculation method is that the control target is gradually loaded first by the specified loading mode to get the load–displacement curve (i.e., the capacity curve) of the structure. And then the capacity curve should be transformed into the capacity spectrum of the acceleration response and the displacement response in the single freedom system. After that, the acceleration response spectrum in acceleration-periodic format is transformed into the acceleration–displacement curve (i.e., the demand spectrum). Finally, the demand spectrum and the capacity spectrum are reflected in the same coordinate system, and the intersection of the two spectrum curves is the point of the ultimate bearing capacity and the deformation capacity to meet the seismic loading level. The basic calculation steps of the capability spectrum are given as follows:

(1) Calculation of the capacity curve. The capacity curve of the structure is obtained by the static pushover analysis method, which is the base shear-top displacement curve ($V_b - u_t$).

(2) Establishment of the capability spectrum curve. The principle is that the structure in the multi-freedom system is transformed into the single freedom system based on the assumption of the pushover analysis, then the influence of the first

vibration mode is considered in the analysis. Therefore, the capacity curve ($V_b - u_t$) is transformed point by point into the capacity spectrum curve ($S_a - S_d$). The specific transformation equations are given as follows:

$$S_a = V_b / M_{1equ} \tag{10.15}$$

$$S_d = u_T / (\gamma_1 x_{n1}) \tag{10.16}$$

$$\gamma_1 = \frac{\sum_{i=1}^{n} m_i x_{i1}}{\sum_{i=1}^{n} m_i x_{i1}^2} \tag{10.17}$$

$$M_{1equ} = \frac{\left(\sum_{i=1}^{n} m_i x_{i1}\right)^2}{\sum_{i=1}^{n} m_i x_{i1}^2} \tag{10.18}$$

where γ_1 is the participation coefficient of the first vibration mode, M_{1equ} is the equivalent mass of the first vibration model, m_i is the mass of the i-th layer, x_{i1} is the amplitude of the first vibration mode on the i-th layer, and x_{n1} is the amplitude of the first vibration mode at the top layer.

(3) Establishment of the demand spectrum curve. In the demand spectrum curve of the Chinese standard for seismic design of buildings (GB50011-2010), the natural vibration period (T) of the structure can be the abscissa and the seismic influence coefficient (α) can be the ordinate, and the substantive content is acceleration response spectrum. Therefore, the format of spectral acceleration-structure natural vibration period ($S_a - T$) of the seismic influence coefficient curve can transform into the format of spectral acceleration-spectral displacement ($S_a - S_b$). The specific transformation equation is given as follows:

$$S_d = (T/2\pi)^2 S_a = (T/2\pi)^2 \alpha(T) g \tag{10.19}$$

where $\alpha(T)$ represents the seismic influence coefficient corresponding to the natural vibration period of the structure.

The response spectrum of the structure in the elastic stage can be obtained by this method. If the structure reaches the plastic condition, the seismic demand needs to be reduced to get the elastic-plastic demand spectrum of the structure, because of the nonlinear energy dissipation of the structure. First, the equivalent damping ratio (ξ_{equ}) in the equivalent single freedom system of the structure is determined, based on the elastic-plastic deformation and hysteretic energy consumption of the structure. Then the equivalent damping ratio (ξ_{eq}) is substituted into the typical elastic demand spectrum to get the reduced elastic demand spectrum and the elastic-plastic demand spectrum of the structure.

(4) Seismic capacity evaluation of the structure. The intersection of the two curves is obtained by the combination between the capacity spectrum curve and the elastic-plastic demand spectrum curve. If one point in the capacity curve of the equivalent single freedom system coincides with the above intersection, this point will be the characteristic point of the node seismic performance, and the corresponding displacement is the spectral displacement of the system under the seismic loading. The fix point displacement of the original structure can be calculated by Eq. (10.4). Then, based on the position of the top displacement on the capacity curve ($V_b - u_t$) of the original structure, the seismic characteristics of the structure can be obtained, including the displacement of each floor, the displacement between floors, and the distribution of plastic hinges. In this way, the seismic performance of the structure can be evaluated.

10.2 Pushover analysis of the large thermal power plant

10.2.1 Project overview

The prototype structure studied in this chapter is the large thermal power plant with a 1000 MW level which belongs to the steel-reinforced concrete structure, and the structure is obtained from the actual engineering. The main power plant adopts the frame-bent structure, including the coal bunker room, the deaeration room, and the steam turbine room. The span of the coal bunker room is 14 m, the span of the deaeration room is 10m, and the span of the steam turbine room is 34 m. The longitudinal column spacing has 11 spans, including 5 spans of 10 m and 6 spans of 12 m. The structure has nine layers, and the elevations of the layers are: 0 m, 8.355 m, 11.275 m, 12.95 m, 16.725 m, 28.755 m, 33.7 m, 41.685 m, 50.255 m, and 58.905 m, respectively.

The layout plan of the frame-bent structure of the thermal power plant has many irregularities in the plane and vertical, including the large volume, the complex shape, and the irregular shape. The structural plane layout is uneven, and the structural layout of each layer is different. And there are large openings in the floor, local floor plane reinforcement, and other situations. The vertical layout of the structure is uneven in the vertical and horizontal directions, and the stiffnesses in the vertical and horizontal directions show many parts, because of the different layouts of each longitudinal frame. The height of each layer of the structure varies, and there are many staggered floors.

Meanwhile, the equivalent replacement method will be used to establish the model. That is, the section steel ratio of the replaced steel frame confined concrete column (SFC) is the same as that of the prototype steel-reinforced concrete column. Because the SFC also belongs to a kind of steel-reinforced concrete column, but has the higher bearing capacity and well seismic performance. Meanwhile, the angle steel is arranged at the four corners and the T-shaped section steel is arranged in the middle of the long side to ensure the good confining effect on concrete because the section form is rectangular. The section parameters of each equivalent column are shown in Table 10.1.

Table 10.1 Section parameters of each equivalent column.

Item	Column 1	Column 2	Column 3	Column 4
Length × width	1200 × 800	1400 × 800	1400 × 800	1000 × 800
Original section steel size	800 × 450 × 18 × 24	800 × 400 × 16 × 24	1000 × 450 × 18 × 24	600 × 400 × 18 × 24
Original steel ratio (%)	3.66	2.79	3.46	3.94
New angle steel size	180 × 14	180 × 14	180 × 18	160 × 16
New T-steel size	150 × 300 × 15 × 15	150 × 300 × 10 × 15	170 × 350 × 10 × 16	150 × 300 × 10 × 15
New steel ratio (%)	3.72	2.82	3.52	3.96

10.2.2 Model establishment

In order to get the stress characteristics of the SFC frame-bent structure, the SAP2000 analysis software is used to establish a three-dimensional calculation model to present the static elastic-plastic analysis in this section. The analysis contents mainly include the mechanical characteristics of the structure under earthquake, failure criteria, and the emergence and development law of the plastic hinges. In the establishment process of the model, the following points should be considered: (1) the frame/cable elements are adopted for the beams, columns, and supporting members in the structure; (2) the shell element is adopted for the shear wall and the floor slab; (3) the $P-\Delta$ effect should be considered; (4) the top displacement of the structure can be the calculated target displacement, and the value is 1/50 of the total height of the structure.

It has been known that the lateral loading mode has a significant influence on the calculation results. In the process of the pushover analysis in this section, two loading models are used to analyze the structure in SAP2000, including the acceleration loading mode and the modal loading, which consider multiple vibration modes.

10.2.3 Calculation parameters

(1) Setting of the plastic hinges

In the SAP2000, the discrete plastic hinges need to be used to simulate the nonlinear behavior of members. Meanwhile, the arrangement of the plastic hinges makes it quite flexible. These plastic hinges can be assigned to any position of the member, and several characteristics can be adopted, including the uncoupled bending moment, the torque, the axial force, and the shear hinges. For example, a beam element can have both a moment hinge and a shear hinge at one end.

Therefore, based on the mechanical characteristics of each member in the large thermal power plant, the hinge in each member should be adopted the following form: the column element adopts the compression-bending plastic hinge (P–M–M) which considers the coupling effect of axial force and bending moment; the beam element adopts the moment hinge (M–M) which considers the bending moment yield at both ends; the supporting unit adopts the axial force hinge (P).

(2) Loading sequence and loading order

In the pushover analysis process based on the SAP2000, the gravity load should be applied first. Then, the nonlinear analysis of the structure under the gravity load should be presented to get the internal force and deformation of the structure. After that, the lateral load is applied, and the internal force and deformation are calculated until the target displacement requirements are met. Therefore, in this section, four working conditions are used to calculate and analyze the prototype structure under 8-degree rare seismic loading. The four working conditions are as follows:

(1) The gravity + X-direction mode. The operation method is that the nonlinear analysis condition under gravity load is operated first, and then the lateral modal loading on the X-direction of the structure is applied.

(2) The gravity + Y-direction mode. The operation method is that the nonlinear analysis condition under gravity load is operated first, and then the lateral modal loading on the Y-direction of the structure is applied.

(3) The gravity + acceleration in X-direction. The operation method is that the nonlinear analysis condition under gravity load is operated first, and then the lateral acceleration loading on the X-direction of the structure is applied.

(4) The gravity + acceleration in Y-direction. The operation method is that the nonlinear analysis condition under gravity load is operated first, and then the lateral acceleration loading on the Y-direction of the structure is applied.

10.3 Analysis and discussion of calculation results

10.3.1 Static elastic-plastic curve

The static elastic-plastic curves of the structure under modal loading and acceleration loading are shown in Fig. 10.1. Curves *a*, *b*, *c*, and d represent the demand

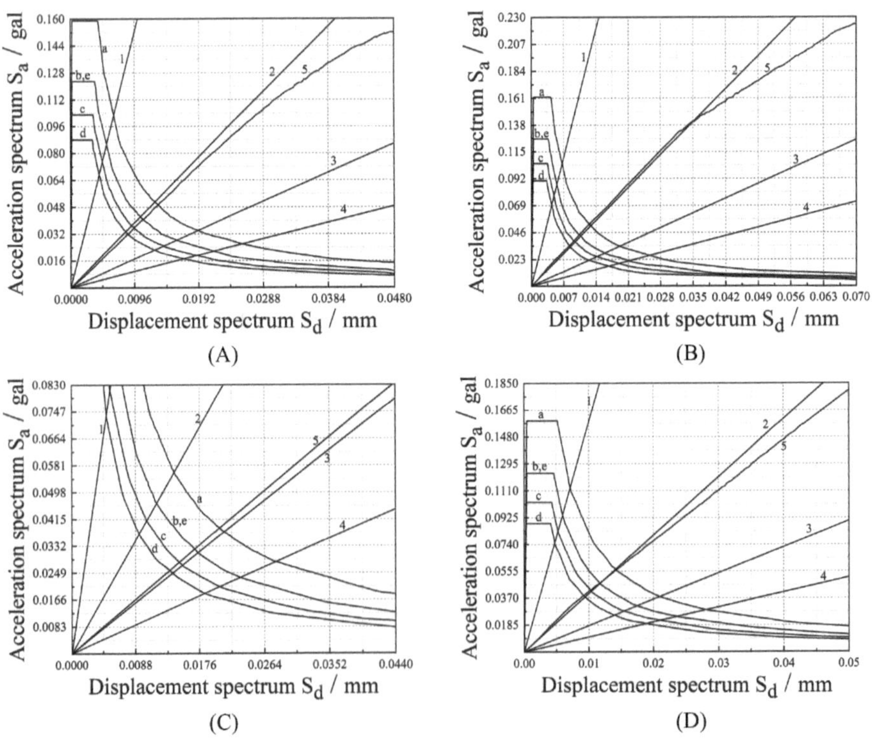

Figure 10.1 Static elastic-plastic curves of structure. (A) Modal loading in X-direction; (B) acceleration loading in X-direction; (C) modal loading in Y-direction; (D) acceleration loading in Y-direction.

spectrum curves with the damping ratios of 5%, 10%, 15%, and 20%, while the curve e represents the demand spectrum curve with the effective damping ratio of the structure. Curves 1, 2, 3, and 4 represent periodic straight lines with periods of 0.5 s, 1 s, 1.5 s, and 2 s, while curve 5 represents the capacity spectrum curve of the structure. The intersection of curve e and curve 5 is the performance point of the structure, which can get the corresponding base shear force and the top displacement. The two parameters show the deformation and the internal force that the structure can bear, and can evaluate the seismic performance and the performance level of the structure. Based on the static elastic-plastic curve of the structure in Fig. 10.1, it can be seen that the structure can reach performance points under the seismic loading in X and Y directions, which shows that the structure can resist earthquakes of this level and achieve the seismic performance index of "no collapse in the large earthquake".

10.3.2 Base shear-top displacement curve

The analysis in SAP2000 can fetch the calculation results under different loading modes. The relationships between the base shear force and the top displacement are shown in Fig. 10.2.

Based on the curve comparisons in Fig. 10.2, it can be found that:

(1) Based on the base shear force-top displacement curve in the X-direction, the base shear force-top displacement curve is basically a straight line under the small base shear, which shows that the structure is at the linear elastic stage and the stress state of each member does not reach the yield stress. With the continuous increase in the base shear, some members of the structure reach the yield state one after another. And the increasing rate of top displacement is sped until the structure becomes a mechanism. Meanwhile, the base shear force and the top displacement reach the maximum values.

(2) Based on the base shear force-top displacement curve in the Y-direction, the base shear-top displacement curve is basically a straight line, which shows that the structure is at

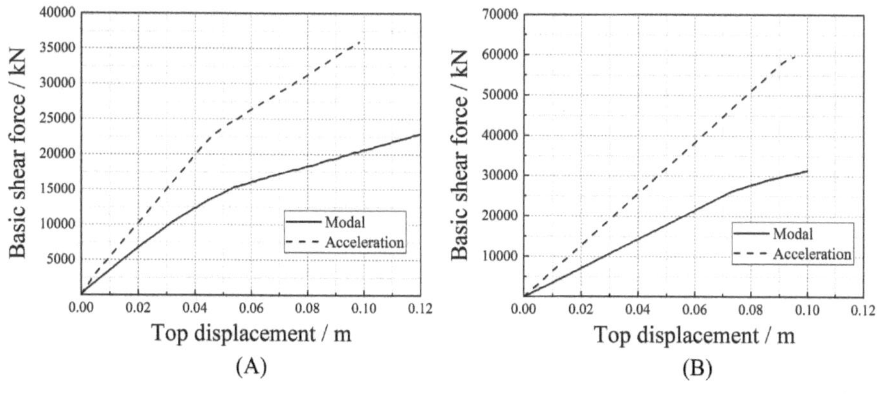

Figure 10.2 Relationships between base shear force and top displacement. (A) X-direction; (B) Y-direction.

the linear elastic stage. Because there are numerous reinforced concrete members in the structure, which mostly belong to the brittle failure and lead to no obvious inelastic stage in the base shear-top displacement curve of the structure.

(3) The mechanical properties of the structure in X and Y directions are quite different. This is mainly because the longitudinal span of the structure differs from the transverse span, so there are more longitudinal supports and shear walls to increase the lateral stiffness, so the stiffness is also greater, resulting in great differences in the stiffness distribution and bearing capacity of the structure in two directions. And this kind of structure is extremely complex, the distribution along the two directions is irregular, and the layer distribution along the height direction is more irregular. This feature also affects the order and state of the structural plastic hinges, resulting in significant differences in the mechanical properties in the two directions.

(4) The mechanical performances of the structure in X and Y directions are quite different. This is mainly because the longitudinal span of the structure differs from the transverse span. And it is necessary to set more longitudinal supports and shear walls to increase the lateral stiffness. These measures lead to the significant differences in the stiffness distribution and the bearing capacity of the structure in two directions. Meanwhile, the structure is extremely complex, the distribution along the two directions is irregular, and the layer distribution along the height direction is more irregular. These features also influence the order and state of structural plastic hinges, which lead to significant differences in mechanical properties in the two directions.

(5) Based on the curve in Fig. 10.2, we can find that the response of the structure under different lateral loading modes is also very different. Therefore, the response of the structure under different loading modes should be integrated to evaluate the seismic performance of the structure in the static elastic-plastic analysis.

10.3.3 Layer shear force

Based on the calculation in SAP2000, the analysis results of the structural layer shear force under different loading modes can be obtained. The change of structural layer shear force with the structural height is shown in Fig. 10.3. Fig. 10.3 shows that the layer shear force under several loading modes is not an oblique

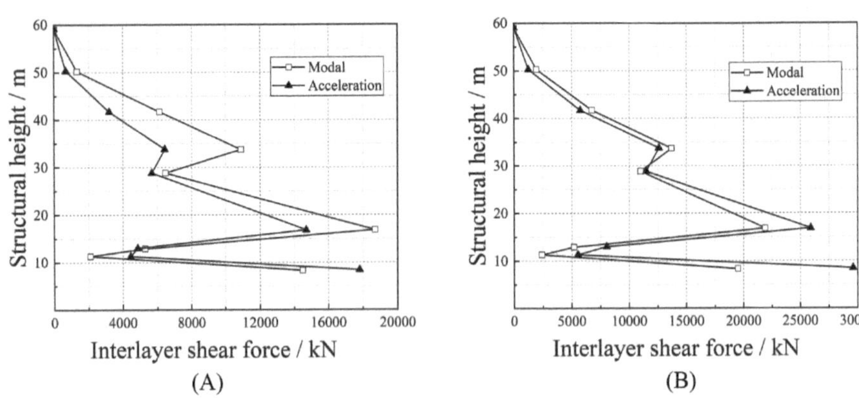

Figure 10.3 Structural layer shear force. (A) X-direction; (B) Y-direction.

straight line, which shows that the structural layer quality is different and the vertical direction of the structure is irregular. Meanwhile, the curve reflects that the shear force of the 2-layer, the 3-layer, and the 5-layer show a decreasing trend, because of the prominent layer and the small layer area. The shear force of the first layer is smaller than that of the 4-layer under the two modal loading modes, due to the small area. The shear force of the first layer under the acceleration loading mode is the largest, mainly because the acceleration loading mode emphasizes the stress at the bottom of the structure.

We should note that the structural layer shear force of the upper part is large under the modal loading mode. And the following conclusions can be drawn: the modal loading mode focuses on the upper part of the structure, and the structure under this analysis shows the low base shear force and the good ductility; the acceleration loading mode focuses on the lower part of the structure, and the structure under this analysis shows the high base shear force and the poor ductility. Therefore, the acceleration loading mode and modal loading mode should be considered comprehensively for the analysis of the seismic performance of the structure.

10.3.4 Interlayer displacement angle

The calculation in SAP2000 can get the structural analysis results of the interlayer displacement angle under different loading modes. The change of structural interlayer displacement angle with the structural height is shown in Fig. 10.4. We can find that under the seismic loading corresponding to the performance point, the deformation under modal loading mode is small, and the deformation under acceleration loading mode is large.

The interlayer displacement angle curve in the X-direction shows that: under the acceleration loading mode, the maximum interlayer displacement angle in the X-direction occurs in the middle of the structure, and the value is 1/710; under the modal loading mode, the maximum interlayer displacement angle in X-direction occurs in the middle and the lower part of the structure, and the value is 1/310; the

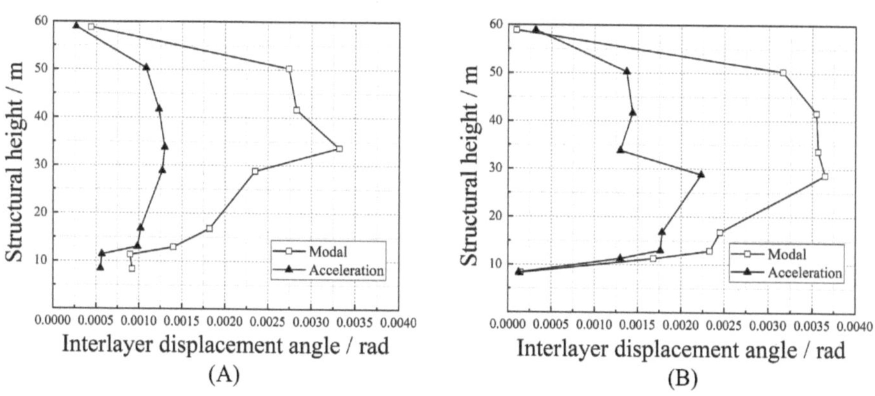

Figure 10.4 Structural interlayer displacement angle. (A) X-direction; (B) Y-direction.

maximum interlayer displacement angles under the two loading modes are less than 1/250, which is the limit value of elastic-plastic interlayer displacement angle of the multilayer and high-rise structures specified in the seismic standard.

It can be seen from the curves that the maximum interlayer displacement angles under the two loading modes are different, which is mainly caused by the uneven stiffness, due to the uneven floor distribution. Moreover, the structural interlayer displacement angle in the modal loading mode is greater than that in the acceleration loading mode. And the upper interlayer displacement angle in this mode is larger, mainly because the structural displacement in the modal loading mode is larger. Meanwhile, there is a large sudden change in the curve, which is mainly because of the complex structure and the significant difference in the distribution of each layer.

The interlayer displacement angle curve in the Y-direction shows that: the maximum interlayer displacement angle in Y-direction occurs in the middle of the structure, and the value is 1/444; under the modal loading mode, the maximum interlayer displacement angle in the X-direction occurs in the middle and the upper part of the structure, and the value is 1/280; the maximum interlayer displacement angles under the two loading modes are less than 1/250, which is the limit value of elastic-plastic interlayer displacement angle of the multilayer and high-rise structures specified in the seismic standard.

It can be seen from the curves that the maximum interlayer displacement angles under the two loading modes mainly occur in the middle and upper part of the structure, mainly because the upper part of the structure has more locally prominent layers and the lateral stiffness is small. Under the loading in the Y-direction, the structural interlayer displacement angle in the modal loading mode is greater than that in the acceleration loading mode. Meanwhile, there is a large sudden change in the curve, because the structure of the thermal power plant is complex, and the uneven layer distribution leads to the uneven stiffness.

10.3.5 Cooperation relationship in the frame-bent structure

For the structural failure under the modal loading mode and the acceleration loading mode in X and Y directions, the ratio of the shear borne by the bent column of each axis in the transverse frame structure to the total transverse base shear is shown in Fig. 10.5.

It can be seen from Fig. 10.5 that the ratio in the modal loading mode is slightly higher than that in the acceleration loading mode. Meanwhile, the shear ratio borne by some bent columns on axis numbers (2, 4, 6, 8, 10, and 12) is large. Because the frame structure bears most of the shear force after the arrangement of the shear wall, we can use the frame as an independent seismic defense line. Although the bent structure can work together with the frame structure, it cannot be used as an independent seismic defense line. In order to make the frame column bear more shear force, more shear walls or supports should be arranged in the frame structure to further improve the shear ratio and seismic capacity of the frame structure.

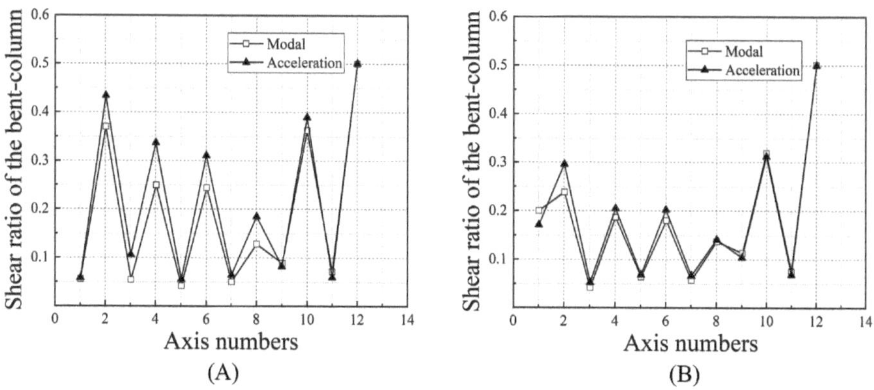

Figure 10.5 Ratio of shear force borne by the bent-column to base shear force. (A) X-direction; (B) Y-direction.

10.3.6 Structural plastic hinge

In the finite element nonlinear analysis of SAP2000, we present the pushover analysis of the structure in the X and Y directions using two different loading modes. Then the distribution of plastic hinges can be obtained when the structure is damaged under an earthquake (see Fig. 10.6), where the distribution of beams, columns, and supports are shown, and the red points are the hinge points.

It can be seen from Fig. 10.6 that plastic hinges basically appear at the beam end, and the distributions of the structural plastic hinges are basically the same under the two loading modes. The differences are that there are more plastic hinges distributed in the bottom layer under the acceleration loading mode, and more plastic hinges distributed in the higher layers under the modal loading mode, mainly because the modal loading emphasizes the weak parts are at the upper part of the structure. Therefore, the combination of modal loading and acceleration loading can better analyze the overall performance of the structure.

Under both loading modes, a large number of plastic hinges are distributed in the beams of the 5th and 6th layers. And the interlayer displacement angle of the 5th layer is large, which is mainly because the 5th layer is the protruding layer of the structure and the 6th layer is the coal hopper beam. In order to meet the production requirements, a series of coal hopper beams with super large sections are arranged, so the stiffness of the coal hopper beam differs from that of an ordinary frame beam. As a result, there are many plastic hinges on these two parts, and a large interlayer displacement angle occurs. This phenomenon is more obvious in pushover analysis in the Y-direction.

In order to study the structural plastic hinge distribution under the two loading modes in detail, we select the representative structural facades for observation.

The structural plastic hinge distribution in the X-direction under modal loading is shown in Fig. 10.7. As shown in Fig. 10.7, the plastic hinge is completely distributed

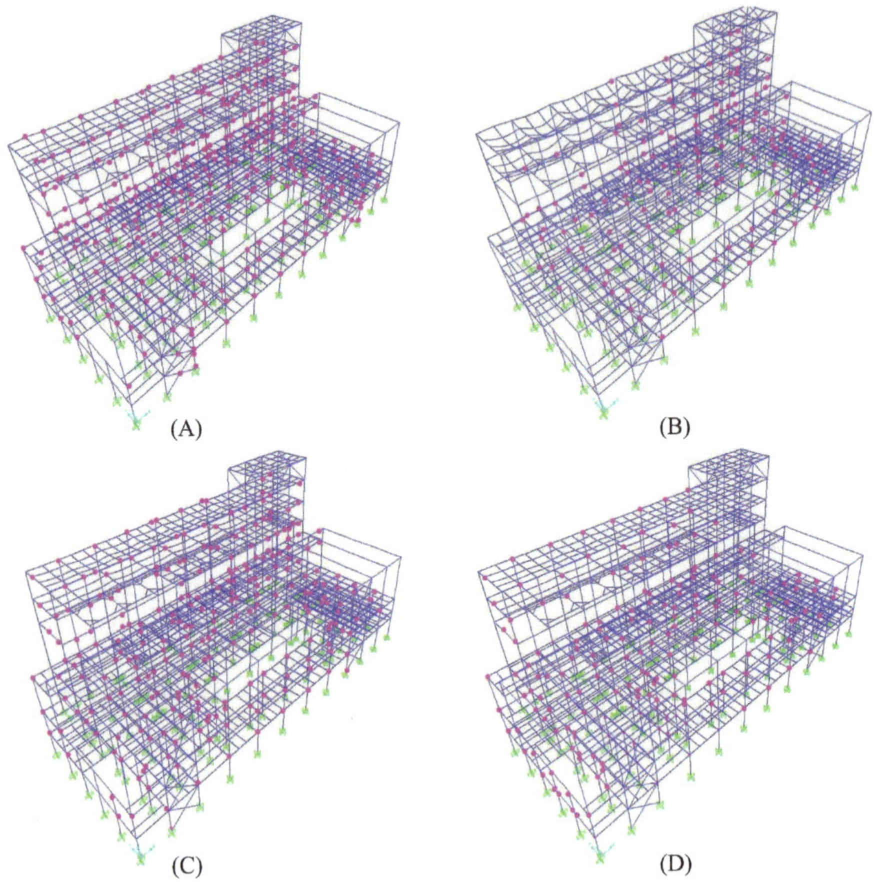

Figure 10.6 Distribution of plastic hinges. (A) Modal loading in X-direction; (B) acceleration loading in X-direction; (C) modal loading in Y-direction; (D) acceleration loading in Y-direction.

at the beam end in the frame structure. In the bent structure, the plastic hinge mainly appears at the lower part of the structure, and the plastic hinge basically appears on the beam, only a few plastic hinges appear on the column and support. Based on the development of the plastic hinge, it can be seen that the plastic hinge first appears in the upper part of the structure, and then develops into the overall structure.

The structural plastic hinge distribution in the X-direction under acceleration loading is shown in Fig. 10.8. As shown in Fig. 10.8, there are fewer structural plastic hinges under acceleration loading mode, and the plastic hinges are all distributed on the beam. The plastic hinges of the bent structure mainly appear in the lower part of the structure, and the plastic hinges of the frame structure are mainly distributed at the beam end of the middle area. Based on the observation of the

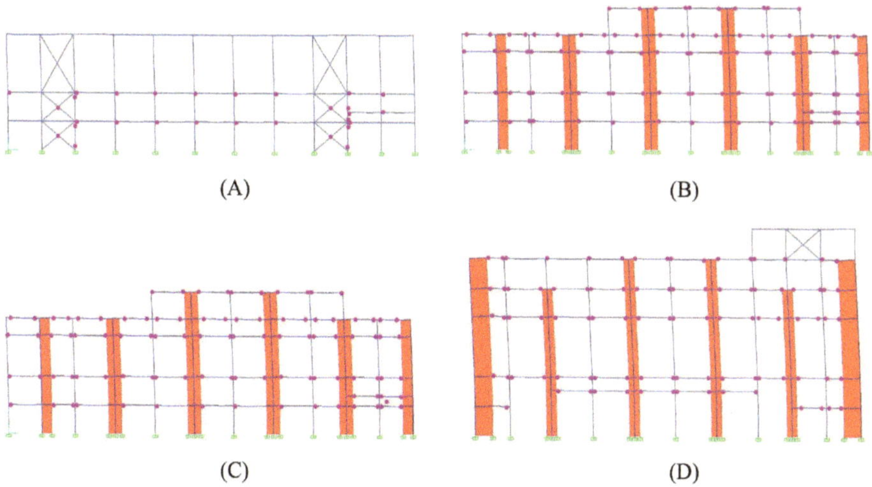

Figure 10.7 Structural plastic hinge distribution in the X-direction under modal load. (A) A-axis structure; (B) B-axis structure; (C) C-axis structure; (D) D-axis structure.

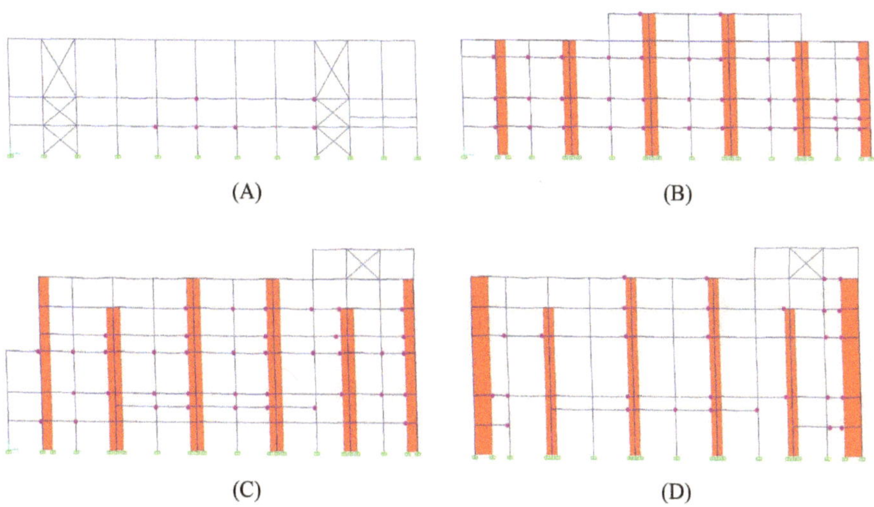

Figure 10.8 Structural plastic hinge distribution in the X-direction under acceleration loading. (A) A-axis structure; (B) B-axis structure; (C) C-axis structure; (D) D-axis structure.

development of the plastic hinges, it can be seen that the plastic hinges first appear at the bottom of bent structure, and then gradually increase from bottom to top in the frame structure. Among them, there are more plastic hinges in the protruding parts of the second layer, third layer, and fifth layer, and a large number of plastic hinges are distributed in the structure between 1-axis and 6-axis.

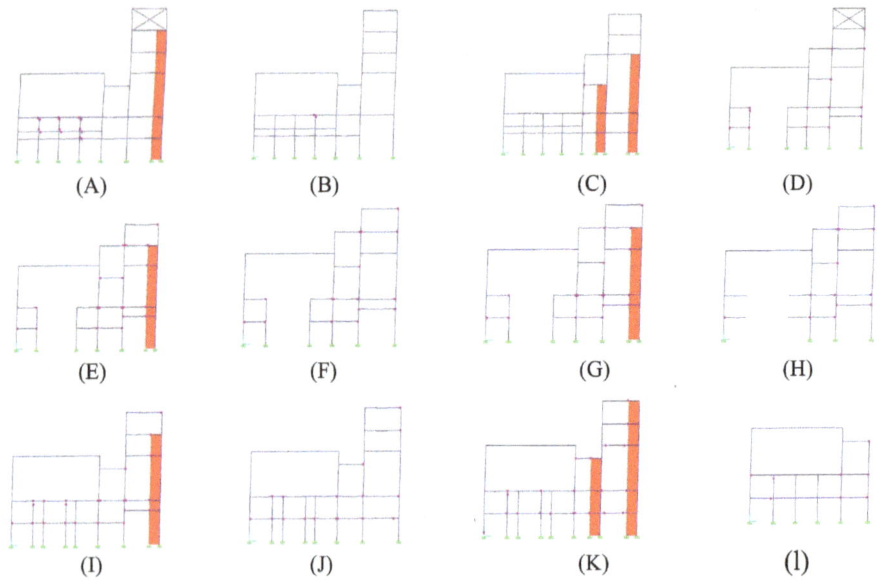

Figure 10.9 Structural plastic hinge distribution in the Y-direction under modal loading. (A) 1-axis; (B) 2-axis; (C) 3-axis; (D) 4-axis; (E) 5-axis; (F) 6-axis; (G) 7-axis; (H) 8-axis; (I) 9-axis; (J) 10-axis; (K) 11-axis; (L) 12-axis.

The structural plastic hinge distribution in the Y-direction under modal loading is shown in Fig. 10.9. As shown in Fig. 10.9, most of the structural plastic hinges are distributed on the beam, and parts of the plastic hinges are distributed on the bent columns under the mode loading mode in the Y-direction. Based on the observation of the development of the plastic hinge, it can be seen that plastic hinges first appear on the bent structure, and then develop into the overall structure. It can be also found that there are more plastic hinges distributed on the beams and columns of the bent structure. Therefore, more shear walls or supports should be arranged to improve the seismic performance of the structure in the Y-direction.

The distribution of structural plastic hinges in the Y-direction under acceleration loading is shown in Fig. 10.10. As shown in Fig. 10.10, there are fewer structural plastic hinges under acceleration loading mode, and some plastic hinges also distributed on bent columns. Based on the observation of the development of plastic hinges, it can be seen that a few plastic hinges appeared at the bottom of the structure, and then gradually increase from bottom to top in the frame structure. Among that, there are more plastic hinges in the structural protruding parts of the second layer, third layer, and fifth layer. Meanwhile, there are many plastic hinges distributed on the bent structure. Therefore, more shear walls or supports should be arranged to improve the seismic performance of the structure in the Y-direction.

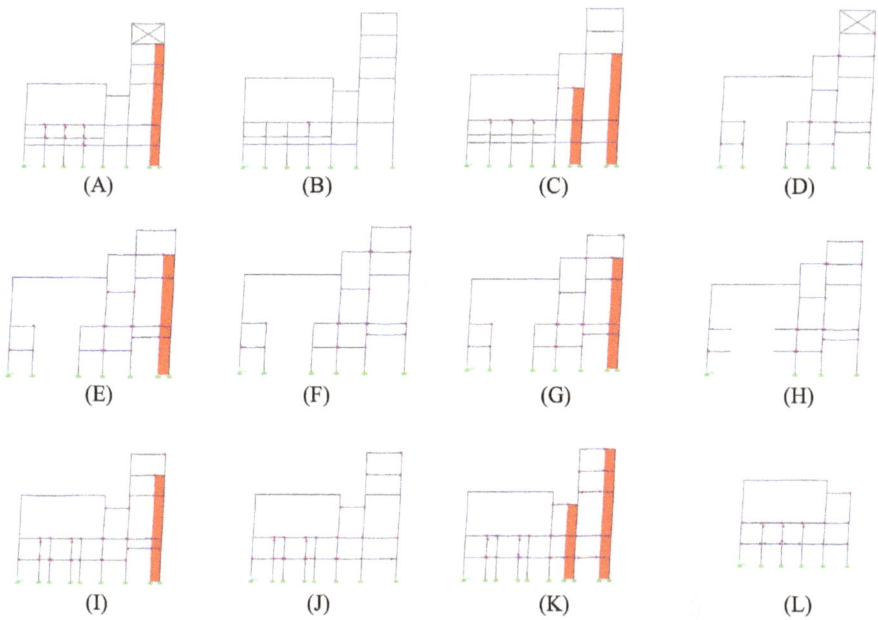

Figure 10.10 Structural plastic hinge distribution in the Y-direction under acceleration loading. (A) 1-axis; (B) 2-axis; (C) 3-axis; (D) 4-axis; (E) 5-axis; (F) 6-axis; (G) 7-axis; (H) 8-axis; (I) 9-axis; (J) 10-axis; (K) 11-axis; (L) 12-axis.

10.4 Introduction of the performance-based seismic design method of the frame-bent structure of the large thermal power plant

The structure of frame-bent structure of the large thermal power plant is complex, and the layout is irregular. The current standard cannot be used in seismic design directly. The seismic performance index of the structure needs to integrate various factors, such as the selection of plant site, the function of structure and equipment, the investment and income, the loss and restoration after the earthquake, and so on. Therefore, the performance-based seismic design method is suitable for the design of the large thermal power plant. The performance-based seismic design method makes the seismic design transition from macro qualitative objectives to multiple specific and quantifiable objectives. It can focus on different performance objectives to pay more attention to the in-depth analysis and demonstration of the implementation of performance objectives. The seismic design of the large thermal power plant should be analyzed and studied carefully, according to the actual engineering conditions, and the joint agreement of the designers and the owner (Cheng & Xue, 2003).

10.4.1 Seismic fortification target

The seismic performance target of the structure is the expected seismic performance level for each seismic fortification intensity. The seismic fortification goal stipulated in the Chinese standard for seismic design (GB 50011-2001, 2008) requires that buildings should have different resistance to earthquakes of different frequencies and intensities in usage period, that is "the building has no damage after small earthquakes, the building can be repaired after the medium earthquakes, and the building does not collapse after the large earthquakes". The specific meaning is that when suffering from an earthquake with the fortification intensity lower than that specified in the region, the buildings are not damaged or repaired so that they can continue to be used; when suffering from an earthquake with the fortification intensity specified in the region, the buildings may damage to a certain extent, but they will not endanger people's lives and the safety of production equipment, and can still be used after repair; when suffering from an earthquake with the fortification intensity higher than that specified in the region, the buildings will not collapse or cause serious life-threatening damage.

Based on the seismic fortification target specified in the Chinese standard for seismic design of buildings, the designed building structure should meet the following conditions: under the frequent earthquake intensity, the structure is basically in the elastic stage and will not damage; under the basic earthquake intensity, the structure will enter the elastic-plastic stage, but there will be no serious damage; under the rare earthquake intensity, the structure will show serious damage, but the structure will not collapse.

10.4.2 Basic content of the performance-based design method

Performance-based seismic design of structures is to put forward corresponding seismic fortification standards according to the determined seismic performance objectives of structures. It mainly includes four aspects: the seismic fortification level, building the performance level, the structural seismic performance target, and the design method to achieve the structural seismic performance target. We mainly divided it into three steps: clarification of the structural performance objectives, the structure design, and structural performance evaluation. Performance-based seismic design methods mainly include the comprehensive design method, the energy-based design method, and the displacement-based design method.

(1) The comprehensive design method

The comprehensive design method is that the minimum total investment of the project on the premise that the building reaches the basic performance goal. This design method considers the important factors of seismic design comprehensively, so as to maximize the performance-based design idea. This design idea can provide the best design scheme, but it has many uncertain factors, a wide range, and a complex design process.

(2) The energy-based design method

The energy-based design method considers that the damage to the structure and the internal facilities are determined by the input energy of the earthquake and the energy consumption of the structure. The advantage of this method is that it can directly estimate the potential damage degree of the structure, and the setting of energy dissipation elements can effectively control the loss. However, this method is not simple enough and has many uncertain factors.

(3) The displacement-based design method

The displacement-based design method uses displacement as an index to evaluate the failure of the structure, which can get the performance level of the structure directly. It can be divided into the ductility coefficient method roughly, the capacity spectrum method, and the direct displacement-based method.

The capacity spectrum method evaluates the structural performance by comparing the ductility demand with the ductility. The capacity spectrum method is to transmit the force–displacement curve obtained by the static elastic-plastic analysis into the acceleration–displacement relationship capacity spectrum of the equivalent single freedom system. The structural seismic performance is evaluated by the capacity spectrum with the demand spectrum, based on the structural ductility coefficient.

The direct displacement-based method focuses on the realization of target displacement. This method is intuitive and simple to implement. Therefore, the direct displacement-based method is selected to study the seismic design method of the frame-bent structure of the large thermal power plant.

10.4.3 Direct displacement-based method

The basic ideas of the direct displacement-based seismic design method are as follows: (1) the original structure is equivalent to the "alternative structure" of the single freedom system, the stiffness adopt the secant stiffness at the maximum displacement, and the damping ratio adopt the viscous damping ratio in the inelastic response which has the same hysteretic energy consumption; (2) the pre-determined design displacement and expected ductility are used to predict the damping ratio; (3) the effective period is determined by the displacement response spectrum, based on the estimated damping ratio and design displacement; (4) the equivalent stiffness at the maximum displacement can be obtained from the single freedom vibration period equation, so that the design base shear force is the product of the equivalent stiffness at the maximum displacement and the design displacement; (5) the seismic loading and the seismic effect of the original structure are calculated from the base shear force.

The direct displacement-based seismic design method mainly adopts the secant stiffness at the maximum displacement and the equivalent single freedom structure to replace the original structure to calculate the seismic response. The key characteristics are as follows: (1) The displacement is taken as the design variable; (2) the seismic effect is calculated by using the expected displacement under a certain level of seismic loading, so as to make the members in the building reach the expected deformation and make the building reach the expected displacement; (3) designers can easily and freely control the damage to the structure.

10.5 Member design method of the large thermal power plant

10.5.1 Shear wall design

The shear wall design in the design of steel-reinforced concrete frame structure mainly refers to the study (Kang, 2009), because SFC is focused in this chapter. The number and location of the shear wall layout have a significant influence on the dynamic characteristics of the structure, which leads to the different seismic shear and overturning moments borne by the structure. The small number of shear walls make the structure can bear large shear force under an earthquake. A large number of shear walls will increase the stiffness of the structure and the seismic effect borne by the overall structure, which will bring difficulties to the section design. Therefore, the layout of the shear wall should follow the following principles:
(1) The two-way lateral force resistance requirements of the structure should be considered simultaneously, and the shear walls should be set in both principal axes of the structural plane.
(2) The shear wall should run through the full height of the structure to ensure the uniformity and continuity of the structure along the height.
(3) The shear wall should be arranged around the building plane as far as possible to increase its torsion resistance arm and improve the torsion resistance capacity of the structure.
(4) The shear wall should be arranged symmetrically on the plane of the structural unit, and the horizontal force line should be close to the stiffness center as far as possible to avoid excessive torsion.
(5) The shear walls should be arranged as T-shaped or L-shaped in two directions, and the shear walls between the two directions are wing walls to improve the torsional stiffness.
(6) The overturning moment borne by the shear wall should be greater than 50%, and the seismic shear borne by the bottom shear wall should be less than 75% of the total shear force.
(7) The processing pipeline can pass through the middle of the shear wall, but it cannot pass through the member at the structural edge, nor should it pass through the reinforced section at the bottom of the shear wall. When the pipeline passes through the wall, whole compensation measures should be taken.

10.5.2 Steel frame confined concrete column design

(1) Dividing axial compression ratio

The failure modes of the ordinary reinforced concrete columns can be divided into the large eccentric compression failure and the small eccentric compression failure. Large eccentric compression failure can be called tensile failure. The tensile reinforcement reaches the yield condition first, and then the concrete edge in the compression area reaches the ultimate compressive strain. This failure has a certain omen and belongs to ductile failure.

The small eccentric compression failure can be called compression failure. The failure is that the concrete edge in the compression area reaches the ultimate compressive strain, but the tensile reinforcement does not reach the yield condition. This failure has no certain omen and belongs to brittle failure. The structural design must meet the requirements of ductility design, so it must be designed according to the large eccentric compression members. And the ductility control of the compression members is considered by the dividing axial compression ratio in the Chinese standard.

Based on the experimental and theoretical studies in the above chapters, it has been known that the SFC has better performance than ordinary steel-reinforced concrete under the same axial compression ratio, including better hysteretic characteristics, better ductility, and better seismic performance. Under the different axial compression ratios, the hysteretic curves of composite columns are also significantly different. With the increase in the axial compression ratio, the hysteretic curve area is larger, and the energy dissipation capacity of the column is enhanced. However, with the increase in the cycle times and deformation, the greater the axial compression ratio leads to the more serious reduction of bearing capacity and the worse ductility of the composite columns.

Based on the Chinese technical specification for the steel-reinforced concrete composite structures (JGJ 138-2001, 2001), the steel-reinforced concrete columns have better hysteretic performance and ductility than reinforced concrete columns under the same axial compression ratio. Therefore, the advantages of steel should be considered in the calculation of the axial compression ratio. The axial compression ratio (μ_N) of steel-reinforced concrete frame columns is calculated by the following equation:

$$\mu_N = N/(f_cA_c + f_aA_a) \tag{10.20}$$

where N is the design value of axial pressure of composite column, f_c is the design value of concrete strength, A_c is the sectional area of the concrete part, f_a is the design value of compressive strength of section steel, and A_a is the sectional area of section steel. For frame columns with different seismic grades and different structural systems, the dividing value of the axial compression ratio should be adjusted. Because SFC has better hysteretic behavior and ductility than the ordinary steel-reinforced concrete column, which leads to a higher value of the dividing axial compression ratio. Furthermore, the value of dividing axial compression ratio is shown in Table 10.2.

(2) Steel ratio

SFC is improved from the steel-encased reinforced concrete column and the steel-reinforced concrete column. It is a novel type of composite column that is based on the theory of the confined concrete and considers the influence of section form. Based on the study results of the steel-reinforced concrete column (Chen et al., 2007; Feng, 2014; Shi et al., 2013; Tang, 2011; Wang & Guo, 2006; Wang et al., 2009), the following conclusions can be drawn: (1) the steel-reinforced concrete column with a larger steel ratio has the better the seismic performance; (2) if the steel ratio is

Table 10.2 Dividing axial compression ratio of SFC.

Structure type	Hooping form	Seismic grade		
		First level	**Second level**	**Third level**
Frame structure	Compound hooping	0.7	0.8	0.9
Frame-shear wall structure Frame tube structure	Compound hooping	0.75	0.85	0.95
Bearing frame structure	Compound hooping	0.65	0.75	0.85

small, the confining effect on the core concrete will be too small to reflect the characteristics of the steel-reinforced concrete column; (3) if the steel ratio is large, it will be difficult to pour concrete, and the section size of section steel is too large to affect the bonding effect between the section steel and the concrete, so it is difficult to make the section steel and the concrete to work together and easy to occur bonding failure. Therefore, almost all countries have regulations on the steel ratio in steel-reinforced concrete columns. The American Institute of Steel Construction (AISC) stipulates that the steel ratio in steel-reinforced concrete members should not be less than 4%. The Japanese standards stipulate that the upper limit of steel ratio in steel-reinforced concrete members should not be over 8%. In Chinese standards, the range of steel ratio is from 2.5% to 7.0%.

Based on the experimental studies of SFC in Chapter 8, all SFC specimens show great mechanical behavior and have very full hysteretic curves. The higher steel ratio leads to the slower decreasing stage after the maximum seismic load, which shows that the seismic performance is better and reflects the superior seismic performance. Moreover, all composite columns show a small reduction in the bearing capacity, which shows that the bearing capacity of the new composite column is relatively stable. It should be noted that the components should have a sufficient concrete cover thickness within the range of reasonable steel ratio. In order to ensure the convenience of construction, service conditions and economy, the steel ratio should be controlled at a range from 3% to 7%, which is the applicable range of the reasonable steel ratio of the large thermal power plant.

10.6 Performance level and performance target of SFC frame-bent structure in the large thermal power plant

10.6.1 Division of the structural performance level

The structural performance level is used to describe the damage degree of the buildings under certain conditions of seismism. It is a combination of structural performance, the performance of non-structural components, and the system performance of indoor facilities. However, the seismic design of building structures mainly takes "human" as the major goal at present, which focuses on protecting life safety during seismism, and rarely considers the influence of key equipment in the structure.

However, for the special structure of the main powerhouse of the large thermal power plant, the layout of the processing pipeline is complex, the plane and vertical layout of the structure are irregular, and equipment safety is more important than structural safety. The loss and secondary disasters caused by the earthquake to the equipment in the structure are huge. It is unreasonable to use the original structural performance design level.

Based on numerous experimental and theoretical existing studies, in order to consider comprehensively the structural characteristics of the frame-bent structure in the large thermal power plant, the seismic performance of the SFC frame-bent structure can be divided into four levels: normal use, basic use, life safety, and near collapse. The seismic performance level of the structure and its divisions are shown in Table 10.3. The seismic performance objectives of the structure are shown in Table 10.4.

Table 10.3 Seismic performance level of the structure.

Performance level	Description of failure mode and structural performance	Damage degree
Normal use	The structure and equipment have not been damaged and still be safe without repair.	Intact
Normal use	The structure is damaged slightly and can use after repair. The equipment can operate normally after simple maintenance.	Minor damage
Life safety	The structure and equipment are seriously damaged and can be used after repair. But the repair cost is high and does not affect life safety.	Moderate damage
Near collapse	The structure is seriously damaged without collapse and cannot be repaired. The equipment is scrapped basically.	Serious damage

Table 10.4 Seismic performance objectives.

Seismic level	Structural seismic performance target			
	Normal use	Normal use	Life safety	Near collapse
Frequent earthquake	A	—	—	—
Fortification earthquakes	B	A	—	—
Rare earthquake	C	B	A	A

In Table 10.3, the performance index A represents that the structure meets the requirements of "normal use" of the performance level under frequent earthquakes, meets the requirements of "basic use" of the performance level under fortified earthquakes, and meets the requirements of "life safety" or "near collapse" of the performance level under rare earthquakes. The performance index B represents that the structure meets the requirements of "normal use" of the performance level under frequent earthquakes and meets the requirements of "basic use" of the performance level under rare earthquakes. The performance index C represents that the structure meets the requirements of "normal use" of the performance level under rare earthquakes.

10.6.2 Quantification of the structural performance objectives

Displacement-based seismic design needs to determine the relationship between structural performance level and structural displacement. The experimental study shows that the interlayer displacement angle can be used as an effective index to reflect the deformation of each member in the frame structure and the influence of layer height. And it can directly reflect the damage degree of the structure, so it has been widely used.

We take the SFC frame-bent structure of the large thermal power plant with a 1000 MW level as an example to describe the seismic design. In the condition of class II site in degree 8 area, when encountering frequent earthquakes, performance target A requires that the structure meet the elastic design, and the seismic bearing capacity and interlayer displacement angle of the structural members meet the requirements (Chu et al., 2018; GB 50011-2001, 2008). Based on the experimental and theoretical studies, it can be concluded that when the limit value of interlayer displacement angle is controlled within the range of 1/750−1/650, the $P-\Delta$ effect is less than 5%. Therefore, the second-order effect of gravity can be ignored in the frame-bent structure. In addition, industrial buildings with non-structural components are looser than ordinary civil buildings. Based on the practical application of the project and the economic comprehensive indicators, the limit value of the elastic interlayer displacement angle of the SFC frame structure can be taken as 1/700−1/600. The deformation of process equipment and the displacement of the process pipeline should meet the corresponding deformation requirements. For the "basic service" performance level, we can suggest the limited value of the structural interlayer displacement angle is 1/600.

10.7 Displacement-based seismic design of the SFC frame-bent structure in the large thermal power plant

10.7.1 Target displacement mode

The target displacement mode of the structure is the way to realize the equivalent transformation from a multi-freedom to a single-freedom system. Therefore, determining the target displacement mode is an important step in displacement-based seismic design.

There are two major methods to determine the target displacement curve: one assumes that the displacement of each layer of the structure reaches the target displacement; another assumes that the displacement of the weak layer of the structure reaches the target displacement, and the lateral displacements of the other layers are less than that of the weak layer. The former is an ideal state, while the latter is more suitable for the actual situation. The design floor displacement curve shown in Eq. (10.21) is used as the initial displacement curve of the structure, and then the pushover analysis is used to correct the curve.

$$\phi_c u_i = \phi_i u_c \tag{10.21}$$

where u_i is the displacement at the i-th layer, u_c and ϕ_c are the displacement and vibration mode value of the floor that firstly reaches the control limit displacement condition, ϕ_i is the shape of vibration mode, and u_c can be calculated by the following equations:

$$\Delta u_i = [\theta] h_i \tag{10.22}$$

$$u_i = \phi_i u_t \tag{10.23}$$

where Δu_c represents the interlayer displacement of different layer heights, $[\theta]$ represents the limit value of interlayer displacement angle, u_i is the absolute displacement of different layer heights, u_t is the absolute displacement of the top layer, and $h_i$is the layer height of the structure.

The dimensionless vibration mode shape can be calculated by the following equation:

$$\phi_i = \left(3\zeta_i - \zeta_i^3\right)/2 \tag{10.24}$$

where, ζ_i is the layer height ratio, $\zeta_i = H_i/H$, H_i is the height of the i-th layer, and H is the total height of the structure.

10.7.2 Equivalent parameters of the equivalent single-freedom system

For the displacement-based seismic design, the first step should be to solve the problem of transformation from the multi-freedom system to the equivalent single-freedom system. The parameters in the equivalent single-freedom system can be set as follows: the equivalent stiffness is K_{equ}, the equivalent mass is M_{equ}, the equivalent damping ratio is ζ_{equ}, the equivalent displacement is u_{equ}, the equivalent height is H_{equ}, the base shear is V_b, and the lateral displacement of each particle is u_i and the horizontal seismism of each particle is F_i.

In order to transmit the structural system with N-degree-freedom into an equivalent single-freedom system, the following assumptions should be set: (1) the multi-freedom system produces seismic response according to the assumed displacement shape; (2) the base shear force of the equivalent system is equal to that of the original system; (3) the works by the lateral seismic force on the two systems are equal.

According to the first assumption, the displacement vector of the multi-freedom system can be expressed by the following equation:

$$u(\xi, t) = \phi(\xi) z(t) \tag{10.25}$$

where $\phi(\xi)$ is the displacement shape function of the multi-freedom system, which can adopt a certain lateral displacement mode according to the distribution of the mass and the stiffness of the structure along the height; the function $z(t)$ is a time-dependent function.

The vibration of particles along the horizontal direction is assumed as a simple harmonic vibration, the above equation can be transmitted into the following expression:

$$u(\xi, t) = \phi(\xi) Y_o \sin(\omega t) \tag{10.26}$$

Based on the above equation, the acceleration vector of the multi-freedom system can be obtained as:

$$a(\xi, t) = -\phi(\xi)\omega^2 Y_o \sin(\omega t) = -\omega^2 u(\xi, t) \tag{10.27}$$

Meanwhile, the relative displacement value of each layer relative to the equivalent displacement can be obtained as follows:

$$c_i = u_i / u_e qu = a_i / a_{equ} \tag{10.28}$$

where a_{equ} is the equivalent acceleration of the equivalent single-freedom system.

Based on the above equation, the lateral seismism of i-th particle of the multi-freedom system can be expressed as:

$$F_i = m_i a_i = m_i c_i a_{equ} \tag{10.29}$$

According to the second assumption, the calculation equation of the base shear force of the structure can be obtained as follows:

$$V_b = \sum_{i=1}^{n} F_i = \left(\sum_{i=1}^{n} m_i c_i \right) a_{equ} = M_{equ} a_{equ} \tag{10.30}$$

Based on above equation, the calculation equation of the equivalent mass of the original structure can be obtained as follows:

$$M_{equ} = \sum_{i=1}^{n} m_i u_i / u_{equ} \tag{10.31}$$

Based on Eqs. (10.28), (10.29), (10.30), and (10.31), the calculation equation of lateral seismism (F_i) of each particle of the original structure can be obtained as follows:

$$F_i = \frac{m_i u_i V_b}{\sum_{j=1}^{n} m_j u_j} \tag{10.32}$$

According to the third assumption, it can be concluded that the works by the left and right lateral earthquakes on the two systems are equal, which can be expressed as follows:

$$V_b u_{equ} = \sum_{i=1}^{n} F_i u_i \tag{10.33}$$

Based on the combination between Eqs. (10.32) and (10.33), the equivalent displacement of the structure is:

$$u_{equ} = \frac{\sum_{i=1}^{n} F_i u_i^2}{\sum_{i=1}^{n} F_i u_i} \tag{10.34}$$

Based on the basic principle of structural dynamics, the calculation equations of the equivalent stiffness and the base shear of the equivalent single-freedom system can be obtained as follows:

$$K_{equ} = \left(2\pi / T_{equ}\right)^2 M_{equ} \tag{10.35}$$

$$V_b = K_{equ} u_{equ} \tag{10.36}$$

where the equivalent stiffness (K_{equ}) of the equivalent single-freedom system is the secant stiffness corresponding to the maximum displacement, and the maximum displacement is the equivalent displacement; T_{equ} is the equivalent period of the equivalent single-freedom system.

For the seismic design of the SFC frame-bent structure in the large thermal power plant, the influence of high vibration mode on the structure needs to be considered. Therefore, the lateral seismism needs to be corrected, which mainly includes the correction of particles on the top layer and the correction of particles on other layers. It is mainly considered that the top layer has a protruding part.

Then the lateral seismism of particle of the top layer can be calculated as follows:

$$F_n = 0.1 V_b + \frac{0.9 V_b m_n u_n}{\sum_{j=1}^{n} m_j u_j} \tag{10.37}$$

The lateral seismism of particles of other layers can be calculated as follows:

$$F_i = \frac{0.9V_b m_n u_n}{\sum_{j=1}^{n} m_j u_j} \tag{10.38}$$

At present, there are many calculation methods for the equivalent damping ratio of structure ξ_{equ}. Most of the calculation methods are mainly based on the equivalent linearization principle. The maximum displacement demand of the actual inelastic system can be calculated by an equivalent linear elastic system with a low lateral stiffness and a high damping ratio. The equivalent damping ratio (ξ_{equ}) in this chapter adopts the empirical calculation equation as follows:

$$\xi equ = \xi_0 + \left(1 - 1/\sqrt{\mu}\right)/\pi \tag{10.39}$$

where ξ_o is the viscous damping ratio at the elastic stage of the structure, the value can be taken as 0.04 for the steel-reinforced concrete frame structure; μ is the displacement ductility coefficient of the structure.

10.7.3 Displacement response spectrum

Based on the analysis, the acceleration response spectrum (S_a) in the current standard can be transmitted into the displacement response spectrum (S_d), and the conversion equation is as follows:

$$S_d = \left(T/2\pi\right)^2 S_a \tag{10.40}$$

where T is the natural vibration period of the structure.

In order to make the seismism of medium period structure and long period structure not too small, the existing seismic standard amplifies the spectral acceleration of the medium period part and long period part of the acceleration response spectrum when developing the acceleration response spectrum. Therefore, the displacement response spectrum is replaced directly by the acceleration response spectrum in Eq. (10.40) will produce large errors for medium period and long period structures. However, for the frame-bent structure of the large thermal power plant, the elastic period is small, and its corresponding spectral acceleration is close to the short period part. Therefore, the displacement response spectrum determined by Eq. (10.40) can be used in the structural system directly in this section, which can be the foundation for calculating the seismic displacement response.

Based on the acceleration response spectrum in the standard, the expression equation of the acceleration response spectrum can be obtained by Eq. (10.40), which is given as follows:

$$S_d = \begin{Bmatrix} \dfrac{0.45 + 10T(\eta_2 - 0.45)}{4\pi^2}\alpha_{max}gT^2 & T \le 0.1\text{ s} \\ \dfrac{\eta_2\alpha_{max}g}{4\pi^2}T^2 & 0.1s \le T \le T_g \\ \dfrac{T_g^\gamma\eta_2\alpha_{max}g}{4\pi^2}T^{2-\gamma} & T_g \le T \le 5T_g \\ \dfrac{\left[0.2\gamma\eta_2 - \eta_1\left(T - 5T_g\right)\right]\alpha_{max}g}{4\pi^2}T^2 & 5T_g \le T \le 6\text{s} \end{Bmatrix} \quad (10.41)$$

The attenuation index (γ) of the decreasing stage in the curve can be calculated as follows:

$$\gamma = 0.9 + (0.05 - \xi)/(0.3 + 6\xi) \quad (10.42)$$

The slope change coefficient (η_l) of the straight decreasing stage can be calculated as follows:

$$\eta_1 = 0.02 + (0.05 - \xi)/(4 + 32\xi) \quad (10.43)$$

The damping change coefficient (η_2) can be calculated as follows:

$$\eta_2 = 1 + (0.05 - \xi)/(0.08 + 1.6\xi) \quad (10.44)$$

where if the damping change coefficient η_2 is less than 0.55, the value can take 0.55 to calculate.

In the above equations, T_g is the natural vibration period of the structure, α_{max} is the maximum value of the lateral seismic influence coefficient. When the fortification intensity is 7 degrees, 8 degrees, and 9 degrees of seismic with basic intensity, α_{max} takes 0.23, 0.45, and 0.9, respectively. While for frequent and rare seismic corresponding to the basic intensity, the value is taken in the existing standard. When the equivalent displacement (u_{equ}), the fortification level, the site type, the damping ratio, and other parameters of the equivalent single-freedom system are known, the corresponding equivalent period (T_{equ}) can be calculated by Eq. (10.40).

10.7.4 Displacement-based seismic design steps

Based on the basic principle of the displacement-based seismic design method in the above section, combined with the mechanical and seismic characteristics of the SFC members and the frame-bent structure in the large thermal power plant, the seismic design steps of displacement-based seismic design can be concluded as follows:

(1) When the structure is designed preliminarily, the layout of the structure should be carried out by considering the needs of the process production. Then the material

strength grade should be selected according to the engineering experience. And the column grid size, layer height, and member section size should be reasonably determined.

(2) The performance target is determined and the corresponding limit value of interlayer displacement angle is given, based on the seismic fortification intensity, the installed capacity of the large thermal power plant, and the owner requirements.

(3) The initial target lateral displacement curve is determined by Eq. (10.26), and the equivalent mass (M_{equ}) and the equivalent displacement (u_{equ}) of the equivalent single-freedom system are calculated by Eqs. (10.31) and (10.34), respectively.

(4) The equivalent damping ratio (ξ_{equ}) and the equivalent period (T_{equ}) are calculated in Eqs. (10.29) and (10.31), respectively. We should note that the displacement ductility coefficient (μ) can be determined by the fortification target: for the SFC structures, when the structure is in "normal use", take $\mu = 1.0$; when the structure is in "near collapse", take $\mu = 2.0 \sim 4.0$.

(5) Based on equations from Eqs. (10.35) to (10.38), the following parameters are determined: the equivalent stiffness (K_{equ}) in the equivalent single-freedom system, the total bottom shear (V_b) of the original structure, and the lateral seismism of each particle.

(6) The lateral seismism and the corresponding gravity effect of the structure are computed. The internal force design value of the member section is obtained after combination, then the section bearing capacity of the member is calculated to take necessary structural measures.

(7) The static pushover analysis is carried out on the designed structure. And the lateral displacement shape pushed to one or several layers of the structure, which reaches the corresponding interlayer displacement angle limit, is compared with the initial target lateral displacement curve. If they are basically consistent, the above design results are effective. Otherwise, the lateral displacement curve obtained from static pushover analysis is recalculated as the modified lateral displacement curve.

10.8 Analysis of practical examples

We take the actual project as an example to carry out the seismic design in this section. One span of the frame-bent structure of the large thermal power plant is shown in Fig. 10.11. And the seismic fortification intensity is 8 degrees (0.2 g), the site category is Class II, and the design earthquake group is Group II. The building is a building of Class I. The preliminary section sizes of members are shown in Table 10.5. The steel materials are Q345 steel. The interlayer displacement angle is used to quantify the performance level of the frame-bent structure. Based on the performance target A, the calculation results meet the "normal use" under the small seismism, the calculation results meet the "basic use" under the medium seismism, the calculation results meet the "life safety" under the large seismism, and the calculation results meet the "near collapse" under the large seismism. The corresponding interlayer displacement angle limits are 1/1000, 1/600, 1/250, and 1/110, respectively.

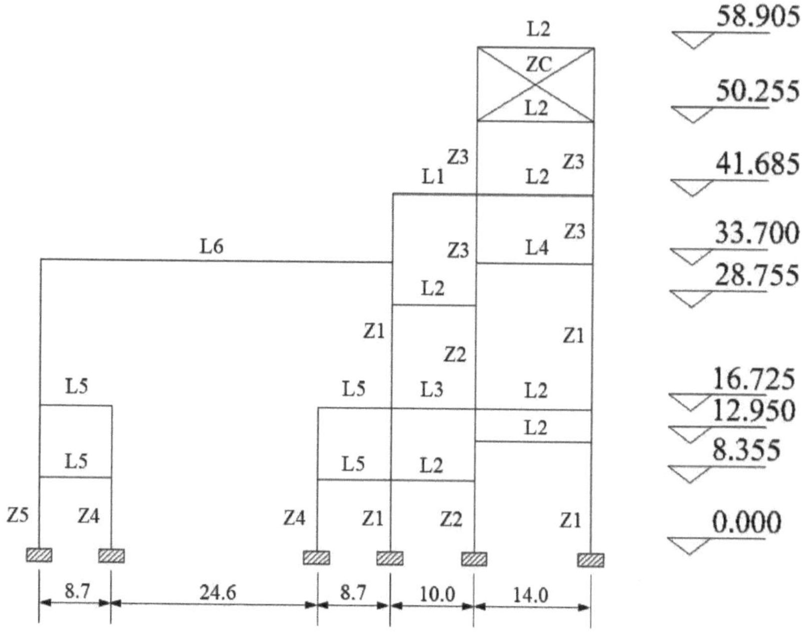

Figure 10.11 Structural calculation diagram (unit: m). A sketch.

Table 10.5 Member section size.

Item	Section size (mm)	Section form
Z1	800Ã—1200	L180 × 14′T150 × 300 × 15 × 15
Z2	800 × 1400	L180 × 18′T170 × 350 × 10 × 16
Z3	800 × 1000	L160 × 16′T150 × 300 × 10 × 15
Z4	700 × 700	Reinforced concrete
Z5	700 × 1200	Reinforced concrete
L1	600 × 1000	Reinforced concrete
L2	600 × 1500	Reinforced concrete
L3	600 × 1800	Reinforced concrete
L4	600 × 3600	Reinforced concrete
L5	400 × 800	Reinforced concrete
L6	H-section steel	HM900 × 500 × 16 × 24
ZC	Square steel tube	400 × 400 × 14

10.9 Seismic design based on the "normal use" condition

The first floor of the structure is assumed as a weak floor in this section, which first reaches the corresponding limit value of the interlayer displacement angle of the

Table 10.6 Lateral displacement of each floor.

Floor	Floor height H_i (m)	H (m)	Floor height ratio ζ_i	Lateral displacement u_i (mm)
8	58.905	58.905	1	39.535
7	50.255		0.853	38.319
6	41.685		0.708	34.961
5	33.7		0.572	30.226
4	28.755		0.488	26.650
3	16.725		0.284	16.385
2	12.95		0.220	12.827
1	8.355		0.142	8.355

performance level of "normal use". The interlayer displacement angle takes 1/1000. Based on the lateral displacement equation of the constant section shear cantilever column, the corresponding interlayer displacement (u_i) is calculated by Eqs. (10.27) and (10.28), then brought into Eq. (10.26) to calculate the lateral displacement of each layer. The results are shown in Table 10.6.

Based on the lateral displacement of and the mass of each layer of the structure, the equivalent displacement $u_{equ} = 28.556$ mm and the equivalent mass $M_{equ} =$ 2138.702 t of the structure, which is calculated by Eqs. (10.31) and (10.34). The displacement ductility coefficient of the structure has $\mu = 1.0$ because the structure is at the performance level of "normal use". Then, the equivalent damping ratio of the structure can be calculated by Eq. (10.39), $\xi_{equ} = 0.04$.

The equivalent displacement, the equivalent mass, the maximum value of seismic influence coefficient (the seismic intensity is 8 degrees in frequent seismic, $\alpha_{max} = 0.16$) and characteristic period $T_g = 0.35$ s can be substituted into the equations from Eqs. (10.41) to (10.44) to get the equivalent period calculation parameters and calculation results of the equivalent single-freedom system:

$$\gamma = 0.9 + (0.05 - \xi)/(0.3 + 6\xi) = 0.9185$$

$$\eta_1 = 0.02 + (0.05 - \xi)/(4 + 32\xi) = 0.0219$$

$$\eta_2 = 1 + (0.05 - \xi)/(0.08 + 1.6\xi) = 1.0694$$

Therefore, the equivalent period of the structure is $T_{equ} =$ 1.74 s. The equivalent mass (M_{equ}) and the equivalent period (T_{equ}) of the equivalent single-freedom system are substituted into Eq. (10.35), and the equivalent stiffness is:

$$K_{equ} = \left(2\pi/T_{equ}\right)^2 M_{equ} = 27.89 \text{ kN/mm}$$

Then, the equivalent stiffness (K_{equ}) and the equivalent displacement (u_{equ}) are substituted into Eq. (10.36), and the layer shear force (V_b) of the structure can be calculated as follows:

$$V_b = K_{equ}u_{equ} = 796.3\ \text{kN}$$

Finally, the lateral seismic force (F_i) at each particle of the structure can be calculated by Eq. (10.38). The calculation parameters and calculation results are shown in Table 10.7.

Based on the above calculated horizontal seismism of the structure, the representative effect of the gravity load, and the wind load effect, the internal force design value of the member section can be calculated. Finally, the bearing capacity and stability requirements of the member section are checked.

In this section, the SAP2000 is used to carry out the pushover analysis on the structure, then the pushover curve of the SFC frame-bent structure and the predicted interlayer displacement angle curve of each floor are obtained. The curve comparisons are shown in Fig. 10.12. In Fig. 10.12, the initial curve is the lateral displacement curve or interlayer displacement curve adopted in the performance

Table 10.7 Lateral displacement and layer shear force under "normal use" condition.

Floor	H_i (m)	M_i (t)	u_i (mm)	m_iu_i (t·mm)	$m_iu^2_i$ (t·mm^2)	F_i (kN)	V_b (kN)
8	58.905	108	39.535	4270	168,807	129.7	129.7
7	50.255	201	38.319	7702	295,136	90.4	220.1
6	41.685	333	34.961	11,642	407,016	136.6	356.7
5	33.7	565	30.226	17,078	516,188	200.4	557.1
4	28.755	215	26.650	5730	152,693	67.2	624.4
3	16.725	544	16.385	8914	146,055	104.6	729.0
2	12.95	177	12.827	2270	29,124	26.6	755.6
1	8.355	415	8.355	3467	28,970	40.7	796.3

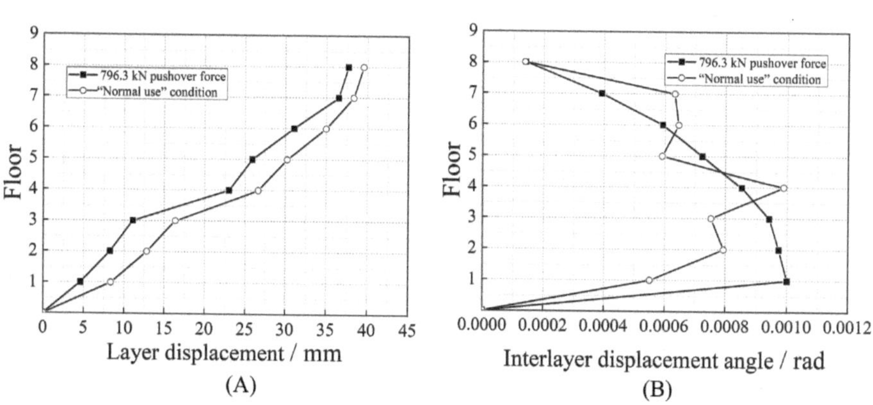

Figure 10.12 Curve comparisons between "normal use" condition and 796.3 kN pushover force. (A) Layer displacement curve; (B) Interlayer displacement angle curve.

level design. The pushover curve is the corresponding lateral displacement curve and the interlayer displacement angle curve when the base shear is V_b = 796.3 kN.

It can be concluded from Fig. 10.12 that when the structure pushes to the performance level of "normal use", the lateral displacement curve of the structure is consistent with the initially assumed lateral displacement curve. However, there are still significant differences in the two interlayer displacement angle curves. The main reason is that the lateral stiffness of the structure is unevenly distributed along the height, so it is not the largest interlayer displacement angle of the first layer of the structure.

It is obviously seen from Fig. 10.12 that the interlayer displacement angle of the structure does not occur in the first layer when the structure reaches the performance level of "normal use", but in the fourth layer of the structure which is also in line with the previous static pushover analysis results. Therefore, the initial displacement curve of the structure is not completely consistent with the practice condition of the structure, and the lateral displacement curve should be corrected.

10.10 Seismic design in "normal use" after correction

This section assumes the fourth layer of the structure is a weak layer. The fourth layer of the structure first reaches the limit value of the interlayer displacement angle required by the design performance level $[\theta] = 1/1000$. When the fourth floor of the structure reaches the limit value of the interlayer displacement angle, the maximum interlayer displacement is 12.03 mm, and the corresponding layer displacement is 38.11 mm. Based on the calculation method in the previous section, the equivalent displacement has $u_{equ} = 26.38$ mm, the equivalent mass has $M_{equ} = 1945.84$ t and the equivalent period has $T_{equ} = 1.58$ s. Meanwhile, the corresponding base shear force is 811.6 kN, and the calculation results of the corresponding layers are summarized in Table 10.8.

Table 10.8 Lateral displacement and layer shear force under "normal use" performance level after correction.

Floor	H_i (m)	M_i (t)	u_i (mm)	m_iu_i (t·mm)	$m_iu^2_i$ (t·mm^2)	F_i (kN)	V_b (kN)
8	58.905	108	38.11	4116	156,856	139.7	139.7
7	50.255	201	36.90	7417	273,686	105.6	245.3
6	41.685	333	31.42	10,464	328,790	148.9	394.2
5	33.7	565	26.21	14,810	388,230	210.8	605.0
4	28.755	215	23.26	5000	116,279	71.2	676.2
3	16.725	544	11.22	6106	68,540	86.9	763.1
2	12.95	177	8.35	1478	12,340	21.0	784.1
1	8.355	415	4.66	1933	9002	27.5	811.6

Based on the comparisons of the base shear force in Tables 10.7 and 10.8, it can be concluded that there is little difference between the two values before and after correction. Therefore, the design results still use the design results before correction, and the bearing capacity and stability of the section do not need to be designed again. At this time, the structural design has not been changed, so the lateral displacement curve after structural correction can coincide with the pushover curve. Therefore, the strength and lateral displacement of the SFC frame-bent structure meet the corresponding requirements under the performance level of "normal use", and the design is completed.

10.11 Seismic design based on the "basic use" condition

Based on the design process of the performance level of "normal use", it can be concluded that the weak layer of the structure is the fourth layer, which means that the interlayer displacement angle of the fourth layer is the largest. Therefore, the interlayer displacement angle for the fourth layer first reaches the required value of the performance level of "basic use". The interlayer displacement angle is taken as 1/600, so the interlayer displacement of the fourth layer is 20.5 mm, and the corresponding layer displacement is 38.76 mm. Based on the lateral displacement and the mass of each layer, the equivalent displacement is $u_{equ} =$ 43.96 mm and the equivalent mass is $M_{equ} =$ 1945.84 t can be calculated by Eqs. (10.31) and (10.34). Because the structure is at the performance level of "basic use", the displacement ductility coefficient of the structure is $\mu = 1.0$. The equivalent damping ratio of the structure can be calculated by Eq. (10.39), $\xi_{equ} =$ 0.04. The equivalent displacement, the equivalent mass, the maximized value of seismic influence coefficient ($\alpha_{max} = 0.23$), and characteristic period $T_g = 0.35$ s can be substituted into equations from Eqs. (10.41) to (10.44) to get the equivalent period calculation parameters and the calculation results of the equivalent single-freedom system:

$$\gamma = 0.9 + (0.05 - \xi)/(0.3 + 6\xi) = 0.9185$$

$$\eta_1 = 0.02 + (0.05 - \xi)/(4 + 32\xi) = 0.0219$$

$$\eta_2 = 1 + (0.05 - \xi)/(0.08 + 1.6\xi) = 1.0694$$

Thus, the equivalent period of the structure is $T_{equ} =$ 1.78 s. The equivalent mass (M_{equ}) and the equivalent period (T_{equ}) of the equivalent single-freedom system are substituted into Eq. (10.35), and the equivalent stiffness is given as follows:

$$K_{equ} = \left(2\pi/T_{equ}\right)^2 M_{equ} = 24.25\ \text{kN/mm}$$

Then, the equivalent stiffness (K_{equ}) and the equivalent displacement (u_{equ}) are substituted into Eq. (10.36) to get the bottom shear (V_b) of the structure, then it is given as follows:

$$V_b = K_{equ}u_{equ} = 1065.72 \text{ kN}$$

Finally, the lateral seismic force (F_i) at each particle of the structure can be calculated by Eq. (10.38). The calculation parameters and the calculation results are shown in Table 10.9

Then, the SAP2000 is used to carry out the pushover analysis on the structure, and the pushover curve of the SFC frame-bent structure is obtained. The pushover curve is the corresponding lateral displacement curve and the interlayer displacement angle curve when the base shear force reaches $V_b =$ 1065.7 kN. The comparisons of the lateral displacement curve of the structure under the performance level of "basic service", the lateral displacement curve obtained by the pushover analysis, and the interlayer displacement angle curve are shown in Fig. 10.13.

Table 10.9 Lateral displacement and layer shear force under "basic use" condition.

Floor	H_i (m)	M_i (t)	u_i (mm)	m_iu_i (t·mm)	$m_iu^2_i$ (t·mm²)	F_i (kN)	V_b (kN)
8	58.905	108	63.51	6859	435,620	183.5	183.5
7	50.255	201	61.49	12,360	760,079	138.6	322.1
6	41.685	333	52.36	17,438	913,115	195.5	517.6
5	33.7	565	43.68	24,682	1,078,191	276.8	794.4
4	28.755	215	38.76	8332	322,930	93.4	887.9
3	16.725	544	18.71	10,176	190,349	114.1	1002.0
2	12.95	177	13.91	2463	34,270	27.6	1029.6
1	8.355	415	7.76	3221	25,000	36.1	1065.7

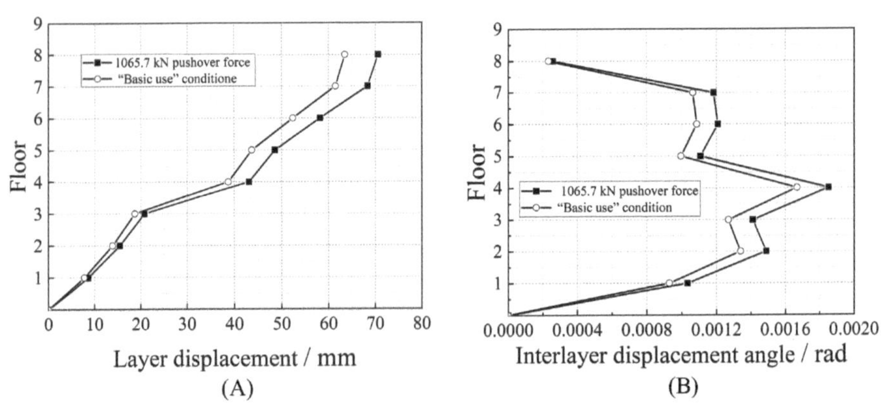

Figure 10.13 Curve comparisons between "basic use" condition and 1065.7 kN pushover force. (A) Layer displacement curve; (B) Interlayer displacement angle curve.

Based on the curve comparison in Fig. 10.13, it can be concluded that the structure curve agrees with the assumed lateral displacement curve and the interlayer displacement angle. Meanwhile, the lateral displacement and interlayer displacement angle of the structure meet the requirements of standard. It means that the deformation capacity of the structure is greater than the displacement demand of the structure under the lateral seismic force corresponding to the performance level of "basic service", $V_b =$ 1065.7 kN. Therefore, the designed SFC frame-bent structure can meet the requirements of the performance level of "basic use" under the frequent seismism of 8 degrees.

10.12 Seismic design based on the "life safety" condition

Based on the design process of the performance level of "normal use", it can be concluded that the weak layer of the structure is the fourth layer, so the interlayer displacement angle of the fourth layer is the largest. Therefore, the interlayer displacement angle for the fourth layer first reaches the required value of the performance level of "life safety". The interlayer displacement angle takes 1/250, so the interlayer displacement of the fourth layer is 48.12 mm, the corresponding layer displacement is 152.42 mm. And the equivalent displacement is $u_{equ} =$ 105.49 mm and the equivalent mass is $M_{equ} =$ 1945.84 t can be calculated by Eqs. (10.31) and (10.44). Because the structure is at the performance level of "life safety", the displacement ductility coefficient of the structure is $\mu = 2.0$. The equivalent damping ratio of the structure can be calculated by Eq. (10.39), $\xi_{equ} =$ 0.2787. The equivalent displacement, the equivalent mass, the maximized value of seismic influence coefficient ($\alpha_{max} = 0.45$), and characteristic period $T_g = 0.35$s can be substituted into equations from Eqs. (10.41) to (10.44) to get the equivalent period calculation parameters and the calculation results of the equivalent single-freedom system:

$$\gamma = 0.9 + (0.05 - \xi)/(0.3 + 6\xi) = 0.7840$$

$$\eta_1 = 0.02 + (0.05 - \xi)/(4 + 32\xi) = 0.0023$$

$$\eta_2 = 1 + (0.05 - \xi)/(0.08 + 1.6\xi) = 0.5651$$

Thus, the equivalent period of the structure is $T_{equ} =$ 2.44 s. The equivalent mass (M_{equ}) and the equivalent period (T_{equ}) of the equivalent single-freedom system are substituted into Eq. (10.35), and the equivalent stiffness is given as follows:

$$K_{equ} = \left(2\pi/T_{equ}\right)^2 M_{equ} = 12.90 \text{ kN/mm}$$

Then, the equivalent stiffness (K_{equ}) and the equivalent displacement (u_{equ}) are substituted into Eq. (10.36) to get the bottom shear (V_b) of the structure, then it is given as follows:

$$V_b = K_{equ}u_{equ} = 1361.14\ \text{kN}$$

Finally, the lateral seismic force (F_i) at each particle of the structure can be calculated by Eq. (10.38). The calculation parameters and the calculation results are shown in Table 10.10.

Then, the SAP2000 is used to carry out the pushover analysis on the structure, and the pushover curve of the SFC frame-bent structure is obtained. The pushover curve is the corresponding lateral displacement curve and the interlayer displacement angle curve when the base shear force reaches $V_b =$ 1361.1 kN. The comparisons of the lateral displacement curve of the structure under the performance level of "life safety", the lateral displacement curve obtained by the pushover analysis, and the interlayer displacement angle curve are shown in Fig. 10.14.

Table 10.10 Lateral displacement and layer shear force under "life safety" condition.

Floor	H_i (m)	M_i (t)	u_i (mm)	m_iu_i (t·mm)	$m_iu^2_i$ (t·mm^2)	F_i (kN)	V_b (kN)
8	58.905	108	152.42	16,461	2,509,040	234.4	234.4
7	50.255	201	147.58	29,664	4,377,826	177.0	411.4
6	41.685	333	125.67	41,849	5,259,267	249.8	661.1
5	33.7	565	104.84	59,234	6,210,055	353.5	1014.6
4	28.755	215	93.01	19,997	1,859,981	119.3	1134.0
3	16.725	544	44.89	24,422	1,096,351	145.7	1279.7
2	12.95	177	33.39	5911	197,387	35.3	1315.0
1	8.355	415	18.63	7730	143,990	46.1	1361.1

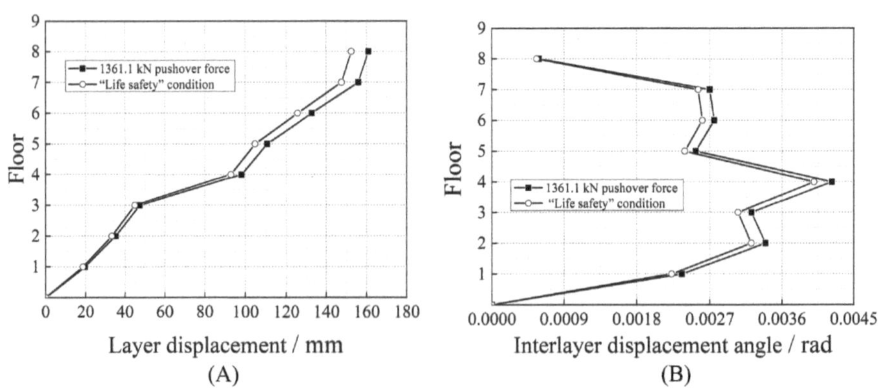

Figure 10.14 Curve comparisons between "life safety" condition and 1361.1 kN pushover force. (A) Layer displacement curve; (B) Interlayer displacement angle curve.

Based on the curve comparison in Fig. 10.14, it can be concluded that the structure curve agrees with the assumed lateral displacement curve and the interlayer displacement angle. Meanwhile, the lateral displacement and the interlayer displacement angle of the structure meet the requirements of standard. It means that the deformation capacity of the structure is greater than the displacement demand of the structure under the lateral seismic force corresponding to the performance level of "life safety", $V_b = 1361.1$ kN. Therefore, the designed SFC frame-bent structure can meet the requirements of the performance level of "life safety" under the frequent seismism of 8 degrees.

10.13 Seismic design based on the "near collapse" condition

Based on the design process of the performance level of "normal use", it can be concluded that the weak layer of the structure is the fourth layer, so the interlayer displacement angle of the fourth layer is the largest. Therefore, the interlayer displacement angle for the fourth layer first reaches the required value of performance level of "near collapse". The interlayer displacement angle takes 1/110, so the interlayer displacement of the fourth layer is 109.36 mm, the corresponding layer displacement is 211.40 mm. And the equivalent displacement is $u_{equ} =$ 239.76 mm and the equivalent mass is $M_{equ} =$ 1945.84 t can be calculated by Eqs. (10.31) and (10.34). Because the structure is at the performance level of "near collapse", the displacement ductility coefficient of the structure is $\mu = 4.0$. The equivalent damping ratio of the structure can be calculated by Eq. (10.39), $\xi_{equ} =$ 0. 3384. The equivalent displacement, the equivalent mass, the maximized value of seismic influence coefficient ($\alpha_{max} = 0.90$), and characteristic period T_g = 0.35 s can be substituted into equations from Eqs. (10.41) to (10.44) to get the equivalent period calculation parameters and the calculation results of the equivalent single-freedom system:

$$\gamma = 0.9 + (0.05 - \xi)/(0.3 + 6\xi) = 0.7840$$

$$\eta_1 = 0.02 + (0.05 - \xi)/(4 + 32\xi) = 0.0023$$

$$\eta_2 = 1 + (0.05 - \xi)/(0.08 + 1.6\xi) = 0.5651$$

Thus, the equivalent period of the structure is $T_{equ} =$ 2.64 s. The equivalent mass (M_{equ}) and the equivalent period (T_{equ}) of the equivalent single-freedom system are substituted into Eq. (10.35), and the equivalent stiffness is given as follows:

$$K_{equ} = \left(2\pi/T_{equ}\right)^2 M_{equ} = 11.02 \text{ kN/mm}$$

Then, the equivalent stiffness (K_{equ}) and the equivalent displacement (u_{equ}) are substituted into Eq. (10.36) to get the bottom shear (V_b) of the structure, then it is given as follows:

$$V_b = K_{equ}u_{equ} = 2642.62\ \text{kN}$$

Finally, the lateral seismic force (F_i) at each particle of the structure can be calculated by Eq. (10.38). The calculation parameters and the calculation results are shown in Table 10.11.

Then, the SAP2000 is used to carry out the pushover analysis on the structure, and the pushover curve of the SFC frame-bent structure is obtained. The pushover curve is the corresponding lateral displacement curve and the interlayer displacement angle curve when the base shear force reaches $V_b =$ 2642.6 kN. The comparisons of the lateral displacement curve of the structure under the performance level of "near collapse", the lateral displacement curve obtained by pushover analysis, and the interlayer displacement angle curve are shown in Fig. 10.15.

Table 10.11 Lateral displacement and layer shear force under "near collapse" condition.

Floor	H_i (m)	M_i (t)	u_i (mm)	m_iu_i (t·mm)	$m_iu^2_i$ (t·mm²)	F_i (kN)	V_b (kN)
8	58.905	108	346.42	37,413	12,960,736	455.0	455.0
7	50.255	201	335.42	67,420	22,614,164	343.7	798.7
6	41.685	333	285.63	95,114	27,167,349	484.9	1283.6
5	33.7	565	238.28	134,627	32,078,752	686.3	1969.9
4	28.755	215	211.40	45,450	9,607,944	231.7	2201.6
3	16.725	544	102.03	55,505	5,663,327	283.0	2484.6
2	12.95	177	75.90	13,434	1,019,626	68.5	2553.1
1	8.355	415	42.34	17,569	743,799	89.6	2642.6

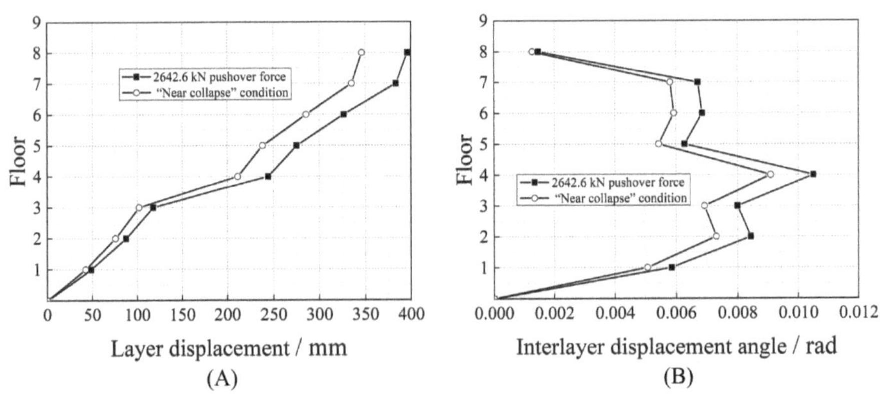

Figure 10.15 Curve comparisons between "near collapse" condition and 2642.6 kN pushover force. (A) Layer displacement curve; (B) Interlayer displacement angle curve.

Based on the curve comparison in Fig. 10.15, it can be concluded that the structure curve agrees with the assumed lateral displacement curve and the interlayer displacement angle. Meanwhile, the lateral displacement and the interlayer displacement angle of the structure meet the requirements of standard. It means that the deformation capacity of the structure is greater than the displacement demand of the structure under the lateral seismic force corresponding to the performance level of "near collapse", $V_b = 2642.6$ kN. Therefore, the designed SFC frame-bent structure can meet the requirements of the performance level of "near collapse" under the frequent seismism of 8 degrees.

10.14 Conclusions

The SAP2000 software is used to carry out the static elastic-plastic analysis of the SFC frame-bent structure of the large thermal power plant under four loading conditions. The calculation results of the whole structure are analyzed in many aspects, including the structural static elastic-plastic curve, the base shear-vertex displacement curve, the interlayer shear force, the interlayer displacement angle, the cooperative working performance of the frame-bent structure, and the structural plastic hinge. Then, the displacement-based seismic design method of the SRC frame-bent structure is proposed, which is used to design and analyze the practical project. The static elastic-plastic analysis of the designed structure is carried out in SAP2000. The design results are verified by the comparisons of the interlayer displacement and the interlayer displacement angle. Based on the above studies, the following conclusions can be drawn:

(1) The performance points can be obtained under the two loading modes in the X and Y directions of the structure, which shows that the structure can resist the seismism level and realize the seismic performance index of "no collapse in a large seismic".
(2) Based on the base shear vertex-displacement curve, the mechanical properties of the structure in the X and Y directions are quite different. This is mainly because the longitudinal span of the structure differs from the transverse span, so there are more longitudinal supports and shear walls to increase the lateral stiffness, which leads to significant differences in the stiffness distribution and the bearing capacity in two directions.
(3) Based on the analysis of interlayer shear force and interlayer displacement angle, we can find that the corresponding curves are not a diagonal line distribution, because of the different layer quality and the vertical irregularity of the structure,
(4) The modal loading mode focuses on the upper part of the structure, so the base shear is small, but the ductility is good. The acceleration loading mode focuses on the lower part of the structure, so the base shear of the structure is large, but the ductility is poor. The acceleration loading mode and modal loading mode should be contemplated in the analysis of structural seismic performance.
(5) During the seismic, the frame structure bears most of the shear force and can be used as an independent seismic defense line. Although the bent structure can work together with the frame structure, it cannot be an independent seismic defense line. In addition, more

shear walls or braces should be arranged in the frame structure to improve the shear ratio borne by the frame structure.

(6) Plastic hinges basically appear at the beam end, and the distribution of structural plastic hinges is basically the same under the two loading modes. However, the plastic hinges of the structure are more distributed in the higher layer in the modal loading model. And the plastic hinges of the structure are more distributed in the bottom layer in the acceleration loading mode.

(7) The structures under the four performance levels of "normal use", "basic use", "life safety" and "near collapse" are designed and analyzed. The results show that the structure designed by the displacement-based design method can well meet the multistage seismic requirements.

Acknowledgments

This work was supported by the National Natural Science Foundation of China [grant numbers 52108171; 51878540]. The authors gratefully acknowledge two Master's students (Mr. Qv Yunsong and Mr. Tian Wenkai) of Xi'an University of Architecture and Technology for the drawing works.

References

Chen, CH, Wang, CK, & Sun, HZ (2007). Experimental study on seismic behavior of steel reinforced concrete compression-bending members. *Earthquake Resistance Engineering and Retrofitting*, *29*, 68–72.

Cheng, B, & Xue, WC (2003). Studies on performance-based seismic design of frame structures. *Earthquake Engineering and Engineering Vibration*, *23*(4), 50–55.

Chu, LS, Liu, J., Wang, SW, & Zhao, J (2018). Direct displacement-based seismic design method of SRC column-steel beam hybrid frames. *Engineering Mechanics*, *35*(8), 100–110.

Feng, JN (2014). *Study on seismic behavior of concrete column steel beam composite frame*. Zhengzhou University.

GB 50011-2001. (2008). Code for Seismic Design of Buildings. China Architecture & Building Press.

JGJ 138-2001. (2001). Technical specification for steel reinforced concrete composite structures (SSRC). China Architecture & Building Press.

Kang, LG (2009). Study on seismic performance test and design method of steel reinforced concrete frame main powerhouse with few walls in large thermal power plant. *Xi'an University of Architecture and Technology*.

Shi, QX, Wang, P, & Wang, QW (2013). Study of factors influencing seismic behavior of SRC columns. *Industrial Construction*, *43*(10), 134–140.

Tang, LJ (2011). Theoretical analysis on quantified performance index for steel reinforced concrete frame structures. *Journal of Huazhong University of Science and Technology (Urban Science)*, *28*(3), 193–195.

Wang, MF, & Guo, ZX (2006). Discussion on seismic performance levels and limit-states about steel reinforced concrete columns. *Earthquake Resistance Engineering and Retrofitting*, *28*(3), 31–37.

Wang, QW, Shi, QX, & Yang, K (2009). Study on seismic performance levels and tolerant deformation values about SRC structures. *Journal of Xi'an University of Architecture and Technology*, *41*(1), 82–87.

Index

F

H

L

M

R

MIX
Papier | Fördert gute Waldnutzung
FSC® C083411

Kolibri 360 GmbH
Ferdinand-Jühlke-Straße 7
99095 Erfurt, Deutschland
produktsicherheit@kolibri360.de